Advances in Seafood Byproducts
2002 Conference Proceedings

Peter J. Bechtel, Editor

Proceedings of the 2nd International
Seafood Byproduct Conference
November 10-13, 2002
Anchorage, Alaska, USA

Alaska Sea Grant College Program (Publisher)
University of Alaska Fairbanks
AK-SG-03-01

Elmer E. Rasmuson Library Cataloging-In-Publication Data

Advances in seafood byproducts : 2002 conference proceedings / Peter J. Bechtel, editor. — Fairbanks, Alaska : Alaska Sea Grant College Program, 2003.

566 p. : ill. ; cm. – (Alaska Sea Grant College Program ; AK-SG-03-01)

Note: "Proceedings of the 2nd International Seafood Byproduct Conference, November 10-13, 2002, Anchorage, Alaska, USA."

Includes bibliographical references and index

1. Fisheries—By products—Congresses. 2. Fishery products—Congresses. 3. Seafood—Congresses. 4. Fish meal—Congresses. 5. Fish oils—Congresses. I. Title. II. Bechtel, Peter J. Series: Alaska Sea Grant College Program report ; AK-SG-03-01.

SH335.8.A38 2003

ISBN: 1-56612-082-9

Citation for this volume is: 2003. P.J. Bechtel (ed.), Advances in Seafood Byproducts: 2002 Conference Proceedings. Alaska Sea Grant College Program, University of Alaska Fairbanks, Fairbanks. 566 pp.

Credits

This book is published by the Alaska Sea Grant College Program, University of Alaska Fairbanks, supported by the U.S. Department of Commerce, NOAA National Sea Grant Office, grant NA 16RG2321, A/151-01 and A/161-01; and by the University of Alaska Fairbanks with state funds. The University of Alaska is an affirmative action/equal opportunity institution. The Alaska Fisheries Development Foundation supported the publication; see pages viii-ix. The views expressed herein are not necessarily the views of the above organizations.

Sea Grant is a unique partnership with public and private sectors combining research, education, and technology transfer for public service. This national network of universities meets changing environmental and economic needs of people in our coastal, ocean, and Great Lakes regions.

Alaska Sea Grant College Program
University of Alaska Fairbanks
P.O. Box 755040
205 O'Neill Bldg.
Fairbanks, Alaska 99775-5040
Toll free (888) 789-0090
(907) 474-6707 Fax (907) 474-6285
http://www.uaf.edu/seagrant/

Contents

Products I—Animal Feeds

Products II—Plant Fertilizers, Alternative Energy, and Other Industrial Uses

Products III—Human Food, Supplements, and Pharmaceuticals

Available and Innovative Technology

Problem Solving—Alaska Model

Product Practices and Logistics

Remarks and Acknowledgments

For us in Alaska, we have seen many changes in our fishing industry since the first International Conference on Fish By-Products held in 1990. Our white fish industry is fully developed and Alaska has emerged as one of the world's leaders in producing surimi and fillets. The fish meal technology presented during the first conference was adopted by our industry. However, our industry has made modifications. For example, most shoreside processors are using a screening system between the drying and milling steps to remove some of the bone fragments to maintain a protein content of at least 65% in the fish meal. Thus, we would like to share with you what we have accomplished as well as learn from you what you have experienced. This sharing of information and needs is good because we all share the same concerns as we strive to increase the utilization of our fishery resources.

It is important to make some acknowledgments. Planning for this conference began in November 2001. Those individuals who served on the organizing committee were:

- Jerry Babbitt (conference chair), National Marine Fisheries Service, Kodiak, Alaska

- Brenda Baxter (conference coordinator), University of Alaska Fairbanks, Alaska Sea Grant College Program, Fairbanks, Alaska

- Peter J. Bechtel, U.S. Department of Agriculture, Agricultural Research Service, Fairbanks, Alaska

- Robin Cababa, The Oceanic Institute, Waimanalo, Hawaii

- Ian Forster, The Oceanic Institute, Waimanalo, Hawaii

- Ronald W. Hardy, University of Idaho, Hagerman Fish Culture Experiment Station, Hagerman, Idaho

- Marc Jones (conference financial manager), Alaska Fisheries Development Foundation, Anchorage, Alaska

- Donald Kramer, University of Alaska Fairbanks, Marine Advisory Program, Anchorage, Alaska

- Scott Smiley, University of Alaska Fairbanks, Fishery Industrial Technology Center, Kodiak, Alaska

Additional members on the program committee included:

- Anthony Bimbo, Technical Consultant, Kilmarnock, Virginia

- John Kilpatrick, Marine Protein and Oil Consultant, West Vancouver, British Columbia

In addition, we have been blessed with the generosity of many groups who willingly sponsored this conference. These sponsors contributed speakers' travel and meeting expenses, and gave us cash for unrestricted use in supporting the conference.

- Alaska Fisheries Development Foundation

- The Oceanic Institute

- National Marine Fisheries Service (Pacific Rim)

- U.S. Department of Agriculture, Agricultural Research Service

- University of Alaska Fairbanks

    Marine Advisory Program

    Alaska Sea Grant College Program

    Fishery Industrial Technology Center

    School of Fisheries and Ocean Sciences

- World Aquaculture Society

- Pacific Seafood Processors Association

Welcoming remarks were made by Dr. Charles Hocutt, University of Alaska, Fairbanks, Alaska, for Dr. Craig Dorman, vice president for research, University of Alaska Statewide System and Dr. Robert Matteri, USDA, Agriculture Research Service, Pacific West Area, Albany, California.

Session chairs were Ronald Hardy, Robbin Cababa, John Kilpatrick, Scott Smiley, Peter Bechtel, Anthony Bimbo, Marc Jones, Donald Kramer, Ian Forster, and Chris Mitchell. Donald Kramer coordinated the poster session.

Conference coordinator was Brenda Baxter who received assistance from Beverly Bradley, Tim Sullivan, and Charles Crapo.

Sue Keller copy-edited and arranged production of this proceedings, Kathy Kurtenbach formatted the text pages, and Tatiana Piatanova designed the cover, all of the Alaska Sea Grant College program.

At the conclusion of the conference I think that all were in agreement that the 2nd International Seafood Byproduct Conference was a great success. The interactions among the international and national seafood industry and the Alaskan processing sector has been exciting and meaningful.

Indeed, this was truly an international conference. For the record, there were 125 participants from 17 countries.

Significant advances have been made in our Alaskan processing sector since the first International Conference on Fish By-Products that was held in 1990. Alaska now makes high quality meals which are highly accepted in international markets. However, we realize that we have a long way to go to more fully utilize our harvested marine resources, and we are thankful to the speakers and participants for sharing with us their knowledge and experiences as we continue to develop economically viable options to utilize our resources.

The true test of the success of this conference will be seen in the coming years as we employ the ideas that were discussed during this conference to increase the utilization of our marine resources. One thing that I am sure of is that, like the proceedings from the first conference, the published proceedings from this conference will serve as the reference book in this field for many years to come.

Jerry Babbitt, Conference Chair
National Marine Fisheries Service, Utilization Research Laboratory, Kodiak, Alaska

Fish Processing Waste: Opportunity or Liability

Keynote Address

John S. Kilpatrick
West Vancouver, British Columbia, Canada

The first International Seafood Byproducts Conference was held in Anchorage from April 25 to 27, 1990, over 12 years ago. Many predictions made then have proved accurate, but many problems identified then still await solutions. Problems were opportunities then and that is even more true today. New concerns have arisen, or old ones have returned. New key words that were little used in 1990 are now commonplace.

The world seafood industry is still producing large quantities of "waste," which I prefer to call secondary process streams, and which can take many forms. Except for whole frozen round fish, which is mainly produced by very large freezer trawlers, all fish processing operations produce "waste" or offal, as secondary process streams. The highest proportion of fish "waste" consists of heads, tails, viscera, and backbones from filleting operations. As the industry moves toward more "value-added" products, which require skinless, boneless fillets, additional waste is produced. When these two streams separate, they are of equal quality—that is, they are at the level of freshness suitable for human consumption. Some "waste" products can have a very high value if they can be separated economically. Many people have tried halibut cheeks or—the Newfoundland delicacy—cod tongues. White fish "v" cuts can be made into high quality minced fish with deboning equipment. Some consumer studies have shown that children tend to prefer fish sticks (fish fingers, as the Europeans call them) made from mince, to those from whole fillets.

The great bulk of fish processing waste, which can amount to over 60% of landed weight, leaves the human consumption stream at equivalent freshness, and is ideal raw material for the highest quality fish meal and fish oil, if the economics and logistics are right. If they are not, there are other profitable approaches.

One of the new key words I referred to is "organic"; another is "sustainable." There are many different definitions of "organic," but the Soil

Association in the United Kingdom, one of the bodies approved by the U.K. government to certify food as "organic," defines "organic" fish meal as only that made from offal and trimmings from processing for human consumption of fish from sustainable fisheries. So, people wanting to produce "organic" farmed fish have to use fish meal and fish oil certified as "organic" in their fish feeds. The largest U.K. fish meal producer has this certification, and the discussion now is how large the premium shall be.

Let us now consider further how things have changed since 1990. Figure 1 shows world fish landings from 1990 to 1999 (FAO figures). For 2000 the estimate is 95 million metric tons, and for 2001 only 90 million t. In 1993, landings reached 105 million t and were as low as 87 million t in 1998. There has been a downtrend from 1990, and no significant growth is foreseen. In contrast, world aquaculture production has doubled since 1990. In 2000, 27.4% of world seafood supplies were from aquaculture and 72.7% were from the capture fisheries. This year the ratio may be close to ⅓:⅔.

Figure 2 shows world fish meal production, which includes fish by-product meals, since 1990. The 12-year average is 6.53 million t, so it has been remarkably stable, except for the El Niño year 1998, when production fell to 5.21 million t, and Peru produced only 815,000 t compared with 1.74 million t the year before.

Figure 3 shows world fish body oil production. This is much more variable than that of fish meal, because oil yields can vary so much. The average yield in Peru for 2002 as of November was only 2.6%; in 2001 it was 4.3%. As with fish meal, there is no prospect of any significant increase in production, and as the 12-year average from 1990 is only 1.24 million t, this is of tremendous significance to the world seafood byproducts industry.

Figure 4 shows world fish meal plus world fish body oil production. You will note the maximum was in 1994 with 9 million t and the minimum in 1998, the El Niño year, with just over six million t.

Figure 5 shows world population growth since 1990. The total increase since 1990 is 850 million people. The annual rate of increase has been declining—from 1.36% in 1998 to 1.25% (estimated) in 2002. This still means 77 million more people to feed in one year. If those 77 million people took a one-gram fish oil capsule each day, that amounts to 28,105 t. And if half the world's population took a capsule a day, that is 1.13 million t of fish oil!

The point is that production of fish meal and fish oil, and other fish byproducts in general, has very limited potential for increase, and replacements for some of their unique properties are very hard to find and tend to be very expensive. On the other hand, production of vegetable proteins and oils continues to increase. It is fortunate that it does, as this trend must exceed population growth to feed the world.

Figure 6 shows world consumption of 12 major protein meals in

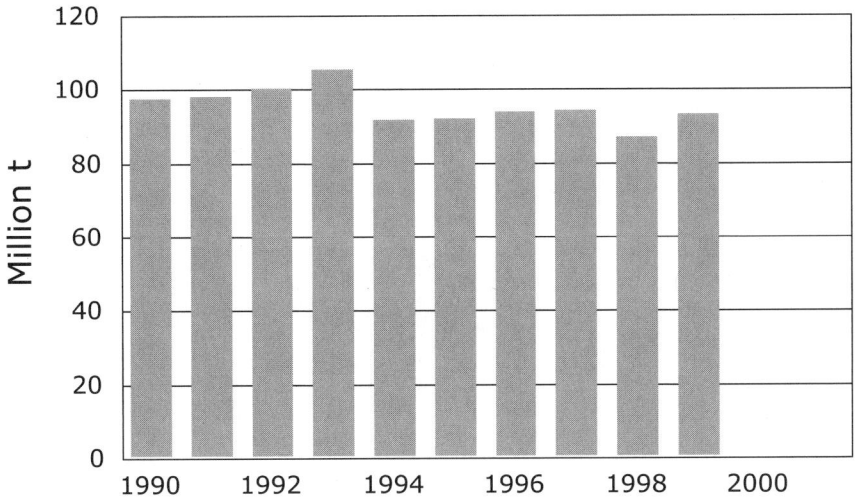

Figure 1. World fish landings from 1990 to 1999 (FAO figures).

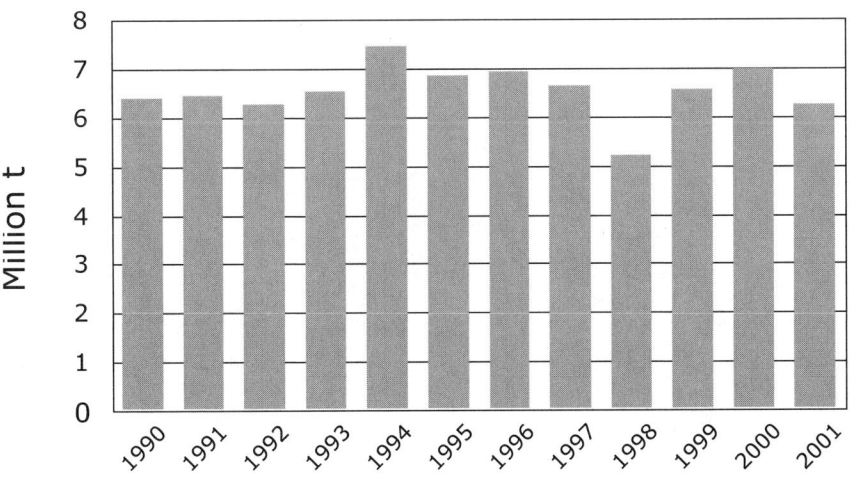

Figure 2. World fish meal production, including fish byproduct meals, 1990-2001.

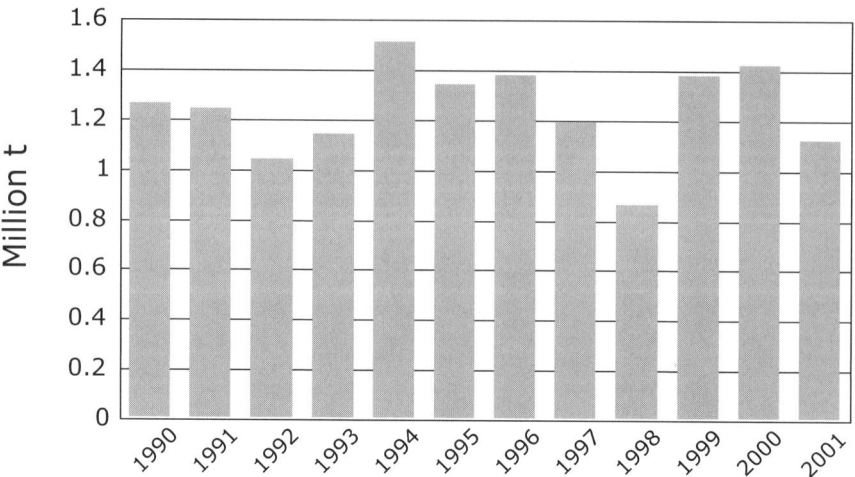

Figure 3. World fish body oil production, 1990-2001.

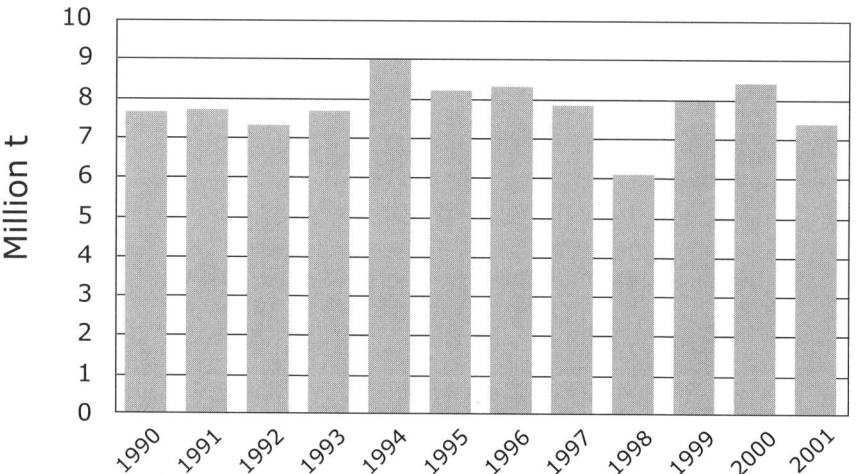

Figure 4. World fish meal plus body oil production, 1990-2001.

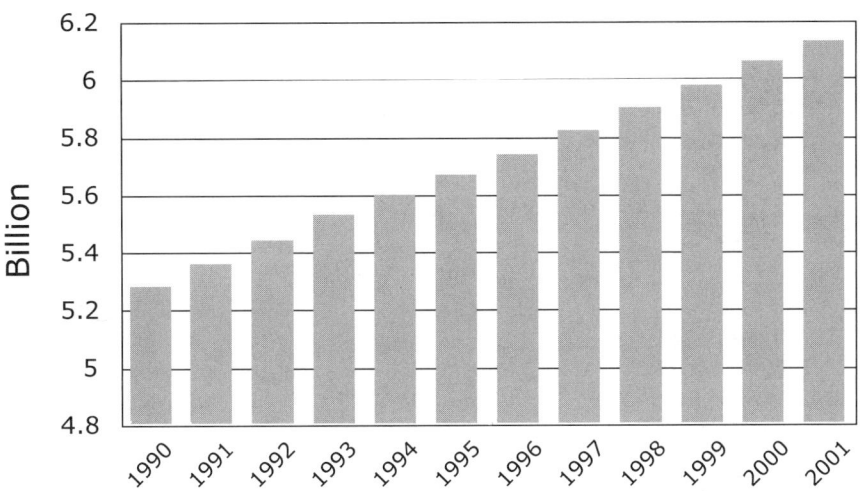

Figure 5. World population, 1990-2001.

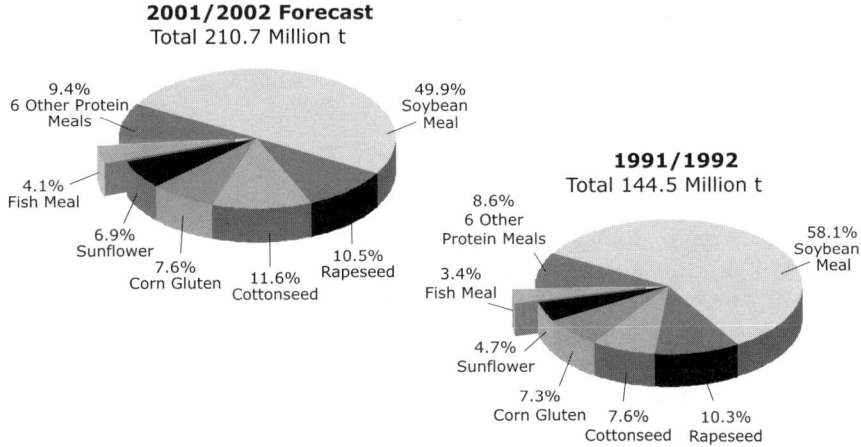

Figure 6. World consumption of twelve major protein meals, 1991-1992 and 2001-2002, showing an increase of 66.2 t (46%) for those years.

1991-1992. The total was 144.5 million t, of which 4.1% was fish meal. The figure also shows world consumption of these same 12 meals in 2001-2002 (forecast). The total is 210.7 million t—or increase of 46%—while fish meal's share of this total has decreased to 3.4%. Note the growing importance of soybean meal-up from 50% in 1991-1992 to 58% in 2001-2002. All the other meals also lost "market share." Corn gluten meal lost only 0.3% and rapeseed meal only 0.2%. However, total production of these two meals increased by 10.9 million t or by 1.7 times the average total world fish meal production. This is important because these two meals both have considerable importance as protein ingredients for aquaculture feeds.

Figure 7 shows that in 1986 fish meal contributed 7% of total world protein meal supply. Soybean meal's market share was at the same level as today-so it has recovered all that it had lost in 1991-1992.

Figure 8 shows world fish meal use in 1995 and 2001. In 1995 aquaculture use was 15% and poultry plus pigs 75%. In 2001 aquaculture was up to 40%, and poultry plus pigs down to 53%. The decline in fish meal use for ruminants reflects European Union (E.U.) restrictions on the use of fish meal in ruminant diets-ostensibly because of the fear of adulteration of fish meal with meat and bone meal (prohibited because of BSE [bovine spongiform encephalopathy] concerns).

Another new key word is traceability. Prohibition of the use of fish meal in ruminant diets will be lifted, but only when E.U. bureaucrats can be convinced that fish meal producers can satisfy zero tolerance for any mammalian protein in fish meal.

Chloramphenicol is also a new key word. This is a good example of new emphasis on safety, traceability, and accountability. It is an E.U. example, but as E.U. bureaucracy goes, so goes the rest of the world tomorrow. Safe food is the new motherhood issue. It is good politics, impossible to argue against, and it will not go away-nor should it.

Chloramphenicol is an antibiotic used only as a last resort because of severe side effects. It has been used illegally in shrimp aquaculture in the Far East, and the authorities there have now tightened inspections and penalties for illegal use to an extreme degree. In November 2001 the Dutch authorities detected chloramphenicol in a container of imported frozen shrimp. They ordered the product destroyed but, instead, the shrimp was added to herring offal and sent to a fish meal plant in Cuxhaven. When the Dutch authorities discovered this, they notified the German authorities--unfortunately by mail and over the Christmas period! By the time the German authorities took action it was mid-January and the original 20 t of shrimp had been incorporated into about 1,000 t of fish meal and distributed to customers as far apart as Italy and Russia. When the fish meal lots were recalled and analyzed, chloramphenicol could not be detected but the regulators argued that this was immaterial. Chloramphenicol was detected in the shrimp (which would make about

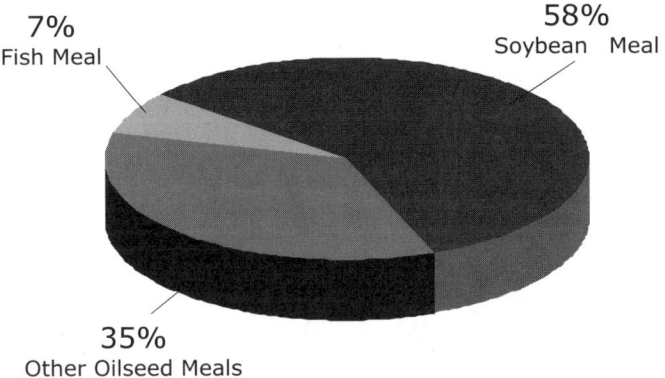

Figure 7. Major protein meals, 1986.

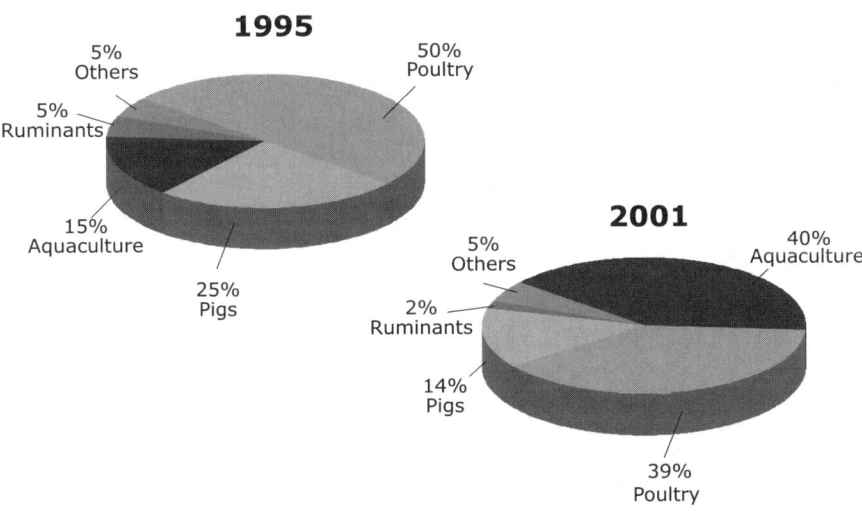

Figure 8. World fish meal use, 1995 and 2001.

4 t of meal), and thus it must be present in the fish meal, which must be destroyed. I include this cautionary tale to illustrate the new regulatory climate, and to emphasize that, as with any other food processor, manufacturers of fish byproducts of any kind need to have absolute product security and safety.

Figure 9 shows world consumption of 17 major oils and fats in 1990-1991 and 2000-2001. Note that fish oil comprised 1.4% of the total, and soybean oil was 20% in 1990-1991. Ten years later fish oil has fallen to 0.9% of total world oil and fat consumption, and soybean oil has risen to 23.7%. Total consumption of oil and fat rose by 39.9 million t (49%) in the 10-year period. World fish body oil production was 1.26 million t in 1990 and 1.12 million t in 2001.

Figure 10 shows world fish oil use in 1995 and 2001. In 1995 aquaculture used 18% and hydrogenation 70%. By 2001 the proportions were almost exactly reversed. Aquaculture used 70% and hydrogenation 19%. I predict that in the very near future, use of fish oil for hydrogenation will virtually have disappeared and aquaculture and nutraceutical use will have achieved absolute dominance.

It is perhaps ironic that this conference [2nd International Seafood Byproducts Conference] is in Alaska, where aquaculture, apart from salmon ranching, is prohibited by law, and I believe that the most important market for a major fishery byproduct, fish oil, is aquaculture feed. Aquaculture is not going to disappear; it comprised 27% of the world supply of fish in 2000 compared with 73% from capture fisheries. The latest estimate for 2001 is 70:30 (38.25 million t from aquaculture and 90 million t from capture fisheries).

The conventional fish meal and fish oil process is well suited to handle the major industrial fisheries of Peru, Chile, Iceland, Norway, Denmark, and others. And aquaculture and animal feeds need all their production.

Conventional fish meal plants can handle filleting waste very well when there is a steady year-round supply. Excellent examples of this success are the large fish meal plants associated with major Alaskan fish processing operations, and plants such as those at Aberdeen in Scotland.

There are many other ways of handling fish processing waste efficiently. They include (and should include) maximizing human consumption use, as in the example I gave of deboning "v" cuts. Much filleting waste is high in bone content, and thus in ash, so that byproduct fish meals may have over 20% ash compared with a low value of 14% (including salt) in conventional whole fish meals.

The expertise is here in this room to describe these alternatives. They include autolysis and enzymatic hydrolysis, where the bone can be screened out of the liquefied product. Salt can be removed from stickwater by reverse osmosis, concentrating the stickwater at the same time, and reducing ash (salt is recorded as "ash" in proximate analysis).

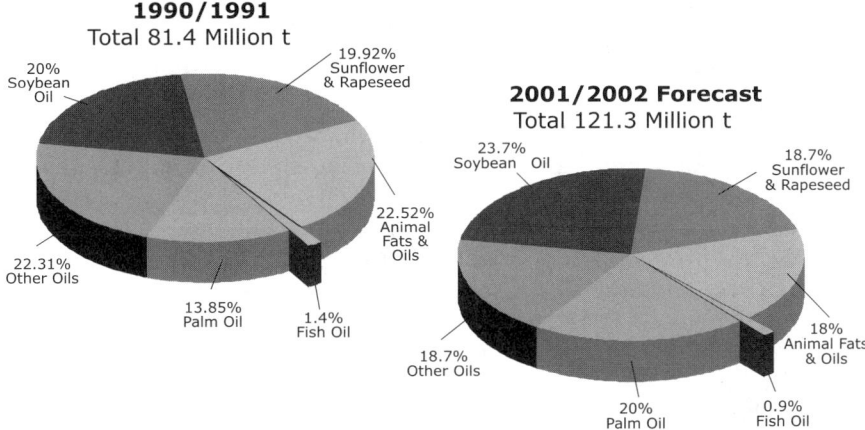

Figure 9. World consumption of seventeen major oils and fats, 1991-1992 and 2001-2002, showing an increase of 39.9 million t (49%) for those years.

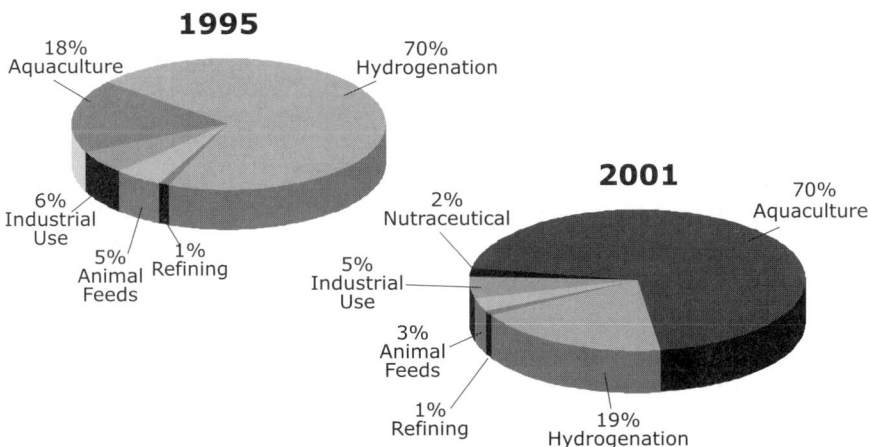

Figure 10.   World fish oil use, 1995 and 2001.

In my opinion, co-drying is much neglected as an economical and efficient way of incorporating marine protein into feed for pigs, poultry, and ruminants. In the early 1980s I visited the National Marine Fisheries Service in Seattle and was most impressed by a hands-on researcher who was co-drying fish hydrolysates with various vegetable proteins and cereals. His name was Ron Hardy and he is also presenting at this conference.

Many niche products can provide profitable sidelines when derived from fish byproducts. I have referred to fish oil capsules earlier. The potential is frightening! Fish bones are an obvious source of phosphorus in animal feeds and fertilizers. Poultry can utilize high ash fish meals to advantage. Also, fish bones are softened in the canning process, and are an excellent digestible source of calcium and phosphorus. I enjoy canned salmon skin, bones, oil, and all. I remember a market survey which found that many people used canned salmon by draining off all the liquid, including the oil with its EPA (eicosapentaenoic acid) and DHA (docosahexaenoic acid), separating and discarding the skin and bones, mashing the salmon up with lots of mayonnaise and voila—a salmon sandwich. This seems almost criminal to me. What about extra-bone canned salmon (compare "extra pulp" orange juice)?

The conference has provided an excellent overview of the 2002 potential for fish byproducts. In my presentation at the 1990 conference, I acknowledged the help of the International Association of Fish Meal Manufacturers in providing industry statistics. Since then the organization has changed its name, first to IFOMA (the International Fish Meal and Fish Oil Association); then in 2001, it merged with the Fish Meal Exporters Organisation, to form IFFO (the International Fishmeal and Fish Oil Organisation). Its Director General then and now is Dr. Stuart Barlow, and we are fortunate to have him here today. I would like to thank him, Ian Pike, and Jean-François Mittaine for providing excellent statistics. Oil World is also an excellent source of information on all edible oils and vegetable protein meals. I also owe a particular debt of gratitude to Dr. Ulf Wijkström of FAO.

World Market Overview of Fish Meal and Fish Oil

Stuart Barlow
IFFO, St. Albans, Hertfordshire, U.K.

Introduction

IFFO (International Fishmeal and Fish Oil Organisation) is the international non-governmental trade organization representing fish meal and oil producers worldwide. It has more than 200 member companies in 38 countries. Two-thirds of the world's producers of fish meal and fish oil are members of IFFO, and 95% of the exporters of fish meal and oil are also part of IFFO. So it is pleasing to be here and to see some of the IFFO members, producers, and associate members who will be very familiar with what I am going to say. The others I hope will learn a little something.

Growth of fishing industry since the 1950s

I want to take you on a fifty-year historic trip of the world's capture fisheries from 1950 through to the end of the 1990s. Capture fisheries grew almost continuously year-by-year until the 1990s, and from the 1990s onward they have been almost static. The increased demand has been met by an increase in aquaculture production. One-third of the world's catch is going into manufacturing fish meal and fish oil.

The per capita supply of fish was almost static over the period from 1950 to the end of the 1990s. But we are coming to a watershed now with capture fisheries almost plateauing. What is going to happen as the population continues to increase? The supply of fish for per capita consumption inevitably is going to go down. What role can aquaculture play in that and what role do fish meal and fish oil play in aquaculture?

Figure 1 shows the fish meal use in the year 2002, and projected for 2010. For 2002 we estimate that the biggest proportion (about 34%) is going into aquaculture, 27% into poultry, 29% into pigs, and 10% into the rest. This means that the majority of fish meal is still going into non-aquaculture species. Therefore there is room for growth in aquaculture without taxing the availability of fish meal, because poultry and pigs can

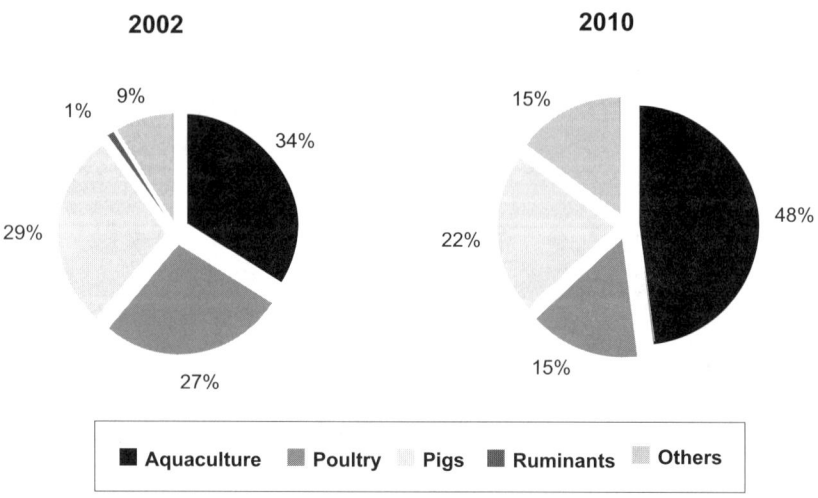

Figure 1. Summary of world fish meal use for 2002, and projections for 2010. (Zero neglible use by ruminants is projected for 2010.)

use vegetable proteins much more readily than many carnivorous fish. Carnivorous fish have great difficulty with vegetable proteins.

Figure 2 shows use of fish oil in aquaculture in 2002, and projections for 2010. For 2002 we have an absolute figure of fish oil consumption in aquaculture which we have then divided by the annual average production of fish oil. This is about 1.25 million metric tons, resulting in an estimated 56% going into aquafeed. In reality in 2002 we are expecting global fish oil production to be about 1 million t, not 1.25 million t, which means that share of aquaculture use will increase to about 75%.

Figure 3 shows the production of fish meal over the past 11 years— 1990 to 2001. Production totals for the entire world decreased in 1998 because of the El Niño in South America. The IFFO-5 countries dominate the world fish meal scene. The combination of Chile, Peru, and Scandinavia make up the IFFO-5, Scandinavia being three countries—Norway, Denmark, and Iceland.

Figure 4 shows catches of fish in Peru for fish meal production from 1970 to 2001. In 1970 Peru was catching 12 million t, and then during the severe El Niño of 1972-1973 the number dropped. There was some recovery but it was very slight. There was a change of species—when anchovy numbers were reduced, sardines started to become quite significant. Again in 1982-1983 there was another El Niño, which caused another reduction, and then slowly the stocks built up again until a big drop at the time of the 1998 El Niño.

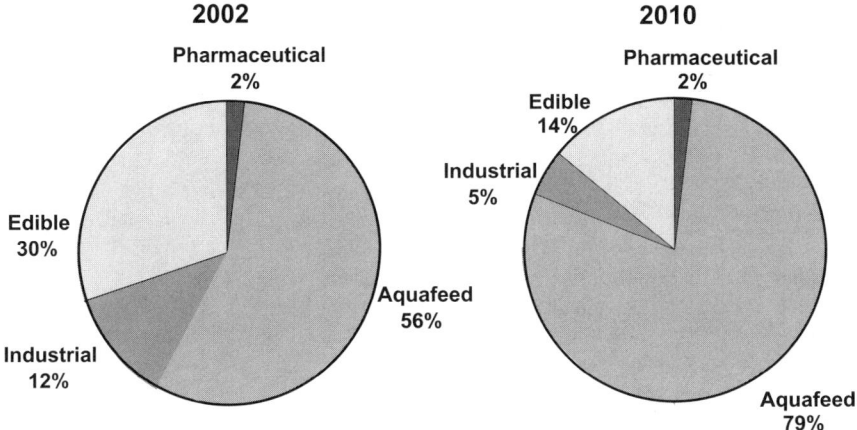

Figure 2. Summary of world fish oil use for 2002, and projections for 2010.

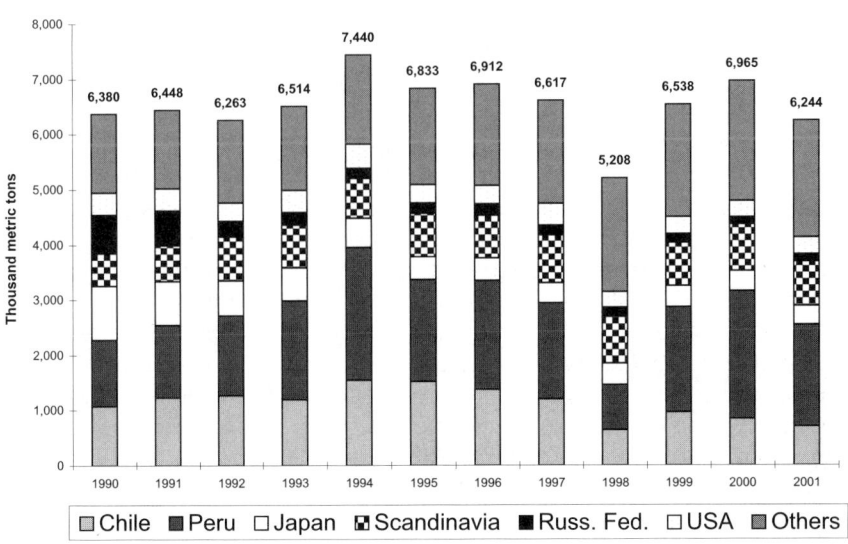

Figure 3. Major fish meal producers worldwide. Scandinavia represents Denmark, Norway, and Iceland.

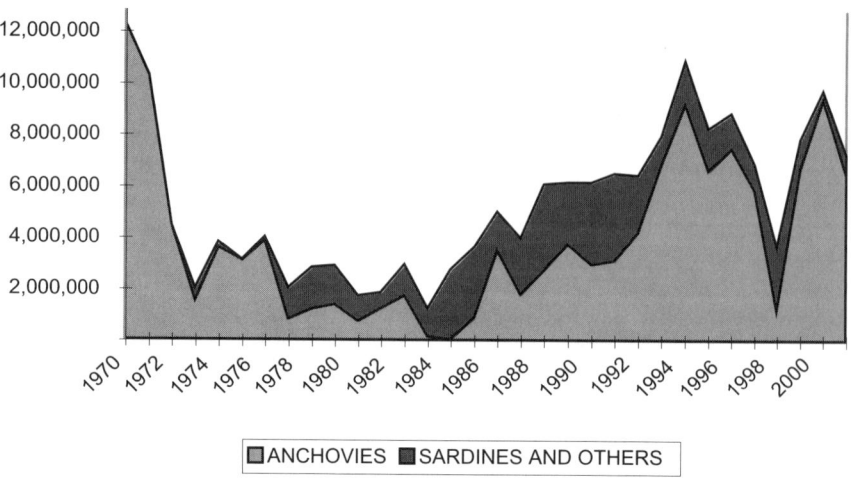

Figure 4. Peru fisheries harvest for meal production, 1970-2001.

Interestingly, at our IFFO Annual Conference, which took place in late 1997 in Rome, FAO biologists were extremely concerned with the forthcoming El Niño. They knew it was going to be one of the biggest on record and were concerned about the effect on Peru and the Peruvian fisheries, since in 1972-1974 El Niño reduced stocks for quite some time. For me one of the key factors is that the fishing industry has matured in terms of management of the fish resources. The Peruvian government in 1998 was very proactive in anticipating the El Niño. NOAA and other organizations are now much better at predicting El Niño. The Peruvian government took precautionary measures including a reduction in fishing. The fisheries were able to recover afterward. This is a major point that we have been making as an industry to anybody who will listen—the media, the supermarkets, etc.—that sustainability is very important to our industry. We are not out there to kill the fish and not worry about the resource next year. We are worried about the resource next year and were pleased to see this recovery of the fisheries.

Figure 5 shows Peruvian fish meal production and export for the same period, 1970-2001. Peruvians do not use very much of the resource at home; it is mainly exported.

Figure 6 shows fish meal production in Chile for the period 1970-2001. In 1974 Chile hardly produced any fish meal. Then production went up, particularly with increase in the catch of jurel in southern Chile. Production peaked about 1994 and has decreased since that time. Chile has gone through a major change of management in their fisheries. Now

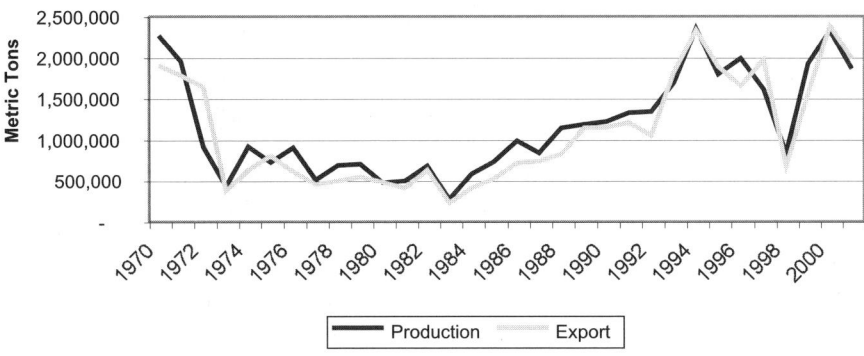

Figure 5. Peru production and export of fish meal, 1970-2001.

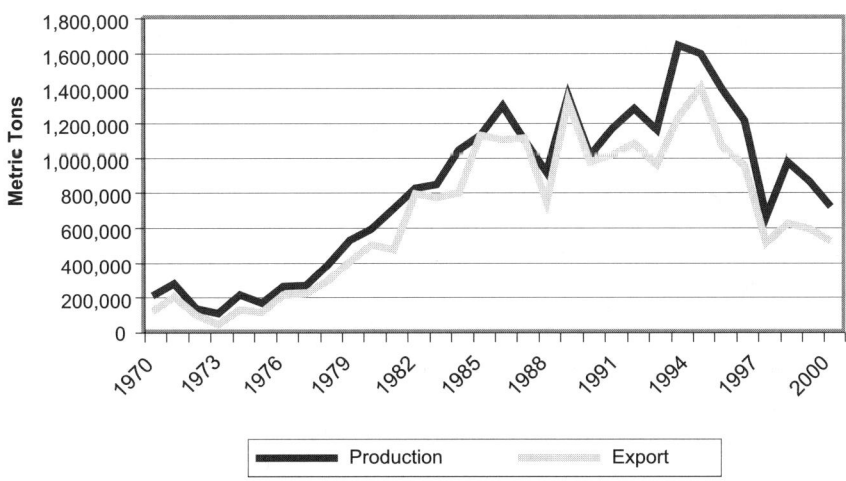

Figure 6. Chile production and export of fish meal, 1970-2001.

they have individual quotas for companies, as opposed to the Olympic style of "shooting the gun" and going out to fish as much as they can get. So again management has played a key role in making this fishery sustainable. For many years Chile exported as much as it could produce. Then from about 1990 onward exports have dropped below production because of the growing salmon aquaculture industry within Chile using much more of their fish meal and fish oil.

So what do we learn from this knowledge? The El Niños of 1972-1973, 1982-1983, and 1997-1998 did have an impact on fish meal and fish oil production. We are much more aware of the importance of these events and we have the management in place to ensure sustainability.

Figure 7 shows the exports of fish meal from 1990 to 2001 coming from the three principal areas—Chile, Peru, and Scandinavia—with the rest of the world playing a small part in the trade of fish meal. In 2001 a trade of about 4 million t of fish meal took place. Figure 8 shows a breakdown of exports in 2000 in a different way: 52% came from Peru, and 13% from Chile.

Figure 9 shows fish meal export destinations principally from the IFFO-5 in the year 2000: 28% go to China, 29% go to other Far East countries, 32% to Europe, and some goes to the Americas, Middle East, and Eastern Europe. Europe has decreased—in 2001 it came down slightly and it has decreased again in 2002 because of controls that have been put in place as a result of European food policy. Politicians are determined to rule the food industry with a very strong hand. Food scares have created all sorts of mayhem within Europe, and it has resulted in a slight reduction of fish meal use within Europe.

Figure 10 shows worldwide fish meal consumption. This is not only exported meal but also domestically produced meal. The Far East is the largest consumer of fish meal, with China consuming 1.5 t.

Fish meal outlook in 2003

At the IFFO Annual Conference, in October 2002 in Mexico, we talked about the outlook for fish meal for the years 2002 and 2003. The principal market maker is the IFFO-5, but mainly Peru. In the beginning of November 2002 no stocks were available in the IFFO-5 countries, and Peru had just started fishing after an unusually long fishing halt during the spawning season. IFFO estimated at its conference that the production of meal for the six months from October 1, 2002, through March 31, 2003, would be 1.28 million t, with Peru producing about 500,000 t, Chile 420,000 t, and Scandinavia 360,000 t. This would be only a 1.5% decrease in production over the same period in 2001-2002. That is in spite of mild to moderate El Niño anticipated to start sometime in the first quarter of 2003 and gaining momentum in the second quarter.

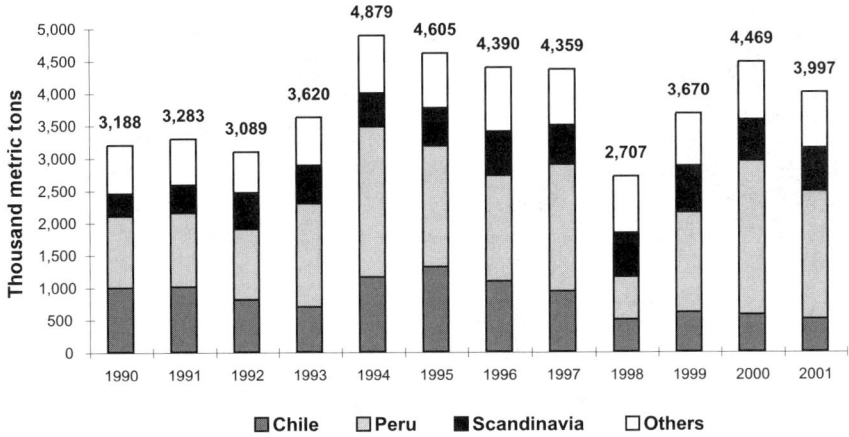

Figure 7. Major fish meal exporters worldwide, 1990-2001.

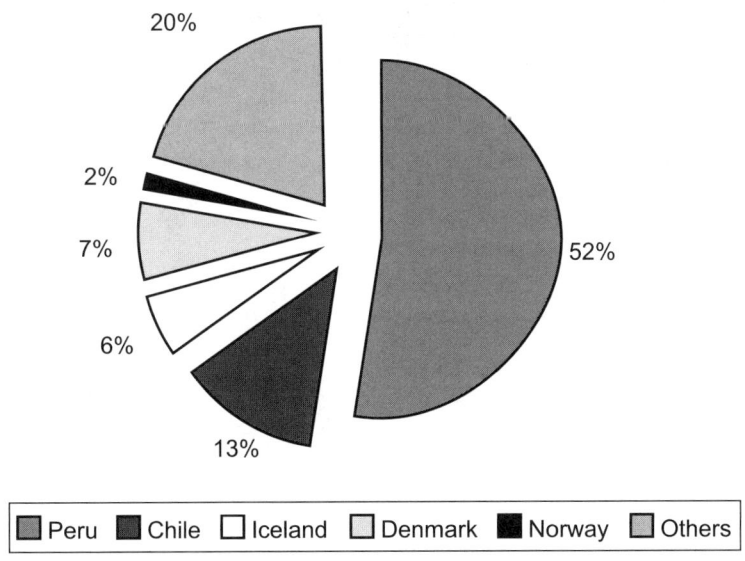

Figure 8.   World exports of fish meal, 2000. (Total is 4.5 million metric tons.)

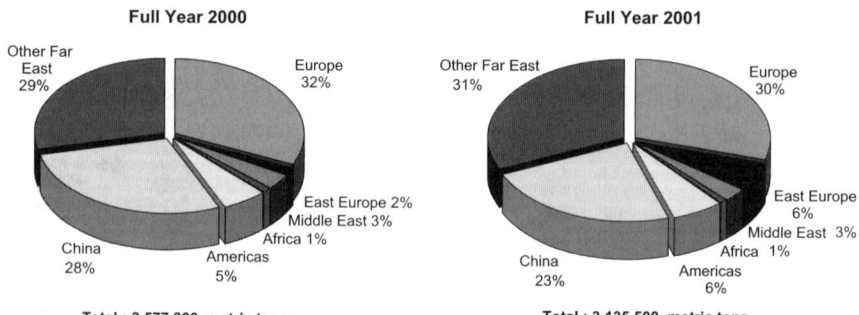

Figure 9. Fish meal exports by country of destination, 2000 and 2001. Information represents exports by IFFO-5.

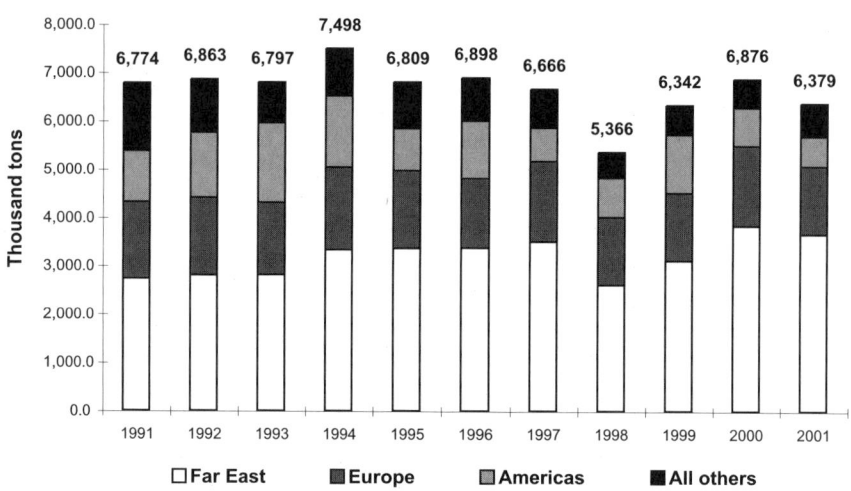

Figure 10. World fish meal consumption and stocks, 1991-2001.

We learned at the conference that experts predict that this El Niño will increase the sea temperature by about 1.5°, which should have little effect on the catch of anchovies. We do not expect to see the El Niño having too much effect on fish meal production. We emphasize that this 1.2 million t of production is a maximum potential as fishing is regulated directly or indirectly through quotas.

Exports for the same six months, October 1, 2002, through March 31, 2003, are estimated at 1.1 million t, which would leave an anticipated stock of about 350,000 t on March 31, 2003. Table 1 shows that this would be the lowest recorded stock in the past six years or more, with an average stock for the past five years at about 600,000 t on that date. These numbers refer to stocks within the IFFO-5 countries. Stocks of fish meal in China have been labeled as excessive at 200,000 t, which might have an impact on the market. It should be recognized that this represents less than two month's supply for Chinese consumption. While we cannot rule out a limited short-lived price correction, the markets fundamentally continue to look healthy with the industry expecting stable prices.

Impact of El Niño

Figure 11 shows prices for fair average quality (FAQ) fish meal from 1997 to the present, in European prices delivered into Europe. Also shown is the price ratio against soybeans. Not too long ago we regarded the ratio of 1.8:2 as being the normal price ratio of FAQ fish meal to soybeans. Then the 1997-1998 El Niño came, fish meal production went down, and the prices went up. The ratio against soybeans went up because soybean production remained relatively static over this period. After the El Niño, prices came down again, but started to increase because aquaculture demanded more and more of the fish meal. As a result ratios against soybeans have increased and therefore value against soybeans has become better.

Fish oil

Figure 12 shows world fish body oil production. Chile, Peru, Scandinavia, and the United States are the main suppliers of fish oil. More fluctuations are seen in fish body oil production that for fish meal, due to the El Niño. Averaging figures over the past ten or eleven years comes to a value of about 1.25 million t of fish oil produced annually.

Figure 13 shows marine oil exporters. Japan used to be an exporter, but is now a net importer. Scandinavia, the United States, Chile, and Peru are the major exporters. Peru is quite variable, depending on what I call the "unknown factor." Production can be enormous and it can be very small. It was very small during the El Niño periods.

Table 1.   Physical stocks of fish meal on March 31, for years 1992-2002, for Chile, Peru, Iceland, Norway, and Denmark (IFFO-5).

1992	1993	1994	1995	1996	1997	1998	1999	2000	2001 (est.)	2002 (est.)
478	1,035	824	623	556	442	357	571	1,039	604	430

Units are thousand metric tons.
Average for 1996-2000 is 593,000 t.

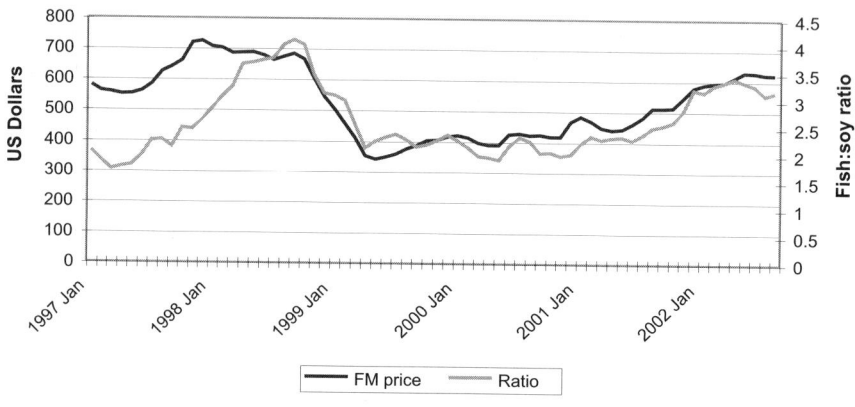

Figure 11.   European fish meal prices, and ratios of fish meal price to soy-bean meal price, 1997-2002.

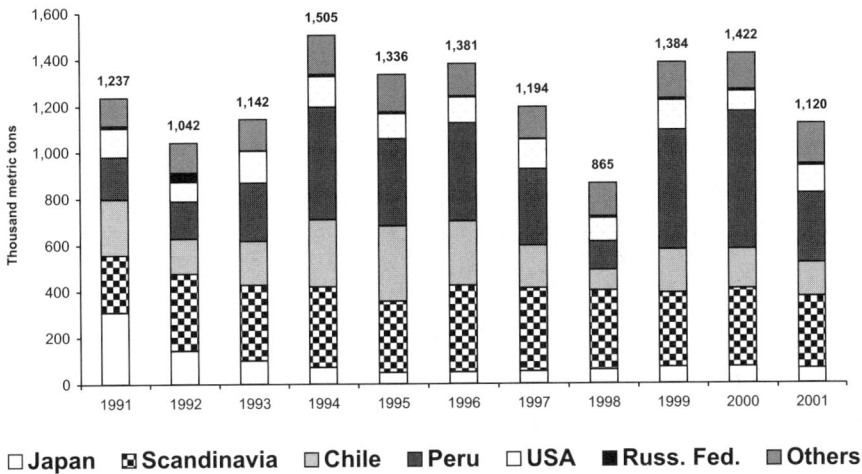

Figure 12. Major worldwide producers of fish body oil, 1991-2001.

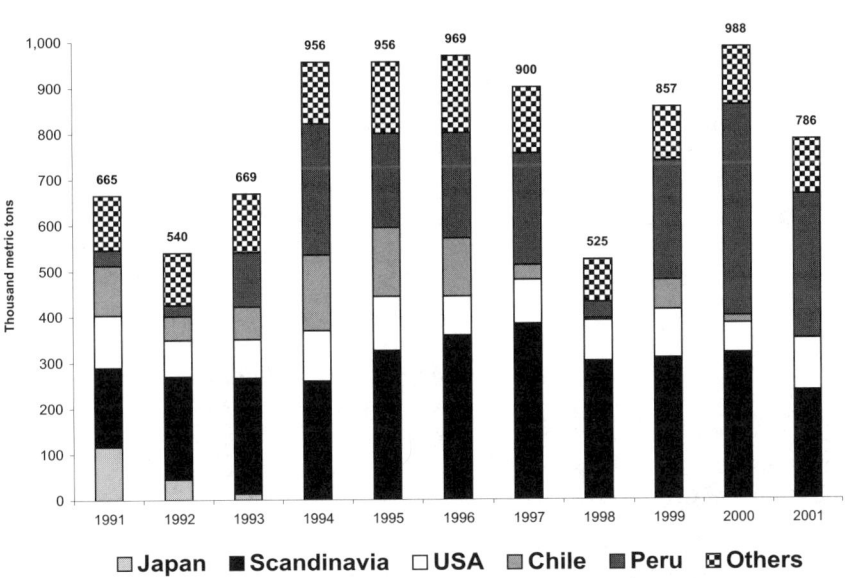

Figure 13. Major worldwide exporters of marine oils and fats, 1991-2001.

Figure 14 shows marine oil consumption—principally in the Americas, Europe, Scandinavia, and the Far East. The Far East is not the dominant consumer of fish oil. Most of this fish oil is going into salmonid consumption in Norway, Chile, Canada, and European countries.

Fish oil outlook for 2003

If current prices continue, together with the recovery of the farmed salmon industry, demand is estimated at between 900,000 and one million t of fish oil for 2003 (Fig. 15).

The big unknown is the supply. Historically a stable production of 600,000 t of fish oil is expected from the rest of the world, excluding Peru and Chile. This means that we would expect Peru and Chile during 2003 to produce between 300,000 and 400,000 t to meet the demand. But what will be the effect of the El Niño? In terms of capture fisheries it is anticipated to have little effect. But in terms of the oil yield of those fisheries it may have a significant effect. In 1998, which was the big El Niño year, production in these two countries reached only 210,000 t, while the average over the last five years was 520,000 t. We could see a considerable hike in the prices of fish oil if Peru and Chile are adversely affected beyond expectations by the El Niño.

Sustainability

Many of us remember the article in *Nature* by Naylor et al. (2000). The authors implied that the growth in aquaculture could put stresses on the fish meal and oil industry to meet increased demand, by ignoring fishing quotas imposed by responsible governments. In reality it has not been the case. In our reply to the *Nature* article, we plotted the growth of aquafeed production during 1985-2000, and fish meal production during the same period (Fig. 16). Fish meal production has not responded in any way to the growth in aquaculture production. However, Naylor et al. have raised the sustainability issue, and it has become an issue for the supermarkets, the customers, the clients, and the media.

Table 2 summarizes all of the management controls that are in place on the fish manufactured into fish meal. There are considerable controls. The sustainability role of our industry is an important image which we have to project to the world at large.

Projections for fish meal and fish oil use in 2010

Figure 1 shows our projections for fish meal use in 2010. Assuming that 6.5 million t of fish meal is still being produced in 2010, and about half of it goes into aquafeed, that still leaves considerable opportunity for more to

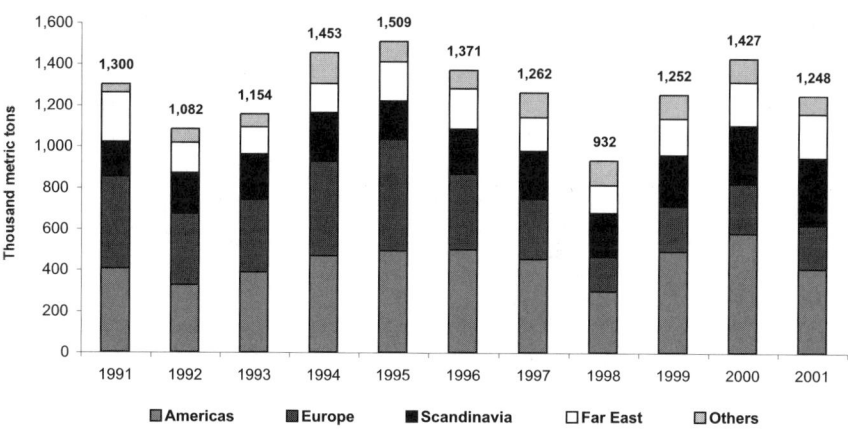

**Figure 14. Major worldwide consumers (and stocks) of marine oils, 1991-
2001.**

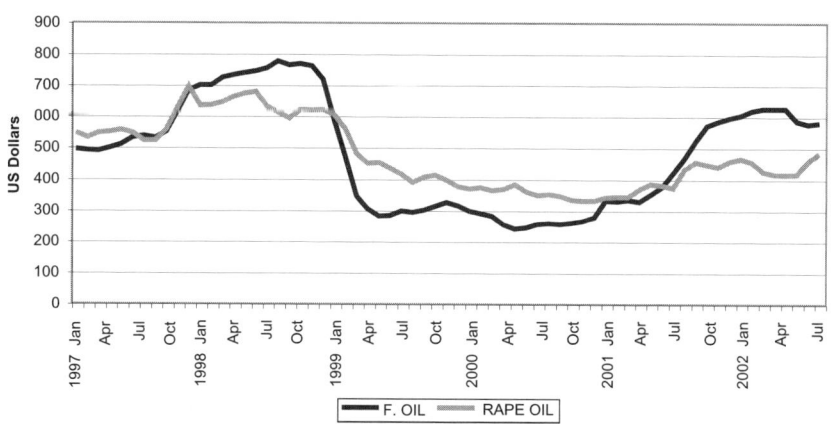

Figure 15. European fish oil and rapeseed oil prices, 1997-2002.

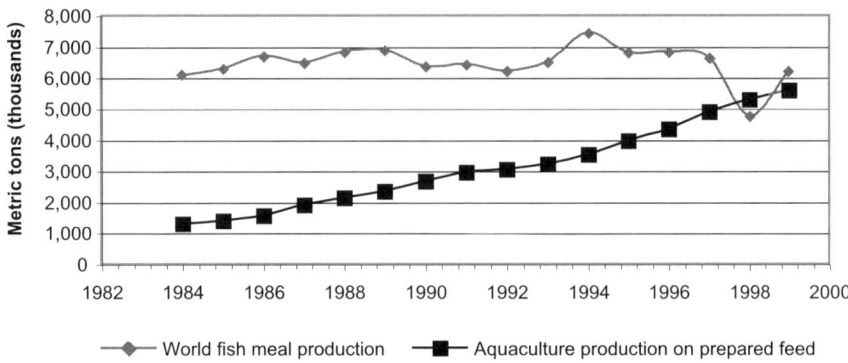

Figure 16. Worldwide fish meal production and growth of aquaculture, 1980-2000. Aquaculture species include salmon, shrimp, eels, marine fish, diadromous fish, and tilapia (carp, bivalves, and plants are excluded). 1999 numbers are projected.

Table 2.   Fishery controls for sustainability.

	Total catch limit	Area catch limit	Min. mesh sizes	Fleet cap controls	Closed areas	Seasonal bans	Min. landing sites
Anchovy	√	√	√	√	√	√	√
Sardine	√	√	√	√	√	√	√
Jack mackerel	√	√	√	√	√	√	√
Horse mackerel	√	√	√	√	√	√	√
Sandeel	√	√	√	√	√	√	
Sprat	√	√	√	√	√	√	
Norway pout	√	√	√	√	√		
Blue whiting			√				
Capelin	√	√	√	√	√	√	√
Herring	√	√	√	√	√	√	√

go into aquafeed as aquafeed grows. Therefore we do not see fish meal being a constraint on the growth of the aquaculture industry for some years.

But fish oil is a different picture. The projections in Fig. 2 are based on an annual world production of 1.25 million t of fish oil. If aquafeed demand is at about 1 million t of fish oil in 2010, depending on the production of fish oil, aquafeed use could be around 80% or even close to 100%. We believe aquafeed composition is going to be impacted by a lack of fish oil.

Concluding remarks

We are as an industry concerned about the future availability of raw material for fish meal and fish oil production. We would like to see more of it going into fish meal and fish oil, but at the present levels of fish landings it does not look likely. FAO through one of its agencies did a survey which showed that about 25-30 million t of fish are still being discarded or wasted and not produced into any usable product. In other words, it is equal in amount to the current quantity of raw material going into the fish meal industry. My plea to you, particularly those of you who are part of the Alaska fishing industry, is that I understand there is some scope for reducing the amounts of wastage, and increasing the amount of recycling. We need as much material as we can get.

I am still grappling with the concept of the cost of diesel oil here, and why you burn fish oil. Serious thought, some energetic thinking, is required on how you can organize your industry so that fish oil is not burned. We need it in the aquafeed business and we need you to come up with some clever solution to help us in meeting the aquafeed demand.

Reference

Naylor, R.L., R.J. Goldburgh, J.H. Primavera, N. Kautsky, M.C.M. Beveridge, J. Clay, C. Folkes, J. Lubchenco, H. Mooney, and M. Troells. 2002. Effect of aquaculture on world fish supplies. Nature 405:1017-1024.

Growing Requirements for Fish Meals and Fish Oils

Albert G.J. Tacon
Aquatic Farms Ltd., Kaneohe, Hawaii

Need for better definitions and information

Fish meal and fish oil production represent one of the oldest and most profitable segments of the marine capture fisheries sector. However, the terms *fish meal* and *fish oil* are very broad generic terms for those products derived from the processing of whole fish and/or fish/shellfish waste that has been processed through cooking, pressing, drying, and milling; fish oil usually being a valuable byproduct of the fish meal manufacturing process.

The nutritional quality and subsequent economic value of a *fish meal* or *fish oil* is dependent on a variety of factors, including the fishing method employed (including temperature and duration of storage prior to processing), the nutrient composition of the fish/shellfish processed (depending upon species mix, fish age, fishing season, and body parts processed), and last but not least the cooking, drying, and/or stabilization methods employed to produce the meal or oil. Fish meals having the highest nutritional quality and market value are those produced from rapidly processed, fresh, uncontaminated whole small pelagic fish species and dried indirectly at low temperatures.

However, at present there is no segregation of global fish meal and fish oil production data with respect to quality or price (Table 1; fair average quality—FAQ, standard, prime, super prime, low temperature—LT). Moreover, the bulk of available statistical information concerning global fish meal and fish oil production is still not being reported down to a species level (Tables 2-3); 79.4% and 43.3% of global fish meal and fish oil production not being reported as non-species specific in 2000, respectively (FAO 2002; Table 4).

Clearly, this situation must be remedied with improved statistical information gathering and reporting from the major fish meal and fish oil (Table 5) producing countries. By so doing, the sector would also be viewed as being much more open and transparent in eyes of the public and consumers.

Table 1. Examples of quality criteria of some Peruvian fish meals (% by weight, dry as-fed basis).

	VLT[a]-steam hot air dried			Direct dried	
	Super prime	Prime	Standard	FAQ[b]	Flame select
Protein, % min	68	67	66	65	67
Fat, % max	10	10	10	12	10
Moisture, % min-max	7-10	7-10	10	10	10
Ash (salt free), % max	13	14	14.5	15.5	14
Salt, % max	3	3	3.5	–	3
Sand, % max	1	1	1	2	1
Salt and sand, % max	–	–	4.5	5	–
TVN mgN/100 g[c]	100	120	150	–	120
Histamine, ppm max	400	1,000	–	–	1,000
Torry modified digestibility, % min	94	92	–	–	90
Antioxidant, ppm min	150	150	150	150	150

[a]VLT = very low drying temperature.
[b]FAQ = fair average quality.
[c]TVN = total volatile nitrogen.
Fish meal prices range from US $415-420/ton for FAQ to US $605-615/ton for super prime.
Sources: AUSTRAL, http://www.frozenfish.com/fishmeal.com, and Fish Meal Market Reports, July 10, 2001,

Increasing demand and rising costs

Due to the limited supplies of fish meal and fish oil (total global production fluctuating between 5 to 7 million t for fish meal and 1 to 1.4 t for fish oil from 1976 to 2000) the demand for these products by the rapidly growing aquaculture sector continues to increase (Hardy and Tacon 2002). This is particularly the case for those major aquaculture producing countries culturing marine and diadromous shrimp and finfish species, including China (shrimp, eels, marine finfish), Thailand (shrimp), Norway (salmonids), and Chile (salmonids).

For example, Figs. 1-4 show the total production, imports, and exports of fish meals and fish oils in Chile, Norway, Thailand, and mainland China. Countries that were once net fish meal and/or fish oil exporting countries have now become major importers of fish meal and/or fish oil to satisfy to a large extent the domestic demand of the aquaculture sector as it continues to grow. In general, the demand for the higher quality fish meals and oils is greatest for the species that have more carnivorous feeding habits and are reared under more intensive culture conditions.

Table 2.   Reported total world fish meal production 1976-2000 (thousand metric tons; dry, as-fed basis).

	1976	1980	1985	1990	1995	2000
Total fish meal production	4,999	4,969	6,313	6,355	6,855	6,943
Fish meal derived from pelagic fish	4,089	4,110	5,541	5,745	6,332	6,649
Oily-fish meal (unspecified)	2,489	2,531	3,333	3,693	3,939	5,309
Anchoveta meal	960	379	291	576	804	417
Capelin meal	109	154	181	119	138	223
Jack mackerel meal	66	114	349	563	959	216
Menhaden meal	193	246	279	203	204	197
Pilchard meal	102	499	965	362	51	153
Tuna meal	38	47	39	41	49	50
Herring meal	36	30	25	43	66	42
Mackerel meal	20	36	7	45	30	21
Clupeoid fish meal (unspecified)	67	56	64	88	90	20
Sardine meal	9	16	7	12	<1	0
Fish meal derived from demersal fish	590	567	455	371	255	123
White fish meal (unspecified)	552	528	426	334	208	101
Hake meal	2	<1	<1	21	36	16
Blue whiting meal	<1	13	5	5	<1	6
Redfish meal	4	6	4	2	2	<1
Cod meal	<1	0	<1	<1	<1	<1
Gadoid fish meal	31	20	15	7	10	0
Sandeel meal	0	0	4	3	0	2
Other marine meals	290	242	290	200	257	161
Fish solubles[a]	283	236	288	196	94	80
Fish meal (unspecified)[b]	<1	<1	1	<1	151	78
Fish silages[c]	6	6	2	3	11	3
Crustacean meals	21	46	26	40	12	10
Crustacean meal (unspecified)	14	40	20	33	4	5
Shrimp meal[d]	1	1	<1	1	2	3
Crab meal[e]	6	5	7	5	5	3

[a]Dried or condensed fish solubles are derived from the drying or evaporation of the aqueous liquid fraction (stickwater) resulting from the wet rendering (cooking) of fish into fish meal, with or without removal of the oil.

[b]Fish meal is defined as the clean, dried, ground tissue of undecomposed whole fish or fish cuttings (processing waste), either or both, with or without the extraction of part of the oil.

[c]Dried or wet fish silages are derived either by ensiling fish with inorganic/organic acids or through microbial fermentation.

[d]Shrimp meal is the undecomposed ground dried waste of shrimp and usually contains parts and/or whole shrimp.

[e]Crab meal is the undecomposed ground dried waste of the crab and usually contains the shell, viscera, and part or all of the flesh.

Source: FAO 2002.

Table 3.   Reported total world fish oil production 1976-2000 (thousand metric tons; dry, as-fed basis).

	1976	1980	1985	1990	1995	2000
Total fish oil production[a]	1,024	1,217	1,481	1,412	1,382	1,352
Pelagic body oils	285	465	735	780	584	762
Anchoveta oil	177	96	154	200	383	597
Menhaden oil	85	132	126	124	108	87
Capelin oil	27	80	114	64	66	68
Herring oil	5	7	5	10	24	10
Pilchard oil	39	133	335	381	3	0
Cod liver oil[b]	25	15	9	11	11	3
Demersal body oils[c]	3	2	2	2	<1	0
Other fish liver oils (unspecified)	<1	2	8	14	13	18
Other marine oils	610	713	720	604	770	569
Fish body oils (unspecified)	582	673	640	482	768	567
Animal oils and fats (unspecified)	26	39	80	122	<1	1
Squid oil	1.6	2.0	0.2	0.5	1.7	0.5

[a]Fish oil is the oil from rendering whole fish or cannery waste.
[b]Demersal fish liver oil.
[c]Demersal body oils include Alaska pollock oil and redfish oil.
Source: FAO 2002.

Table 4.   Global fish meal and fish oil production by major species groups in 2000.

	% production	Million metric tons[a]
Fish meals		
Species specific	20.6	1.43
Non-species specific fish meals	79.4	5.51
Total reported		6.94
Fish oils		
Species specific	56.7	0.766
Non-species specific	43.3	0.586
Total reported		1.35

[a]As-fed basis.
Source: FAO 2002.

Table 5.   Reported major fish meal and fish oil producing countries in 2000.

	Fish meal (%)[a]	Fish oil (%)[a]
Peru	32.2	43.4
Chile	12.7	13.3
China	11.6	3
Japan	5.6	4.4
Thailand	5.6	–
Denmark	5.6	10.4
United States	4.8	6.4
Norway	3.8	6.5
Iceland	3.6	5.8
Others	14.4	8.6
Total	6,943,078 t	1,352,430 t

[a]Percent total production, as-fed basis.

Source for total reported production is FAO 2002.

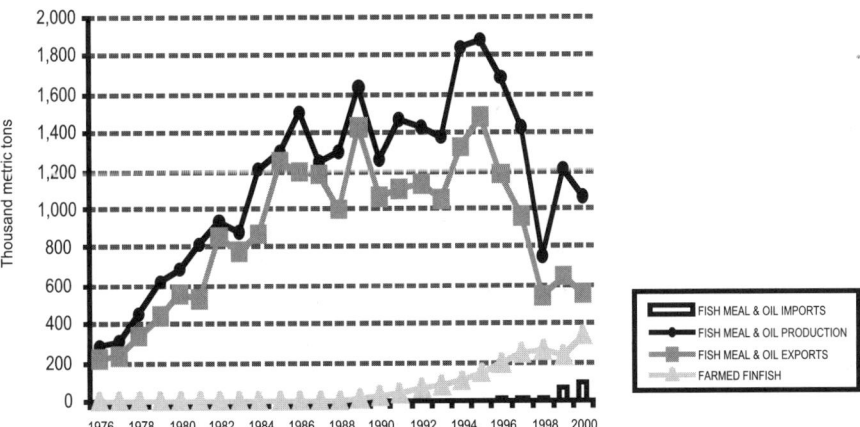

Figure 1. Total production, imports, and exports of fish meals and fish oils in Chile, including farmed finfish production.

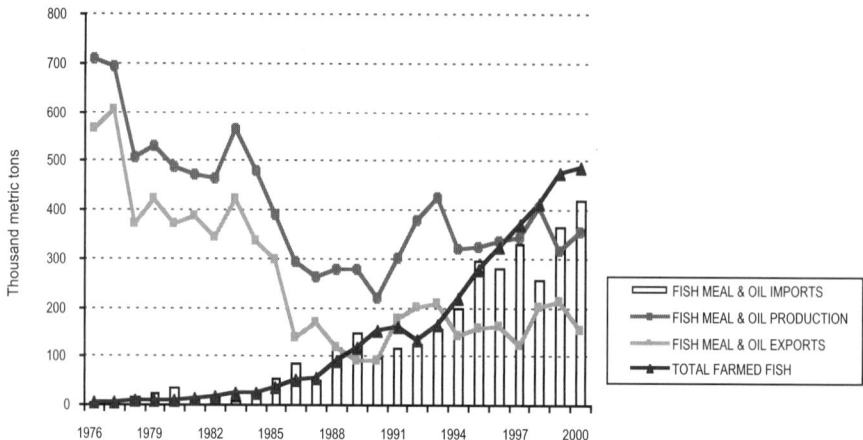

Figure 2. Total production, imports, and exports of fish meals and fish oils in Norway, including farmed fish production.

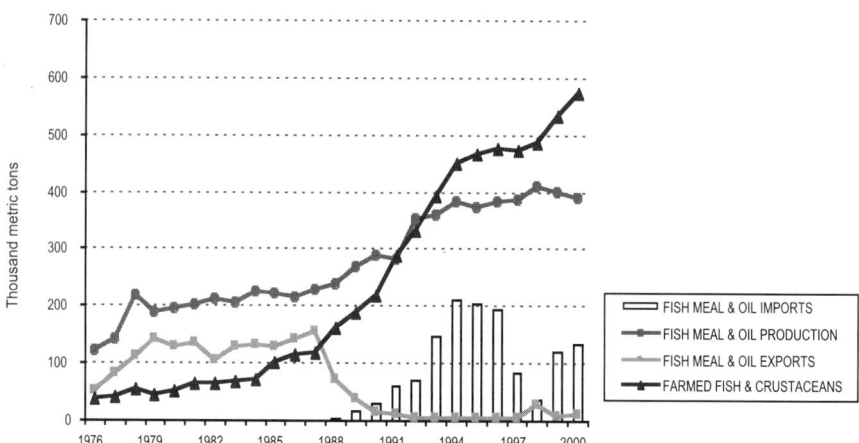

Figure 3. Total production, imports, and exports of fish meals and fish oils in Thailand, including farmed fish and crustacean production (FAO 2002).

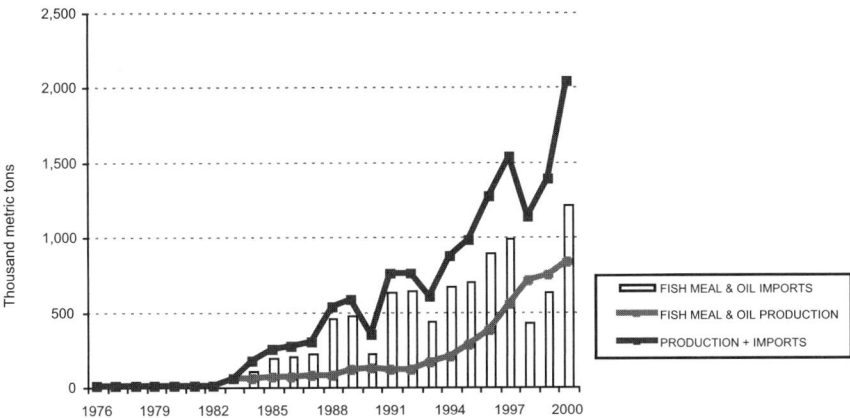

Figure 4. Total production and imports of fish meals and fish oils in China (FAO 2002).

Clearly, the countries that have their own domestic fish meal and fish oil supplies and/or have the greatest market buying power (for importing these commodities in large quantities, including possible tax incentives), will have the upper hand in the long run. Figure 5 shows the estimated imports of fish meal and soybean into mainland China from 1995 to 2001. These imports largely support the rapidly growing resident domestic aquaculture and livestock production sector. After the United States, China has the second largest compound animal feed industry in the world (Tacon and Forster 2000).

Coupled with the limited supplies of fish meal and fish oil, and increased fishing restrictions placed by major producing countries, it is perhaps not surprising that the increasing demand placed by the aquaculture sector for higher quality fish meals and fish oils has resulted in marked increases in the prices of these valuable commodities. For example, in Chile alone the price of fish oil has more than doubled from under $300 to over $600 per ton and the price of fish meal increased from $450 to $600 per ton from 2001 to 2002 (Fig. 6). Similarly, following the El Niño year of 1998, fish meal and fish oil costs (FAQ, FOB Peru) have increased from $349 in 1999 to $555 in 2002 per ton (Fig. 7), and from $222 to $495 per ton, respectively (Fig. 8). By comparison, the price of soybean meal and soy oil over the same period, has increased from only $163 to 182 per ton (Fig. 7) and $401 to 448 per ton (Fig. 8), respectively.

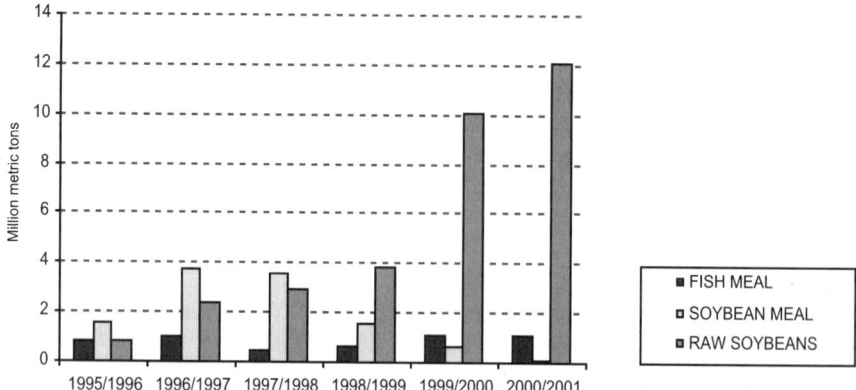

Figure 5. Estimated imports of soybean and fish meal into China from 1995 to 2001. Total soybean imports increased from 0.8 million t for the 1995-1996 growing season to 12.1 million t in 2000-2001, 24% of the total world raw soybean imports. By contrast, soybean meal imports have decreased from a high of 3.61 million t in 1997-1998 to a low of 0.12 million t in 2000-2001. Fish meal imports have fluctuated from 0.45 to 1.12 million t, and for 2000-2001 amounted to 1.08 million t or 29% of total world fish meal imports (Mielke 2001).

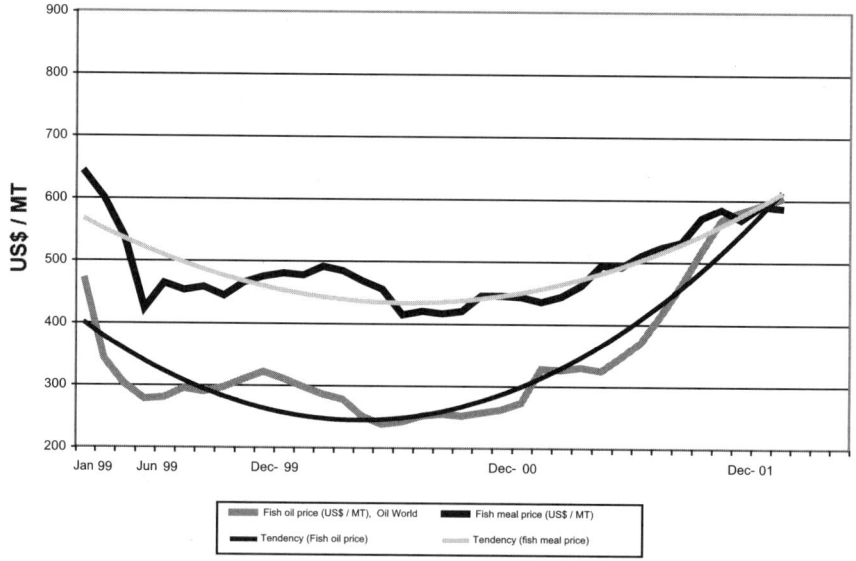

Figure 6. Average fish meal and fish oil price in Chile from 1999 to 2002 (modified from Hinrichsen 2003).

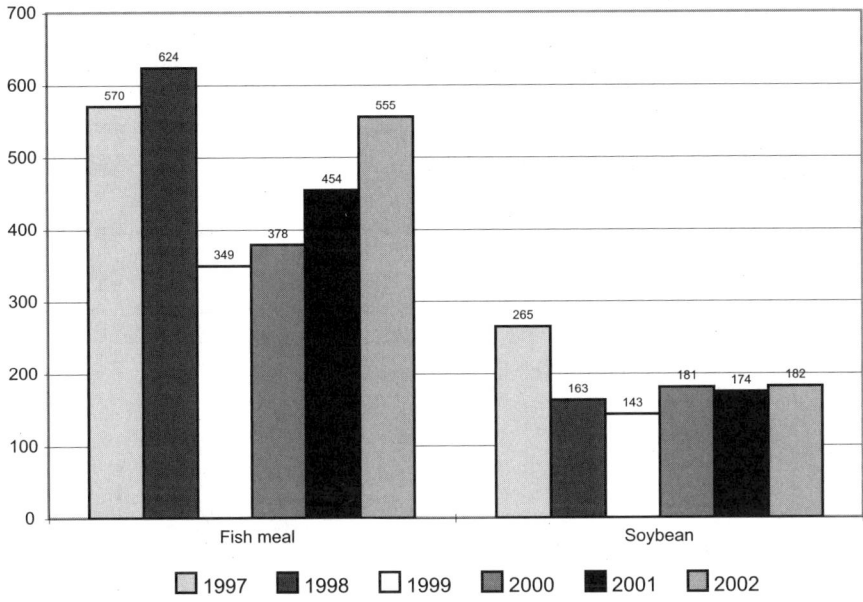

Figure 7. Evolution of yearly average prices for fish meal FAQ, FOB Peru) and soybean meal (FOB Brazil). Values are US $ per metric ton. Source: J.F. Mittaine, International Fishmeal and Fish Oil Organisation (IFFO), pers. comm., Jan. 2003.

U.S. seafood market, international trade, and static/decreasing prices

The United States is currently the world's largest seafood market by value, with consumers spending over $55.3 billion for fishery products in 2001, including $38.2 billion or 69% from food service establishments (restaurants, carry-outs, caterers), $16.8 billion or 30.4% in retails sales for home consumption, and $276.3 million for industrial products (NMFS 2002). To satisfy seafood demand, the United States imported a record quantity of edible seafood products in 2001 (Fig. 9), totaling 1.86 million t and valued at $9.9 billion; the United States being the second largest importer of seafood after Japan (valued at $15.7 billion). These two countries accounted for over 42% of the total global fisheries import trade in 2000 (FAO 2002). In addition to edible seafood products, the United States is one of the world's largest importers and exporters of nonedible seafood products (Fig. 9; the main nonedible seafood products including pearls, and to a lesser extent fish meals and fish oils). The total value of nonedible seafood imports and exports in 2001 was $8.7 billion and $8.6 billion, including $43 million and $109 million of fish meal/fish oil

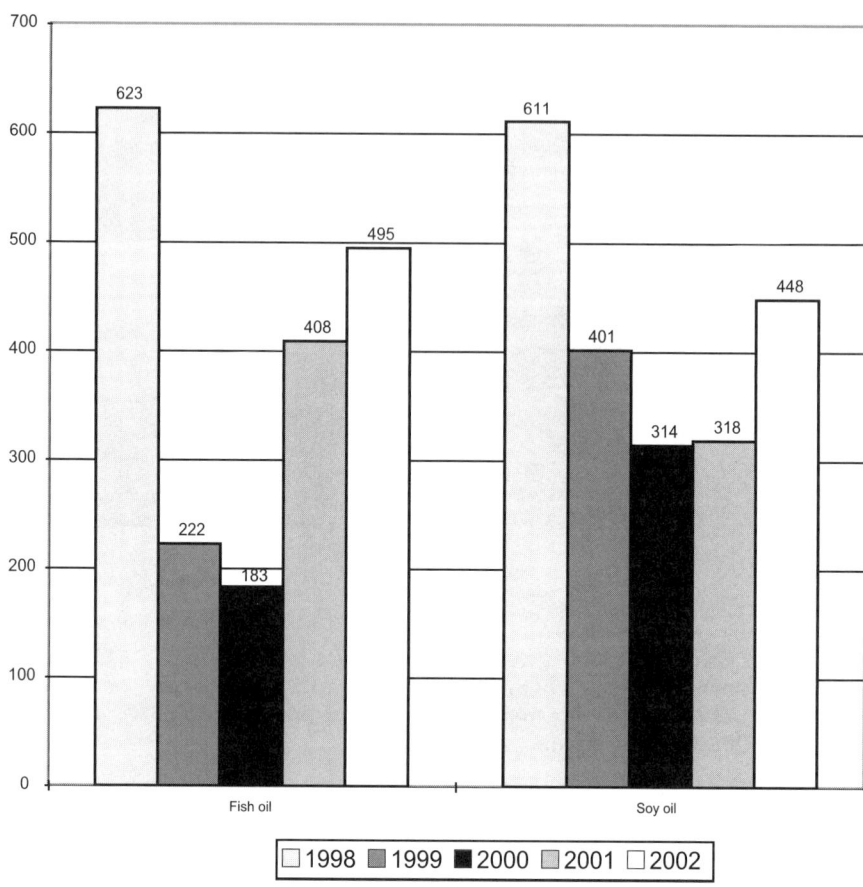

Figure 8. Evolution of yearly average prices for fish oil (FOB Peru) and soy oil (FOB Brazil). Values are US $ per metric ton. Source: J.F. Mittaine, International Fishmeal and Fish Oil Organisation (IFFO), pers. comm., Jan. 2003.

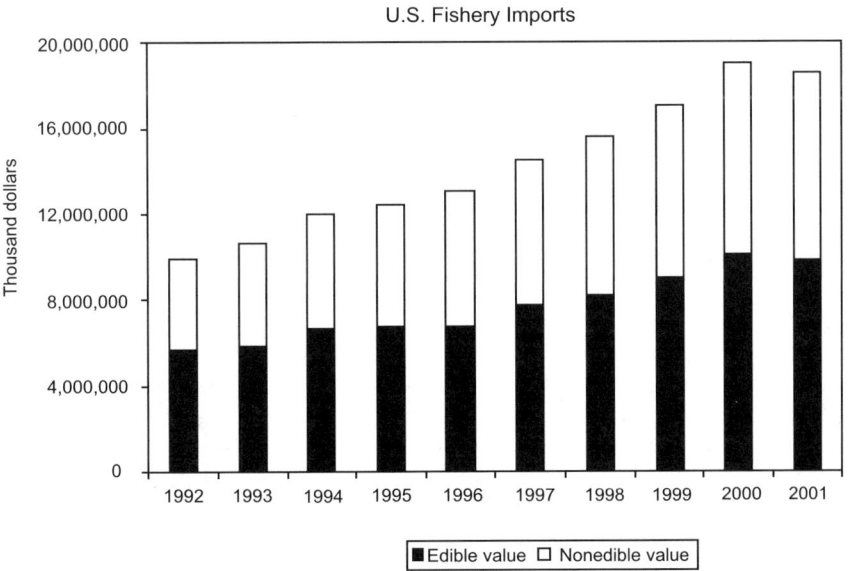

Figure 9. U.S. Fishery product imports and exports, 1992-2001 (NMFS 2002).

imports and exports, and \$40.8 million and \$6.9 million of ornamental fish imports and exports, respectively (NMFS 2002, USDA 2002).

Although total U.S. landings from capture fisheries and aquaculture amounted to 5.2 million t (Fig. 10) in 2000 and was ranked fifth globally (after China, Peru, Japan, and India), the United States had a trade deficit in edible and nonedible fishery products of \$6.8 billion in 2001, including 3.6 billion in shrimp imports or 37% of total edible fishery imports by value (NMFS 2002). The deficit in seafood products represents the second largest trade deficit after petroleum products for any natural resource product. Fresh and frozen fish and seafood are the top U.S. food import, followed by meat products, alcoholic beverages, processed fishery products, and canned fruits and vegetables.

During recent years commodity fish and shrimp prices have remained relatively static, and in some cases have actually decreased within many seafood importing countries. To a large extent these price decreases have been due to increased global production and supplies of the major seafood commodity species such as farmed shrimp and salmon. For example, according to Harvey (2002) in the United States the first six months of 2002 have shown a 37% decrease in catfish imports, a 22% increase in tilapia imports, a 15% increase in shrimp imports, a 27% increase in Atlantic salmon imports, and a 22% increase in mussel imports and with consequent static/decreasing domestic seafood prices.

Increased consumer pressure for improved feed and food safety

Recent fears concerning the possible transfer of mammalian infectious agents such as bovine spongiform encephalopathy (BSE) through animal feeding or concerning the possible contamination of animal feeds and human foods with dioxins, PCBs, and other environmental contaminants (i.e., heavy metals, antibiotic residues) have placed food safety issues on the radar screen of consumers, especially within developed countries (Fiedler et al. 1998, Howgate 1998, WHO 1999, Cooke and Sellers 2000, Hottlet 2001, IFOMA 2001, Josupeit et al. 2001). In this respect it is important to remember that the consumer sits at the top of the food chain, and consequently has the power to determine whether a food product is consumed or not, and thereby determine the ultimate market success or failure of a food product. However, the food safety hazards can be effectively controlled and eliminated through the application of the principles of the hazard analysis and critical control point (HACCP) system as a management tool from "farm to fork" (Reilly and Kaferstein 1997, Spencer Garrett et al. 1997). This approach implies that HACCP must be applied throughout the entire supply chain from the newly hatched egg and the water in which the animals or plants are reared and fed, to the handling and processing of the farmed produce, and continue to the education of the consumer.

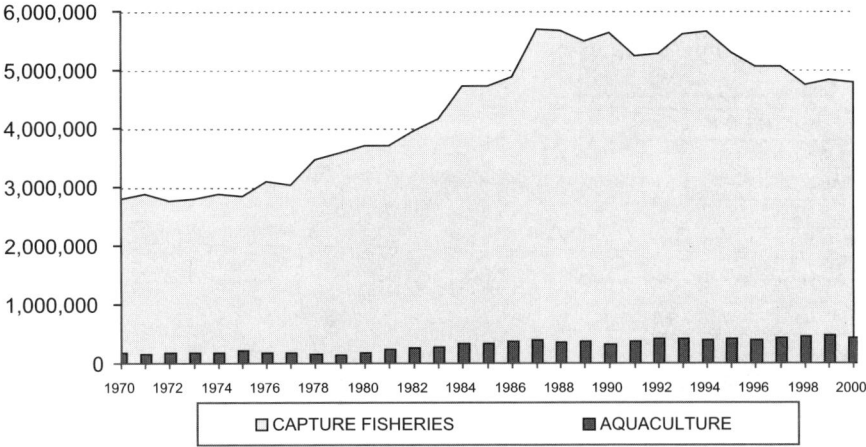

Figure 10. Total capture fisheries and aquaculture production in the United States, 1970-2000. Total capture fisheries production has increased from 2,794,298 t in 1970 to 4,787,683 t in 2000, growing at an average rate of 1.9% per year (1970-2000). Total aquaculture production has increased from 168,681 t in 1970 to 428,262 t in 2000, growing at an average rate of 3.1% per year (1970-2000). Contribution of aquaculture toward total fisheries landings in the United States has increased from 5.7% in 1970 to 8.2% in 2000 (FAO 2002).

Clearly, as the world's food and aquaculture supply becomes more global, food quality and safety issues will become more prominent, and so the need for traceability throughout the entire supply chain and development of food safety assurance schemes will become the norm rather than the exception. It is also important to note that as a direct result of the food safety scares and scandals in Europe, the fastest growing food sector globally has been the organic agriculture sector, including within the United States (Scialabba and Hattam 2002). For example, based on current estimates of certified organic aquaculture production, it is estimated that production will increase 240-fold from 5,000 t in 2000 to 1.2 million t by 2030 (Tacon and Brister 2002). These estimates are primarily based on existing organic aquaculture production levels from developed countries, and the assumption that the major markets for certified farmed aquatic products will be North America and Europe in the West, and Singapore, Japan, Australia, and New Zealand in the East. The latter is fuelled by the growing awareness within these countries of environmental pollution and the safety of aquatic products for human consumption, and awareness

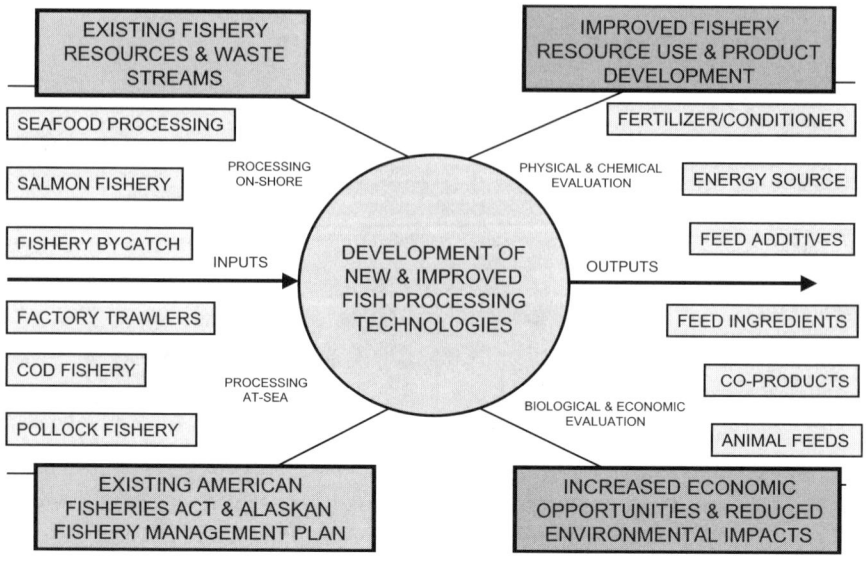

Figure 11. Alaska fisheries byproduct utilization.

of the state of global fishery resources and long-term sustainability of current aquatic food production systems.

The upshot of the above is increasing retailer and consumer pressure for improved feed and food assurance schemes, including the greater need for traceability and transparency in food production systems, and demand for more environmentally sustainable or *greener* aquafeeds and aquaculture produce (Tacon and Barg 2001). It follows therefore, that fish meal and fish oil will be the focus of increased pressure from aquafeed manufacturers and consumers alike, in terms of both nutrient and possible contaminant content, as well as the sustainability of the fisheries from which they are derived.

Concluding statement

Last, but not least, it is important to mention here that our growing requirements for fish meal and fish oil will also be directly influenced by environmental factors beyond our immediate control, namely our climate. In particular, there is also growing concern about the uncertainty of our changing climate and the direct effect of our changing ocean currents and weather upon marine fisheries landings and agricultural food production.

On a final note concerning Alaska and the subject of this conference,

Fig. 11 summarizes the advantages of improved fishery resource utilization within Alaska.

References

Cooke, M., and R. Sellers. 2000. Fair warning: New FDA surveys on dioxin in feed. Feed Management 51(3):19-21.

FAO. 2002. Fishstat Plus: Universal software for fishery statistical time series, 1970-2000. FAO Fisheries Department, Fishery Information, Data and Statistics Unit. Vers. 2.30. http://www.fao.org.

Fiedler, H., K. Cooper, S. Bergek, M. Hjelt, C. Rappe, M. Bonner, F. Howell, K. Willet, and S. Safe. 1998. PCDD, PCDF, and PCB in farm-raised catfish from Southeast United States: Concentrations, sources, and CYP1A induction. Chemosphere 37(9-12):1645-1656.

Hardy, R.W., and A.G.J. Tacon. 2002. Fish meal: Historical uses, production trends and future outlook for supplies. In: R.R. Stickney and J.P. McVey, (eds.), Responsible marine aquaculture. CABI Publishing, New York, pp. 311-325.

Harvey, D.J. 2002. Aquaculture outlook. Electronic outlook report from the Economic Research, USDA, LDP-AQS-16, Oct. 10, 2002. http://www.ers.usda.gov.

Hinrichsen, J.P. 2003. Influence of salmon and shrimp price on the quality of aquafeeds and feed ingredients used in Latin America. In: International aqua feed directory and buyers guide 2003. Turret West Ltd., Bartham Press, Luton, U.K., pp. 30-24.

Hottlet, O. 2001. Heightened surveillance of chloramphenicol residues in shrimp. Global Aquaculture Advocate 4(6):10.

Howgate, P. 1998. Review of the public health safety of products from aquaculture. Int. J. Food Sci. Technol. 33:99-125.

IFOMA (International Fishmeal and Oil Manufacturers Association). 2001. EU publishes further proposals on dioxins in feeds and foods. IFOMA Update 111, May 2001. St. Albans, U.K.

Josupeit, H., A. Lem, and H. Lupin. 2001. Aquaculture products: Quality, safety, marketing and trade. In: R.P. Subasinghe, P. Bueno, M.J. Phillips, C. Hough, and S.E. McGladdery (eds.), Aquaculture in the third millennium. Technical proceedings of the Conference on Aquaculture in the Third Millennium, Bangkok, Thailand, 20-25 Feb 2000, pp. 249-257.

Mielke, T. (ed.). 2001. Oil World annual 2001. ISTA Mielke GmbH, Hamburg, Germany. http://www.oilworld.de.

NMFS. 2002. Fisheries of the United States 2001. NOAA National Marine Fisheries Service, Office of Science and Technology, Fisheries Statistics and Economics Division, Silver Spring, Maryland. http://www.st.nmfs.gov/stl.

Reilly, A., and F. Kaferstein. 1997. Food safety hazards and the application of the principles of the hazard analysis and critical control point (HACCP) system for their control in aquaculture production. Aquac. Res. 28:735-752.

Scialabba, N.E., and C. Hattam (eds.). 2002. Organic agriculture, environment and food security. FAO, Sustainable Development Department, Environment and Natural Resources Service, Rome, Italy. 252 pp.

Spencer Garrett, E., C. Lima dos Santos, and M.L. Jahncke. 1997. Public, animal, and environmental health implications of aquaculture. Emerg. Infect. Dis. 3(4):453-457.

Tacon, A.G.J., and U.C. Barg. 2001. Responsible aquaculture for the next millennium, In: L.M.B. Garcia (ed.), Proceedings of the Seminar-Workshop on Responsible Aquaculture Development in Southeast Asia, 12-14 October 1999. Southeast Asian Fisheries Development Center, Iloilo City, Philippines, pp. 1-26.

Tacon, A.G.J., and D.J. Brister. 2002. Organic aquaculture: Current status and future prospects. In: N.E. Scialabba and C. Hattam (eds.), Organic agriculture, environment and food security. FAO, Sustainable Development Department. Environment and Natural Resources Service, Rome, Italy, pp. 163-176.

Tacon, A.G.J., and I.P. Forster. 2000. Global trends and challenges to aquaculture and aquafeed development in the new millennium. International Aquafeed: Directory and buyers' guide 2001. Turret RAI, Uxbridge, Middlesex, U.K., pp. 4-25.

USDA. 2002. Aquaculture outlook/LDP-AQS-15, March 6, 2002. U.S. Department of Agriculture, Economic Research Service. http://www.ers.usda.gov.

WHO. 1999. Food safety issues associated with products from aquaculture. Report of a joint FAO/NACA/WHO study group. WHO Tech. Rep. Ser. 883. 55 pp.

Utilization of Fish Byproducts in Iceland

Sigurjon Arason
Icelandic Fisheries Laboratories, and University of Iceland, Department of Food Science, Reykjavik, Iceland

Abstract

Fisheries are the single most important industry in Iceland, and will continue to play an important role in the economy of Iceland for a long time to come. In 2001 the total catch was around 2 million tons, accounting for 62% of the country's merchandise exports. The living marine resources are, however, limited and it is important to utilize these resources in a sustainable way. It is also important to maximize their value by producing high-priced products from the raw material, which is currently being used for fish meal or simply discarded. For example, today all cod heads from land-based processing plants are being utilized and lately the freezing trawlers have begun freezing them onboard for processing on shore. Fortunately, most of the byproducts are no longer regarded as waste but are used as raw material for fish processing like roe, liver, mince, viscera, etc.

The byproducts from salting, freezing, and canning fresh fish and other processes have different qualities and potentials. Therefore, quality management is important and new technologies are emerging that will allow a new range of products to be made from byproducts which will, for example, benefit the pharmaceutical, cosmetics, and food industries worldwide.

Introduction

The living marine resources in Icelandic waters are the most important natural resources in the country. In 2001, the total catch was around 2 million tons, (Fig. 1), accounting for 62% of the value of exported products and around 48% of the foreign currency earnings that year. The most important fish species in Icelandic waters belong to the gadoids: cod, haddock, pollock, and blue whiting. Other important species are ocean perch, Greenland halibut, herring, capelin, and shellfish like shrimp, lobster, and scallop (Fig. 2).

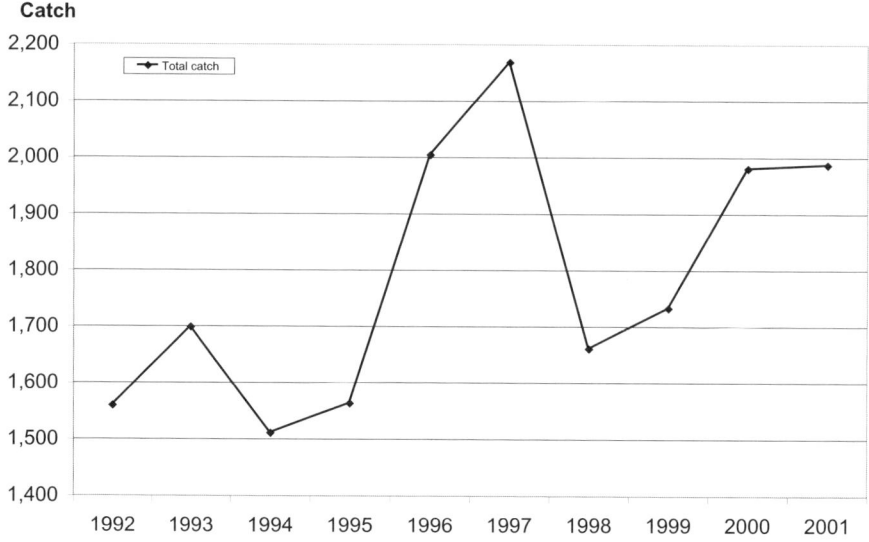

Figure 1. Fish catch on Icelandic catching grounds, 1992-2001.

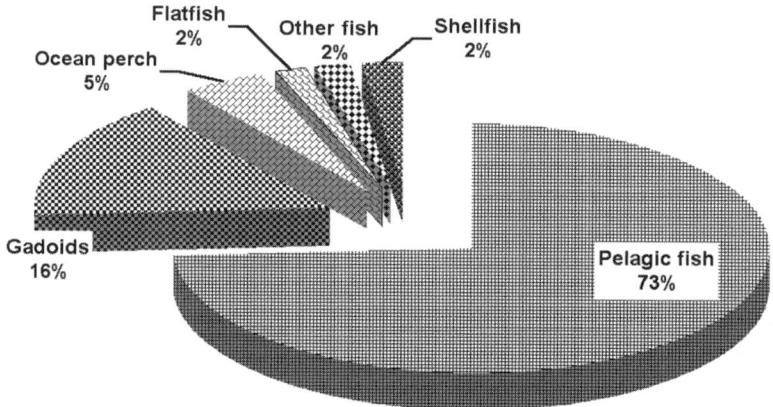

**Figure 2. Catch of Icelandic vessels by fish species fishing in 2001;
the total catch was 1,941,905 tons.**

Fish meal and oil constitute the bulk of the volume of products from fisheries in Iceland or 63% of total, but their value is far less or only about 14% of the total value of exported seafood products (Fig. 3).

The Icelandic fishing fleet and the processing plants are highly mechanized as can be seen by the fact that only about 10% of the work force is employed in fishing and fish processing. Icelanders, like many other nations, have realized that their fish resources are limited and that a collapse of any of the major stocks would be economically disastrous. In 1984 fixed quotas for each vessel were introduced in order to control exploitation of the fish stocks (Valdimarsson 1990). The main objective of the quota legislation was to prevent overfishing and to encourage responsible handling of all catches and exploitation of under-utilized marine life. There is no doubt that the quota system has had a major effect on changing the attitude toward full utilization of the catches. The fishermen and the processing industries are becoming more aware of the possibilities of making marketable products from raw materials that currently are either used for fish meal or simply discarded. Through research and development, publicly funded institutions assist the industry to increase utilization of seafoods.

During the last decade, the annual fish catch has stabilized at about 1.5-2 million tons, and according to fish biologists, no further growth is expected in the near future. Hence, optimal utilization of fishery byproducts is becoming increasingly important to provide more marine fish raw material for various industrial purposes.

The various fish processes give different byproducts and the production yield varies greatly. Calculation of yield in the fish processing industry is generally based on gutted fish with head. Some species, like ocean perch, herring, and other pelagic fish, will reach the factory ungutted and evaluation of processing yield for these species will be based on ungutted fish. Utilization of byproducts is highly dependent on the processing method. Cod can be frozen at sea or on shore and salted cod may be filleted or butterfly split, which will influence the utilization of the fish (Table 1) (Arason 2001b). The "by-raw materials" are better controlled in an onshore operation than at-sea processing. The utilization of the "by-raw materials" from freezing trawlers has increased, however, during the last five years. In 2001 byproducts brought to shore by Icelandic freezing trawlers were:

Fish cut-offs 1,946 tons

Fish heads   3,161 tons

Roe   87 tons

A.

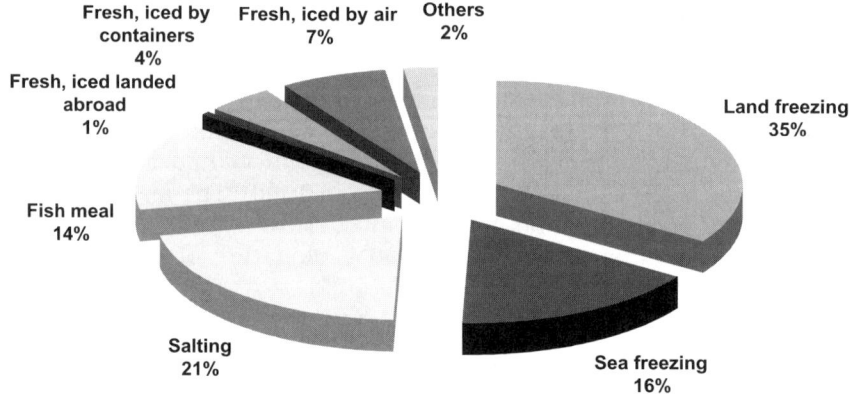

B.

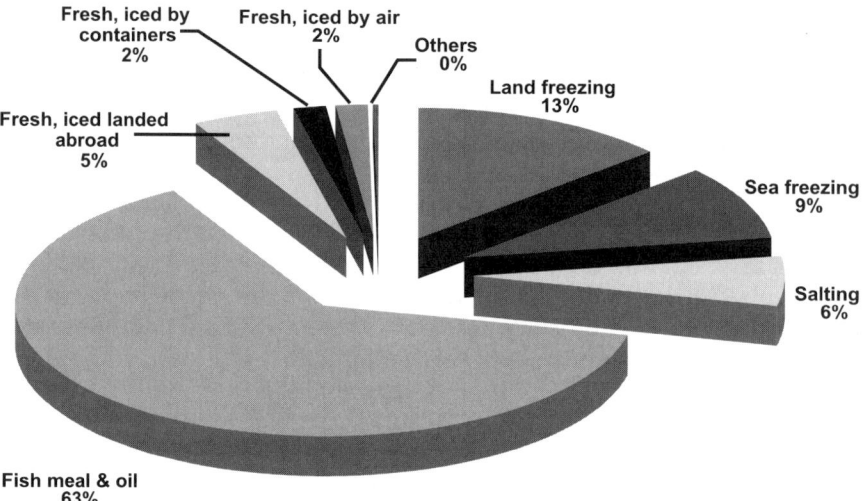

Figure 3. The main Icelandic fishery products area, 2001. A = export value, B = quantity of export.

Table 1.   Processing yield percent of cod at different production.

	Filleting			Splitting
	Frozen on shore	Frozen at sea	Salting	Salting
Product	48	45	60	72
Head	30	34	23	22
Backbone	15	16	16	6
Skin	3	3		
Cut	4	2	1	
	100	100	100	100
Viscera	18	18	18	18
By-raw material total	70	73	58	46

Utilization of fish byproducts

The discussion on byproducts from groundfish will be divided into three categories. In the first we will look at utilizing the viscera, in the second we will look closely at frames, cut-offs, etc., and we will give an overview of the utilization of heads, which represent the main volume and value of byproduct export from Icelandic fisheries (Fig. 4).

Fish viscera

Viscera (including liver and roe or milt) constitute between 10 and 25% of the net weight of fish. In Iceland, most of the intestines are discarded at sea. It is well known that the intestines, the stomach, and the pyloric caecum contain large quantities of digestive enzymes. For a number of years the University of Iceland, the Icelandic Fisheries Laboratories (IFL), and others have worked toward producing crude enzyme mixtures, containing high concentrations of cod enzymes. At the university further work is being carried out by purifying the specific enzymes, especially trypsin and chymotrypsin (Asgeirsson et al. 1989).

Hydrolytic enzymes, especially proteinases, have many uses and potential applications are in industry, medicine, and research. Among these applications are detergent production, leather processing, chemical modifications, and food processing. Enzymes, isolated from cold water marine organisms, may prove to be especially useful for these purposes. The cold-active or psychrophilic enzymes are frequently more active at low temperatures than their mammalian or bacterial counterparts, a characteristic that could be beneficial in many industrial processes, as

A)

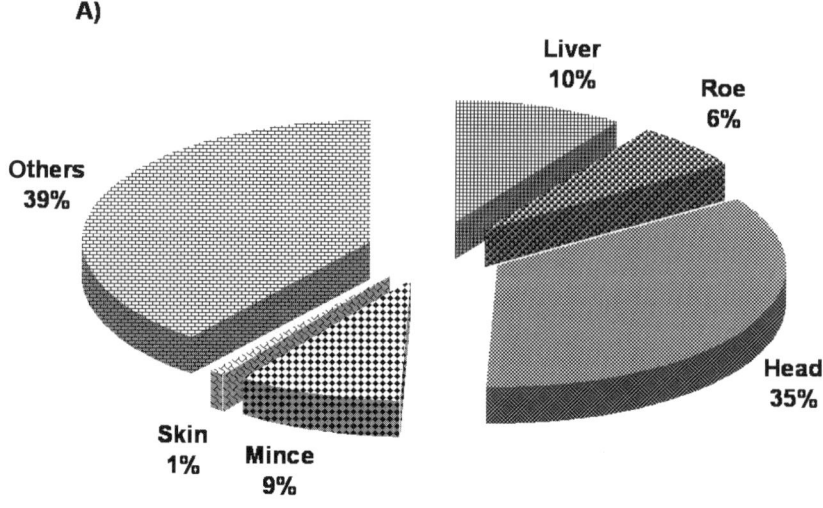

B)

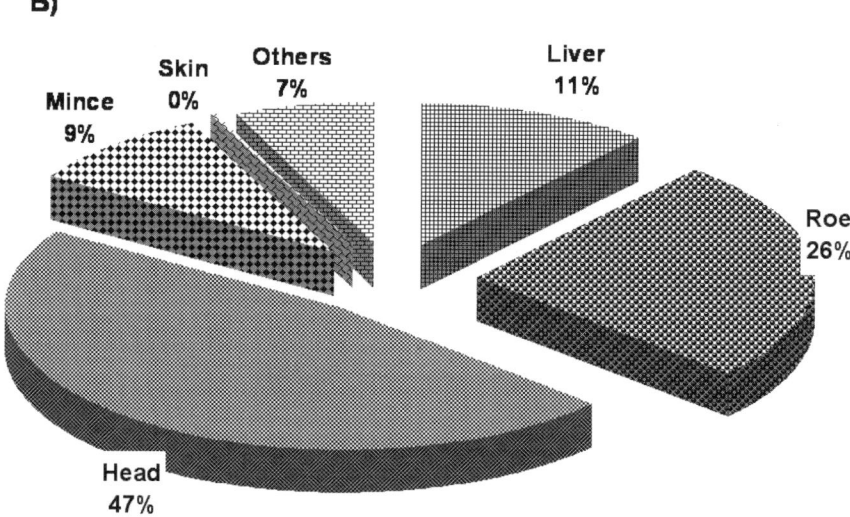

Figure 4. Export of byproducts from groundfish processing in Iceland, 2001.
A. Quantity of export byproducts 2001 was 45,543 tons. B. Export
value for 2001 was $73.5 million

well as in medical, pharmaceutical, hygienic, and cosmetic applications (Bjarnason 2001).

Various food processing applications are also being considered, such as in the chill-proofing of beer, biscuit manufacture, tenderizing of meats, preparation of minimally treated fruit and vegetable beverages, and hydrolysis of various food proteins, such as gelatin, vegetable proteins, and collagens. Currently, cryotin is being prepared on a pilot plant scale for marketing trials and application tests, both in-house and in collaboration with external partners. Purification of the individual proteinases, trypsin, chymotrypsin, elastase, and collagenase is also being scaled up to allow large-scale tests to be conducted. Cryotin is now being used in a patented process to prepare high quality, all-natural flavorings for food processing and innovative cooking (Bjarnason 2001).

Penzyme, a pure, super-active proteinase from cod, is currently being sold in Iceland as an enzyme ointment called PENZIM gel or lotion. The PENZIM ointment is a soothing, moisturizing, cleansing, and nourishing skin healing treatment for dry or chapped skin. It also appears to have good qualities as an "age-specialist" product for facial skin, and rejuvenates whole body skin by removing the outermost layer of dead skin cells (Bjarnason 2001).

There is also a growing market for fish hydrolysates with defined qualities. This is one area where fish enzymes might become valuable in producing a variety of fish protein hydrolysates from scrap fish.

The "by-raw materials," which have been utilized from viscera, include liver, roe, milt, and stomach. The best-known Icelandic fish byproducts are undoubtedly the cod liver oil and cod roe. From 1901 to 1950 the annual production of cod liver oil was about 5,000 tons per year and around 2,000 tons of salted cod roe.

Roe

There is a high demand for frozen and salted cod roe for smoking, canning, and the production of various kinds of spreads, such as "caviar" spread. The roe from cod, haddock, and pollock are only available for a period of three months each year. The ripening stage is very important in terms of product quality.

In Iceland, all iced fish must be gutted if the individual fishing trip is longer than 24 hours. Traditionally, only roe from fish that has been brought to shore ungutted for salting or freezing have been utilized. Increased roe prices have changed this and today roes from most of the harvested groundfish are being utilized. Vessel workers, who are eviscerating the fish at sea, collect the roe in insulated plastic tubs, using salt for preservation; or the roes are frozen onboard the freezing trawlers. Table 2 shows the different categories of exported roe from gadoid species.

The lumpfish roe industry is now well established in Iceland, and most of the roe is packed as caviar under various foreign labels and sold

Table 2. Roe production from gadoid species in Iceland, 2001.

	Quantity (tons)	Value (thousand $)
Industry frozen roes	483.6	2,947
Special quality frozen roes	603.6	5,677
Salted roes	287.0	4,881
Sucker salted roes	1,181.7	387
Canned roes	27.9	8,742
Others	213.0	464
Total	2,796.8	23,099

on the European market. The export value was $10.8 million in 2001 and the quantity was 1,184 tons.

Capelin roe has almost exclusively been sold in Japan, but some producers are now producing caviar or spreads for the European market. The export value was $10.4 million in 2001 and the quantity was 6,165 tons.

Liver

One part of cod liver is used for making medicinal liver oil. The University of Iceland, Lysi Ltd., the Icelandic Fisheries Laboratory (IFL), and others have been working on the product development. The main effort has been toward producing cod liver oil with a higher proportion of omega-3 fatty acids. Icelanders have long been loyal consumers of cod liver oil, and since the media started reporting on the alleged beneficial effects of the omega-3 fatty acids on human health the consumption has increased even more. Generally there is now more demand for cod liver than can be supplied from the uneviscerated catches alone.

Canned cod liver has always been in demand, but over the last few years only 220 tons have been produced annually. One reason for this is the short operating period. Table 3 summarizes the various categories produced from liver.

Although most of the emphasis has been placed on the health benefits of fish oil, it is known that fish oil and fish liver oil contain other interesting compounds. With improved separation techniques and more gentle processing methods, these oils might play an even more important role in the pharmaceutical and health food industry in the near future.

Fish frames and collars

Considerable quantities of fish flesh can be recovered from the remaining collars and the frames after filleting groundfish. Bone separator

Table 3. Production of different products from liver of gadoid species in Iceland, 2001.

	Quantity (tons)	Value (thousand $)
Medicinal liver oil	1,573	4,258
Groundfish oil	2,146	2,438
Canned liver	220	1,109
Frozen and fresh liver	238	277
Other liver products	168	149
Total	4,344	8,232

technology is well known and widely used in many countries. In Iceland, practically no fish mince is produced except from the cut-offs rendered by the production of boneless fillets (Table 4). The mince from the collar and the frames is darker in color than the cut-off mince and therefore sells at lower prices.

From one ton of gutted cod some 100-180 kilograms of mince can be produced from the different parts of the fish. Following is the possible mince production out of various parts from the processing of cod fillets:

Cut-off	3-4%
Collar	4-5%
Head	3-4%
Belly flap	5-6%
Frame	4-5%
Total mince	15-18%

MESA is an Icelandic company that has developed machinery for scraping flesh and cutting belly flaps from the backbones after splitting or filleting of cod. These products have been exported and used as value-added products on the salted fish market.

Fish skin

Fish skin is collected and exported to Canada and Spain by an Icelandic fish processor. Research in Iceland indicates that fish skin can be stored refrigerated and frozen for a short period of time without negative effect on the functional properties of the gelatin (Gudmundsson and Hafsteinsson 1997). Fish skins are also being processed into leather (Table 5).

Fish offers new applications as a food ingredient, both because it has properties that are different from mammalian gelatin and also because it

Table 4. Production of minces from gadoid species in Iceland, 2001.

	Quantity (tons)	Value (thousand $)
Cod mince	3,526	5,990
Other groundfish mince	654	809
Total	4,180	6,798

Table 5. Icelandic exported products from skin, 2001.

	Quantity (tons)	Value (thousand $)
Frozen skin	425	65
Leather from fish skin	0.2	57
Total	425.2	122

can be used in food where mammalian gelatin is not desired (e.g., in kosher and halal food, because of the bovine spongiform encephalopathy [BSE] scare, etc.). Fish gelatin is soluble in cold water, which is an important quality in frozen products. There is also a market for non-gelling gelatin, such as in the cosmetic industry as an active ingredient (i.e., shampoo with protein).

Fish heads

In Iceland, as in many other countries, fish tongues and cheeks are considered delicacies, although some cheeks and tongues are also exported. Until recently an obstacle for processing these products was the relatively high manpower requirement, but now the Icelandic company MESA has a machine for this process. The machine uses fish heads and removes head cheeks and tongues. Another machine is used for splitting fish heads and tearing the gills out. The machine can process all sizes of heads, from the smallest to the largest, without special adjustments. As the product from one of the MESA machines is new on the market, time has to be allowed for further product development (Sigurdsson 1992, 1993). In Iceland the tongues and "double cheeks," products from the head splitter, have mostly been salted and sold on the salt fish markets in Portugal and

Table 6. Products from cod heads in Iceland, 2001.

	Quantity (tons)	Value (thousand $)
Dried heads	11,432	24,948
Salted head products	4,277	9,146
Frozen head products	186	501
Total	15,895	34,595

Spain. Most of the cod heads from the Icelandic onshore processing line are utilized into different products (Table 6).

Fish heads contain relatively little meat. However, the meat from the fish head is considered a delicacy due to its taste and excellent texture. If cod heads are taken as an example it can be assumed that the tongue constitutes approximately 1-4% of the weight of the head, cheeks 5-15%, collar 15-20%, and upper head meat 5-15% (Fig. 5).

Drying cod heads indoors

In Iceland, indoor drying has been tested in regions where geothermal energy is found. The reason is that the cost of oil or electricity for heating during the drying process is considerably higher than the cost of hot water or geothermal steam. It is more profitable to locate the processing near inexpensive hot water and steam sources, and collect the raw material and transfer it to the processing plant.

The price of energy for heating varies much from one energy source to another and from one location to another. The price of oil has fluctuated; the price of hot water and electricity has changed less, although it tends to follow the price of oil (Fig. 6).

The energy cost of heating air for drying one kilogram of dried cod heads in Iceland is lowest using geothermal resources (Fig. 6). The prices are extrapolated using the price in March 2001. The main conclusions are that the energy required for evaporating one kilogram of water from a substance is 4,400 kJ (1,100 kcal), the efficiency of oil boilers is estimated at 90%, coefficient of performance of heat pumps 2.5, and it is assumed that the hot water is cooled from 80°C to 30°C (Arason 2001a).

Weather conditions limit outdoor drying in Iceland. Indoor drying of fish, such as cod heads, stockfish, or small fish, is done in such a way that hot air is blown over the fish and the moisture from the raw material subsequently removed.

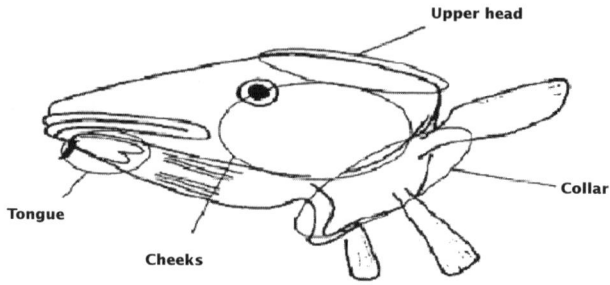

Figure 5. Location of the muscle on a cod head.

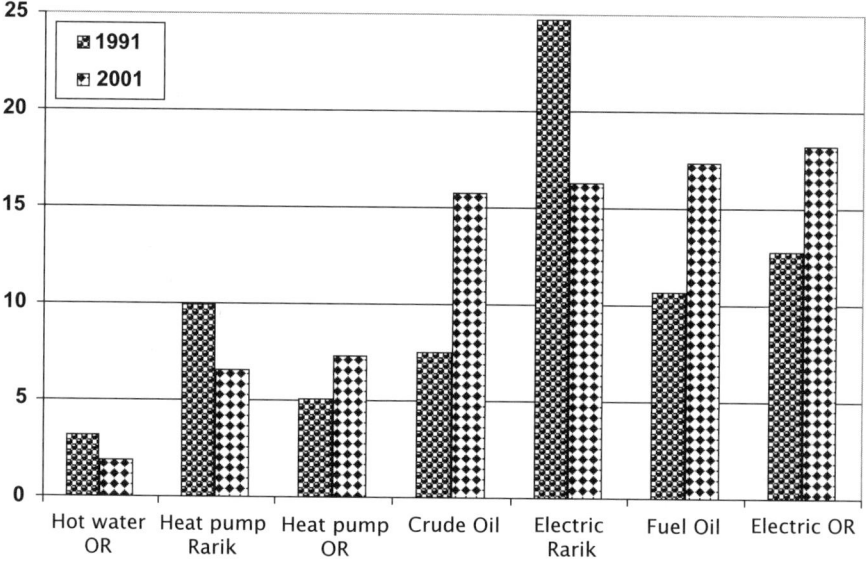

Figure 6. Comparison of prices for different types of energy for heating, ISK for drying one kilogram of dried cod head, based on cost in March 2001 ($1 US = 90 ISK). (OR = Reykjavik Energy, Rarik = Icelandic State Electricity, ISK = Iceland Kroner.)

Value (thousand $)   Quantity (tons)

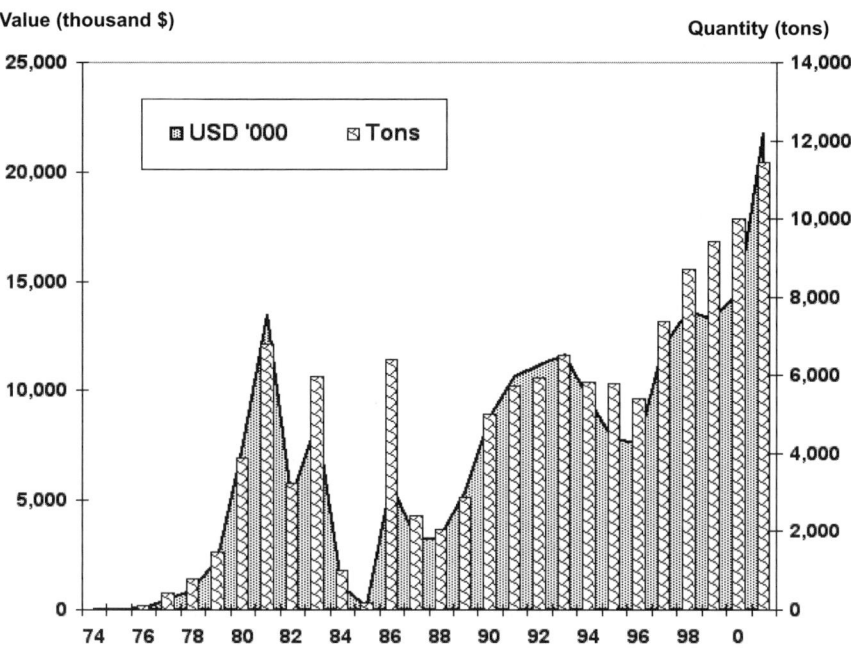

Figure 7. Quantity and value of exported cod-head products from Iceland, 1979-2001.

There is an extensive advantage in drying fresh raw material year-round compared to being dependent on the weather conditions. The process is faster and the drying time is shortened from weeks to days. The main advantages of indoor drying are:

- Shorter drying time than outdoor drying.

- Drying year-round and more even export shipments.

- The product is more consistent in quality and water content.

- Flies and insects are prevented from contaminating the product.

- Utilization of local energy sources.

Traditionally, cod heads used to be dried by hanging them on outdoor stock racks, but about 20 years ago indoor drying was begun. In Iceland the production of dried cod heads increased from 1,000 tons to about 14,000 tons per year which equals about 70,000 tons of undried heads. This practice has grown and now there are several drying factories in Iceland. The largest drying factories are Laugafiskur, Samherji, Hnotskurn, and Thorungavinnslan. The last one specializes in drying seaweed and kelp. In all, about twenty factories air-dry cod head products, all except three using geothermal energy. One drying plant is using oil and another is using a heat pump system. A third one uses a heat pump extracting the heat from coolants from the freezing machinery. One plant uses geothermal steam for drying but most companies use geothermal water. Most of the drying cabinets are constructed for batch drying where cod heads are arranged on trays. Only two cod-head drying plants, including Thorungavinnslan, use conveyor-belt dryers.

There are also about fifty small drying plants for producing dried fish snack, which is a very popular delicacy in Iceland. Most of the small dryers are using geothermal energy. Full drying of cod heads indoors has been successful and the drying is divided into two stages, primary drying and secondary drying (Fig. 8). The cod heads are treated in three different ways prior to the drying (Arason et al. 1992).

Primary drying is done in a rack cabinet or a conveyor-belt cabinet. The rack cabinet is the most common with cod heads arranged in one layer on the racks where about 25 kg of heads can be arranged per square meter. The water content of the heads at the end of this stage is about 50-55%.

Secondary drying of semidried cod heads is conducted in drying containers with hot air blown through it. The water content of the cod heads after drying is about 15%, or the water activity of the products must be lower than 0.6, which is achieved after about 3 days in the drying container.

The greatest advantage of dividing the drying process is that relatively large quantities of cod heads may be placed in the secondary drying facilities than in the primary drying cabinets. The initial and operational costs of secondary drying are much lower than for primary drying, so the production cost is lower in a divided process.

It is also possible to extract the heat from the drying air, which is either blown out or recycled. The recycling of the heat is important, in particular in locations outside the geothermal regions. Results from a preliminary study at IFL indicate that up to 35% energy savings can be achieved by using heat exchangers, and up to 70% savings through the use of heat pumps.

Primary air-drying: batch dryer, rack type

The most common equipment for indoor drying in Iceland is a rack cabinet, the cabinets most frequently consisting of two tunnels with a pyramid in the center. The pyramid can be moved in such a way that if the cabinet is only half full all the airflow is directed through one tunnel. Air valves are inserted in the inlet and recycling outlets, but the regulation of the valves is controlled by the air humidity, measured at the opening of the cabinet (Fig. 9). A regulating valve on the hot water inlet connected to a thermometer, which is located at the same place as the humidity sensor, controls the temperature in the drying cabinet.

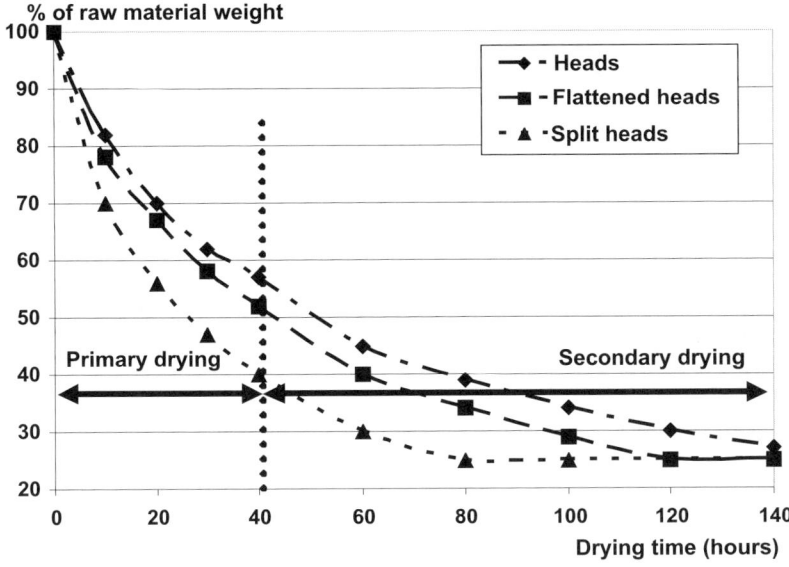

Figure 8. The figure shows how the weight of cod heads changes with time in indoor drying (Arason and Arnason 1992).

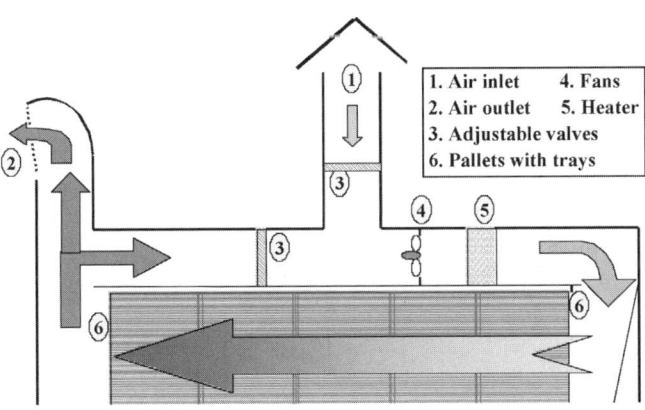

Figure 9. The construction of the rack drying cabinet.

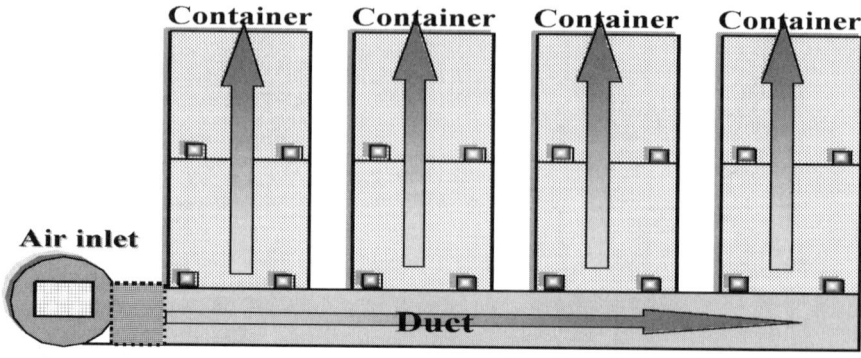

Figure 10. Secondary air-drying unit.

One rack cabinet dryer with heat pumping system is in use in Iceland. The air is heated in the condenser and then blown through the cabinet. In the evaporator, the chilled air and the moisture, which was absorbed in the cabinet, is condensed before the air is heated again in the condenser. About 40% of the energy needed for heating is supplied by electricity and the other 60% comes through reuse of the condensing heat, which is released in the evaporator. The heat pumping systems can be of much use in warming where geothermal energy is not available. On the other hand, the initial capital cost is high, and since there is not much experience from these types of cabinets people have been hesitant to experiment with them.

Secondary air-drying

When a fish is dried in rack cabinets or in conveyor-belt cabinets, it is removed from the cabinet when the water content is about 50-55%. The fish is then placed in a drying container of 1-4 m³ volume. The container is located on top of an air tunnel duct and the air is blown up through it. It is possible to pile 3-4 containers on top of each other (Fig. 10).

Shrimp offal

Shrimp offal constitutes about 50% of the shrimp. It contains protein, chitin, and the coloring agent astaxanthin, which is a necessary ingredient for salmon feeds. One company, Primex, produces chitin products from shrimp shell in Iceland. Chitosan is a biodegradable polysaccharide with a great application potential, ranging from flocculants in wastewater treatment and additives in foods and cosmetics to numerous technical and medical applications (Sikorski et al. 1995). Two companies produce shrimp meal in Iceland. Table 7 shows the different export categories from shrimp shell.

Table 7.   Icelandic exports of products from shrimp shell, 2001.

	Quantity (tons)	Value (thousand $)
Chitin	174	1,911
Shrimp meal	612	328
Total	787	2,239

The Icelandic AVS project

The Icelandic Ministry of Fisheries' priority project on added-value for marine catch is called the AVS project. Although the government intends to play an important part in the AVS project of adding value to the catch, the industry will play the most important role and must lead the way. The most important areas of that growth are likely to be in the areas that are relatively small now. For potential growth in seafood value in 2007 and 2012, see Fig. 11.

The future

One thing that all agree on is that there is not an endless amount of fish in the sea and therefore we need to manage our fisheries better than in the past. There are many ways in which it is possible to optimize fisheries management. It is, however, unlikely that we can increase the catch of wild fish but we can increase fish farming. Furthermore, there are many ways to increase the value of our catches.

Annual discard from the world fisheries has been estimated to be approximately 25% of the total catch. Utilization is driven by economic factors. The conditions onboard the vessels are currently not optimized for a cost effective utilization. Therefore, the main challenge will be to build knowledge of (1) the market potential of the different byproducts and (2) which processes, for practical reasons as well as preservation, should be done onboard the vessel and which on shore.

Increasing the proportion of the catch intended for human consumption and other value added products (pharmaceuticals, feed ingredients, etc.) would increase the profitability and reduce the amount of "waste."

In order to achieve this it is necessary to:

• Develop systems to sort and handle the byproducts on board.

• Find safe and cost-effective preservation methods.

• Improve logistics to get the byproducts from the vessels to the processing plants.

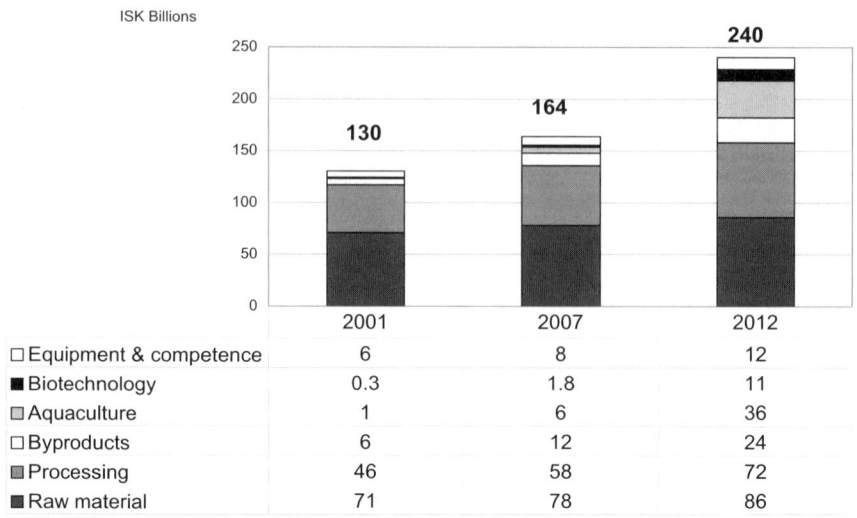

ISK Billions	2001	2007	2012
□ Equipment & competence	6	8	12
■ Biotechnology	0.3	1.8	11
□ Aquaculture	1	6	36
□ Byproducts	6	12	24
■ Processing	46	58	72
■ Raw material	71	78	86

Figure 11. Icelandic potential growth in seafood value (ISK billions) 2007 and 2012 (Palsson 2002).

In order to upgrade the byproducts, more knowledge is needed on the chemical composition, including seasonal variation, of the fishery stocks. This is necessary in order to find effective and safe preservation and storage procedures and to find biomolecules with possible application in the food, feed, and pharmaceutical industries (Fig. 12).

More knowledge is also needed on processing methods to extract the interesting fractions/biomolecules. Finally, there is a need to study the market for these compounds. Byproducts contain components with applications in food, healthcare products, pharmaceuticals, and cosmetics. Screening of the valuable components and an evaluation of the market situation for each component are needed.

The use of geothermal energy for drying of fish products is likely to increase in the future. The interest in Iceland is mainly focused on the use of geothermal energy in low-heat regions. The fish meal industry is likely to use geothermal steam in the processing within a few years. It can be expected that the price of oil will increase more than the local energy in the future, and therefore it is worth paying attention to the use of locally available energy sources for the fishing industry. New, feasible alternative uses of geothermal energy are within sight, such as in freeze drying of food. Equipment designed for drying fish can also be used for drying other industrial products.

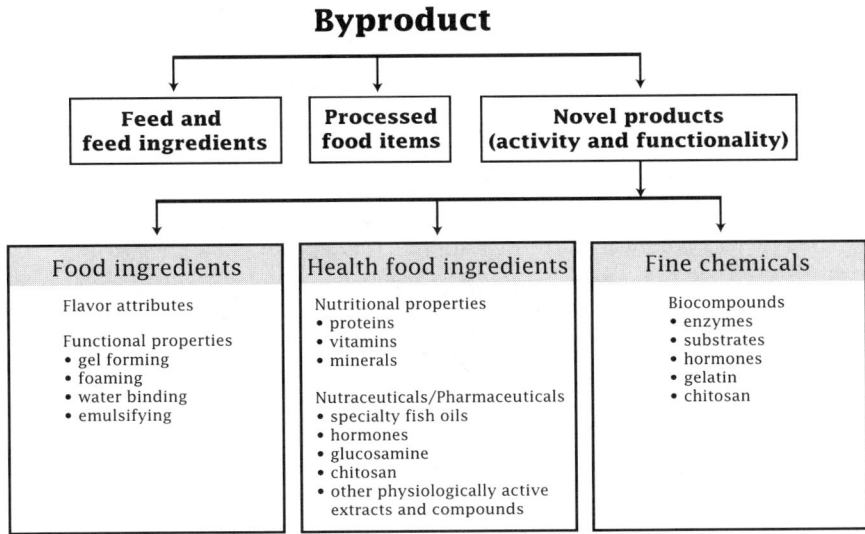

Figure 12. There are many interesting possibilities for byproducts from fish.

References

Arason S. 2001a. The drying of fish and utilization of geothermal energy: The Icelandic experience. Keynote lectures, 1st Nordic Drying Conference, Trondheim, 27-29 June 2001. CD-disk.

Arason, S. 2001b. Status i Norden-mengder/utnyttelse, ver diskaping, politiske mål, reguleringer, virkemidler. Conference Verdiskaping av marine biprodukter etter år 2000. Trondheim, Jan. 2001.

Arason, S., and H. Arnason. 1992. Utilization of geothermal energy for drying fish products. Geothermics 21(5/6):745-75.

Arason, S., G. Thoroddsson, and G. Valdimarsson. 1992. The drying of small pelagic fish: The Icelandic experience. In: J.R. Burt (ed.), Pelagic fish: The resource and its exploitation. Fishing News Books, Surrey, England, pp. 291-298

Asgeirsson, B., J.W. Fox, and J.B. Bjarnason. 1989. Purification and characterization of trypsin from the poikilotherm *Gadus morhua*. Eur. J. Biochem. 180: 85-94.

Bjarnason, J. 2001. Biotechnological applications of fish offal in Iceland. Conference Verdiskaping av marine biprodukter etter år 200. Trondheim, Jan. 2001.

Gudmundsson, M., and H. Hafsteinsson. 1997. Gelatin from cod skins as affected by chemical treatments. J. Food Sci. 62(1):37-39.

Palsson, P.G. (ed.). 2002. The Icelandic Ministry of Fisheries' priority project on added-value for marine catch. AVS-report. 97 pp. (In Icelandic.)

Sigurdsson, A.M. 1992. Apparatus for scraping fish flesh off the backbone after splitting. Icelandic patent no. 1491. (In Icelandic.)

Sigurdsson, A.M. 1993. Machine for cutting tongues, cheeks and belly flaps from fish heads. U.S. patent no. 5,226,848.

Sikorski, Z.E., A. Gilberg, and A. Ruiter.1995. Fish products. In: A. Ruiter (ed.), Fish and fishery products. CAB International, Wallingford, U.K., pp. 315-346.

Valdimarsson, G. 1990. Utilization of selected fish byproducts in Iceland: Past, present and future. In: S. Keller (ed.), Making profits out of seafood wastes. Alaska Sea Grant College Program, University of Alaska Fairbanks, Fairbanks, pp. 71-78.

Byproducts from Chile and the Antarctic

Max Rutman
Marine Biotechnology Consultant, Santiago, Chile

Luciano Diaz
Sea Food Technologies, Santiago, Chile

Juan Pablo Hinrichsen
Hinrichsen S.A., Santiago, Chile

What is a byproduct?

There are several ways to define a byproduct. The first is economically: it has a lower total value than the main product. However, after experience with byproducts, a new more practical definition emerges. A byproduct has no significance to company executives and management often does not consider the relevance of byproducts. It is not in the management mind. It is a constant struggle to attract attention to the role of byproducts in the market, their processing, and what to do with the raw materials.

Chile began as a producer of byproducts, as did Peru. The surrounding area has an estimated 50 million tons of fish, much of which goes to fish meal and oil. In Antarctica, according to the Russians, there are 3 billion tons of raw material available; others claim that number is 500 million. In either case, there are plenty of raw materials. If one is able to capture whatever is available, then the capacity is 5-20 times what is available in the remainder of the world. There is plenty of potential byproduct for this century.

Fishing in Chile

Many of the raw materials from Chile (Fig. 1) are used for fish meal, but some of it is going into human consumption, and the fish processing waste goes to fish meal. The amount of fish meal coming from the aqua industry is growing, but the total fish meal production is decreasing. The

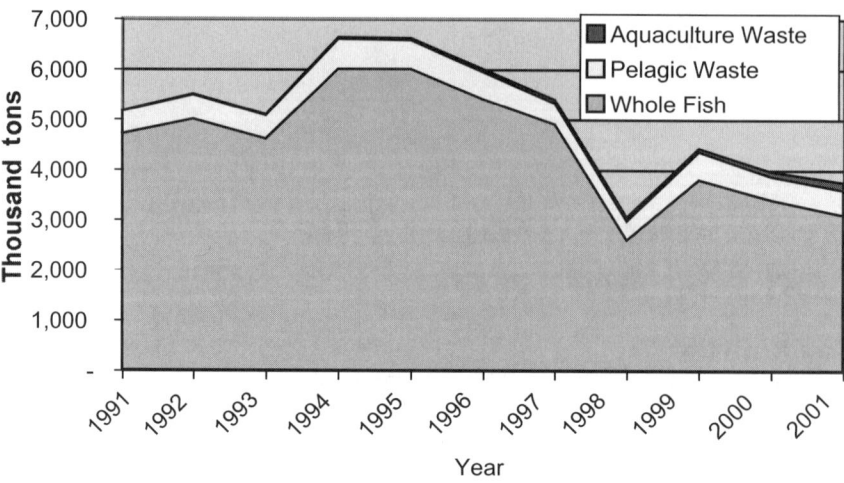

Figure 1. Raw material for byproducts in Chile.

percentage of pelagic fish consumed by humans has grown up to 35%, which would have been unusual 20-30 years ago (Fig. 2).

The aquaculture industry, which includes salmon and trout, is growing as well, so the waste levels in the industry are growing in proportion (Fig. 3). There is a special way to handle waste products. Waste from the fish meal industry is used to feed the salmon, providing a method of eliminating one form of waste while creating another (salmon processing wastes), which has different characteristics. Both wastes are fresh, low in protein, and high in bones, because they are consistent in composition. Salmon products cannot be used for salmon feed, but can be used for other purposes.

The history of Chilean byproducts

The list of byproducts includes fish meal (Sernapesca 2002), which is produced from a whole fish. Chile produced 30 different types of fish meal for chicken, salmon, and Japanese and Taiwanese eels. Many types of fish meal were available, but they were all essentially similar. Fish oil was initially produced primarily for margarine but now much of the oil is used in aquaculture diets. In the 1960s people learned of fish protein concentrate for human consumption. Efforts to develop it were unsuccessful because there was no market for this product. Fish hydrolysates have been developed and are produced in a plant that generates "BioCP,"

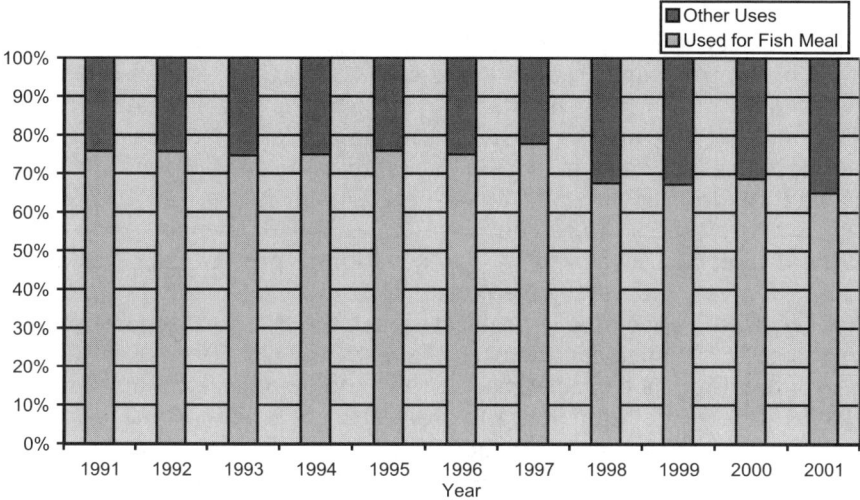

Figure 2. Pelagic fish utilization.

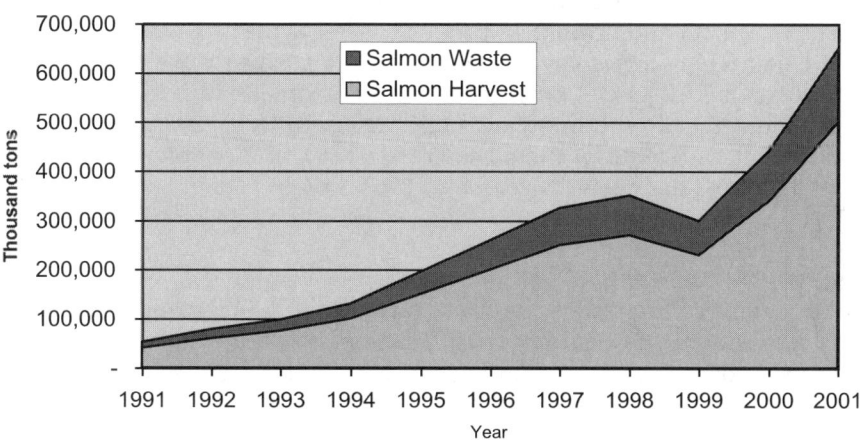

Figure 3. Salmon production and waste in Chile.

a sophisticated fish hydrolysate rich in lipids. The plant produces 4,000 tons per year which is sold primarily as a feed ingredient for piglets.

The future is uncertain, but it is easy to imagine some products appearing in the market. The first uses of fish processing wastes are fish meal and oil; they are the most marketable products. However, this may not be the best approach. There is great interest in hydrolysates, although having a hydrolysate is irrelevant if there is no market for it. The processing of hydrolysates can be complicated, but it will develop slowly into a larger category of byproducts. The question at present concerns price. If a hydrolysate market price is economically equivalent to producing fish meal, why develop a hydrolysate when the fish meal is easier to sell and is an established commodity? At present hydrolysates are produced only when there is a special market for them. Bone meal is another area that needs consideration. When hydrolysates are produced, the separated bones can be fed to either animals or humans.

The idea of making a protein-oil feed ingredient product for enhanced performance is attractive, and the market could become large; however people in these markets should be consulted on this issue

Utilizing enzymes from fish byproducts has always been an interesting topic, and there are now markets and technologies that are suited to utilizing these products. However, once an enzyme attracts attention, the molecular biologist will clone the gene, insert it in an appropriate microorganism and ultimately produce the enzyme at a lower cost.

Other technologies have been introduced, such as those that produce amino acids and gels from wasted meat. Many other uses are being developed, but they are still in the planning stages. In terms of the nutraceutical market, one-tenth of the Alaska resources could supply everyone on the planet for the next two years!

Krill

In the Antarctic four to five ships are catching krill. In Japan, it is used for bait and Russia and other countries have attempted to use it with little success. The U.S. ship is producing krill meat and krill meal.

The krill harvest peaked during the period from 1978 to 1982 and 1986 to 1992 at 500,000 tons and there were 40 factory ships, mostly from the Ukraine (Ichii 2000). These vessels left when a better business proposition arose, but returned when the proposition failed. Later, the entire fleet collapsed due to the economic collapse of the Ukraine. The current harvest is now approximately 100,000 tons of krill.

In the past krill was used for human consumption as a canned product or as krill meat in the Soviet Union. At present it is used mostly for bait and krill meal, and small amounts are utilized in a sophisticated technology to make a meat product. It is not difficult to catch krill, but it is hard to process and even harder to sell. Current allowed catch is 4 million tons of krill. There are plenty available for catch. Entering the business requires a $50 million ship.

When meat or canned product is produced, only 25% is utilized. What to do with the unutilized material becomes the most important issue in krill fisheries both from the ecological as well as economic perspective.

Hydrolysates are one option, and also there is the option to work with the solubles

Useful enzymes can be produced from krill, but the process approaches a pharmaceutical area. There is a patent on the enzyme used for wound healing, and a group in Sweden is testing the krill enzyme in toothpaste. Krill oil is rich in pigments and is an ideal product for nutraceuticals. It can be used in aquaculture diets when and if fish farmers are willing to pay more for it. Other products can be developed with krill, such as krill chitin, depending on the market demands.

Krill meat has a good flavor and is being tested in numerous countries to determine whether it will be accepted (Dimitri Sclabos, Tharos Ltda., Santiago, Chile, pers. comm., 2002). In order for krill meat to be successful, something must be done about the 75% remainder that is discarded into the sea. Krill byproducts are mandatory in order to have a successful fishery.

References

Ichii, T. 2000. Krill harvesting. In: I. Everson (ed.), Krill biology, ecology and fisheries. Blackwell Science, Oxford, U.K.

Sernapesca. 2002. Anuario Estadistico de Pesca (Chile). Servicio Nacional de Pesca, Valparaíso, Chile.

What Is the Potential Market for Seafood Byproducts in Southeast Asia for Aquaculture and Livestock Feeds?

Suchart Thanakiatkai
Protector Nutrition (Thailand) Co. Ltd., Bangkok, Thailand

Introduction

In Asia, we are hungry for raw material for our aquaculture and livestock feed industries and are looking for whatever you have to sell.

Table 1 lists the estimated protein needs of Asian countries for swine, poultry, and fish. China has the largest protein need. Most of the protein is in the form of low-grade quality fish meal, which is only used at 2-4% in the feeds. Table 2 lists the estimated animal protein needed for shrimp by various Southeast Asia countries. In Asia, Thailand has the largest protein need for shrimp raised in aquaculture systems. In Southeast Asia fish oil is often used in the feeds of shrimp, fish, and livestock (sows, piglets, day old chicks).

The major exporters of fish meal and fish oil to Asia are Peru (60%), and Chile (30%). Local production of fish meal is estimated to meet 8% of shrimp needs and 20% of livestock needs.

Product specifications

You don't have to have sophisticated products to meet the fish meal needs of much of Asia. As shown in Table 3 a fish meal with 55% protein and 8% fat can be used for livestock feeds, and this product can be either flame dried or steam dried. For shrimp a steam dried product having 60% protein, 8% fat, and a minimum of 95% digestibility would be readily used. As shown the ash content of fish meal for shrimp feed is much lower than that destined for livestock feeds. For shrimp it would be desirable to have a higher protein content; however, 60% will suffice, with lower ash products for shrimp and higher ash fish meals for livestock feeds. Estimated fish oil specifications for export to Asia and used in

Table 1.   Animal protein requirements for swine, poultry, and fish (metric tons), 2001-2003.

	Livestock	Fish
China	80,000	500,000
Thailand	50,000	20,000
Indonesia	60,000	NA
Vietnam	10,000	NA
Malaysia	10,000	NA
Philippines	10,000	NA
India	10,000	NA
Total	230,000	520,000

The total is 750,000 based on 2-4% in feed.

Table 2.   Animal protein requirements for shrimp feed in Asia and Southeast Asia, 2001-2003.

China	50,000 tons
Thailand	100,000 tons
Vietnam	50,000 tons
Following 4 countries combined	50,000 tons
Malaysia	
Indonesia	
India	
Philippines	
Total	250,000 tons

Table 3.   Usable seafood byproduct meals for shrimp and livestock feeds, 2001-2003.

	Shrimp	Livestock
Protein %	60	55
Fat %	8	8
Ash %	12	20
Moisture %	12	12
TVN mg	120	1,000
Digestibility %	95	90
Histamine ppm	150	1,000
Salt (NaCl) %	2	4
Antioxidant ppm	250	400

TVN = total volatile nitrogen.

shrimp feed are listed on Table 4. As show the minimum omega-3 fatty acid content would be 18-20%.

Requirements for selling fish byproducts to Asia

I want to draw your attention to the fact that if you want to sell your fish oil to Asia, it has to be packed in a drum. Fish meal has to be packed, because this avoids the biggest problem, namely *Salmonella*. Regarding shipping charges, I can tell you that today, to ship our fish hydrolysate from Vancouver or Seattle in a 20-foot container that holds 20 tons costs us $500. The remaining issue is how to get something from Alaska down through Canada. The documents normally required in Asia include certificate of origin and analysis, health certificate, packing list, and production process flow chart. The production process flow chart is required because of the Asian taxation system (e.g., single species and mixed seafood meals and oils can have different tax rates).

Fish hydrolysate and bone meal

Table 5 shows the composition of fish hydrolysates and fish solubles. Both are 50% water and similar in protein and ash content. Many people think fish bones are a real problem when making fish hydrolysates. However, the dried bone can contain 20% protein, 16% calcium, and 8% phosphorus. The dried bone can be used to replace dicalcium and rock phosphate in feed formulations.

Table 4.   Specifications for seafood oil to be used in Asia for shrimp feed.

Moisture and impurities %	1 maximum
Omega-3	18-20 minimum
Acid value, KOH/gm	10 maximum
FFA %	3-5 maximum
Iodine value, Hanus	160-200 maximum
Unsaponifiable matter %	1 maximum
Antioxidant ppm	250-350 minimum

FFA = free fatty acids.
Source is Peru, Chile, and Norway.

Table 5.   Composition of fish hydrolysate and fish solubles.

	Fish hydrolysate	Fish solubles
Protein %	36	38
Moisture %	50	50
Ash %	2	1
Salt (NaCl) %	6	10
Oil %	2	5
Solid matter %	40	Liquid

Fish hydrolysate is salmon from U.S. and Scotland.
Fish solubles are tuna from Japan and Korea.

Tuna byproduct process

The processing of tuna byproducts in Thailand is shown in Fig. 1. Thailand is the biggest tuna exporter in the world, with a minimum of half a million tons of tuna per year. The raw material is cooked and heated to 90°C. It then undergoes pressure separation into solids on one side and oil and water on the other side. We centrifuge the liquid into crude oil and a water-soluble fraction. In order to make fish hydrolysates, we add enzyme and heat at 50°C for three hours. We heat the fish solubles at 100°C for four hours. On the solids side, after separation of the liquid the solids are dried, bones are removed by sieving, and the solids are ground to make fish meal.

Comparison of fish meal by country of origin

Fish meal is made in many countries and the protein content does show considerable variation. High protein content does not always mean high

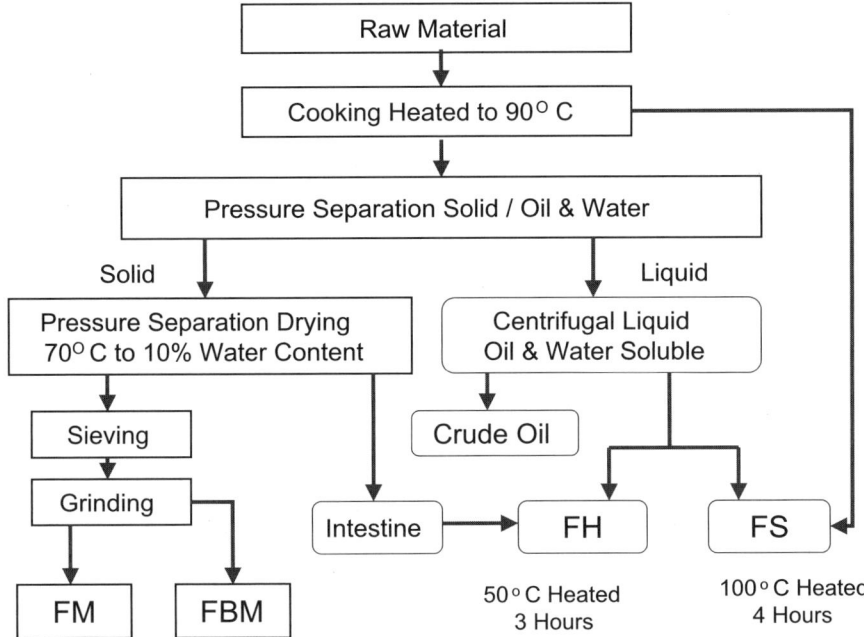

Figure 1. Production process for tuna byproducts in Thailand. FM = fish meal, FBM = fish bone meal, FH = fish hydrolysates, and FS = fish solubles.

quality as the amino acid composition varies greatly (Feed International 2002). The amounts of limiting amino acids such as lysine and methionine have been shown to vary in meals from different countries of origin.

Summary

As stated earlier, I am here to tell you that in Asia we need these things and are willing to buy whatever you can sell us. I would like to thank Dr. Ron Hardy, and the one who pushed me into the fish meal and oil business, Mr. Paul Gusman of H.A. Baker.

Reference

Feed International. 2002. July 2002, pp. 13-14.

Seafood Byproduct Production in the United Kingdom

Michaela Archer
Sea Fish Industry Authority, Hull, United Kingdom

Introduction

The fish industry developed around fishing ports at a time when landings were plentiful and there was little concern about environmental impacts. Nowadays, it is widely recognized that natural resources and the environment are under threat and are increasingly protected by law. Government policy is now focusing on the protection of resources, promoting sustainable utilization, and reducing emissions to the environment. Fishing opportunities are reduced and waste generation and disposal are increasingly penalized. A further problem faced by the fish industry is the increasing concern about farm animal diseases, particularly those transmissible to humans such as bovine spongiform encephalopathy (BSE). This is resulting in the closure of some of the existing routes of seafood byproduct utilization and adding further restriction and costs to disposal.

While the U.K. fishery resources are reducing, the costs of our industry are increasing. Imported seafood products remain available, often at low cost, and now take the major share of the retail market. Retail price competition is fierce and the industry is squeezed to very low profit margins between low retail prices and high operating costs. It is therefore essential that the industry minimizes waste and maximizes the value of the material available to it.

In recent years, the U.K. Sea Fish Industry Authority has carried out a survey of the seafood byproducts produced by the U.K. seafood industry and identified potential uses for that material (Archer et al. 2001). This paper summarizes the findings of the report. It includes an outline of relevant legislation, describes the current utilization of seafood byproducts in the United Kingdom, and identifies potential utilization options that may be suitable for commercial development.

Legislation

The utilization and disposal of animal byproducts is now highly regulated in the United Kingdom following recent problems such as BSE and foot and mouth disease. The Animal Byproducts Order 1999 (as amended) restricts how, what, and where animal byproducts are processed and how they can be utilized. It applies to whole fish, parts of fish, or products of fish origin that are not intended for human consumption. It does not apply to skins, shells, blood, and similar products that are not used in the manufacture of feedstuffs, providing the animal from which they came shows no clinical signs of any disease communicable to humans or animals.

Animal byproducts are currently categorized into high-risk or low-risk material. High-risk material is that which poses a risk if allowed back into the food chain, whereas low-risk material poses no risk in the food chain, providing it is processed according to specified, approved methods. Fish and shellfish are classed as low-risk unless (1) they show clinical signs of disease communicable to humans or fish, (2) are spoiled in such a way that they pose a risk to human or animal health, or (3) contain residues that may pose a danger to human or animal health.

The options for the disposal of animal byproducts are restricted according to the level of risk. High-risk material must only be sent for or disposed of by rendering or part-rendering in approved premises, incineration, burning other than in an incinerator, or burying if it is a place where access is difficult or the quantity of byproduct and distance to disposal premises do not justify transporting it. It can also be used for diagnostic, educational, or research purposes, or be exported from Great Britain. There are further options for the disposal of low-risk material including production of pet food; production of pharmaceutical and technical products; and feeding to zoo, circus, and fur animals, recognized packs of hounds, and maggots farmed for fishing bait at registered premises.

The legislative framework is further complicated by additional controls on what can be utilized in animal feed. Recent examples of these include restrictions on feeding swill to animals and a ban on feeding fish meal to ruminants. Unfortunately seafood is caught up in these, even though the controls were primarily implemented to prevent the risks that are associated with warm-blooded animals.

Environmental legislation is also becoming more stringent, with the introduction of landfill taxation and increasing controls over what is acceptable for landfilling. For example, the landfilling of biodegradable waste is being phased out in favor of other waste management options such as composting.

The overall effect of all these restrictions is to increasingly add difficulty and cost to the disposal and utilization of seafood byproducts. It has now become a problem with a considerable financial impact.

Table 1.   The U.K. seafood trade for 1999.

	Official landings[a] (t)	Imports[b] (t)	Exports[b] (t)	Balance of trade[b] (t)
Demersal	236,398	249,404	72,640	413,162
Pelagic	107,277	147,363	114,970	139,670
Shellfish	147,891[c]	87,652	81,866	153,677
Total	491,566	484,419	269,476	706,509

[a]Live weight.

[b]Product weight of seafood quantity used/consumed in the U.K.

[c]Includes aquaculture and fishery order landings. In the U.K. an individual has the right to apply for exclusive rights to harvest one species in one area of the seabed, known as fishery order landings.

Estimates of seafood byproduct quantities produced in U.K.

Different types and quantities of seafood byproducts are generated at various stages between capture and consumption. This section summarizes the main types of byproduct produced, by quantity and origin wherever possible. Byproducts produced at sea include discards, gutting, and *Nephrops norvegicus* (Norway lobster) heads and claws. Byproducts generated onshore are produced during commercial processing, retailing, catering, and also by the consumer. The main sources of seafood byproducts are separately identified but the data is not available to quantify byproducts from consumers and catering businesses.

The calculation of the quantities of byproduct generated are based on official landings, import, and export statistics collected by government departments (Sea Fish Industry Authority 1999; Table 1).

In the context of this paper, seafood byproducts are defined as captured and discarded fish and shellfish, comprising un-utilized, undersized or non-quota species, and components of the fish or shellfish that are not usually used directly for human consumption (i.e., viscera, frames, skin, heads, shell, lugs, and flaps, etc). Estimations of the quantities of these byproducts were made using data from Sea Fish Industry Authority discard studies and previously published data on processing yields from Waterman (undated).

Estimation of discards at sea

For every haul of fish and shellfish a proportion will be discarded at sea. In demersal fisheries, this is estimated at about 50% of the weight of the catch. In pelagic and shellfish fisheries, discard rates vary significantly but average about 13% of the catch. Table 2 shows the estimated quantities of fish and shellfish discarded in 1999.

Table 2. Estimate of fish and shellfish discarded at sea from U.K. vessels landing into the U.K.

	Landings (t)	Discards per catch (%)			Discards (t)		
		Min.	Max.	Ave.	Min.	Max.	Ave.
Demersal	236,398	40	60	50	94,559	141,839	118,199
Pelagic	107,277	5	20	12.5	5,364	21,455	13,410
Shellfish	110,929[a]	5	20	12.5	5,546	22,186	13,866
Total	448,604				105,470	185,480	145,475

[a]Excludes aquaculture and fishery order landings.

Table 3. Estimate of the byproducts produced from processing at sea.

Sector	Metric tons
Demersal	36,966
Pelagic	—
Shellfish (*Nephrops norvegicus*)	8,796
Total	45,762

Byproducts from basic processing at sea

A significant proportion of fish are processed at sea by basic processing operations such as gutting; however, this does not apply to small-sized fish or some species (e.g. *Squalus acanthias* [spurdog shark]). Pelagic fish and shellfish are generally not processed at sea, with the exception of damaged or small-sized *Nephrops norvegicus*. Table 3 summarizes the quantities of processing/gutting byproducts at sea.

Byproducts from onshore processing

After fish and shellfish are landed, they normally receive some form of processing in a shore-based facility. The amount of byproduct produced varies from 50 to 70% of the gutted weight of fish, although for shellfish this can constitute up to 88% of their live weight. Table 4 summarizes the quantities of byproduct produced during shore-based processing.

If this material is to be commercially utilized, it is essential to consider its production on a regional basis. An estimate of the quantity of byproduct generated in each region of the United Kingdom is derived from relating the total fish supply to data on the processing capacity in the regions, Table 5. (Joseph and Findlater 1996).

Table 4. Estimated quantities of seafood byproducts produced during onshore processing.

Sector	Metric tons
Demersal	154,143
Pelagic	50,269
Shellfish	
Mollusks	67,166
Crustaceans	29,459
Total	301,037

Summary of seafood byproducts produced by U.K.

Of a total available U.K. fish and shellfish resource of approximately 850,000 metric tons, it is estimated that 43% ends up as products for human consumption and the remainder is classed as byproduct (Fig. 1). This material is produced at different stages of the distribution chain. The majority of byproducts are produced in the onshore processing sector (35% of the resource) whereas discards and basic processing at sea produce smaller quantities (17% and 5% of the resource respectively).

Current utilization of fisheries resource

A significant quantity of seafood byproducts is already utilized in the production of other products. Fish meal production is the biggest outlet for seafood byproduct material in the United Kingdom. It currently generates an income of up to $50 per metric ton for material from demersal species but this is variable between regions. High-grade pelagic material can generate income up to $115 per metric ton but is more typically less than $50 per metric ton. Other uses include pet food and maggot farms which may pay up to $65 per ton. Small-scale uses include fishing bait which may or may not generate revenue. However, in many regions of the United Kingdom, including the southwest, the south and the southeast, and in remote regions of Scotland, there are no or few outlets for seafood byproducts and so they are disposed of in landfills. This can cost about $100 per ton in some regions but it can vary dramatically.

To put these figures into context, one metric ton of *Gadus morhua* (cod) currently costs about $2,400 per ton at the auction market. Approximately 50% of this resource is discarded as byproduct. If this is used in other products, it could earn up to $65 per ton but if it is discarded as waste

Table 5.   Quantities of seafood byproducts produced by region.

Region	Product processed (%)	Estimate of byproduct produced (t)
Humberside	44%	132,456
Grampian	29%	87,302
Northern England	6%	18,062
Southwest England	5%	15,052
South, Midlands, Wales	5%	15,052
Highlands and Islands	4%	12,041
Other Scotland	4%	12,041
Northern Ireland	3%	9,031
Total	100%	301,037

it can cost an additional $100 per ton. The U.K. industry cannot sustain the loss of this increasingly valuable resource and so must look to ways of maximizing the use of this material.

Alternative uses for seafood byproducts

Before considering the utilization of seafood byproducts it is important to consider how to reduce the overall quantity produced. For the catching sector, this would include the introduction of more selective fishing gear to reduce discards. For the processing sector there are options for maximizing yield by utilizing fish mince, tongues, cheeks, fins, fish heads, stomachs, and roe/milt, etc.

Further to these options, seafood byproducts are rich in potentially valuable proteins, minerals, enzymes, pigments, chemicals, and flavors, etc. that have many uses in a range of different industrial applications. Potential uses for seafood byproducts can be separated into two broad areas:

1. Aquacultural, agricultural, and bulk food uses which include fish meal, fish oils, other animal feeds, fish protein hydrolysates, fish protein concentrate, fertilizer, composting, and silage.

2. Non-nutritional uses which include chitin, chitosan, carotenoid pigments, enzymes, leather, aggregates, cosmetics, pharmaceuticals, and charcoal.

These lists are not exhaustive and research continues to generate new uses for seafood byproducts. However, many of these new uses are still at laboratory or pilot scale and are very slow to reach the commercial

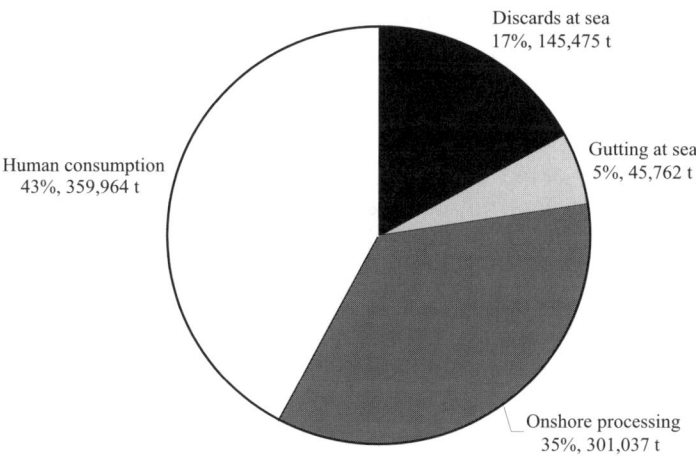

Figure 1. Utilization of the U.K. seafood resource.

stage. The United Kingdom in particular has been slow in utilizing this potentially valuable resource.

Currently, however, a number of projects are developing saleable products from seafood byproducts. These include the proposed development of an anaerobic digestion facility in Peterhead, Northeast Scotland. The resulting product will be a liquid plant feed which can be certified for use in the U.K. organic sector. The United Kingdom is a major importer of organic plant feed so the potential for this product is significant. The facility was scheduled to be operational in 2001 but unfortunately the opening has been delayed.

In addition, a plant has opened on the Scottish west coast that takes minced salmon, demersal, and pelagic material to convert into a range of different products including fertilizer, fish oils, etc.

A recent Ph.D. project at Queen's University Belfast in Northern Ireland studied the biological extraction of high quality chitin from *Nephrops norvegicus*. Traditional chemical based methods of extracting chitin can be expensive and yield a lower quality product, which has made it economically unattractive in the United Kingdom. By using biological methods of extraction, the *Nephrops* byproduct was broken down into three components: chitin, calcium carbonate, and proteinaceous liquor. By looking at different ways to utilize these products, the research team identified that the liquor had significant potential as an aquaculture feed. It was initially tested on juvenile lobsters and showed promising results. Other species will be considered. As a result of this work, a collaborative project was established in Northern Ireland. The intention is to install the technology in Northern Ireland ports, for use by local processors.

Perhaps one of the most promising potential uses for seafood by-products is composting, which is practiced in many different countries throughout the world. This low technology use of seafood byproducts may have significant potential for areas in the United Kingdom that currently have to dispose of seafood byproducts in landfills. The Sea Fish Industry Authority has completed a feasibility study on seafood byproduct composting, which concluded that it offers a relatively simple, adaptable, low technology solution for seafood byproduct disposal in areas remote from fish meal plants. Seafood-derived compost offers a number of unique benefits which, if produced according to national standards and certified as organic, would maximize its value. Potentially it offers a range of solutions to the industry and thus requires further technical investigation. We intend to set up a demonstration project which could lead to commercial uptake.

Conclusion

Seafood byproduct disposal in the United Kingdom is becoming increasingly difficult and costly, and it is important to identify better ways of fully utilizing the increasingly valuable fishery resource. There are a number of potentially valuable ways of utilizing seafood byproducts. Further effort should focus primarily on facilitating a greater commercial uptake of these ideas, as there are currently few outlets for seafood byproducts in this country. The Sea Fish Industry Authority is currently liasing with academia, industry, producers of byproduct utilization equipment, and other interested parties to try to turn the research into commercial reality.

References

Archer, M., R. Watson, and J.W. Denton. 2001. Fish waste production in the U.K. Sea Fish Industry Authority, Hull, U.K.

Joseph, M., and A. Findlater 1996. A survey of the U.K. sea fish processing industry. Sea Fish Industry Authority, Hull, U.K.

Sea Fish Industry Authority. 1999. Key indicators: Winter 1999. Sea Fish Industry Authority, Statistics Department, Hull, U.K.

Waterman, J.J. (undated). Measures, stowage rates and yields of fishery products. Torry Advisory Note 17, Torry Research Station. http://www.fao.org/wairdocs/tan/x5898e/x5898e00.htm.

Future of Alaska's Fisheries Resources

Loh-Lee Low
*National Marine Fisheries Service, Alaska Fisheries Science Center, Seattle,
Washington*

Abstract

Fishing has always played an important role in the history of Alaska.
Alaska has the bragging rights for the most productive fisheries in the
nation. Alaska accounts for 51% of the nation's recent catches and has
55% of its long-term potential yield. Alaska has sustainable fisheries for
salmon, groundfish, herring, crab, shrimp, mollusks, and many other
inshore and offshore species. They have all been under active management. This paper will review the major fisheries resources, their management arrangements, and outlook. The issues of resource utilization and
management are complex but the outlook is cautiously optimistic. There
is a tendency toward more active management, more considerations of
ecosystem and climatic concerns, and more legal and administrative requirements in the management process.

Introduction

The waters off Alaska are very productive for fisheries and other living
marine resources. Alaska accounts for 51% of all fisheries production
in the United States (NMFS 1999, Fig. 1). An update of this report is in
preparation. Recent average production has been more than 2.5 million
metric tons (t) annually. There is even more room to grow as the long-
term fisheries production potential off Alaska is 55% of that for the Nation. The waters off Alaska are home and migration area to a diverse mix
of seabirds, marine mammals, salmon, groundfish, pelagic fish, shellfish,
and other animals. These resources have sustained traditional human use
for centuries. In modern times, the resources have come under increasingly higher degrees of management, and intensity of utilization is actually decreasing in most cases. The entire fisheries management process

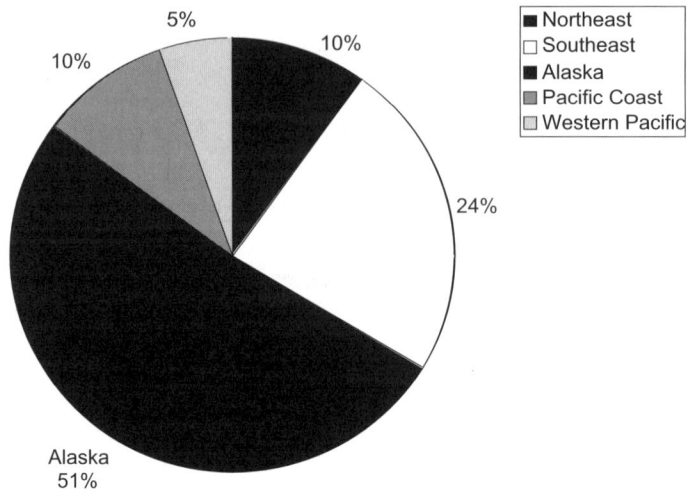

Figure 1. Regional apportionment of recent average production of fisheries resources from the United States, 1998-2001.

is now geared toward conservative use and achieving balanced sustained use of all of the resources and of the ecosystems.

There are reportedly more than 400 species of fish and shellfish alone, and significant numbers of species of marine mammals, seabirds, and other living marine resources. This paper covers only the major commercially exploited species. They are the five species of Pacific salmon, more than fifteen major species of groundfish, Pacific herring, and six species of crabs, scallops, sea snails, and shrimps.

Salmon resources

Alaska's commercial salmon harvests generally have increased over the last three decades. Long-term trends show salmon catches to be relatively high in recent years for most species (Figs. 2-6). Only chinook salmon (*Oncorhynchus tshawytscha*) has not faired as well as the other salmon species, although catch of chinook salmon can be considered to be in the historical average range. The figures that show the long-term trend indicate that salmon resources have fluctuated over the past 102 years, and have been relatively high in recent years (Fig 7).

Alaska salmon catches are known to vary widely with area and from year to year. Variation in catches from district to district and in short time frames have caused fishermen to experience much anxiety. The

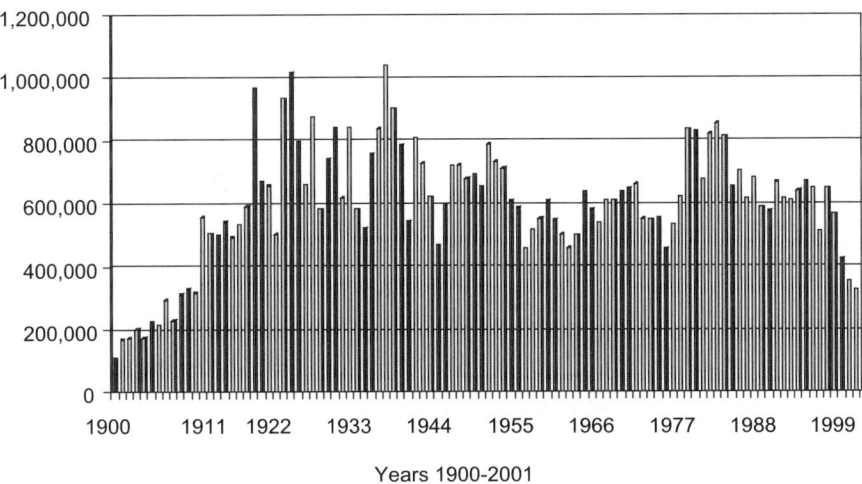

Figure 2. Catch trend of chinook salmon from Alaska, 1900-2001.

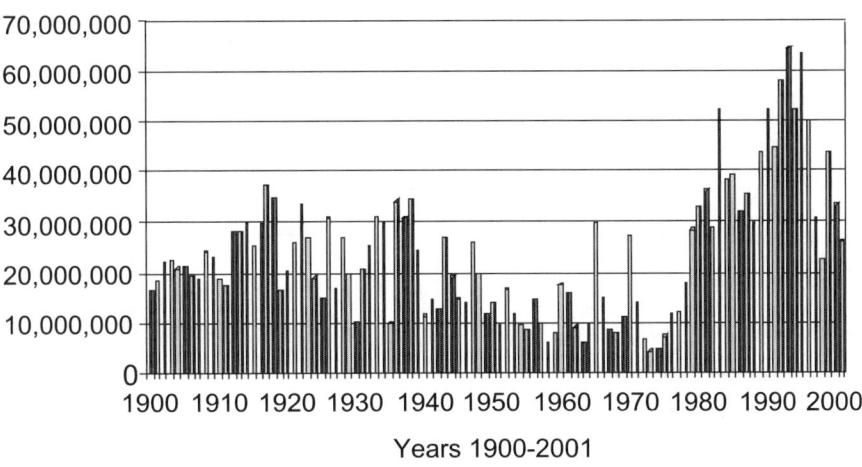

Figure 3. Catch trend of sockeye salmon from Alaska, 1900-2001.

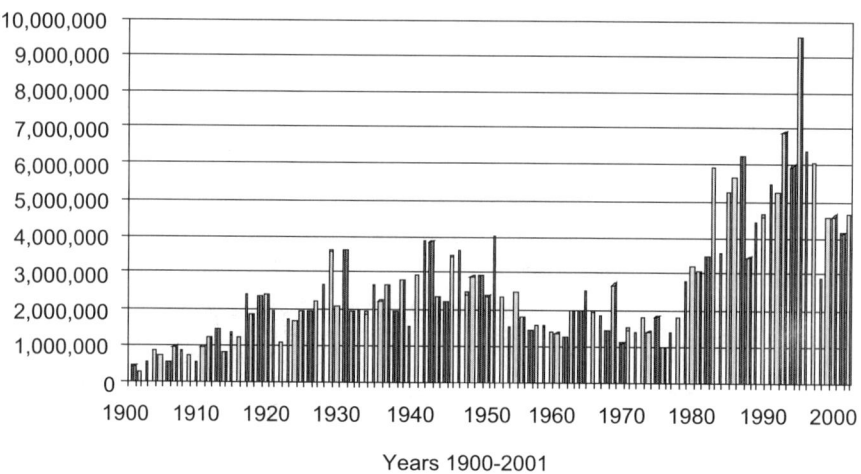

Figure 4. Catch trend of coho salmon from Alaska, 1900-2001.

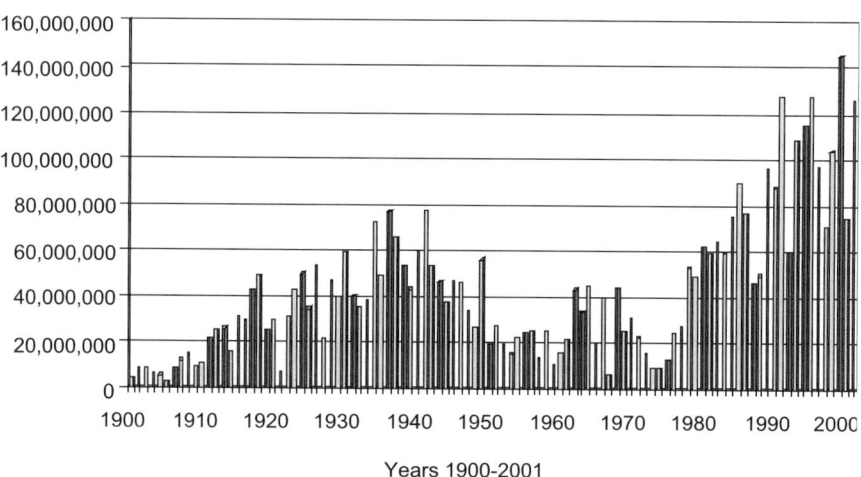

Figure 5. Catch trend of pink salmon from Alaska, 1900-2001.

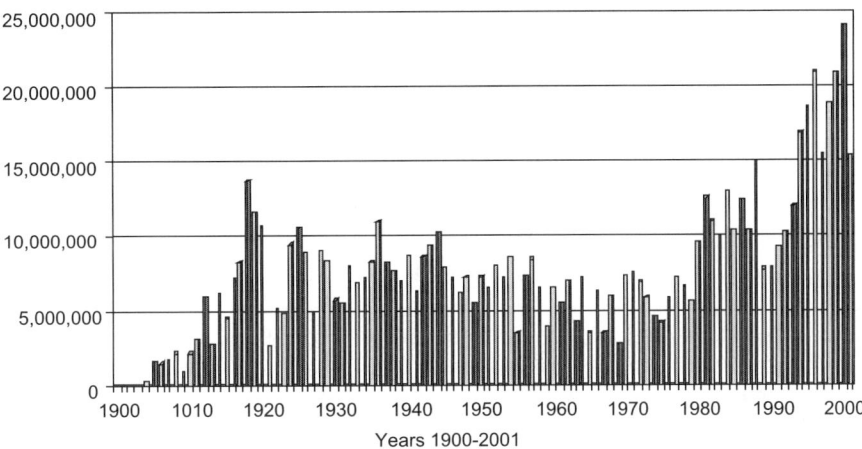

Figure 6. Catch trend of chum salmon from Alaska, 1900-2001.

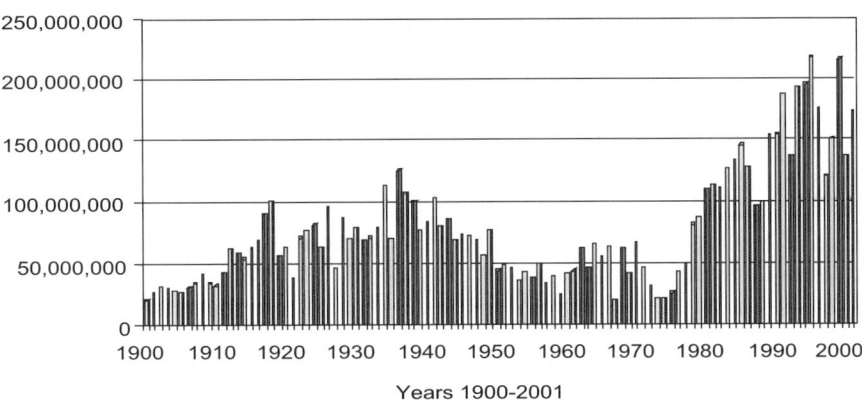

Figure 7. Catch trend of all species of salmon combined from Alaska, 1900-2001.

negative and bad experiences tend to get emphasized while good times are often taken for granted. While there is no denial that social problems of low salmon returns have been encountered for Bristol Bay sockeye salmon (*Oncorhynchus nerka*) and Yukon chinook in recent years, the fact remains that for the whole of Alaska, salmon catches have been on the high side.

What is the main problem for the salmon fishing industry in recent years? It is price. The prices paid to the fishermen have been declining steadily due to increasing competition from farm-raised salmon. Aquaculture salmon production has grown steadily and the price received for wild salmon catches has steadily declined (Fig. 8). Some prices paid for the lesser quality salmon are so low that the cost of catching far exceeds the value of the fish!

From the point of view of stock status, all five species of Pacific salmon—pink (*Oncorhynchus gorbuscha*), sockeye (*O. nerka*), chum (*O. keta*), coho (*O. kisutch*), and chinook (*O. tshawytscha*) are fully utilized, and stocks in most regions of the state generally have rebuilt to near or beyond previous high levels (Table 1). Research has been extensive into all aspects of life histories of the Pacific salmon, and the information has been used to regulate the fisheries.

The Alaska Department of Fish and Game manages salmon fisheries, while the Alaska Board of Fisheries has responsibility for allocating the yield of salmon among user groups. This clear separation of management authority is one of the strengths of the Alaska management system. The Web site for the Alaska Department of Fish and Game (ADFG) (http://www.ak.gov/adfg/geninfo/special/sustain/managmnt.pdf) describes Alaska's salmon management as "A Story of Success." It states that the management system is science-based, is in-season abundance-based, and it lets its managers manage.

Groundfish resources

The groundfish complexes are the most abundant of all fisheries resources off Alaska. The biomass of all the groundfish species combined total more than 26.4 million t and 80% of the biomass is found in the Bering Sea–Aleutian Islands region. The rest are located in the Gulf of Alaska. The percentage distribution of the 1999-2001 average harvest of groundfish was 88% from the Bering Sea–Aleutians, 10% from the Gulf of Alaska, and 2% was Pacific halibut (*Hippoglossus stenolepis*, Fig. 9). The rate of fisheries exploitation, however, varies species by species. The North Pacific Fishery Management Council (NPFMC) manages groundfish resources beyond the 12-mile territorial seas, while ADFG manages the inshore groundfish resources. Prior to implementation of the Magnuson Fisheries Conservation and Management Act (MFCMA) of 1976, the only groundfish species of significant commercial value harvested by domestic

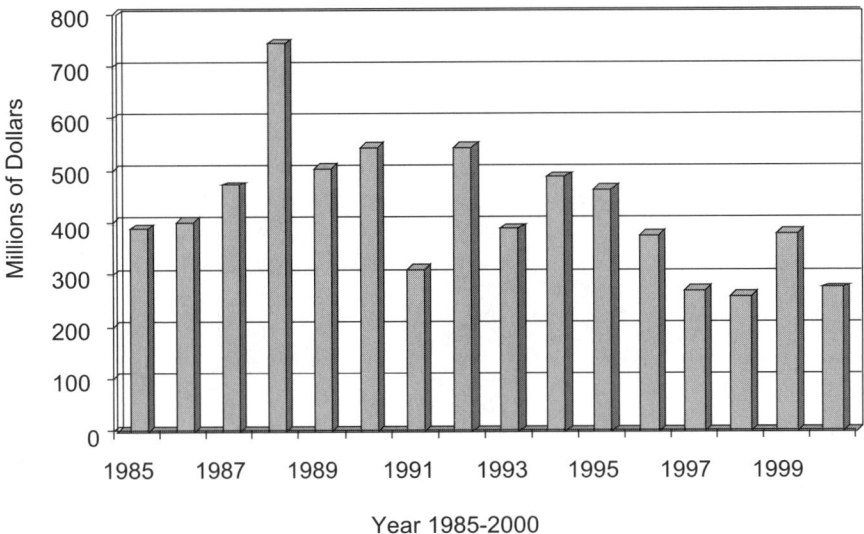

Figure 8. Value of salmon catch from Alaska, 1985-2000.

Table 1.   Status of salmon resources off Alaska.

Species	Recent average yield (RAY)	Current potential yield (CPY)	Long-term potential yield (LTPY)	Utilization	Stock level relative to LTPY
Pink	156,700	129,300	129,300	Full	Above
Sockeye	108,400	113,000	113,000	Full	Below
Chum	78,000	50,400	50,400	Full	Above
Coho	16,300	17,700	17,700	Full	Below
Chinook	4,300	5,300	5,300	Full	Below
Total all salmon	363,700	315,700	315,700		

Numbers metric tons.
LTPY and CPY from 1980-2000 averages; RAY is for 1995-2000.

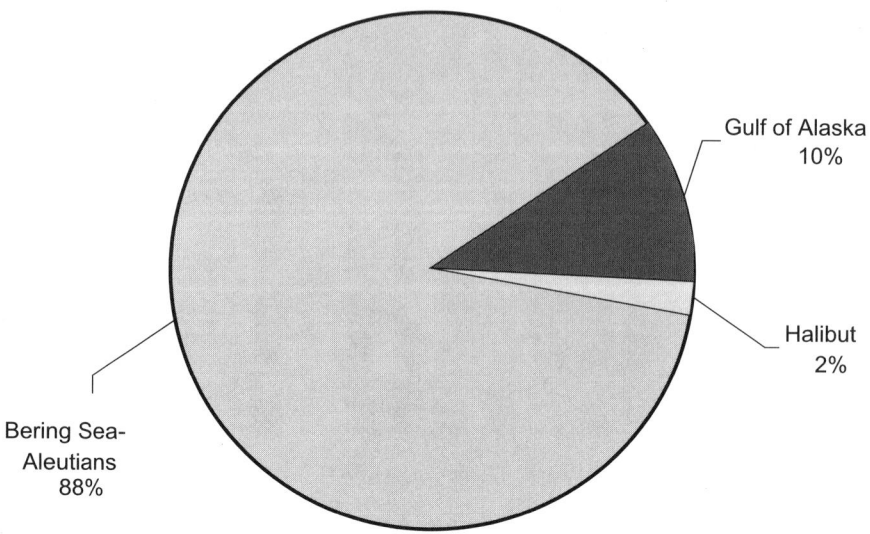

Figure 9. Regional apportionment of average groundfish catch from Alaska, 1998-2001.

fisheries was Pacific halibut. The other groundfish species were harvested mostly by foreign fisheries before that time. The Magnuson Act stimulated growth of the domestic fisheries, initially in joint-venture operations with foreign partners and then rapidly replacing them by 1983.

Pacific halibut

The Pacific halibut resource is found from the Bering Sea to California, with the center of abundance in the Gulf of Alaska. The resource is managed by a bilateral treaty between the United States and Canada and through research and regulation recommendations from the International Pacific Halibut Commission (IPHC).

The halibut resource has been healthy, and the total catch has been near record levels (Fig. 10), totaling 52,413 t in 2001. The breakdown by fishery was 43,217 t for commercial fisheries; 4,740 t for recreational fisheries; 440 t for personal use; 3,877 t as bycatch in other fisheries; and 139 t as mortality due to fishing by lost gear and discards. Overall the Pacific halibut resource, including stocks off Canada, is fully utilized each year and the stock level is near its long-term potential yield (Table 2).

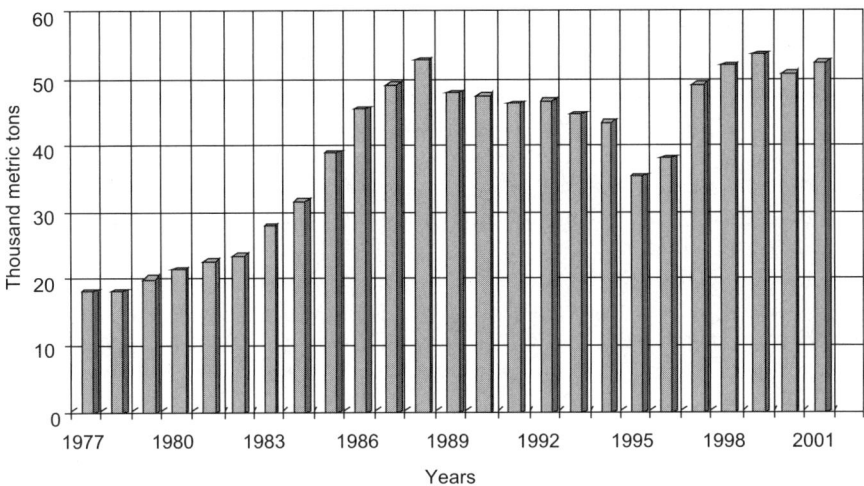

Figure 10. Catch trend of the Pacific halibut resource, 1977-2001.

Table 2.   Status of Pacific halibut resources.

Species	RAY	CPY	LTPY	Utilization	Stock level relative to LTPY
Bering Sea	10,097	21,105	15,000	Moderate	Above
Gulf of Alaska	33,756	42,644	40,000	Near full	Above
U.S. Pacific Coast	908	1,115	1,000	Full	Near
Canada	7,616	7,972	7,000	Full	Near
Total resource	52,377	72,836	63,000		

Numbers are metric tons.
LTPY and CPY from 1980-2000 averages; RAY is for 1995-2000.
RAY = recent average yield; CPY = current potential yield; LTPY = long-term potential yield.

Bering Sea–Aleutians groundfish

The dominant species of groundfish harvested from the Bering Sea–Aleutians groundfish complex are walleye pollock (*Theragra chalcogramma*, 73%), Pacific cod (*Gadus macrocephalus*, 11%), yellowfin sole (*Pleuronectes asper*, 4%), rock sole (*Pleuronectes bilineatus*, 3%), and atka mackerel (*Pleurogrammus monopterygius*, 3%). The rest of the species make up 1% or less of the total catch. The catch trend since 1977 is shown in Fig. 11.

The groundfish catch in recent years (1999-2001) was about 1.55 million t, taken from the groundfish complex exploitable biomass of about 21 million t. The long-term potential yield is almost 5 million t while current potential yield is 3.18 million t (Table 3). This current potential yield, however, has not been allowed to be fully harvested because catch quotas have been capped at a 2 million t optimum yield limit set in the fishery management plan. The economically more valuable species, like pollock and Pacific cod, however, have been harvested closer to their full biological potential while many less valuable species are relatively lightly harvested.

Pollock produce the largest catch of any single species inhabiting the U.S. Exclusive Economic Zone. The three main stocks, in decreasing order of abundance, are eastern Bering Sea stock, Aleutian Basin stock, and the Aleutian Islands stock. Other species of great commercial importance are Pacific cod, flatfishes (yellowfin sole, rock sole *(Pleuronectes bilineatus)*, Pacific halibut, etc.), sablefish *(Anoplopoma fimbria)*, rockfishes, Atka mackerel *(Pleurogrammus monopterygius)*, and other species. Many of the groundfish stocks are not fully utilized as there is need to protect ecologically related or incidentally impacted species while fishing for groundfish.

The Bering Sea–Aleutian Islands groundfish fishery is managed by the NPFMC under a Federal Fishery Management Plan. ADFG manages the inshore portion of the groundfish resource within its 12-mile territorial seas. Federal management is promulgated under the Magnuson Act of 1976. The management of groundfish resources is shrouded with ecosystem-related issues because the fishery takes a large amount of catch. The gear used also takes incidental catches of various species. The trawl gear impacts sea bottom profiles and biological communities. Thus, the legal demands to mitigate all these ecologically related issues when conducting a groundfish fishery can be rather demanding.

Gulf of Alaska groundfish

Groundfish abundance in the Gulf of Alaska increased since 1977, peaking at 5.3 million t in 1982 and 1988, and most recently in 1997 at 5.49 million t. Abundance since then has remained relatively stable, fluctuating between about 4 and 5 million t. The estimated long-term potential yield is 451,440 t (Table 4). The current potential yield for the groundfish complex totaled 394,780 t and recent average yield is 203,512 t. The

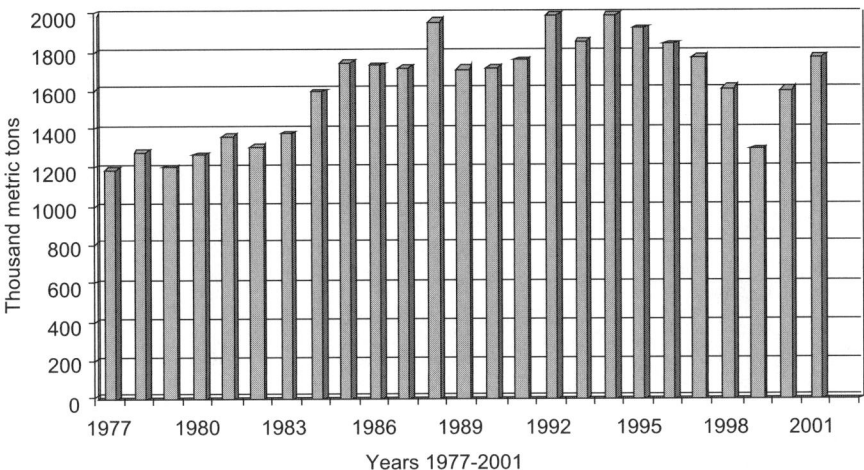

Figure 11. Catch trend of groundfish from the Bering Sea–Aleutians region, 1977-2001.

Table 3.   Status of Bering Sea–Aleutians groundfish resources.

Species	RAY	CPY	LTPY	Utili- zation	Stock level relative to LTPY
Pollock	1,135,650	2,138,110	3,608,000	Under	Below
Pacific cod	169,470	223,000	294,000	Under	Near
Yellowfin sole	68,580	115,000	136,000	Under	Near
Greenland turbot	5,970	8,100	36,500	Under	Below
Arrowtooth flounder	12,420	113,000	137,000	Under	Near
Rock sole	39,740	225,000	268,000	Under	Near
Flathead sole	18,000	82,600	101,000	Under	Below
Alaska plaice	11,100	143,000	172,000	Under	Near
Other flatfish	1,800	18,100	21,800	Under	Near
Sablefish	1,630	4,480	6,750	Full	Below
Pacific ocean perch	10,000	14,800	17,500	Full	Near
Other red rockfish	210	7,800	10,400	Under	Below
Other rockfish	800	1,040	1,380	Full	Near
Atka mackerel	54,120	49,000	82,300	Full	Below
Other fish	21,200	39,100	78,900	Under	Below
Total resource	1,550,690	3,182,130	4,971530		

Numbers are metric tons.
LTPY and CPY from 1980-2000 averages; RAY is for 1995-2000.
RAY = recent average yield; CPY = current potential yield; LTPY = long-term potential yield.

Table 4.   Status of groundfish resources in the Gulf of Alaska.

Salmon Species	RAY	CPY	LTPY	Utili- zation	Stock level relative to LTPY
Pollock	78,912	58,250	84,090	Full	Below
Pacific cod	59,024	57,600	77,100	Full	Above
Flatfish	31,009	232,850	231,060	Under	Above
Sablefish	12,202	12,820	19,350	Full	Above
Atka mackerel	166	600	6,200	Unknown	Unknown
Slope rockfish	16,805	24,830	30,530	Full	Above
Thornyhead rockfish	1,304	1,990	2,330	Full	Above
Pelagic shelf rockfish	3,799	5,490	8,220	Under	Above
Demersal shelf rockfish	291	350	480	Full	Above
Total groundfish	203,512	394,780	509,360		

Numbers are metric tons. LTPY and CPY from 1980-2000 averages; RAY is for 1995-2000.
RAY = recent average yield; CPY = current potential yield; LTPY = long-term potential yield.

wide disparity between the current potential yield and the recent average yield is due to underutilization of groundfish species, particularly for flatfish, that could not be fully harvested without exceeding incidental catch limits of Pacific halibut set by the North Pacific Fishery Management Council.

Gulf of Alaska groundfish catches have ranged from a low of 129,640 t in 1978 to a high of 352,800 t in 1984 (Fig. 12). The groundfish catches are dominated by pollock, followed by Pacific cod, flatfish, and rockfish. Groundfish catches since 1989 have fluctuated around 200,000 t.

Pollock abundance has been decreasing in recent years due to a lack of strong recruitment since the 1989 year class. Pollock are considered fully utilized. Pacific cod are abundant and fully utilized. The Pacific cod stock has been declining for the past several years due to a lack of significant recruitment. Flatfishes are in general very abundant, largely due to great increases in arrowtooth flounder *(Atheresthes stomias)* biomass. Flathead sole *(Hippoglossoides elassodon)*, rex sole *(Errex zachirus)*, and arrowtooth flounder are managed as separate categories, and the rest of the flatfish are managed as deepwater or shallow-water groups. Flatfishes are underutilized due to Pacific halibut bycatch considerations. Sablefish are considered to be at a low but stable population level and are projected to increase in the near future due to improved recruitment. They are fully utilized. Sablefish have been harvested under an individual fishing quota system since 1995. This has significantly changed the dynamics of the fishery. The Atka mackerel stock occurs mainly in the Aleutian Islands region. Its abundance in the Gulf of Alaska is much lower and highly variable.

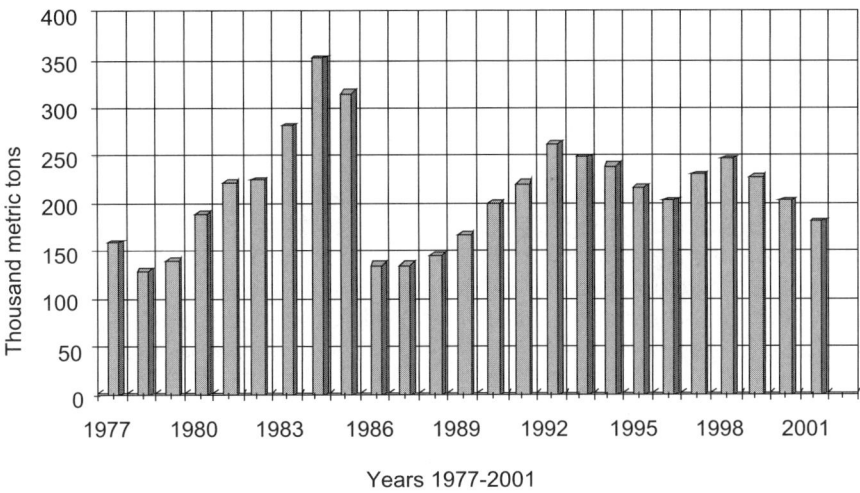

Years 1977-2001

Figure 12. Catch trend of groundfish from the Gulf of Alaska, 1977-2001.

For management purposes, rockfishes in the Gulf of Alaska are divided into four assemblages or species groups: slope rockfishes, pelagic
shelf rockfishes, thornyhead rockfish, and demersal shelf rockfish. Slope
rockfish are at low levels of abundance and fully utilized. Within this
group, Pacific ocean perch *(Sebastes alutus)*, shortraker *(S. borealis)* and
rougheye *(S. aleutianus)* rockfish, and northern *(S. polyspinus)* rockfish are
managed as separate categories. The principal species of the slope group,
Pacific ocean perch and shortraker and rougheye rockfish, are highly valued. Slope rockfish, particularly Pacific ocean perch, were intensively exploited by foreign fleets in the 1960s. In recent years, Pacific ocean perch
have rebounded from the heavy exploitation of three decades ago due to
good recruitment from a series of year classes. Thornyhead rockfish are
highly valued and believed to be at average levels of abundance. Dusky
rockfish *(S. ciliatus)* is the dominant species in the pelagic shelf rockfish
group, but its abundance estimate is variable due to problems assessing this species with current trawl survey methodology. Demersal shelf
rockfish assessment and management are focused on the target species,
yelloweye rockfish *(S. ruberrimus)*. Traditional population assessment
methods (e.g., trawl surveys) are not considered useful for surveying
demersal shelf rockfish because of their affinity for rough terrain. Rockfishes in general are conservatively managed due their long life spans
and consequential sensitivity to overexploitation.

Shellfish resources

Alaska's major offshore shellfish fisheries in the Bering Sea are king crabs, tanner crabs, sea snails, and shrimps. The king and tanner crab fisheries are managed primarily by the State of Alaska, with advice from a federal fishery management plan. The sea snail fishery falls under management of a federal preliminary fishery management plan. Shrimp and other near-shore fisheries are managed by the Alaska Department of Fish and Game. The status of the major shellfish resources is shown in Table 5.

Three king crab species (red, blue, and golden or brown—*Paralithodes camtschaticus, P. platypus*, and *Lithodes aequispina*) and two Tanner crab species (Tanner crab and snow crab—*Chionoecetes bairdi* and *C. opilio*) have been the traditionally harvested species. Exploratory fisheries on new deep-water stocks of other species of king and Tanner crabs have begun producing minor landings.

The northern pink shrimp *(Pandalus borealis)* is the most important of the Alaskan shrimp resources. For 30 years, shrimp abundance has been too low in the Bering Sea to support a commercial fishery. Bering Sea shrimp catches by Russia and Japan peaked at 32,000 t in 1963, declining quickly thereafter, until the fishery ended in 1973 (Fig. 13).

The Japanese pot fishery for snails, conducted from about 1972 until it ended in 1987, peaked at about 13,000 t in 1974. The snail stocks of the Bering Sea are underutilized, with no reported catch since 1997.

In general, crab and shrimp resources are depressed throughout Alaska. The red king and Tanner crab stocks in Bristol Bay are particularly low (Figs. 14-15). The red king crab fishery was closed in 1994 and 1995, following assessment of the spawning stock, which has declined to a low level. During the 1996 Tanner crab fishery only 800 t was landed, and the fishery has been closed since 1997. Snow crab abundance declined in 1999 to a low level. The fishery remains open; however, harvest rates have been reduced.

The bycatch of crabs in trawl and pot fisheries continues to be a major issue. Not only is bycatch an allocation problem, but unknown mortalities from discards of females and subadult crabs from pot and trawl catches could have an impact on the crab stocks. When crab abundance is low, the unknown bycatch mortality, if high enough, could impose unacceptable risks to stock recovery.

Herring resources

Pacific herring *(Clupea pallasii)* is the major pelagic species harvested in Alaska. The fisheries occur in specific inshore spawning areas of the Gulf of Alaska and the Bering Sea. In the Gulf of Alaska, spawning fish concentrate mainly off southeast Alaska, in Prince William Sound, and around the Kodiak Island–Cook Inlet area. In the Bering Sea, the centers

Table 5. Status of shellfish resources in the Bering Sea.

Species	RAY	CPY	LTPY	Utili-zation	Stock level relative to LTPY
King crabs	8,050	8,050	36,481	Full	Below
Tanner crabs	997	997	21,751	Full	Below
Snow crabs	71,389	71,389	37,202	Full	Below
Shrimps	1,152	1,152	14,722	Full	Below
Sea snails	0	0	3,062	Under	Unknown
Total	81,588	81,588	113,218		

Numbers are metric tons.
RAY = recent average yield; CPY = current potential yield; LTPY = long-term potential yield.

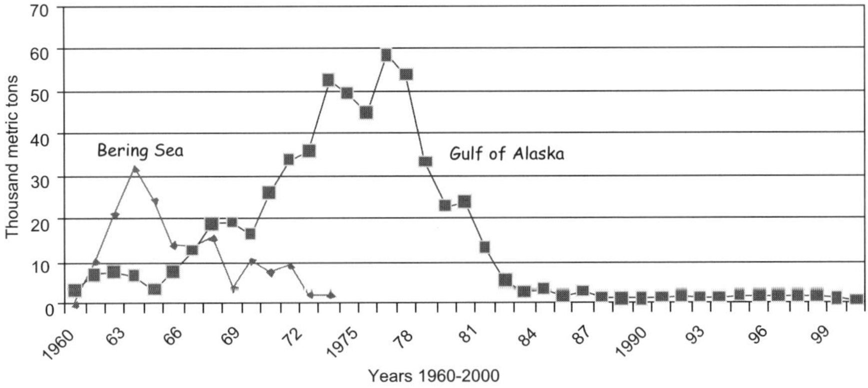

Figure 13. Catch trend of shrimp from the Bering Sea–Aleutians and the Gulf of Alaska, 1960-2000.

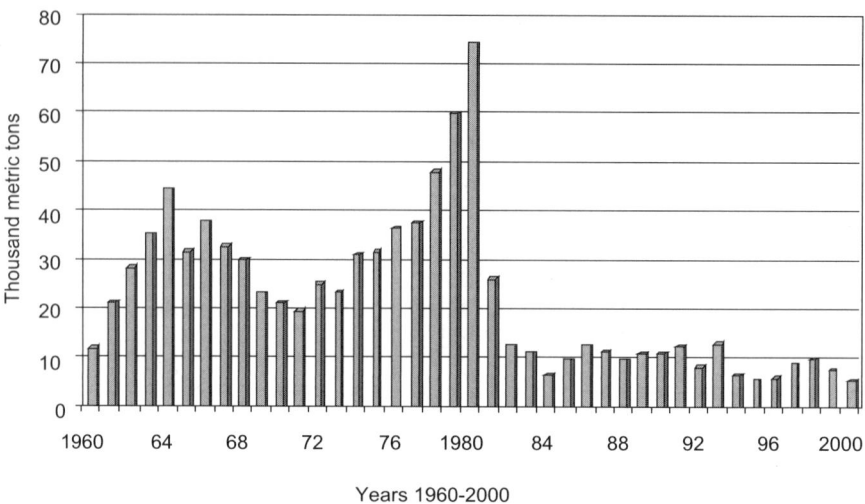

Figure 14. Catch trend of king crab from the Bering Sea, 1960-2000.

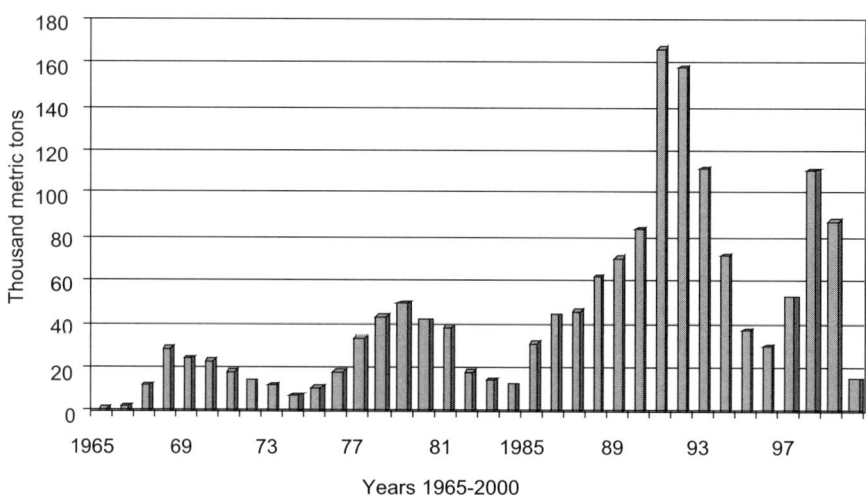

Figure 15. Catch trend of tanner crab from the Bering Sea–Aleutians region, 1965-2000.

of spawning abundance are in northern Bristol Bay and the eastern shore of Norton Sound. This fishery occurs within state waters (3-mile limit), and it is therefore monitored and managed by the Alaska Department of Fish and Game.

Alaska's herring industry began as early as 1878. The Bering Sea fishery began in the late 1920s. Since 1977 Bering Sea herring have been harvested primarily in inshore sac roe fisheries, and catches have since risen slowly but steadily, reflecting better stock conditions. A portion of the Bering Sea harvest is taken as bycatch in the offshore federally managed groundfish fishery. Retention of herring in these fisheries is prohibited, with regulations limiting herring bycatch to no more than about 1,000 t annually. From 1997 to 2001 the actual herring bycatch averaged 751 tons.

From catch records (Fig. 16), it is evident that herring biomass fluctuated widely due to influences of strong and weak year-classes. Currently the herring populations in Alaska remain at moderate levels and are in relatively stable condition, with the exception of Prince William Sound (Table 6). Herring abundance levels typically increase abruptly following major recruitment events, then decline slowly over a number of years because of natural and fishing mortality. Prince William Sound herring continue to be depressed from a disease outbreak in 1993. In more recent years, statewide herring harvests have averaged about 45,000 t. About 10% of the commercial harvest is taken for food and bait, and the rest is taken in the sac roe fisheries. In addition, there is a roe-on-kelp fishery that harvests about 400 t of product annually.

The value of the fishery is on roe-on-kelp products when the herring spawn. Thus the fishing season is very short as spawning events take place over hours or days. ADFG makes pre-season forecasts of the stocks and actual season openings and closures are determined virtually real time as spotter airplanes and helicopters are used to estimate biomass real time. Thus the fisheries can be furious as short-term management calls for fishing seasons may be made over periods of hours to days.

Outlook

The NPFMC looks cautiously to an optimistic future for Alaska's fisheries resources. The outlook is indeed optimistic as the fisheries management process is indeed conservation oriented. To borrow language from the ADFG Web site, "the management is science based, it is in-season abundance based, and it lets its managers manage." While active management is key to assuring Alaska's optimistic future, there is no doubt that Mother Nature plays a greater role.

Periodic weather regime shifts and other aspects of ocean variables (physical, chemical, and biological) affect the health and dynamics of Alaska's fisheries resources. The fisheries management process is set up

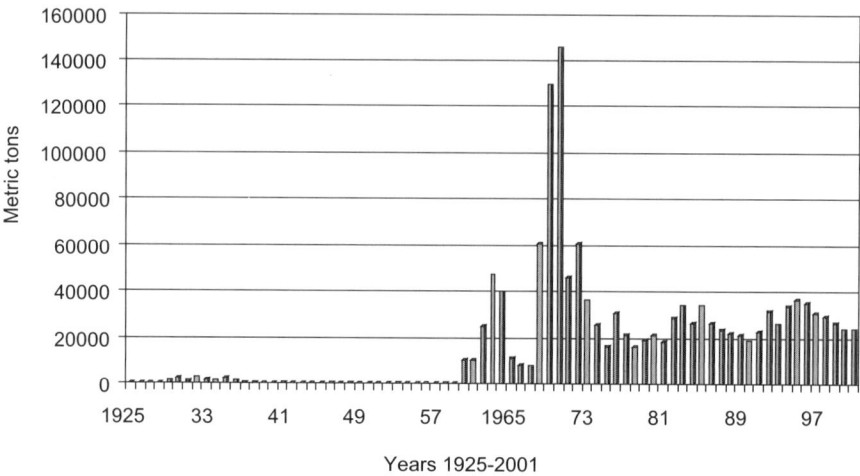

Years 1925-2001

Figure 16. Catch trend of Pacific herring from the Bering Sea–Aleutians region, 1925-2001.

Table 6.   Status of Pacific herring resources off Alaska.

Area	RAY	CPY	LTPY	Utili-zation	Stock level relative to LTPY
Gulf of Alaska	10,030	11,366	19,000	Full	Near
Bering Sea	24,038	27,420	28,000	Full	Near
Total herring	34,068	38,786	47,000		

Numbers are metric tons.
LTPY is based on 10-20 year average catch; CPY is based on 2002 catch quotas; RAY is for 1995-2000.
RAY = recent average yield; CPY = current potential yield; LTPY = long-term potential yield.

to consider as much as possible the known impact of weather on fisheries resources, and to take precautionary actions to mitigate for these factors. Management must also now consider a broader aspect of fishing on ecologically related species and the broader aspect of ecosystem management.

Ecosystem management

Some species in the Bering Sea underwent large changes between the 1950s and the 1980s. Among the best documented are the declines of Steller sea lions *(Eumetopias jubatus)* and northern fur seals (*Callorhinus ursinus)*, and the increase and dominance of groundfish (pollock and flatfishes). A frequently proposed explanation is that human exploitation of top predators and/or a shift in the physical oceanography altered the structure of the eastern Bering Sea ecosystem. While research is continuing, the NPFMC has to mitigate for these ecosystem considerations in its groundfish management process.

Do we have ecosystem management? No, but yes! There is still a general lack of data and understanding about what ecosystem-based management means. The theory of ecosystem-based management is sound, but its practice is subject of much debate. How much ecosystem-based consideration is adequate ecosystem-based management?

The NPFMC has an approach to ecosystem management. It has made significant progress toward incorporating ecosystem considerations into management of groundfish fisheries. Steps have been taken to lessen human impacts on the environment due to fishing, while at the same time providing sustained yields of fishery resources. The council's ecosystem-based management approach involves public participation, reliance on scientific research and advice, conservative catch quotas, comprehensive monitoring and enforcement, bycatch controls, gear restrictions, temporal and spatial distribution of fisheries, habitat conservation areas, and other biological and socioeconomic considerations.

The most basic ecosystem consideration employed is a precautionary approach to harvesting of fish resources. It means setting harvest levels at safe biological levels. Harvest levels are set to provide for adequate spawning biomass. For groundfish management, the trend is toward even safer harvesting strategies that account for (a) bycatch, (b) ecosystem interactions that particularly impact seabirds and Steller sea lions, (c) preservation of essential fish habitat, and (d) vagaries of weather and other uncertainties on the populations and the recruitment process.

Even before ecosystem considerations became a common concept in fisheries management, management of the groundfish resources in the Bering Sea–Aleutians region already started in 1979 with a conservative maximum annual catch of two million metric tons despite higher acceptable biological catch levels.

Input toward ecosystems management is a continuing activity. The NPFMC has an ongoing program to encourage fishermen and coastal community residents to submit observations of unusual occurrences, significant changes, and other items of interest seen during the fishing year so that these observations can be incorporated into the resource assessment process. Each year, the SAFE report (Stock Assessment and Fishery Evaluation) also incorporates a chapter on ecosystem considerations that bring together all relevant data and observations on weather and ecosystem events for consideration into the council's management process.

Outside fisheries impacts

Besides direct management of Alaska's fisheries resources, there is also concern for outside fishing impacts that would affect the management and outlook of Alaska's fisheries resources. Some of Alaska's fisheries resources are transboundary in their migrations and can be intercepted outside of U.S. jurisdiction. The main fishery resources that are transboundary are the five Pacific salmon species, pollock, and Pacific herring in the Bering Sea.

Coordinated management of salmon transboundary problems are pretty much covered by the North Pacific Anadromous Fish Commission (NPAFC) and by a 1998 bilateral agreement between the U.S. and Russia. The NPAFC bans high seas fishing for salmon and the bilateral agreement discourages driftnet fisheries beyond 25 miles off the Russian coasts. Still, there have been occasions for suspicions that salmon resources bound for the U.S. have been intercepted by illegal fisheries in Russian waters.

In the northern Bering Sea, the pollock resources are transboundary across the U.S.-Russian border. There is not yet any coordinated management of the resource, although there has been much coordinated research on the resource. There are great differences in the exploitation rates of pollock resources in the Russian EEZ and the U.S. EEZ, with rates significantly higher than the 10-15% harvest rate in the U.S. zone. The Russian harvest rates in recent years were 35% to rates exceeding 50%.

The U.S. and Russia have a regular forum to address common issues in the Bering Sea, including fisheries enforcement issues. This forum is the "Intergovernmental Consultative Committee" which meets generally once a year to resolve issues about salmon and pollock management.

Perhaps the toughest problem to resolve is an agreement to define the exact coordinates of the boundary between the United States and Russia in the Bering Sea and the Chukchi Sea. From 1886 when Russia sold Alaska to the United States, until 1976-77 when both countries extended fisheries to 200 nautical miles, the exact coordinates of the boundary was not a major issue and there was general acceptance of a treaty line referenced from three coordinate points. Under extended jurisdictions and modern interpretations of international laws, the boundary coordinates are much more difficult to resolve between the two countries. A temporary agreement was reached in 1979 and was ratified by the U.S. Congress. However, it has yet to be ratified by the Russian Duma. In the meantime, both countries have observed the provisions of the temporary agreement.

References

NMFS. 1999. Our living oceans. Report on the status of U.S. living marine resources, 1999. NOAA Technical Memorandum NMFS-F/SPO-41. 301 pp.

Utilization of Alaska's Seafood Processing Byproducts

Chuck Crapo
University of Alaska Fairbanks, Fishery Industrial Technology Center,
Kodiak, Alaska

Peter Bechtel
University of Alaska Fairbanks, USDA Agricultural Research Service
Laboratory, Fairbanks, Alaska

Abstract

Byproducts processing has a long history in Alaska, dating from 1882 when
the Alaska Oil and Guano Co. started producing fish oil and fertilizer from
whole herring. The use of whole herring to produce meals, oils, and other
products was a thriving industry with peak herring meal production in 1936
at 19,000 tons. These plants ceased operations in the late 1950s. Not until
the Federal Water Control Act of 1972 and establishment of the 200 mile
exclusive economic zone were new waste reduction plants constructed
in Alaska for processing of shellfish and finfish wastes. Since then, the
amount of fish meal produced from fish and shellfish harvested from Alaska
waters has increased from 5,000 tons in 1976, to over 70,000 tons in 2002.

The amount of marine fish processing byproducts produced from
the 2000 harvest was approximately 1,007,000 metric tons (t) of which
approximately 53%, 40%, and 7% was processed on shoreside, catcher-processor, and motherships, respectively. Approximately 70% of the total fish
harvest was from the Bering Aleutian region. Of the total fish processing
byproducts produced in Alaska, approximately 70% come from Pollock
and only 9% from salmon. It was estimated that less than 40% of the fish
processing byproducts are utilized for human or animal nutritional or
industrial products.

At present roe is a highly valued byproduct that is always saved,
and depending on economic conditions there can be markets for cod
stomachs, livers, and heads. With the development of the markets for
fish mince there has been a substantial increase in the recovery of muscle
flesh from fish frames. However, most of the fish processing byproducts

including heads, frames, viscera, and skin are disposed of or made into meal and oil for the animal industries. There is opportunity to produce value added products from underutilized fish processing byproducts.

Introduction

Fish and shellfish processing byproducts have been utilized for many years to make meals and oils (Windsor and Barlow 1981). Alaska has produced fish meal and fish oil for over one hundred years (Meehan et al. 1990). The establishment of the Alaska Oil and Guano Company in 1882 started an industry that continues today. The early fish meal industry processed whole herring for edible oil, fertilizer, and animal feeds. Fish meal production varied from year to year depending on the catch and other uses for herring such as pickled product (Fig. 1). Production between the 1890s and late 1910s was small, never exceeding 2,000 tons. After World War I, fish meal production increased substantially due to the westward development of Alaska's resources. Peak production of herring meal occurred in 1936 at 19,000 tons. During World War II, production fell dramatically due to war restrictions and herring quotas. The herring reduction fishery rebounded in the late 1940s, only to be discontinued in the late 1950s because of poor economics and less government regulations.

Fish meal was not produced again in Alaska until the early 1970s. Since that time production has steadily increased until over 70,000 t have been made in 2002 (Fig. 2). The reestablishment of fish meal operations in Alaska was the result of the passage of the 1972 Federal Water Quality Control Act and favorable prices for meals. The act required fish processing wastes in non-remote areas be discharged or dumped in tidal flow areas to prevent buildup. The subsequent Fisheries Conservation and Management Act in 1976 spurred fish meal plant construction as the cod and pollock fisheries developed in the Gulf of Alaska and Bering Sea. These modern fish meal plants have been designed to handle the byproducts from surimi, pollock, and cod fillet and salmon operations. Raw materials include viscera, heads, frames, and skin.

Fish meal production is entering a new era. The fishery management councils are requiring full utilization of resources; no longer can material be discarded (NPFMC 2002). There appears to be opportunity for those processors able to produce products with specific properties (Barlow 2002). Many aquaculture operations need low ash fish meal to meet animal growth and environmental needs (Hardy 2002). The development of nutraceuticals from fish oils and proteins is a tantalizing opportunity (Shahidi 2002). In Alaska, fish meal production has been a volume operation, processing materials with little consideration of their properties.

In order for processors to adapt to the new opportunities or requirements, both environmental and market driven, they must know volumes and types of waste they handle and how much of the total material stream

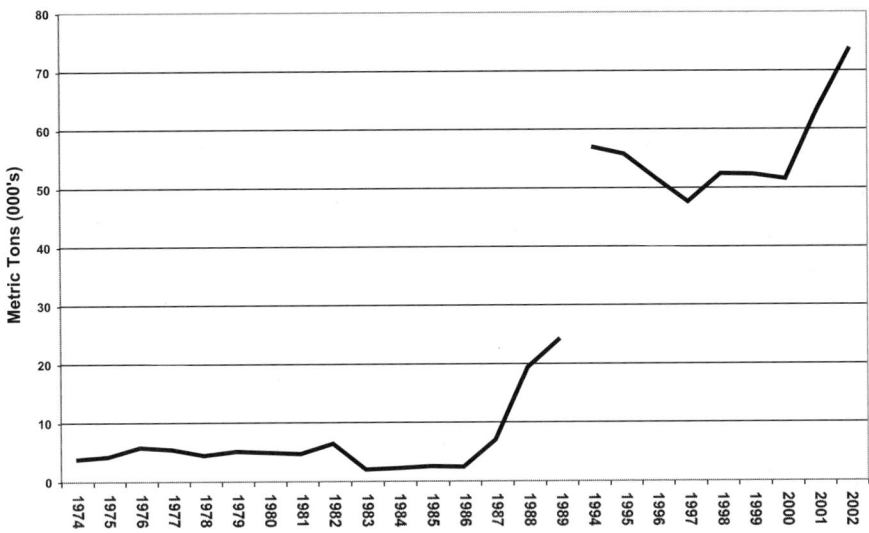

Figure 1. Alaska fish meal production, 1890-1958.

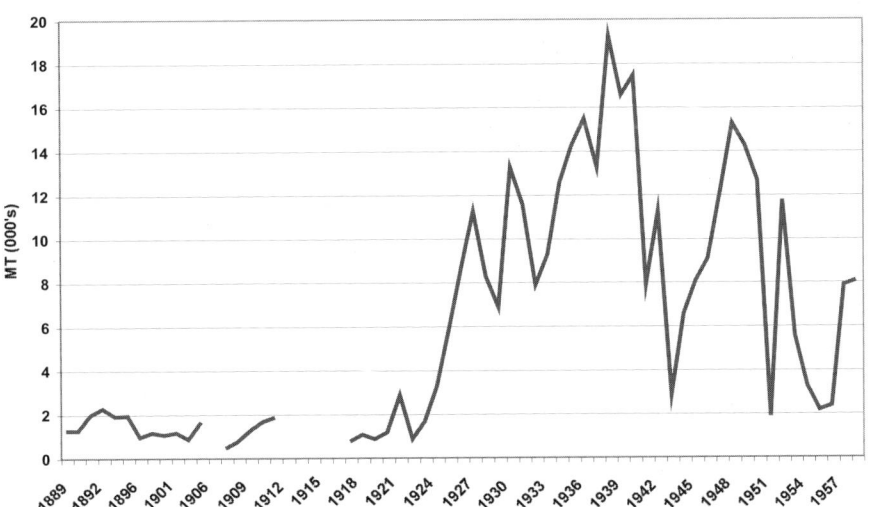

Figure 2. Alaska fish meal production, 1974-2002.

is being utilized. This paper evaluates the production of fish meal and fish oil in Alaska and determines the amounts and types of raw material produced in Alaska's commercial fisheries.

Methods

Using statistics from the 2000 fishing seasons, total catch by species, region, and processing sector were determined (ADFG 1999, 2000a,b,c; NMFS 2000a,b,c,d,e). Reported fish meal and fish oil production was broken down to region and processing sector. Calculations were made using reported values and anecdotal information from processors to estimate the ratio of products and byproducts. Production of the few processing facilities not covered in the NMFS (U.S. National Marine Fisheries Service) database were estimated.

Percent heads, frames, viscera, skin, and fillets by species were obtained from published recoveries (Kizevetter 1971, Babbitt 1990, Crapo et al. 1993). Using these published and estimated values, the amount of individual byproducts were calculated. This calculation took the amount of fish caught from NMFS and ADFG (Alaska Department of Fish and Game) statistics and multiplied them by the percent of the individual waste component (e.g., head is 20% of fish weight). The waste stream from most pollock and cod fillet operations generated similar byproducts. For other species, such as salmon and Pacific halibut processing, heads and viscera were the major byproducts. The data put forth attempts to reflect current commercial processing practices. Estimates were made of the amount and type of wastes generated by each established processing sector, catcher-processor, mothership, and onshore processor.

Finally, total fish byproduct utilization was estimated by converting wet waste volume to dry material using moisture data (Bechtel 2003) and comparing with published fish meal and fish oil production.

Results

In the past several years, the annual harvest of fish and shellfish from Alaska waters has been over 2 million metric tons. In 2000, 2,008,143 t of fish and shellfish were harvested (Table 1). Shellfish and aquacultured mollusks accounted for 102,894 t or 5.1% of the catch. These products do not significantly impact the production of fish meal and fish oil in Alaska and have been excluded from the subsequent analysis. Alaska pollock was the most abundant, comprising 1,067,738 t or 53% of the landings. The next two species, salmon and Pacific cod, contributed 319,472 t and 226,709 t, 16% and 11%, respectively. More than 75% of the total fish harvest in Alaska is pollock, salmon, and cod.

The annual harvest by region is shown in Table 2. The regions are those used by NMFS in reporting their data. Catch by regions showed

Table 1. 2000 commercial seafood harvest in Alaska.

	Metric tons	Data source[a]
Marine finfish		
Alaska pollock	1,067,738	NMFS 2000
Salmon	319,472	ADFG 2000
Pacific cod	226,709	NMFS 2000
Flatfish	141,530	NMFS 2000
Atka mackerel	39,986	NMFS 2000
Perch	17,077	NMFS 2000
Sablefish	13,547	NMFS 2000
Rockfish	10,472	NMFS 2000
Pacific herring	32,509	ADFG 2000
Pacific halibut[b]	32,686	NMFS 2000
Others	3,523	NMFS 2000
Total	1,905,249	
Marine shellfish		
Crab	99,663	ADFG 2000
Shellfish	2,486	ADFG 2000
Squid	333	NMFS 2000
Total	102,482	
Marine aquaculture production		
Oyster	405	ADFG 1999
Clams	12	ADFG 1999
Mussels	2	ADFG 1999
Total	419	

[a]NMFS = National Marine Fisheries Service. ADFG = Alaska Department of Fish and Game.
[b]NMFS report halibut harvest with head and viscera removed; adjusted for 12% viscera, 16% head.

Table 2.   2000 Alaska fish harvest by region and processing sector, in metric tons.

	Total	Shoreside	Catch-Proc.	Mothership
Alaska pollock	1,067,738	557,790	411,663	98,285
Bering Aleutian area	998,063	488,426	411,420	98,217
Gulf of Alaska area	69,675	69,364	243	68
Salmon	319,471	319,471		
Southeast Alaska area	109,677	109,677		
Central Alaska area	150,868	150,868		
Westward Alaska area	56,063	56,063		
AYK Alaska area	2,863	2,863		
Pacific cod	226,709	91,754	123,593	11,362
Bering Aleutian area	173,660	47,599	115,432	10,629
Gulf of Alaska area	53,049	44,155	8,161	733
Flatfish	72,476	11,714	60,676	86
Bering Aleutian area	50,937	1,995	48,859	83
Gulf of Alaska area	21,539	9,719	11,817	3
Atka mackerel	39,986	1	39,985	
Bering Aleutian area	39,826	39,826		
Gulf of Alaska area	160	1	159	
Perch	17,077	4,211	12,866	
Bering Aleutian area	8,073	12	8,061	
Gulf of Alaska area	9,004	4,199	4,805	
Sablefish	13,547	11,352	2,195	
Bering Aleutian area	1,470	631	839	
Gulf of Alaska area	12,077	10,721	1,356	
Yellowfin sole	69,740	1,777	67,949	14
Bering Aleutian area	69,740	1,777	67,949	14
Others	3,523	904	2,616	3
Bering Aleutian area	2,770	157	2,610	3
Gulf of Alaska area	753	747	6	
Pacific herring	32,509	32,509		
Gulf of Alaska area	32,509	32,509		
Pacific halibut	32,686	32,686		
Gulf of Alaska area	32,686	32,686		
Rockfish	10,472	5,437	5,017	18
Bering Aleutian area	1,164	58	1,088	18
Gulf of Alaska area	9,308	5,379	3,929	
Totals	1,895,510	1,069,606	726,560	109,768
Bering Aleutian totals	1,345,703	536,260	696,923	108,964
Gulf of Alaska totals	240,760	144,285	30,476	804
Southeast Alaska area	109,677	109,677		
Central Alaska area	150,868	150,868		
Westward Alaska area	56,063	56,063		
AYK Alaska area	2,863	2,863		

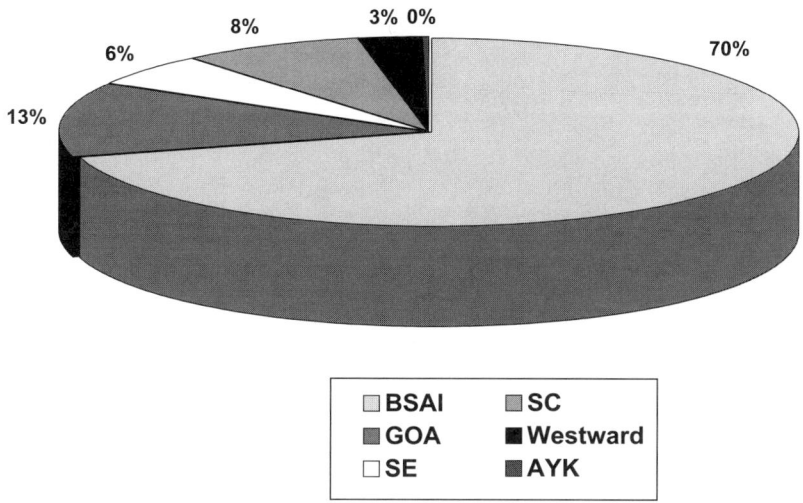

8%   3% 0%

6%   70%

13%

□ BSAI	■ SC
■ GOA	■ Westward
□ SE	■ AYK

Figure 3. 2000 Alaska catch by region, total 1,905,206 t. BSAI = Bering Sea/Aleutians; GOA = Gulf of Alaska; SE = Southeast; SC = South-central; AYK = Arctic Yukon Kuskokwim.

that 70% (1,345,703 t) of all finfish were harvested in the Bering Sea and Aleutian Islands (BSAI), primarily groundfish (Fig. 3). By contrast, the Gulf of Alaska (GOA) landed only 13% or 240,760 t of the Alaska catch. The salmon data listed by area in Alaska represents the data collection and tabulation by ADFG. Viewing the data by processing sector, the shoreside processors handled 56% (1,069,606t) of the catch followed by catcher-processors with 38% or 726,560 t (Fig. 4). The data indicate shoreside pollock processing the BSAI region comprises the largest volume of fish and hence, fish byproducts, in Alaska. All other fisheries, regions and processing sectors are significantly smaller.

During 2000, 83,148 t of fish meal and fish oil were produced (Table 3). These values were tabulated from the data collected and reported by NMFS. The numbers of shoreside plants is small but their volumes are large and easily reported. However, for the catcher processor vessels the values may be less reliable since only 11 vessels reported oil or meal products in 2000. The BSAI accounted for 77% of the production and the GOA for 23%, reflecting the volumes of catch in each region. Processing by sector indicates 83% of the fish meal and fish oil were produced by shoreside facilities. Catcher-processors accounted for only 15% of the production. Comparing the fish byproducts percentages with landed catch, it becomes evident that the shoreside processors are utilizing more

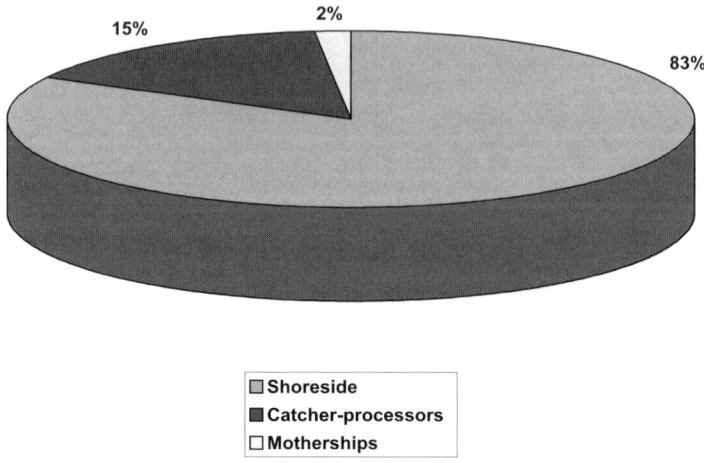

Figure 4. 2000 Alaska catch by sector, total 83,148 metric tons.

Table 3.   2000 production of fish meal and oil in Alaska, in metric tons.

	Fish meal	Fish oil	Meal + oil
Shoreside plants			
Bering Aleutian area	36,785	12,899	49,684
Gulf of Alaska area[a]	15,921	3,268	19,189
Total	52,706	16,167	68,873
Catcher-processors			
Bering Aleutian area	12,571	305	12,875
Motherships			
Bering Aleutian area	814	585	1,400
Grand total	66,091	17,057	83,148

[a]Assumed 13,000 t of meal and 3,000 t of oil from Kodiak shoreside processors.
Calculations from National Marine Fisheries Service 2000 statistics.

raw material than either catcher-processors or motherships. This is due in part to environmental regulations for shoreside processors and the limited space for production and storage of fish meal and fish oil aboard vessels. Of the 84,000 t of fish byproducts, it is estimated that 85% to 90% are produced from groundfish and the remainder are from salmon, primarily in the Gulf of Alaska.

What is the composition of the fishery byproducts available for fish meal and fish oil production? Depending on the fishery and markets, this can vary significantly (Table 4). For example, the majority of Alaska salmon is processed as dressed and headed fish leaving only heads and viscera. Recoveries from this product average 77%, producing only 23% byproducts. In 2000, this calculated to 86,257 t of which 57,505 t was heads and 28,752 t was viscera (Fig. 5). For Alaska pollock, the product forms can include surimi, mince, fillets, and dressed and headed fish. The product mix can vary between processors and seasons and it is difficult to develop an exact recovery rate. Based on anecdotal production information from several processors, the recovery for all forms averaged 34%. The remaining 66% was viscera, heads, frames, and skin. For 2000, the volume of byproducts from pollock processing was 704,707 t. Breakdown of byproduct components was 32% viscera, 26% heads, 33% frames, and 9% skin (Fig. 6). Other groundfish, such as Pacific cod and flatfish, parallel Alaska pollock use while blackcod, rockfish, and yellowfin sole are similar to salmon.

The final objective was to estimate how much of the raw byproducts were converted to fish meal and fish oil in 2000. The total amount of byproducts from all species were calculated as heads, viscera, frames, or skin and then converted to a dry weight basis (Table 5). Each component had a slightly different moisture level ranging from 19% for frames to 25% for viscera. This calculation used overall values from fish processors in Kodiak, Alaska. Overall, about 40% of the available byproducts is being used for fish meal and fish oil production. The shoreside processors use about 60% of their material while catcher-processors use only 15%. Since the biggest concentration of fish meal plants is in the BSAI region, the bulk of the byproducts being used come from groundfish. Much of the remaining shoreside material is salmon heads and viscera which is widely dispersed throughout Alaska and much more difficult to use.

Summary

A considerable volume of processing byproducts is unused. There are many reasons for this including problems with remote processing, short seasons, shipping logistics, and costs and economics. However, for those existing fish meal plants located in specific areas, there may be opportunities to produce products from individual components. The outcome of this analysis is a better quantitative understanding of the types and amounts of fish byproducts available for future use.

Table 4.   Fish waste components by processing sector in 2000, in metric tons.

	% of fish[a]	Total	Shoreside	Catch-proc.	Mother-ship
Alaskan pollock					
Total harvest	100	1,067,738	557,790	411,663	98,285
Total waste[b]	66	704,707	368,141	271,698	64,868
Heads	17	181,515	94,824	69,983	16,708
Viscera	21	224,225	117,136	86,449	20,640
Frames	22	234,902	122,714	90,566	21,623
Skin	6	64,064	33,467	24,700	5,897
Salmon					
Total harvest	100	319,471	319,471		
Total waste	27	86,257	86,257		
Heads	18	57,505	57,505		
Viscera	9	28,752	28,752		
Pacific cod					
Total harvest	100	226,709	91,754	123,593	11,362
Total waste[c]	46	104,286	55,970	45,729	4,204
Heads	18	40,808	16,516	22,247	2,045
Viscera	19	43,075	17,433	23,483	2,159
Frames	18	16,516	16,516		
Skin	6	5,505	5,505		
Flatfish					
Total harvest	100	72,476	11,714	60,676	86
Total waste	72	52,183	8,434	43,687	62
Heads	20	14,495	2,343	12,135	17
Viscera	14	10,147	1,640	8495	12
Frames	30	21,743	3,514	18,203	26
Skin	8	5,798	937	4,854	7
Atka mackerel					
Total harvest	100	39,986	1.0	39,985	
Total waste	32	12,796	0.3	12,795	
Heads	19	7,597	0.2	7,597	
Viscera	13	5,198	0.1	5,198	
Perch					
Total harvest	100	17,077	4,211	12,866	
Total waste	38	6,489	1,600	4,889	
Heads	26	4,440	1,095	3,345	
Viscera	12	2,049	505	1,544	

Table 4.   (Continued.)

	% of fish[a]	Total	Shoreside	Catch-proc.	Mother-ship
Sablefish					
Total harvest	100	13,547	11,352	2,195	
Total waste	32	4,335	3,633	702	
Heads	21	2,845	2,384	461	
Viscera	11	1,490	1,249	241	
Yellowfin sole					
Total harvest	100	69,740	1,777	67,949	14
Total waste	31	21,619	551	21,064	4
Heads	17	11,856	302	11,551	2
Viscera	14	9,764	249	9,513	2
Others					
Total harvest	100	3,523	904	2,616	3.0
Total waste	33	1,163	298	863	1.0
Heads	18	634	163	471	0.5
Viscera	15	528	136	392	0.5
Pacific herring					
Total harvest[d]	100	32,509	32,509		
Pacific halibut					
Total harvest	100	32,686	32,686		
Total waste	28	9,152	9,152		
Heads	16	5,230	5,230		
Viscera	12	3922	3922		
Rockfish					
Total harvest	100	10472	5437	5017	18
Total waste	43	4503	2338	2157	8
Heads	31	3246	1685	1555	6
Viscera	12	1257	652	602	2
Total harvest	1,905,934	1,069,606	726,560	109,768	
Total waste	1,007,490	536,375	403,585	69,147	
Total heads	289,384	182,047	129,345	18,779	
Total viscera	328,962	171,675	135,917	22,815	
Total frames	273,161	142,744	108,769	21,649	
Total skin	75,368	39,910	29,554	5,904	

[a]Crapo et al. 1993. Species harvest values from Table 1.
[b]The 66% total waste value is low, especially when surimi paste is made.
[c]Assume shoreside produced fillets (61% waste) and catcher processors headed and gutted fish (37% waste).
[d]Assume herring are frozen and shipped elsewhere for processing.

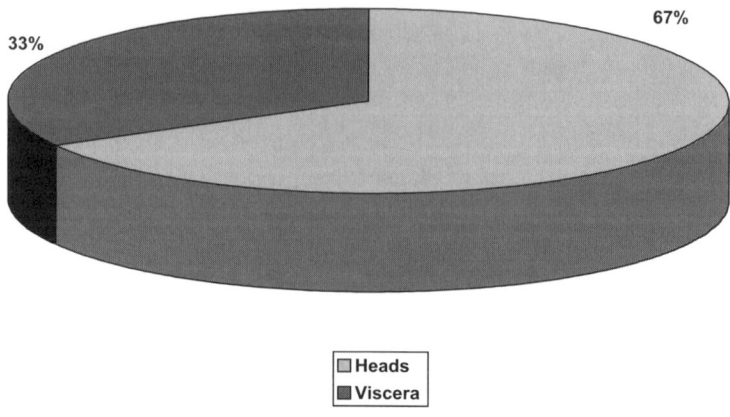

Figure 5. 2000 Alaska salmon byproducts, total 86,257 metric tons.

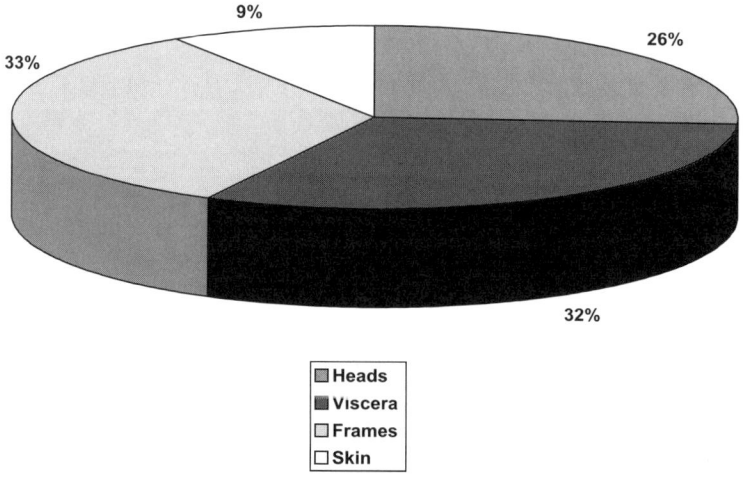

Figure 6. 2000 Alaska pollock byproducts, total 704,707 metric tons.

Table 5. Fish waste utilization in 2000 by processing sector, in metric tons.

	Dry matter[a] %	Total waste[b]	Total DM	Shoreside waste[b]	Shoreside DM	Catch-proc. waste	Catch-proc. DM	Mothership waste	Mothership DM
Heads	20	289,384	57,877	182,047	36,409	129,345	25,869	18,779	3,756
Viscera	25	328,962	82,241	171,675	42,919	135,917	33,979	22,815	5,704
Frames	19	273,161	51,901	142,744	27,121	108,769	20,666	21,649	4,113
Skin	22	75,368	16,581	39,910	8,780	29,554	6,502	5,904	1,299
Total processing waste solids			208,599		115,230		87,016		14,872
Total Alaska meal and oil			83,148		68,873		12,875		1,399
% solids recovered			39.9	59.8			14.8	9.4	

[a]From Bechtel 2003.
[b]Calculations based on values from Table 3, this report.
Not used in the calculation were 32,260 t of solids from fish discards and crab waste solids of approximately 10,000 t.
DM = dry matter.

References

ADFG. 1999. Alaska commercial herring sac roe harvest. Alaska Department of Fish and Game, Commercial Fisheries. http://www.cf.adfg.state.ak.us/geninfo/finfish/herring/catchval/99catch.htm.

ADFG. 2000a. Preliminary Alaska commercial shellfish catches and exvessel values. Alaska Department of Fish and Game, Commercial Fisheries. http://www.cf.adfg.state.ak.us/geninfo/shellfish/99_value.htm.

ADFG. 2000b. Alaska food/bait herring harvest. Alaska Department of Fish and Game, Commercial Fisheries. http://www.cf.adfg.state.ak.us/geninfo/finfish/herring/catchval/2kFBcatc.htm.

ADFG. 2000c. Alaska commercial salmon harvests—exvessel values. Alaska Department of Fish and Game, Commercial Fisheries. http://www.cf.adfg.state.ak.us/geninfo/finfish/salmon/catchval/blusheet/00exvesl.htm.

Babbitt, J.K. 1990. Intrinsic quality and species of North Pacific fish. In: S. Keller (ed.), Proceedings of the International Conference on Fish By-products. Alaska Sea Grant College Program, University of Alaska Fairbanks, Fairbanks, pp. 39-45.

Barlow, S. 2003. World market overview of fish meal and fish oil. In: P. Bechtel, (ed.), Advances in seafood byproducts: 2002 conference proceedings. Alaska Sea Grant College Program, University of Alaska Fairbanks, Fairbanks.

Bechtel, P.J. 2003. Properties of different fish processing byproducts from pollock, cod and salmon. J. Food Process. Preserv. 27:101-116.

Crapo, C., B. Paust, and J. Babbitt. 1993. Recoveries and yields from Pacific fish and shellfish. Alaska Sea Grant College Program, University of Alaska Fairbanks, Fairbanks. 32 pp.

Hardy, R. 2003. Marine byproducts for aquacultural use. In: P. Bechtel, (ed.), Advances in seafood byproducts: 2002 conference proceedings. Alaska Sea Grant College Program, University of Alaska Fairbanks, Fairbanks.

Kizevetter, I.V. 1971. Chemistry and technology of Pacific fish. (Translated in 1973 by Israel Program for Scientific Translations Ltd.). U.S. Department of Commerce. Also available from Coronet Books, ISBN 0-70651-2715. 304 pp.

Meehan, M.J., F.M. Husby, C. Rosier, and R.L. King. 1990. Historic and potential production and utilization of Alaskan marine by-products. In: S. Keller (ed.), Proceedings of the International Conference on Fish By-products. Alaska Sea Grant College Program, University of Alaska Fairbanks, Fairbanks, pp. 31-38.

NMFS. 2000a. Bering Sea and Aleutians groundfish discards in round metric tons. National Marine Fisheries Service, Juneau, Alaska. http://www.fakr.noaa.gov/1993/bdisc93.txt.

NMFS. 2000b. Bering Sea and Aleutians groundfish quotas and preliminary catch in round metric tons. National Marine Fisheries Service, Juneau, Alaska. http://www.fakr.noaa.gov/2000/bsa00b.txt.

NMFS. 2000c. Gulf of Alaska groundfish quotas and preliminary catch in round metric tons. National Marine Fisheries Service, Juneau, Alaska. http://www.fakr.noaa.gov/2000/goa00b.txt.

NMFS. 2000d. Individual fishing quota (IFQ) allocations and landings for 2000. (halibut and sablefish). National Marine Fisheries Service, Juneau, Alaska. http://www.fakr.noaa.gov/ram/00IFQland.htm.

NMFS. 2000e. 2000 pollock and Pacific cod products. National Marine Fisheries Service, Juneau, Alaska. http://www.fakr.noaa.gov/2000/pacpol00.txt.

NPFMC. 2002. Assessment in changes of IRIU [Improved Retention and Improved Utilization] flatfish requirements. North Pacific Fishery Management Council, Anchorage, Alaska. http://www.fakr.noaa.gov/npfmc/Plan%20analysis.htm.

Shahidi, F. 2003. Nutraceuticals and bioactives from seafood byproducts. In: P. Bechtel, (ed.), Advances in seafood byproducts: 2002 conference proceedings. Alaska Sea Grant College Program, University of Alaska Fairbanks, Fairbanks.

Windsor, M., and S. Barlow. 1981. Introduction to fishery byproducts. Fishing News Books, Ltd. Farnham, Surrey, U.K.

The Potential Fate and Effects of Seafood Processing Wastes Dumped at Sea: A Review

Bodil A. Bluhm
University of Alaska Fairbanks, School of Fisheries and Ocean Sciences, Fairbanks, Alaska

Peter J. Bechtel
University of Alaska Fairbanks, USDA Agricultural Research Service Laboratory, Fairbanks, Alaska

Abstract

In Alaskan waters, approximately 0.4×10^6 t of fish processing wastes are currently generated by catcher processing vessels per year of which more than 95% is from the Bering Aleutian area. About 85% of this offal is discharged offshore. Offal and discard transfer organic material from the sea bottom and water column to the surface, changing its availability in the marine food web. This paper aims at summarizing the current knowledge on the fate and effects of fish processing offal and discard dumped at sea. Studies mostly from the North Sea, the Mediterranean Sea, and Australian waters showed that considerable fractions of floating offal and discard are taken by seabirds. Sinking material was partially consumed by midwater scavengers (especially sharks and marine mammals), but more extensively by seafloor opportunists such as fish, crustaceans, and echinoderms. Several seabird and fish species have expanded their population abundances and/or ranges apparently as a result of increased food availability from fishery discards and offal. Based on the results from the literature, scenarios are developed regarding the potential fate and effects of the offshore Alaska fishery wastes on the marine environment, focusing on the southeastern Bering Sea.

Introduction

The eastern Bering Sea continental shelf supports one of the most productive groundfish fisheries in the world (Bakkala 1993) with an annual catch ranging from 1.2 to 2.2×10^6 t since 1970 (North Pacific Fishery Management Council 1998). The two major fisheries in the eastern Bering Sea, walleye pollock and Pacific cod, are highly regulated with the goal of maintaining a sustainable fishing industry. About 70% of the total landings since 1970 have been walleye pollock (Goddard and Walters 1999) in which about 30% of the estimated total biomass were harvested in recent years (Ianelli et al. 2002). Approximately 0.4×10^6 t of fish processing wastes were generated in 2000 by catcher processing vessels in Alaskan waters, of which over 95% were from the Bering-Aleutian area (Crapo and Bechtel 2003). While some fish byproducts are utilized to produce various products such as fish meal, fish oil, and human foods (Keller 1990, Bechtel 2003), approximately 85% of this fish processing waste is returned to the ocean when the vessels are at sea (Crapo and Bechtel 2003). The dumping of seafood processing offal in Alaskan waters is regulated by the EPA Region 10 office under the National Pollutant Discharge Elimination System for Alaskan Seafood Processors (NPDES) (http://yosemite.epa.gov/R10/WATER.NSF). In these regulations, offshore seafood processing is defined as operating and discharging more than one nautical mile from shore. Effects of dumping much farther out at sea are thought to be much less than for nearshore and onshore seafood processing (B. Hill, EPA, Seattle, pers. comm.).

Seafood processing waste and discard disposal at sea return organic matter, especially proteins and fats, in the form of suspended and dissolved organic materials to the sea (Champ et al. 1981). The procedure of discharging is regulated in several of the large fishing nations and discard species composition and amounts are monitored in various fisheries, e.g., by ICES (International Council for the Exploration of the Sea) in the North Atlantic and by the North Pacific Groundfish Observer Program in the North Pacific. However, little attention has been paid to the potential fate of the discard and seafood wastes and their impact on the marine ecosystems (Goñi 2000, Groenewold 2000), especially in offshore areas.

Dumping of large amounts of material in defined areas is known to affect water quality parameters such as biochemical and chemical oxygen demand, total dissolved solids, oil and grease content, nutrient concentrations, pH, and turbidity (Champ et al. 1981). In addition, the biological communities, from bacteria to macrofauna, may be affected either directly by the additional food supply, and/or indirectly through the above listed factors. The few studies conducted on this issue imply that dumping may have a selective effect on communities, generally favoring scavenging species over other feeding types, which in turn may be put at a competitive disadvantage (Wassenberg and Hill 1987, Dayton et al. 1995, Olaso et al. 1998). A "scavenger" is a carnivore that does not kill its own prey but has the ability to actively consume pieces of carrion (Britton and Morton 1994).

Under which conditions and to what extent dumping may affect the structure, diversity, and functioning of marine communities at the dump site remains largely unknown (Kennelly 1995, ICES 1996, Goñi 1998, 2000). There is much less information on the discharge of fish processing wastes from catcher processors at sea, where discharge may have different effects than nearshore. The general assumption made is that offshore dumping of fish processing wastes in the Bering Sea will not concentrate at a distinct site but will be dispersed over a wide geographic area.

This paper aims at reviewing the scarce literature on potential effects of seafood processing waste and discard dumped at sea with a focus on how scavenging and opportunistic feeding guilds at the sea surface, in the water column, and at the seafloor have been found to be impacted by dumped waste with regard to their population density and structure, and feeding behavior. Results from the literature are used to infer assumptions on the potential fate of dumped waste in the southeastern Bering Sea.

For the purpose of this paper, "discard" is defined as fish and invertebrates caught at sea and subsequently discharged unprocessed. "Offal" is used for discharged viscera, heads, tails, and other products resulting from processing. "Waste" is used for the sum of offal and discard. Note, however, that not all referenced sources gave definitions of their usage of these terms, which in those cases are used as given.

Literature review

Queirolo et al. (1995) estimated over 1.3×10^6 t of offal produced for the combined Bering Sea–Aleutian Islands and Gulf of Alaska in 1994. Crapo and Bechtel (2003) calculated 403,585 t of offal to be produced by catcher or processor vessels offshore in 2000, predominantly in the southeastern Bering Sea. Based on a total harvest of about 1.9×10^6 t of marine finfish in 2000 (calculated by Crapo and Bechtel 2003), processing waste makes up a considerable share of total landings. As a reference, in the North Sea, roughly 60,000-90,000 t of offal are produced, and 784,000 t of bycatch are discarded annually (Furness et al. 1992, Garthe et al. 1996). According to Garthe et al. (1996), the sum equals 4% of the total available fish biomass and 22% of the landings. Approximately 100,000 t of bycatch are discarded annually on the northeast coast of India (Fishing News International 1991). The New Zealand west coast hoki fishery alone dumped up to 23,000 t of wastes, 37% of the catch in 1986 (Livinston and Rutherford 1988). What happens to this material once it is discharged?

Factors impacting the fate of dumped fishery waste

The fate of dumped seafood offal and discard is decided by a combination of factors comprising vessel practices, offal characteristics, and environmental and biological characteristics at the dumping site. When discarded, parts of the waste will float for a certain amount of time while others will sink almost immediately (Harris and Poiner 1990, Wassenberg and Hill 1990). Studies on floating/sinking characteristics of seafood offal are to our knowledge not available, although there is anecdotal evidence that much of the offal sinks once discharged; all studies reviewed in this respect were conducted on discard. Apparently, discarded fish with intact gas bladders will float if they could not resorb or release gas fast enough to compensate for the depth change, when hauled to the surface (Harris and Poiner 1990). In fish without an air bladder and in invertebrates, the density of the discarded particles is determined by the individual contributions of fat, muscle, and other tissues. Floating waste is organic matter that was transferred from the bottom of the sea to the surface, thus becoming available to surface scavenger and opportunistic assemblages which otherwise could not access this material at greater water depths (Hill and Wassenberg 1990).

Sinking material spends a certain amount of time in the water column where it is exposed to the midwater scavenging guild before arriving at the seafloor (Hill and Wassenberg 1990). For 27°C water in Australia, Hill and Wassenberg (1990) reported a sinking rate of about 0.1 m per second for fish and crustacean discard ranging from 50 to 100 g in weight. Cuttlefish, in contrast, sank more slowly at 0.05 m per second. Several of the fish and cephalopod species tested in their study had significant correlations between discard weight and sinking rate while others did not. In Chiniak Bay off Kodiak, Alaska, fish offal settled to the bottom at 150 m within an hour of dumping (Stevens and Haaga 1992).

Regarding floating material, Blaber et al. (1995) observed in Australia that part of the discard remained floating for up to six hours. Studies to determine the partitioning of floating and sinking material used the term "floating material" for items floating on the surface after 1 hour (Hill and Wassenberg 1990, 2000), 15 minutes (Evans et al. 1994), and 5 minutes (Harris and Poiner 1990). In terms of fish discard, 80% of the flatfish and 77% of all fish were reported to sink in the North Sea beam trawl fishery and the Australian prawn fishery (Groenewold 2000, Hill and Wassenberg 2000), respectively. Two dab species in the North Sea (Evans et al. 1994) and rays in the Australian prawn fishery (Harris and Poiner 1990) sank at 100%. Regarding invertebrate discard, 90% sank in the North Sea beam trawl fishery (Groenewold 2000) and over 90% sank in the Australian prawn fishery (Hill and Wassenberg 2000). From the latter study, 55% to 75% of the cephalopods, 99% of the crabs and lobsters, and all of the scallops sank (Harris and Poiner 1990; Hill and Wassenberg 1990, 2000; Groenewold 2000).

The size of dumped material will select for the scavenging guild able to utilize the material. Factors such as individual body weight of the scavenger and bill length in birds determine what size discard and offal items a scavenger feeds on (Furness et al. 1992, Garthe et al. 1996). EPA regulations for Alaska dictate grinding waste to 0.5 inch, a size that may be unsuitable for larger scavengers. Continuous as opposed to bulk dumping can have an effect on scavenger utilization rate. Scavenging birds, for example, often cannot swallow or access the whole discard or offal when high amounts are dumped within a short time period (Garthe et al. 1999). In addition, the share of dumped material utilized by scavengers may vary with the time of day and/or the season. A scavenging Australian swimming crab was most active at dusk (Wassenberg and Hill 1987), while several gull species in the North Sea fed on discard day and night (Garthe and Hüppop 1996). Migrating seabirds are most abundant in their breeding areas in the summer, while a considerable amount of fishery in the Bering Sea occurs in the winter.

Oceanographic features such as temperature and currents have an impact on how fast offal is dispersed and degraded and whether oxygen depletion can become a problem. Experiments on decaying, with different fish species, weight ranges, and temperatures, reported temperature to be the main factor determining rate and duration of decay (Groenewold 2000). North Sea discarded fish of about 40 to 150 g weight remained available for scavengers for more than 18 days at 5°C, and around 8 days at 15°C. Bottom temperatures in the southeastern Bering Sea ranged from –0.8 to 7.7°C (mean 3.4°C) as reported in the 1996 bottom trawl survey (Goddard and Walters 1999) and are about 3 to 4°C on average throughout the year (Luchin et al. 1999). The observed 18-day availability for decaying discard in 5°C in the North Sea water may come close to the decaying processes on the southeastern Bering Sea shelf. Oxygen depletion is likely to be most pronounced in areas with low current flow, such as fjords and protected bays, or in sensitive habitats or communities. Small-scale vertical oxygen gradients can be critical such as in the Baltic Sea where fish survived 30 cm above the bottom, but died below (Arntz and Rumohr 1982 as reported in Jones 1992). Changes in the CPUE in the Norwegian lobster fishery were related to reduced oxygen concentrations forcing lobsters out of their burrows (Rosenberg 1985). On the Bering Sea shelf, however, oxygen concentrations on the bottom were above 6 mg per liter in the late 1970s (T. Whitledge, University of Alaska Fairbanks, 2002, pers. comm.), an indication that oxygen depletion may not present a problem offshore.

The faunal composition at the site will eventually determine what species are available to feed on the offal. Thus it is important to know the faunal composition when assessing which species may benefit and which may be negatively impacted by the fertilizing effect of offal. Potentially disadvantaged as well as favored species may comprise commercially harvested species or prey of the latter. The following paragraphs focus on the scavenging macrofauna assemblages at the ocean surface and in the pelagic and benthic realms and their relation to waste material and carrion. Microbial communities are likely affected by discharged fishery waste, as has been documented in a nearshore study close to a seafood processing plant on Kodiak (Himelbloom and Stevens 1994), but is not discussed further in this paper.

Surface scavengers

Floating seafood processing wastes are available to surface scavengers, mainly birds. The Bering Sea breeding seabird population was estimated at 40×10^6 individuals, augmented by non-breeding shearwaters (Gould et al. 1982). Shuntov (1999) estimated an annual prey consumption by seabirds of roughly 0.3×10^6 t in the western Bering Sea with almost half composed of nekton, mainly fish. His extrapolation to the entire Bering Sea resulted in approximately 2×10^6 t total annual prey consumption. The portion of birds in the Bering Sea feeding on offal and discard has not been estimated to our knowledge.

In the North Sea, Camphuysen et al. (1995) reported 1.4 to 3.4×10^6 scavenging birds in the winter that were known to regularly feed on fishery waste. In the summer the number of scavenging birds increased to 3.0 to 6.0×10^6. This equals 66% and 53% of the total North Sea avifauna in summer and winter, respectively. These bird populations consumed about 255,000 t of discard and 55,000 t of offal, equaling 39% of available discard and offal (Garthe et al. 1996). The total amount of available fishery waste in the North Sea was estimated to potentially support 5.9 to $>6 \times 10^6$ birds (Garthe et al. 1996, Camphuysen and Garthe 2000).

Seabird species known to feed on discard and/or offal in different regions of the world include gulls, fulmars, jaegers, terns, shearwaters, petrels, albatrosses, and frigate birds (Gould et al. 1982, Garthe and Hüppop 1994, Blaber et al. 1995, Oro et al. 1995, Hill and Wassenberg 2000, Hüppop and Wurm 2000). At one particular site, 10 out of 16 species feeding in part on discards were significantly more abundant within 6 km of fishing vessels of the continental shelf of Grays Harbor, Washington (Wahl and Heinemann 1979). Garthe and Hüppop (1994) reported that 18 species of birds followed a ship conducting an experiment on discards in the North Sea. Species for which discards have come to play a role in breeding success include the southern Buller's albatross off New Zealand, whose chicks were fed with large amounts (60% of the diet by weight) of gadiform and macruorid discards (James and Stahl 2000). Similarly, fishery offal was reported to account for 80% of the fish and 63% of the total diet of Westland petrels off New Zealand during the Hoki (*Macruronus novaezelandiae*) fishery season, which coincides with the chick-rearing season (Freeman 1998). The importance of discard for some species is also noticeable, when comparing diet during and after fishing seasons. After the New Zealand Hoki season, fishery waste accounted for only 31% of the fish and around 35% of the total diet of the Westland petrels (Freeman 1998). Several tern species in Australia had over 20% of discard taxa in their pellets during fishing season, but shifted to mostly benthic prey during the closed season (Blaber et al. 1995). On the island of Heligoland in the North Sea, approximately 70 and 73% of pellets from wintering herring and great black-backed gull, respectively, contained exclusively fishery discard remains when trawlers were fishing. When no trawlers were operating, the numbers of the two species of gulls present dropped by 86 and 80% and adults had 13 and 24% lower body mass (Hüppop and Wurm 2000), respectively.

Although debate continues concerning whether the abundance and distribution of the northern fulmar in the North Atlantic is constrained by the availability of fishery waste, it is "indisputable" that northern fulmars are major consumers of fishery waste in the southern part of their range (Phillips et al. 1999). The southern range of the northern fulmar in the Atlantic falls into an area equivalent to the southeastern Bering Sea in latitude. While the results from the Atlantic cannot necessarily be transferred to the North Pacific, northern fulmars, along with glaucous-winged gulls, are known to be strongly attracted to ships in the southeastern Bering Sea, where they have been reported to forage on garbage or offal (Gould et al. 1982). Even though evidence for other potentially scavenging seabirds in the Bering Sea such as jaegers, shearwaters, petrels, and albatrosses remains anecdotal for this area, the results from other areas of the world for related species suggest a possibility, but also need for future study, for a similar feeding adaptation.

Birds tend to be size-selective in their feeding behavior (Hill and Wassenberg 1990). Hence, the size and weight of discharged discard and processing waste partially determines its availability to the birds. Most of the available data deal with discards and do not address the issue of fish processing offal released at sea. No such studies were conducted in the Bering Sea.

While seafood processing offal comprises guts, heads, etc. ,with nonstreamlined shapes, discards can include a wide size range of undersized target species and non-target species (both vertebrates and invertebrates) (Queirolo et al. 1995). In Torres Strait, Australia, the discard predominantly weighed less than 70 g and black crested terns and frigate birds were found to feed on prey with mean weights of 19 g (Blaber and Wassenberg 1989) and 6 to 102 g (Diamond 1975), respectively. In the North Sea, fulmars consumed 10-31 cm long herring, 11-43 cm whiting, 8-37 cm cod, but only 13-17 cm dab (Garthe and Hüppop 1994). The preferred size spectrum of the lesser black-backed gull and northern gannet were comparable, while black-legged kittiwakes ate smaller maximum sizes and did not feed on any flatfish. Fish that were offered but not selected were mostly in the size range of 10-42 cm; a comprehensive table can be found in Garthe and Hüppop (1994). Overall, a wide range of discard items seems to be suitable for birds, since, for example, 70-90% of roundfish discarded by shrimpers in coastal North Sea were taken by herring gulls, terns, and black-headed gulls (Berghahn and Roesner 1992, Walter 1997). Similarly, 84% of discarded roundfish, especially Gadidae and Clupeidae, and 8% of flatfish were taken in experimental discard in another North Sea study. The lower rate of flatfish consumption was probably due to their shape (Garthe and Hüppop 1994).

The amounts of discard and offal utilized by seabirds may vary with the time of day and/or the season. Garthe and Hüppop (1996) reported an experiment in which larid gulls fed day and night on discards in the North Sea while Hill and Wassenberg (1990) observed less bird feeding activity on discards at night. The Audouin's gull in the northwest Mediterranean Sea followed purse seiners at night to feed on discarded fish (Arcos and Oro 2002). With regard to seasonal utilization, herring, and black-headed gulls in the North Sea were most actively scavenging in late summer and fall, but were less abundant in June (Garthe et al. 1999). Around the British Isles, gannets mainly utilized discards in spring and were found to partially displace herring gulls from feeding around the boats at that time (Furness et al. 1992). For the southeastern Bering Sea, Gould et al. (1982) compiled distribution and abundance maps of the seabirds at different seasons, showing that scavenging birds such as northern fulmars and glaucous-winged gulls are abundant in this area throughout the year, while shearwaters and other migrating birds are abundant in the summer but practically absent in the winter.

A current version of the North Pacific pelagic seabird distribution is under way and the database can be found at http://www.absc.usgs.gov/research/NPPSD/index.htm. Since several fisheries such as the pollock fishery occur throughout the year, seasonal variability in seabird abundance may make a noticeable difference in scavenging rates.

Midwater scavenging

Little is known about midwater scavenging on discard and processing waste. In two areas in Australia, midwater scavenging was found to be low, the main scavengers being sharks (Hill and Wassenberg 1990, Wassenberg and Hill 1990). As part of the reason for low scavenging rates, the authors discussed the relatively short period of time the material spends in midwater on continental shelf areas. For the Gulf of Alaska, fishery offal was reported to contribute 12% by weight to total shark stomach contents in a study on sleeper shark diet (Yang 1999). This may indicate a potential for offal consumption by sharks in the Bering Sea. There is evidence that shark abundance, mainly salmon sharks, sleeper sharks, and spiny dogfish, has increased in Alaskan waters throughout the 1990s (Hulbert 2002). The reasons for this increase are being debated.

Utilization of seafood waste by marine mammals is poorly studied. Dolphins (*Tursiops truncatus*) followed a research vessel that was releasing experimental discard in Australia while passing through a previously trawled area. In untrawled areas, dolphins were not associated with the same vessel releasing discards (Hill and Wassenberg 1990). The Bering Sea supports one of the richest assemblages of marine mammals in the world, comprising 25 species of marine mammals including sea lions, walrus, seals, sea otters, whales, dolphins, and porpoises (Loughlin et al. 1999). The majority of these occur on the continental shelf coinciding with the main fishery activity (e.g., Moore et al. 2002). There is anecdotal evidence of orcas following trawlers in the Bering Sea (H. Douglas, University of Alaska School of Fisheries and Ocean Sciences, Fairbanks, 2002, pers. comm.) and of sperm whales and orcas feeding on longline catches and discards, but the scavenging behavior of mammals has not systematically been studied to date in the Bering Sea.

Fate of processing waste and discard on the seafloor

The ocean floor has long been used as a disposal site for all sorts of anthropogenic wastes; however, the impacts of such discharges on the ecosystem remain difficult to assess (Stevens and Haaga 1994). Even though the rate of midwater scavenging is undetermined, a fraction of offal and discard always becomes available to benthic scavengers (ICES 1996). In systematic studies in the North Sea and Australia on average at least half of discharged fishery offal and discard sank to the bottom (Harris and Poiner 1990; Hill and Wassenberg 1990, 2000; Groenewold 2000). In some fish and invertebrate species, 100% sank.

In the North Sea, Groenewold (2000) found that 46 benthic macrofauna species responded to baited traps, though catches in most traps were dominated by 2-4 species. Overall, hermit crabs, swimming crabs, and starfish contributed 70 80% to the total catches; other common groups comprised amphipods, shrimp, whelks, crabs, and fish, especially gadoids. Amphipods, particularly lysianassoids, have been documented as one of the most ubiquitous scavenging groups from shallow to deep water and from low to high latitudes (Ingram and Hessler 1983, Sainte-Marie 1986, Slattery and Oliver 1986, Moore 1994, Hargrave et al. 1995). Stomach content analysis from nine demersal fish species caught in recently trawled areas in the southern North Sea showed that two flatfish species, two gurnards, whiting, and dragonet had scavenged on benthos that had been damaged by fishing gear (Groenewold 2000). In the Australian Torres Strait, benthic scavengers were dominated by fish species, in particular nemipterids during the day and saurids and muraenids during the night. Invertebrates, dominated by brachyuran crabs and brittle stars, were overall less attracted to bait in this study (Hill and Wassenberg 1990). Seafood processing waste dumped nearshore in Kodiak was predominantly scavenged on by amphipods, flounder, and sculpins (Stevens and Haaga 1994).

Benthic ecologists who have worked and are currently working on the Bering Sea shelf listed crabs, shrimps, amphipods, sea stars, flatfish, gadoids, and sculpins as likely leading scavengers (S. Jewett and

H. Feder, University of Alaska Fairbanks, School of Fisheries and Ocean Sciences, pers. comm.; J. Grebmeier, University of Tennessee, Knoxville, pers. comm.; pers. obs.). Groups of aggregated starfish, *Asterias amurensis*, were actually observed feeding on offal shortly after trawling in a trawling impact study in the southeastern Bering Sea (Brown 2003). In the same area, a study on groundfish food habits revealed that Pacific cod, walleye pollock, arrowtooth flounder, flathead sole, yellowfin sole, Pacific halibut, and skates also consumed fishery offal (Livingston and deReynier 1996). The total abundance and biomass of scavenging fauna is yet to be estimated, but generally the biomass of benthic fauna is high in the Bering Sea (Grebmeier et al. 1995). Total epifauna biomass in the southeastern Bering Sea ranged from <3 to >10 g per square meter. For the scavenging starfish *Asterias amurensis* biomass ranged from <1.4 to 4.2 g per square meter (Feder and Jewett 1981). Potentially scavenging snow crab *Chionoecetes opilio* had local abundances of >200,000 per square nautical mile (Zheng et al. 2001) and bottomfish biomass ranged from <200 to >600 kg per hectare as estimated in the 1996 trawl survey (Goddard and Walters 1999).

Since a considerable fraction of the Bering Sea fishery is conducted close to and over the shelf break (Fritz et al. 1998), part of the discharged material may end up in the deep Bering Sea, where diversity and abundance of benthic communities is different from the shelves. The scavenging guild in the deep sea across all oceans is dominated by lysianassoid amphipods, shrimp, grenadier fish, zoarcid fish, hagfish, and brittle stars (e.g., Lampitt et al. 1983, Wilson and Smith 1984, Smith 1985, Hargrave et al. 1995). Even though there is a tendency for decreasing macrofaunal abundance and biomass with increasing depth in most taxa (e.g., Piepenburg et al. 2000), numerous scavengers have been documented to arrive within minutes to hours at great depths (Smith 1985; Witte 1999; pers. obs.).

The attraction area of scavengers in the North Sea was measured as >1,000 m^2 for gadoid fish; >100 m^2 for swimming crabs, hermit crabs, starfish and the isopod crustacean *Natatolana borealis*, and 10 m^2 for shrimps and ophiuroids (Groenewold 2000). The Australian swimming crab *Portunus pelagicus* approached discard from about 50 m away and reached trawl discard-simulated bait after 10 min, moving toward it in a zigzag pattern at 8 cm per second on average (Wassenberg and Hill 1987).

This and other studies indicate that benthic scavengers tend to react to organic input on short time scales and arrive at carrion, baited traps and discards within minutes to hours, reaching their maximum abundance at about two to three days, in shallow and deep-sea areas (Smith 1985, Sainte-Marie 1986, Lindeboom and deGroot 1998, Witte 1999; pers. obs.). Grenadier fish in the abyssal North Pacific arrived at baited traps rigged with cameras after 7.5 to 40 minutes (Wilson and Smith 1984). Shrimps and lysianassoid amphipods reached a shark carcass in the deep Arabian Sea within 20-40 min, followed by zoarcid fish after 5 hours (Witte 1999). In the southern North Sea, the numbers of swimming crabs in traps did not change much after 1 to 2 days (Groenewold 2000), and brittle star and starfish abundance reached their maxima after 2 to 3 days, while amphipod abundance already declined after the first day (Lindeboom and deGroot 1998). Similarly, immigration into trawled tracks to feed on trawl-damaged fauna in the southern North Sea lasted for 1-3 days before abundances decreased (Groenewold 2000).

The scavenging rates of benthic organisms seem to vary with food type, food quantity, and region. In the North Sea, scavengers consumed about 150 g fish in 24 hours at two test sites (Lindeboom and deGroot 1998). In situ experiments of discard fish clearance rates ranged from 0.8 to 4.5 g ash free dry weight per day in a southern North Sea study (Groenewold 2000). In deep-sea studies, scavengers ate 6 kg of a shark carcass in 5.35 days in the Arabian Sea (Witte 1999), and about 1-2 kg of teleost fish in 48 hours in the Antarctic (pers. obs.). Smith (1985) observed brittle star scavenging on a 2 kg parcel of carrion off California and calculated a turnover of 13 days.

With regard to food preferences, a study from the North Sea found that fish as bait was most attractive for swimming crabs, while starfish preferred crushed mollusk meat (Groenewold 2000). Most scavengers in this study seemed indifferent to the quality of the bait (fresh or decaying), except for one species of crab, which was more attracted by decaying fish. According to Sainte-Marie (1986), the absolute number of amphipod scavengers increased with increasing bait size in the St. Lawrence Estuary, Canada. During 24 hour trap deployments, Sainte-Marie caught 2 to 474 amphipods with 1 g of bait versus 9,468 to 40,534 individuals with 700 g of bait.

An estimate for the total amount of offal and discard consumed by seafloor inhabitants in a particular sea area is an undertaking that has yet to be accomplished. However, diets of a few individual species have been studied for offal and discard contribution in various seas. In Queensland, Australia, 33% of the diet of the brachyuran *Portunus pelagicus*, the most abundant invertebrate attracted to simulated trawl-discards, consisted of discards (Wassenberg and Hill 1987). In Sendai Bay, northern Japan (138-444 m water depth), discarded Pacific saury (*Cololabis saira*) contributed 21.8% of the total diet of Pacific cod, walleye Pollock, and oilfish, and up to 77.7% in large (>30 cm) oilfish (Yamamura 1997). The discard saury that could still be measured were in a size range of 16-28 cm. Olaso et al. (1998) postulate that many of the blue whiting (*Micromesistius poutassou*), accounting for 20% of the diet of lesser spotted dogfish (*Scyliorhinus canicula*) in the Bay of Biscay, were discards. In the Greenland Sea, redfish (*Sebastes* sp.), along with northern shrimp (*Pandalus* sp.), dominated the food of the starry ray (or thorny skate, *Raja radiata*; Pedersen 1995). The author discusses the possibility that redfish become available to rays as discard from the shrimp fishery. The diet of the same ray species was also investigated in the northwest Atlantic from West Greenland to Georges Bank where more than 30% of the food was composed of discarded fish and fish viscera, mainly from cod and haddock (Templeman 1982).

In the Bering Sea, the most comprehensive study was conducted by Livingston et al. (1993) and Livingston and deReynier (1996), who estimated a mean annual offal consumption of 174,630 t by Pacific cod, walleye pollock, arrowtooth flounder, flathead sole, yellowfin sole, Pacific halibut, and skates in 1990-1992 (Fig. 1). For yellowfin sole, the contribution of fish, mostly discharged offal, to the total diet was estimated as 21% (by weight) in the shallow southeastern Bering Sea (Brown 2003).

Effects on populations and trophic structure

In several sea areas across a wide range of latitudes, populations of scavenging vertebrates and invertebrates have been documented to use discarded fish and invertebrates and/or seafood processing offal as a minor, major, or exclusive food source (Goñi 1998). Although it remains challenging to quantify both the amounts dumped and utilized, and the impact of this food source, several studies indicate that populations changed due to proliferation of scavenger species (e.g., Hill and Wassenberg 1990, ICES 1996, Olaso et al. 1998).

An increase in the numbers of large scavenging seabirds in the North Sea area over the last decades has been reported (e.g., Garthe et al. 1996, Hüppop 1996, ICES 1996, Camphuysen and Garthe 2000). Range expansion and population growth of especially gulls but also northern fulmars in the North Atlantic have been supported in part by an increasing availability of human refuse and fishery discard (Dunnet et al. 1990, Phillips et al. 1999). An impressive example is a gull species, *Larus audouinii*, in the northwest Mediterranean Sea, which has established colonies wholly dependent on discard from the ground fishery (Goñi 2000). Several studies directly linked discard availability to breeding success. In northeastern Spain, the breeding success of the yellow-legged gull, *Larus cachinans*, was negatively affected by a trawling moratorium that coincided with the chick rearing stage (Oro et al. 1995). The egg volume of the lesser black-backed gull, *Larus fuscus*, in the western Mediterranean Sea decreased significantly in a trawling moratorium year versus a fishing year in which the diet was dominated by discard fish (Oro 1996). These examples implicate strong dependence of certain populations on artificial food sources. However, several tropical Australian bird species have not shown any obvious effect of discard availability on their reproductive activity (Blaber et al. 1995). In conclusion, fishery wastes and competition between scavenging species evoked by the supply of discards and offal may lead or have led to changes in the avifauna (Furness et al. 1992) and in seabird food webs (Furness and Ainley 1984, Ryan and Moloney 1988).

Offal and discard sinking beyond the reach of seabirds becomes available to the midwater and benthic communities where it has a potential

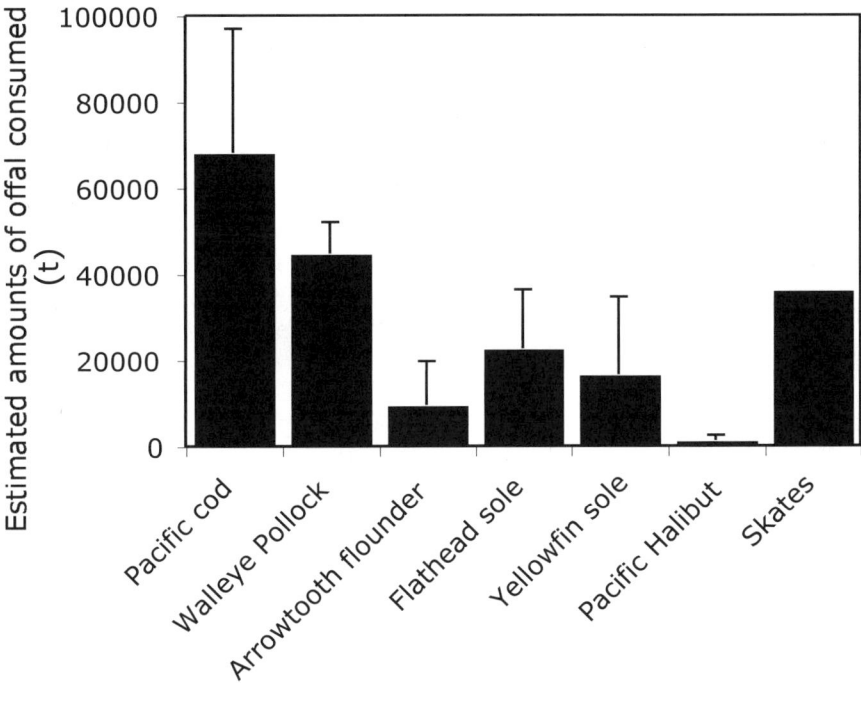

Groundfish predator

Figure 1. Estimated amounts of offal, as means with S.D. from 1990 to 1992, consumed by groundfish on the eastern Bering Sea shelf during May-September, the main feeding season. Skates were sampled only in 1992. Data from Livingston and deReynier (1996).

of affecting the abundance, reproductive success, and patchiness of the fauna, possibly favoring mobile scavengers and opportunists over slow or sessile ones and other feeding types (Dayton et al. 1995). As an example, the proportion of dogfish, *Scyliorhinus canicula*, in catches in the ICES division VIIIc, the Spanish Cantabrian Sea, has increased from 3% in 1983 to 7% in 1994 (ICES 1996, Olaso et al. 1998). This increase is attributed to the extensive feeding by the dogfish on discarded blue whiting. Also, the population increase in the starry ray, *Raja radiata*, in the Greenland Sea was suggested to be a result of increased food availability from discards (Pedersen 1995, ICES 1996) as were the increases of some flatfishes in the North Sea and Gulf of Thailand (Hall 1999). Wassenberg and Hill (1987) argue that swimming crab, *Portunus pelagicus*, population densities may be higher due to the availability of prawn trawl discard.

For the Bering Sea, Conners et al. (2002) documented shifts in benthic biomass and community structure in the early 1980s, consistent with shifts in climate indices in the late 1970s. These changes included increases in walleye pollock, Pacific cod, rock and flathead sole, skate, and non-crab invertebrate abundance. Their analysis suggests higher groundfish biomass at their three study sites during 1980-2000 than during 1960-1980. Based on the discussed findings from other sea areas and on the improbability of one cause–one effect relations, it cannot be excluded that the considerable fishery-related input of organic matter on the Bering Sea shelf supported the suggested biomass increase. The effects of dumping fish processing waste on the Bering Sea biomass remains to be elucidated.

Closing remarks

Trawling results in a movement of organic material from the seafloor to the sea surface. This transfer of organic matter changes its availability in the food web and may affect feeding behavior, reproductive success, abundance, and competitive relations in surface, midwater, and benthic communities.

Studies like the European Community–funded IMPACT II project (Lindeboom and deGroot 1998), attempting a comprehensive approach to assess the effects of different types of fisheries on the North Sea and Irish Sea benthic ecosystem, could help identify and quantify the types and magnitude of effects in the southeastern Bering Sea. They could aid in determining effects of discard and offal dumping on fueling scavenger species populations. Whatever the results may be, a judgment on "desirable," "beneficial," and "adverse" effects will remain a question of perspective.

References

Arcos, J.M., and D. Oro. 2002. Significance of nocturnal purse seine fisheries for seabirds: A case study off the Ebro Delta (NW Mediterranean). Mar. Biol. 141: 277-286.

Arntz, W.E., and H. Rumohr. 1982. An experimental study of macrobenthic colonisation and succession, and the importance of seasonal variation in temperate latitudes. J. Exp. Mar. Biol. Ecol. 64:17-45.

Bakkala, R.G. 1993. Structure and historical changes in the groundfish complex of the eastern Bering Sea. NOAA Tech. Rep. NMFS 114. 91 pp.

Bechtel, P.J. (ed.). 2003. Advances in seafood byproducts: 2002 conference proceedings. Alaska Sea Grant College Program, University of Alaska Fairbanks, Fairbanks.

Berghahn, R., and H.-U. Roesner. 1992. A method to quantify feeding of seabirds on discards from shrimp fishery. Netherlands J. Sea. Res. 28:347-350.

Blaber, S.J.M., and T.J. Wassenberg. 1989. The feeding ecology of the piscivorous birds *Phalacrocorax varius, P. melanoleucos* and *Sterna bergii* in Moreton Bay, Australia: Diet and dependence on trawler discards. Mar. Biol. 101:1-10.

Blaber, S.J.M., D.A. Milton, G.C. Smith, and M.J. Farmer. 1995. Trawl discards in the diets of tropical seabirds of the northern Great Barrier Reef, Australia. Mar. Ecol. Progr. Ser. 127:1-13.

Britton, J.C., and B. Morton. 1994. Marine carrion and scavengers. Oceanogr. Mar. Biol. Annu. Rev. 32:369-434.

Brown, E. 2003. Effects of commercial trawling on essential fish habitat on the Bering Sea shelf. M.S. thesis, University of Alaska Fairbanks.

Camphuysen, C.J., and S. Garthe. 2000. Seabirds and commercial fisheries: Population trends of piscivorous seabirds explained? In: M.J. Kaiser and S.J. de Groot (eds.), The effects of fishing on non-target species and habitats: Biological, conservation and socio-economic issues, chapter 11. Blackwell Science, Oxford, pp. 163-184.

Camphuysen, C.J., B. Calvo, J. Durinck, K. Ensor, A. Follestad, R.W. Furness, S. Garthe, G. Leaper, H. Skov, M.L. Tasker, and C.J.N. Winter. 1995. Consumption of discards by seabirds in the North Sea. Netherlands Institute for Sea Research, Den Burg, Texel. NIOZ-Rapport 1995-05. 202 pp.

Champ, M.A., T.P. O'Connor, and P.K. Park. 1981. Ocean dumping of seafood wastes in the United States. Mar. Pollut. Bull. 12:241-244.

Conners, M.E., A.B. Hollowed, and E. Brown. 2002. Retrospective analysis of Bering Sea bottom trawl surveys: Regime shift and ecosystem reorganization. Prog. Oceanogr. 55:209-222.

Crapo, C.A., and P.J. Bechtel. 2003. Utilization of Alaska fish processing byproducts. In: Bechtel, P.J. (ed.) 2003. Advances in seafood byproducts: 2002 conference proceedings. Alaska Sea Grant College Program, University of Alaska Fairbanks, Fairbanks.

Dayton, P.K., S.F. Thrush, M.T. Agardy, and R.J. Hofman. 1995. Environmental effects of marine fishing. Aquat. Conserv. 5:205-232.

Diamond, A.W. 1975. Biology and behaviour of frigatebirds *Fregata* spp. on Albara Atoll. Ibis 117:302-323.

Dunnet, G.M., R.W. Furness, M.L. Tasker, and P.H. Becker. 1990. Seabird ecology in the North Sea. Netherlands J. Sea Res. 26:387-425.

Evans, S.M., J.E. Hunter, A. Elizal, and R.I. Wahju. 1994. Composition and fate of the catch and bycatch in the Farne Deep (North Sea) *Nephrops* fishery. ICES J. Mar. Sci. 51:155-168.

Feder, H.M., and S.C. Jewett. 1981. Feeding interactions in the eastern Bering Sea with emphasis on the benthos. In: D.W. Hood and J.A. Calder (eds.), The eastern Bering Sea shelf: Oceanography and resources, vol. II. University of Washington Press, Seattle, pp. 1229-1261.

Fishing News International. 1991. Why stocks of large fish fall. Fishing News International, March 1991:1-64.

Freeman, A.N.D. 1998. Diet of Westland petrels *Procellaria westlandica*: The importance of fisheries waste during chick-rearing. Emu 98:36-43.

Fritz, L.W., A. Greig, and R.F. Reuter. 1998. Catch-per-unit-effort, length, and depth distribution of major groundfish and bycatch species in the Bering Sea, Aleutian Islands, and Gulf of Alaska regions based on groundfish fishery observer data. NOAA Tech. Memo. NMFS-AFSC-88. 179 pp.

Furness, R.W., and D.G. Ainley. 1984. Threats to seabird populations presented by commercial fisheries. ICBP (International Council for Bird Preservation) Tech. Rep. 2:701-708.

Furness, R.W., K. Ensor, and A.V. Hudson. 1992. The use of fishery waste by gull populations around the British Isles. Ardea 80:105-113.

Garthe, S., and O. Hüppop. 1994. Distribution of ship-following seabirds and their utilization of discards in the North Sea in summer. Mar. Ecol. Progr. Ser. 106: 1-9.

Garthe, S., and O. Hüppop. 1996. Nocturnal scavenging by gulls in the southern North Sea. Colonial Waterbirds 19:232-241.

Garthe, S., K.C.J. Camphuysen, and R.W. Furness. 1996. Amounts of discards by commercial fisheries and their significance as food for seabirds in the North Sea. Mar. Ecol. Progr. Ser. 136:1-11.

Garthe, S., U. Walter, M.L. Tasker, P.H. Becker, G. Chapdelaine, and R.W. Furness. 1999. Evaluation of the role of discards in supporting bird populations and their effects on the species composition of seabirds in the North Sea. ICES Coop. Res. Rep. No. 232:29-41.

Goddard, P., and G. Walters. 1999. 1996 bottom trawl survey of the eastern Bering Sea continental shelf. NOAA, NMFS-AFSC Report 99-05. 165 pp.

Goñi, R. 1998. Ecosystem effects of marine fisheries: An overview. Ocean Coast. Manag. 40:37-64.

Goñi, R. 2000. Fisheries effects on ecosystems. In: C. Sheppard (ed.), Seas at the millennium: An environmental evaluation, chapter 115. Elsevier Science, Amsterdam, pp. 117-133.

Gould, P.J., D.J. Forsell, and C.J. Lensink. 1982. Pelagic distribution and abundance of seabirds in the Gulf of Alaska and eastern Bering Sea. U.S. Fish and Wildlife Service FWS/OBS-82/48. 294 pp.

Grebmeier, J.M., W.O. Smith Jr., and R.J. Conover. 1995. Biological processes on Arctic continental shelves: Ice-ocean-biotic interactions. In: W.O. Smith Jr. and J.M. Grebmeier (eds.), Arctic oceanography: Marginal ice zones and continental shelves. Coastal and estuarine studies 49, chapter 7. American Geophysical Union, Washington, D.C., pp. 231-261.

Groenewold, S. 2000. The effects of beam trawl fishery on the food consumption of scavenging epibenthic invertebrates and demersal fish in the southern North Sea. Ph.D. thesis, University of Hamburg. 146 pp.

Hall, S.J. 1999. The effects of fishing on marine ecosystems and communities. Fish Biology and Aquatic Research Ser. 1. Blackwell Science, Oxford. 277 pp.

Hargrave, B.T., G.A. Phillips, N.J. Prouse, and P.J. Cranford. 1995. Rapid digestion and assimilation of bait by the deep-sea amphipod *Eurythenes gryllus*. Deep-Sea Res. 42:1905-1921.

Harris, A.N., and I.R. Poiner. 1990. By-catch of the prawn fishery of Torres Strait; Composition and partitioning of the discards into components that float or sink. Aust. J. Mar. Freshw. Res. 41:37-52.

Hill, B.J., and T.J. Wassenberg. 1990. Fate of discards from prawn trawlers in Torres Strait. Aust. J. Mar. Freshw. Res. 41:53-64.

Hill, B.J., and T.J. Wassenberg. 2000. The probable fate of discard from prawn trawlers fishing near coral reefs: A study in the northern Great Barrier Reef, Australia. Fish. Res. 48:277-286.

Himelbloom, B.H., and B.G. Stevens. 1994. Microbial analysis of a fish waste dump site in Alaska. Bioresour. Technol. 47:229-233.

Hulbert, L. 2002. Do Pacific sleeper sharks prey on Steller sea lions? AFSC (Alaska Fishery Science Center) Quarterly Research Reports, April-June 2002. http://www.afsc.noaa.gov/Quarterly/amj2002/divrptsABL3/htm.

Hüppop, O. 1996. Die Brutbestände Helgoländer Seevögel von 1952 bis 1995. Ornithol. Jahresber. Helgoland. 6:72-75. (In German.)

Hüppop, O., and S. Wurm. 2000. Effects of winter fishery activities on resting numbers, food and body condition of large gulls *Larus argentatus* and *L. marinus* in the southeastern North Sea. Mar. Ecol. Progr. Ser. 194:241-247.

Ianelli, J.N., S. Barbeaux, T. Honkalehto, G. Walters, and N. Williamson. 2002. Eastern Bering Sea walleye pollock stock assessment. NMFS-AFSC stock assessment and fishery evaluation documents. http://www.fakr.noaa.gov/npfmc/safes/safe.htm.

ICES. 1996. Report of the working group on ecosystem effects on fishing activities. ICES (International Council for the Exploration of the Sea) Headquarters, 12-21 March 1996, Copenhagen. 89 pp.

Ingram, C.L., and R.R. Hessler. 1983. Distribution and behavior of scavenging amphipods from the central North Pacific. Deep-Sea Res. 30:683-706.

James, G.D., and J.-C. Stahl. 2000. Diet of southern Buller's albatross (*Diomedea bulleri bulleri*) and the importance of fishery discards during chick rearing. N.Z.J. Mar. Freshw. Res. 34:435-454.

Jones, J.B. 1992. Environmental impact of trawling on the sea bed: A review. N.Z.J. Mar. Freshw. Res. 26:59-67.

Keller, S. (ed.). 1990. Making profits out of seafood wastes. Alaska Sea Grant College Program, University of Alaska Fairbanks, Fairbanks. 239 pp.

Kennelly, S.J. 1995. The issue of bycatch in Australia's demersal trawl fisheries. Rev. Fish Biol. Fish. 5:213-234.

Lampitt, R.S., N.R. Merrett, and M.H. Thurston. 1983. Inter-relations of necrophagous amphipods, a fish predator, and tidal currents in the deep sea. Mar. Biol. 74:73-78.

Lindeboom, H.J., and S.J. deGroot. (eds.). 1998. IMPACT II. The effects of different types of fisheries on the North Sea and Irish Sea benthic ecosystems. New Zealand Oceanographic Institute, NIOZ Rapport 1998-1, RIVO-DLO Report C003/98. 404 pp.

Livingston, P.A., and Y. deReynier. 1996. Groundfish food habits and predation on commercially important prey species in the eastern Bering Sea from 1990 to 1992. NOAA, NMFS-AFSC Report 96-04. 214 pp.

Livingston, P.A., A. Ward, G.M. Lang, and M.-S. Yang. 1993. Groundfish food habits and predation on commercially important prey species in the eastern Bering Sea from 1987 to 1989. NOAA, NMFS-AFCS Report 93-05. 92 pp.

Livinston, M., and K. Rutherford. 1988. Hoki wastes on west coast fishing grounds. Catch 15:1-15.

Loughlin, T.R., I.N. Sukhanova, E.H. Sinclair, and R.C. Ferrero. 1999. Summary of biology and ecosystem dynamics in the Bering Sea. In: T.R. Loughlin and K. Ohtani (eds.), Dynamics of the Bering Sea. Alaska Sea Grant College Program, University of Alaska Fairbanks, Fairbanks, pp. 387-408.

Luchin, V.A., V.A. Menovschikov, V.M. Lavrientiev, and R.K. Reed. 1999. Thermo-haline structure and water masses in the Bering Sea. In: T.R. Loughlin and K. Ohtani (eds.), Dynamics of the Bering Sea. Alaska Sea Grant College Program, University of Alaska Fairbanks, Fairbanks, pp. 61-92.

Moore, P.G. 1994. Observations on the behaviour of the scavenging lysianassoid *Orchomene zschaui* (Crustacea: Amphipoda) from South Georgia (South Atlantic). Mar. Ecol. Prog. Ser. 113:29-38.

Moore, S.E., J.M. Waite, N.A. Friday, and T. Honkalehto. 2002. Cetacean distribution and relative abundance on the central-eastern and the southeastern Bering Sea shelf with reference to oceanographic domains. Prog. Oceanogr. 55: 249-261.

North Pacific Fishery Management Council. 1998. Stock assessment and fishery evaluation report for the groundfish resources of the Bering Sea/Aleutian Islands regions. North Pacific Fishery Management Council, Anchorage, Alaska. 635 pp.

Olaso, I., F. Velasco, and N. Perez. 1998. Importance of discarded blue whiting (*Micromesistius poutassou*) in the diet of lesser spotted dogfish (*Scyliorhinus canicula*) in the Cantabrian Sea. ICES J. Mar. Sci. 55:331-341.

Oro, D. 1996. Effects of trawler discard availability on egg laying and breeding success in the lesser black-backed gull *Larus fuscus* in the western Mediterranean. Mar. Ecol. Progr. Ser. 132:43-46.

Oro, D., M. Bosch, and X. Ruiz. 1995. Effects of a trawling moratorium on the breeding success of the yellow-legged gull *Larus cachinnans*. Ibis 137:547-549.

Pedersen, S.A. 1995. Feeding habits of starry ray (*Raja radiata*) in West Greenland waters. ICES J. Mar. Sci. 52:43-53.

Phillips, R.A., M.K. Petersen, K. Lilliendahl, J. Solmundsson, K.C. Hamer, C.J. Camphuysen, and B. Zonfrillo. 1999. Diet of the northern fulmar *Fulmarus glacialis*: Reliance on commercial fisheries? Mar. Biol. 135:159-170.

Piepenburg, P., A. Brandt, K.v. Juterzenka, M. Mayer, K. Schnack, D. Seiler, U. Witte, and M. Spindler. 2000. Patterns and determinants of the distribution and structure of benthic faunal assemblages in the northern North Atlantic. In: P. Schäfer, W. Ritzrau, M. Schlüter, and J. Thiede (eds.), The northern North Atlantic: A changing environment. Springer, Berlin, pp. 179-198.

Queirolo, L.E., L.W. Fritz, P.A. Livingston, M.R. Loefflad, D.A. Colpo, and Y.L. deReynier. 1995. Bycatch, utilization, and discard in the commercial ground-fish fisheries of the Gulf of Alaska, Eastern Bering Sea, and Aleutian Islands. NOAA Tech. Memo. NMFS-AFSC-58. 148 pp.

Rosenberg, R. 1985. Eutrophication: The future marine coastal nuisance? Mar. Pollut. Bull. 16:227-231.

Ryan, P.G., and C.L. Moloney. 1988. Effect of trawling on bird and seal distributions in the southern Benguela region. Mar. Ecol. Progr. Ser. 45:1-11.

Sainte-Marie, B. 1986. Effect of bait size and sampling time on the attraction of the lysianassid amphipods *Anonyx sarsi* Steele & Brunel and *Orchomenella pinguis* (Boeck). J. Exp. Mar. Biol. Ecol. 99:63-77.

Shuntov, W.P. 1999. Seabirds of the western Bering Sea. In: T.R. Loughlin and K. Ohtani (eds.), Dynamics of the Bering Sea. Alaska Sea Grant College Program, University of Alaska Fairbanks, Fairbanks, pp. 651-682.

Slattery, P.N., and J.S. Oliver. 1986. Scavenging and other feeding habits of lysian-assid amphipods (Orchomene spp.) from McMurdo Sound, Antarctica. Polar Biol. 6:171-177.

Smith, C.R. 1985. Food for the deep sea: Utilization, dispersal, and flux of nekton falls at the Santa Catalina Basin floor. Deep-Sea Res. 32:417-442.

Stevens, B., and J.A. Haaga. 1992. Draft manuscript. Ocean dumping of seafood processing wastes: Comparisons of epibenthic megafauna as submersible in impacted and non-impacted Alaskan bays, and estimation of waste decomposition rate. NOAA, NMFS, Kodiak Laboratory. Unpublished.

Templeman, W. 1982. Stomach contents of the thorny skate, *Raja radiata*, from the northwest Atlantic. J. Northw. Atl. Fish. Sci. 3:123-126.

Wahl, T.R., and D. Heinemann. 1979. Seabird and fishing vessels: Co-occurrence and attraction. Condor 81:390-396.

Walter, U. 1997. Quantitative analysis of discards from brown shrimp trawlers in the coastal area of the East Frisian Islands. Arch. Fish. Mar. Res. 45:61-76.

Wassenberg, T.J., and B.J. Hill. 1987. Feeding by the sand crab *Portunus pelagicus* on material discarded from prawn trawlers in Moreton Bay, Australia. Mar. Biol. 95:387-393.

Wassenberg, T.J., and B.J. Hill. 1990. Partitioning of material discarded from prawn trawlers in Moreton Bay. Aust. J. Mar. Freshw. Res. 41:27-36.

Wilson, R.R., and K.L. Smith Jr. 1984. Effect of near-bottom currents on detection of bait by the abyssal grenadier fishes *Coryphaenoides* spp., recorded in situ with a video camera on a free vehicle. Mar. Biol. 84:83-91.

Witte, U. 1999. Consumption of large carcasses by scavenger assemblages in the deep Arabian Sea: Observations by baited camera. Mar. Ecol. Progr. Ser.183: 139-147.

Yamamura, O. 1997. Scavenging on discarded saury by demersal fishes off Sendai Bay, northern Japan. J. Fish Biol. 50:919-925.

Yang, M.-S. 1999. Diet of Pacific sleeper shark, *Somniosus pacificus*, in the Gulf of Alaska. Fish. Bull., U.S. 97:406-409.

Zheng, J., G.H. Kruse, and D.R. Ackley. 2001. Spatial distribution and recruitment patterns of snow crabs in the Eastern Bering Sea. In: G.H. Kruse, N. Bez, A. Booth, M.W. Dorn, S. Hills, R.N. Lipcius, D. Pelletier, C. Roy, S.J. Smith, and D. Witherell. Spatial processes and management of marine populations. Alaska Sea Grant College Program, University of Alaska Fairbanks, Fairbanks, pp. 233-255.

Marine Byproducts for Aquaculture Use

Ronald W. Hardy
University of Idaho, Hagerman Fish Culture Experiment Station,
Hagerman, Idaho

Abstract

Aquaculture production increased by more than 12% per year in the past 15 years and is expected to continue to do so for the foreseeable future, requiring more feeds and thus more fish protein and oil. The percentage of annual world fish meal and oil production used by the aquaculture feed industry was 10% and 6% in 1988 compared to ca. 43% and 75%, respectively, in 2000. Over 70% of the fish meal and oil consumed by aquaculture is used in feeds for four species groups, i.e., marine shrimp, salmonids, marine fish, and carp. Fish meal and oil fulfill specific roles in feeds for each species group, and thus marine byproducts can have different specifications for each. The amino acid profile of fish meal is optimum for aquaculture feeds, but supply and demand concerns will increasingly result in fish meal being combined with grain and oilseed proteins to produce blends suitable for growing fish. Thus, the protein and amino acid content and digestibility of fish meal and marine byproducts will become critically important as proteins from marine sources switch from a primary role in feeds to a specialty role as sources of essential amino acids that are limited in proteins from grains and oilseeds. Likewise, fish oils will switch from being a primary energy source in diets to a specialty role in finishing diets to supply omega-3 fatty acids.

Introduction

On a worldwide basis, aquaculture has become a significant producer of food. Annual growth of 12% or more over the past 15 years is expected to continue, resulting in an increasing proportion of fisheries products being farmed rather than being captured from wild stocks. In 2002, approximately 30% of fish products for human consumption were produced by aquaculture (Kilpatrick 2003, this volume), and this proportion is expected to increase to 50% by 2025 (New 1999). Growth has resulted not only from increases in the number of ponds or marine cages in use but also from increasing the inputs into existing aquaculture systems. Higher inputs support higher productivity of existing aquaculture systems.

Higher inputs mean two things to the aquaculture feed industry: more feed and higher quality feed. Currently, global feed production for farmed fish and crustaceans is approximately 13,000,000 t, and predictions are for feed production to increase to over 37,000,000 t by the end of the decade (Barlow 2000), an increase of 24,000,000 t (Tables 1 and 2). Feeds for salmonids and marine fish have always been complete feeds, i.e., ones that supply all of the nutritional needs of the fish. Complete feeds have improved in quality over the past 10-15 years, mainly by becoming more nutrient and energy dense, and by eliminating poorly digested ingredients from diet formulations. Pond-reared fish, in contrast, obtain a significant proportion of their nutritional needs from pond biota. The degree to which feeds supply essential nutrients to pond-reared fish increases as rearing densities increase beyond the capacity of natural foods in ponds to supply them. Fish farmers around the world have found that as they increase feed inputs, the biomass and economic yields from ponds increase as well. Thus, great areas of low-input, pond-based aquaculture, mainly in Southeast Asia and China, are being converted from low-input systems to high-input systems that depend upon high quality feeds to supply an increasing proportion of nutrients used by the fish.

The effects of the aquaculture industry's growth and of changes in feed input in pond-based aquaculture have been dramatic with respect to the use of marine proteins by the aquaculture feed industry. In the mid-1980s, less than 10% of annual fish meal production was used by the aquaculture feed sector. Today, that proportion is over 40% (Hardy and Tacon 2002). Similarly, aquaculture now uses nearly 75% of annual global fish oil production, up from under 10% 17 years ago, but this change is mainly due to the adoption of high lipid feeds by the salmon farming industry, rather than increasing the inputs into pond-based aquaculture systems.

Role of fish meal and oil in farmed fish diets

Given the dramatic increase in the proportion of annual fish meal and fish oil production used by the aquaculture feed sector, it is appropriate to examine the role of fish meal and oil in diets for farmed fish, past, present and future. First, it is important to note that fish meal and oil are produced from species of fish that are not generally utilized directly for human food. These species include herring and capelin in Norway and Iceland, sand eel in Denmark, capelin in South Africa, anchovies in Peru and northern Chile, jack mackerel in central Chile, sardines in Japan, and menhaden in the United States. Second, it is also important to note that the rapid increase in fish meal and oil use in aquaculture feeds over

Table 1.   World fish feed production by species groups in 2000.

Species group	Feed production (t)	Percent of total
Salmon/trout	1,636,000	13
Shrimp	1,570,000	12
Catfish	505,000	4
Tilapia	776,000	6
Marine finfish	1,049,000	8
Cyprinids (carp)	6,991,000	52
Total	13,106,000	

Table 2.   Estimated fish feed production by species groups in 2010.

Species group	Feed production (t)	Percent of total
Salmon/trout	2,300,000	6
Shrimp	2,450,000	7
Catfish	700,000	2
Tilapia	2,497,000	7
Marine finfish	2,304,000	6
Cyprinids (carp)	27,000,000	73
Total	37,226,000	

the past 15 years has not resulted in over-exploitation of the stocks of fish harvested primarily to produce fish meal and oil. Production of fish meal has averaged between 6,000,000 and 7,000,000 t per year since the mid-1980s, except in El Niño years, when production has been lower. Fish oil production has averaged between 1,200,000 and 1,300,000 t. No significant changes are evident in annual harvest or fish meal and oil production associated with increased aquaculture production or increased intensification of aquaculture and concomitant higher feed inputs. Increased fish meal and oil use in the aquaculture sector has come at the expense of other uses.

Fish meal has been the protein source of choice in diets for farmed fish for several reasons. First, the protein content of fish meals is relatively high, generally 65-72%, depending upon the fish species used to produce the fish meal. Second, the amino acid profile of fish meal closely matches the dietary requirements of most carnivorous fish species. Third, protein and amino acid apparent digestibility is relatively high in good quality fish meal in most farmed fish species (Hajen et al. 1993, Sugiura 2000, Sugiura and Hardy 2000). Finally, fish meal–based diets are highly palatable to most farmed fish. Of these properties of fish meal, plant protein sources are similar with respect to apparent protein and amino acid digestibility, and protein concentrates are similar to fish meal in protein content. However, amino acid profiles of plant protein sources do not match the dietary requirements of carnivorous fish species as well as do the amino acid profiles of marine proteins, and some plant protein sources lower feed intake, presumably by lowering feed palatability, when replacement levels are high. Thus, the role of plant proteins up to now has primarily been to replace a portion of the fish meal protein in diets for carnivorous fish species to lower the price of the feed, extend fish meal supplies when they are tight, or to reduce the total phosphorus level of the diet. Certain protein ingredients from the rendering industry, e.g., blood meal and feather meal, have been used in a similar fashion, and also to increase the total protein level of the diet.

The amount of protein supplied by fish meal in diet formulations for various species of fish differs significantly, depending on whether the species is carnivorous or omnivorous. Salmon and trout, for example, are fed diets that contain 38-44% crude protein during the grow-out stage, where most feed is used during a production cycle (Storebakken 2002). Catfish, in contrast, are fed diets containing 28-32% crude protein, most of which is supplied by soybean meal (Robinson and Li 2002). Members of the carp family are fed diets with protein contents varying from 0 to 35%, depending on species, where they are farmed, and life-history stage (Takeuchi et al. 2002, Shivananda Murthy 2002). Fry and fingerling carp are fed diets containing higher protein levels than are post-juvenile fish. Carp diets intended for use in high-input rearing systems contain 15-25% fish meal. Although this is a relatively low fish meal inclusion level, the tremendous increase in high-input carp culture has dramatically increased the amount of fish meal used by this production sector to about 17% of the total amount of fish meal used in all aquaculture diets in 2000 (Barlow 2000). Altogether, 2,115,000 t of fish meal was used in diets for fish and shrimp in 2000 (Table 3). The percentage of fish meal in the diet of various species groups ranged from 55% for marine flatfish (flounder, turbot, halibut) to 3% for catfish (channel catfish, African catfish). Carp averaged 5%, but this figure includes both high-input and low-input systems. Carp farming is converting to high-input systems, and this will increase the total use of fish meal in this production sector, despite an anticipated reduction in the percentage of fish meal used in diets (Barlow 2000). Carp feed production is anticipated to increase from about 7,000,000 t in 2000 to 27,000,000 t by 2010. Soybean meal will likely supply the bulk of protein in carp diets of the future, but fish meal will continue to be used, especially in diets for fry and fingerling carp.

**Table 3. World fish meal use in fish feeds
(2000 estimate).**

Species group	Fish meal (t)	Percent of total
Salmon	454,000	21.5
Marine finfish	415,000	19.6
Shrimp	372,000	17.6
Cyprinids (carp)	350,000	16.5
Trout	176,000	8.3
Eels	173,000	8.2
Flatfish	69,000	3.3
Other fish	106,000	5.0
Total	2,115,000	

Future protein requirements for aquaculture feeds

Estimates of future protein requirements for aquaculture feeds depend on future production from various segments of the aquaculture industry and on annual production levels of fish meal. Barlow (2000) predicted that aquaculture feed production would increase from 13,098,000 t in 2000 to 37,226,000 in 2010. At today's fish meal use levels in diet formulations for various species groups, the amount of fish meal needed to produce 37,226,000 t of fish and shrimp feed would be 4,081,000 t in 2010, up from 2,115,000 t in 2000. This slightly exceeds the amount of fish meal traded worldwide in non–El Niño years. However, Barlow predicts that the percentage of fish meal in diet formulations will decrease and that total fish meal use by the aquaculture industry will be 2,831,000 t in 2010 (Table 4). Total protein in these diet formulations will not change. The difference between 4,081,000 t and 2,831,000 t, e.g., 1,755,000 t, will be supplied by other protein sources. What sources could supply this protein? The leading candidates are soy products. If we assume that the fish meal used in fish feeds contains 70% crude protein, then 1,228,500 t of protein (1,755,000 × 0.7) from sources other than fish meal will be needed annually in fish feeds by 2010. If soybean meal (48% crude protein) were used to supply this protein, the increase in total use in all aquaculture feeds would be 2,559,375 t. If soy protein concentrate were used, the total would be less, approximately 1,640,000 t, because of its higher protein content. If other protein concentrates from grains or oilseeds, e.g., wheat gluten meal, corn gluten meal, canola protein concentrate, were used, the amounts would be similar.

Table 4.   Estimated fish meal use by species group in 2000 and 2010.

Species group	2000 (%)	2010 (%)	2000 (t)	2010 (t)
Salmon	40	30	454,000	377,000
Trout	30	25	176,000	147,000
Marine fish	45	40	415,000	688,000
Flatfish	55	45	69,000	263,000
Shrimp	25	20	372,000	485,000
Catfish	3	0	15,000	0
Carp	5	2.5	350,000	675,000
Total			2,115,000	2,831,000

Diet formulations will change for salmon, trout, shrimp, and marine fish

Approximately 70% of the fish meal used by the aquaculture feed industry is used in diets for salmon, trout, shrimp, and marine fish (Table 3). These species account for about 32% of total fish feed production, and 15% of total farmed fish production. Barlow (2000) predicts that by 2010, the percentage of annual fish meal production used in diets for these species groups will decrease to 52% of the total amount of fish meal used by aquaculture. A portion of this percentage decrease will result from higher total use of fish meal in diets for other species groups, but most of the decrease will be the result of lower percentages of fish meal being used in diet formulations for salmon, trout, shrimp, and marine fish, and concomitant higher use of alternate protein sources. Examining diet formulations used at present and likely to be used in the future sheds light on which of the properties of fish meal are likely to be of highest value in diet formulations of the future. At present, fish meal constitutes between 40% and 55% of diet formulations for Atlantic salmon (Storebakken 2002). This percentage is expected to decrease to approximately 30% in the next decade, with increasing percentages of plant protein concentrates making up the difference. For rainbow trout, current use levels range from 25% to 40%, depending on fish meal price relative to alternative protein sources (Hardy 2002). Use levels are expected to decrease by 5%, meaning that the percentage of fish meal in the diet will likely be 25% at most in the near future, assuming that there is sufficient demand to keep fish meal prices at the high end of the price range. Using the European sea bass as an example of a farmed marine fish species, current fish meal levels in diets exceed 50% (Kaushik 2002), but future levels are expected to be 40%. Extending these expected trends to other farmed species of carnivorous fish, it is clear that diet formulations will shift from high reliance on fish meal to reliance on blends of fish meal and plant protein concentrates, making it more difficult in some cases to balance diets with respect to limiting essential amino acids. For some fish species, higher inclusion levels of plant protein sources, especially those derived from oilseed meals, will lower the palatability of diets. In some formulations, replacing fish meal with plant protein sources will alter both the mineral balance and bioavailability of minerals in the diet.

Alternative protein sources

Because 70% of the fish meal used in diets for fish and crustaceans is used to produce diets for salmonids, marine fish, and shrimp, these production sectors are the focus of most research with respect to the use of alternate protein sources. Numerous studies have been conducted to evaluate the effects of replacing various percentages of fish meal in diets for these fish, and without exception none has successfully replaced 100% of fish meal without reducing fish performance. At best, 50% of fish meal in diets for salmon and trout can be replaced by soy protein concentrate, and 25-30% with soybean meal (Stickney et al. 1996, Medale et al. 1998, Mambrini et al. 1999, Refstie et al. 2000). Similar findings have been reported in studies of wheat gluten meal, corn gluten meal, and rapeseed protein concentrate (Teskeredzic et al. 1995, Weede 1997). There are many reasons for the failure of diets containing high replacement levels of plant proteins to support fish growth and feed efficiency equivalent to levels supported by diets in which fish meal is the predominate protein source. Most of the poor performance is associated with amino acid profiles of plant protein sources and lower feed palatability when replacement reaches high levels. This has been demonstrated by studies involving amino acid supplementation of diets containing high levels of plant proteins and by pair-feeding studies where differences in feed intake of groups of fish fed diets differing in fish meal replacement levels are removed as sources of variation in fish performance (Weede 1997, Refstie et al. 1997).

Global fish meal supplies are finite and, assuming that the demand for high-protein fish diets will increase dramatically over the next decade, the role of fish meal in diets is likely to shift, causing shifts in the roles of alternative protein sources as well. Alternative protein sources will be used to extend global fish meal supplies, keeping the cost of diets and thus the cost of farmed fish within market expectations. Increasing concern about the environmental impacts of aquaculture will keep pressure on aquaculture to use "environmentally friendly" diets, meaning those low in total phosphorus. Marine proteins will shift from being the primary source of protein in fish diets to being used to balance the amino acid deficiencies in diet formulations that inevitably arise when plant proteins are used to supply the bulk of total dietary protein. Synthetic methionine will augment fish diets containing high amounts of soybean

Table 5.　Predicted protein needs for aquaculture feeds in 2010.

Species group	Feed (t)	Fish meal (t)
2000 (est.)	13,098,000	2,115,000
2010 (with today's diet formulations)	37,226,000	4,586,000
2010 (with lower % in diets)		2,831,000[a]
Difference between two 2010 estimates		1,755,000[b]

[a]Barlow 2000.
[b]Fish meal equivalent to be supplied by other protein sources.

meal or full-fat soybean meal, given the fact that soy protein products are deficient in methionine, but protein concentrates made from grains, e.g., wheat, corn, and other small grains, are deficient in several amino acids for which there are no inexpensive synthetic replacements. These proteins must be blended with other proteins that are relatively rich in the amino acids that are deficient in grain protein concentrates. Marine proteins can fulfill this role, although this represents a paradigm shift in their use in fish diet formulations.

These changes will alter demand for proteins produced from marine byproducts, making byproducts that increase diet palatability more valuable. Amino acid profiles of proteins made from marine byproducts will become a property of increasing value, driving the production of proteins made from marine byproducts away from total recovery by conversion to fish meal and toward removal of seafood byproduct components, e.g., bone, skin, and cartilaginous components, prior to reduction into protein sources (Rathbone et al. 2001). Fish oil will increase in value and the residual level of fish oil in fish meal will enhance fish meal value. The potential of mineral imbalances in fish diets containing high amounts of plant protein concentrates may result in addition of semi-purified bone meal from seafood byproducts being added back at optimum levels to fish diets. Besides being an excellent source of dietary phosphorus, bone recovered from seafood byproduct processing waste contains an array of essential minerals and trace elements that will be needed in fish diet formulations in which plant protein concentrates replace a large proportion of fish meal.

At present, it is impossible to formulate a nutritionally balanced diet for salmonids that contains only plant proteins, unless synthetic amino acids are added to make up for deficiencies of several essential amino acids in plant proteins (Table 6). However, if products from poultry or animal rendering, e.g., blood meal, poultry byproduct meal, and feather meal are used, it is possible to formulate a diet that meets the essential amino acid requirement of salmonids, and presumably marine fish. If soy protein concentrate is used as the main protein source in a rainbow

Table 6. **Amino acid concentration (g/100 g, wet weight) of plant protein ingredients and low-temperature-dried fish meal.**

Amino acid	Soy protein concentrate	Corn gluten	Wheat gluten	Low temperature dried fish meal	RBT dietary requirement
Arginine	4.04	1.34	2.18	3.35	1.5
Histidine	1.442	0.91	1.35	1.54	0.7
Isoleucine	3.17	2.37	2.78	3.15	0.9
Leucine	5.53	10.26	5.40	5.56	1.4
Lysine	3.84	0.91	1.20	4.69	1.8
Methionine	0.81	1.09	0.98	1.88	1.0[a]
Phenylalanine	2.76	2.79	3.00	2.28	1.8[b]
Threonine	3.03	2.06	2.25	3.42	0.8
Valine	5.59	2.85	3.38	4.09	1.2
Crude protein	64.6	65.9	75.5	73.0	44.0

[a]Plus cystine.
[b]Plus tyrosine.

trout diet formulation, including about 25% low-temperature-dried fish meal makes it possible to balance the diet without using synthetic amino acids. If one had a marine protein source that was rich in methionine and lysine, it may be possible to reduce the marine protein source to 15-20%, with soy protein concentrate making up the remainder of dietary protein. This simple example illustrates the approximate limits that now confront the aquaculture industry as it prepares to cope with the fact that in the near future, annual fish meal production will be insufficient to supply the needs of aquaculture using current diet formulations for carnivorous farmed fish species that rely on fish meal to supply the bulk of dietary protein. However, just replacing a portion of the fish meal in current diet formulations with combinations of marine protein and plant protein concentrates will allow continued growth of aquaculture production to supply expected future demand for fisheries products.

Conclusion

In conclusion, diet formulations for farmed fish are expected to change in the future, mainly through a reduction in the percentage of fish meal used to produce grow-out diets. The extent of these changes will vary depending on the species of fish, but in general higher percentages of plant proteins will be used in place of fish meal. This will create several problems. Balancing the essential amino acid content of diets will be more difficult, given the fact that soy products are low in methionine, and grain-derived proteins are low in arginine, lysine, and methionine compared to fish meal (Table 6). Diet palatability may become an important consideration in diet formulation, especially when oilseed-derived proteins are added to diets. Another important issue is associated with dietary minerals, both levels in diets and bioavailability. Fish meal is an excellent source of many essential minerals, and plant proteins are not. Plant proteins contain phytate, the storage form of phosphorus in seeds, and phytate phosphorus is unavailable to monogastric animals, including fish. Further, phytate is known to interfere with the availability of certain trace elements, especially zinc, making it necessary to over-fortify diets to ensure adequate dietary zinc intake in fish fed diets containing high levels of phytate, especially in the presence of high dietary calcium levels (Richardson et al. 1985, Gatlin and Phillips 1989). We can expect the amount of fish meal used in aquaculture feeds to be close to 50% of annual global production, but we can also expect an increase in the demand for specialty marine products produced specifically for use in diets for farmed fish. These products will have special characteristics that overcome problems associated with expanded use of plant-derived protein concentrates. This will necessitate the expanded recovery and utilization of seafood processing waste and bycatch, with the additional refinement of partitioning of the seafood waste stream into segments that can be further processed to produce specialty products designed to enhance palatability, enrich diets with limited amino acids, and increase dietary efficiency, e.g., retention of dietary nutrients to support fish growth.

References

Barlow, S. 2000. Fishmeal and fish oil. The Advocate 3(2):85-88.

Gatlin, D.M.I., and H.F. Phillips. 1989. Dietary calcium, phytate and zinc interactions in channel catfish. Aquaculture 79:259-266.

Hajen, W.F., D.A. Higgs, R.M. Beames, and B.S. Dosanjh. 1993. Digestibility of various feedstuffs by post-juvenile chinook salmon (*Oncorhynchus tshawytscha*) in seawater. 2. Measurement of digestibility. Aquaculture 112:333-348.

Hardy, R.W. 2002. Rainbow trout, *Oncorhynchus mykiss*. In: C.D. Webster and C.E. Lim (eds.), Nutrient requirements and feeding of finfish for aquaculture. CABI Publishing, New York.

Hardy, R.W., and A.G.J. Tacon. 2002. Fish meal historical uses, production trends and future outlook for sustainable supplies. In: R.R. Stickney (ed.), Sustainable aquaculture. CABI Publishing, New York.

Kaushik, S.J. 2002. European sea bass, *Dicentrachus labrax*. In: C.D. Webster and C.E. Lim (eds.), Nutrient requirements and feeding of finfish for aquaculture. CABI Publishing, New York, pp. 28-39.

Mambrini, M., A.J. Roem, J.P. Cravedi, J.P. Lalles, and S.J. Kaushik. 1999. Effects of replacing fish meal with soy protein concentrate and of DL-methionine supplementation in high-energy, extruded diets on the growth and nutrient utilization of rainbow trout, *Oncorhynchus mykiss*. J. Anim. Sci. 77:2990-2999.

Medale, F., T. Boujard, F. Vallee, D. Blanc, M. Mambrini, A. Roem, and S.J. Kaushik. 1998. Voluntary intake, nitrogen and phosphorus losses in rainbow trout (*Oncorhynchus mykiss*) fed increasing dietary levels of soy protein concentrate. Aquat. Living Resour. 11:239-246.

New, M.B. 1999. Global aquaculture: Current trends and challenges for the 21st century. World Aquaculture Magazine 8-13:63-79.

Rathbone, C.K., J.K. Babbitt, F.M. Dong, and R.W. Hardy. 2001. Performance of juvenile coho salmon *Oncorhynchus kisutch* fed diets containing meals from fish wastes, deboned fish wastes, or skin-and-bone by-product as the protein ingredient. J. World Aquacult. Soc. 32(1):21-29.

Refstie, S., S.J. Helland, and T. Storebakken. 1997. Adaptation to soybean meal in diets for rainbow trout (*Oncorhynchus mykiss*). Aquaculture 153:263-272.

Refstie, S., O.J. Korsoen, T. Storebakken, G. Baeverfjord, I. Lein, and A.J. Roem. 2000. Differing nutritional responses to dietary soybean meal in rainbow trout (*Oncorhynchus mykiss*) and Atlantic salmon (*Salmo salar*). Aquaculture 190:49-63.

Richardson, N.L., D.A. Higgs, R.M. Beames, and J.M. McBride. 1985. Influence of dietary calcium, phosphorus, zinc and sodium phytate level on cataract incidence, growth and histopathology in juvenile chinook salmon (*Oncorhynchus tshawytscha*). J. Nutr. 115:553-567.

Robinson, E.H., and M.H. Li. 2002. Channel catfish, *Ictalurus punctatus*. In: C.D. Webster and C.E. Lim (eds.), Nutrient requirements and feeding of finfish for aquaculture. CABI Publishing, New York, pp. 293-318.

Shivananda Murthy, H. 2002. Indian major carps. In: C.D. Webster and C.E. Lim (eds.), Nutrient requirements and feeding of finfish for aquaculture. CABI Publishing, New York, pp. 262-272.

Stickney, R.R., R.W. Hardy, K. Koch, R. Harrold, D. Seawright, and K.C. Massee. 1996. The effects of substituting selected oilseed protein concentrates for fish meal in rainbow trout diets. J. World Aquacult. Soc. 27:57-63.

Storebakken, T. 2002. Atlantic salmon, *Salmo salar*. In: C.D. Webster and C.E. Lim (eds.), Nutrient requirements and feeding of finfish for aquaculture. CABI Publishing, New York, pp. 79-102.

Sugiura, S.H. 2000. Digestibility. In: R.R. Stickney (ed.), Encyclopedia of aquaculture. John Wiley and Sons, Inc., New York, pp. 209-218.

Sugiura, S.H., and R.W. Hardy. 2000. Environmentally friendly feeds. In: R.R. Stickney (ed.), Encyclopedia of aquaculture. John Wiley and Sons, Inc., New York, pp. 299-310.

Takeuchi, T., S. Satoh, and V. Kiron. 2002. Common carp, *Cyprinus carpio*. In: C.D. Webster and C.E. Lim (eds.), Nutrient requirements and feeding of finfish for aquaculture. CABI Publishers, New York, pp. 245-261.

Teskeredzic, Z., D.A. Higgs, B.S. Dosanjh, J.R. McBride, R.W. Hardy, R.M. Beames, J.D. Jones, M. Simell, T. Vaara, and R.B. Bridges. 1995. Assessment of undephytinized and dephytinized rapeseed protein concentrate as sources of dietary protein for juvenile rainbow trout. Aquaculture 131:261-277.

Weede, N. 1997. Low phosphorus plant protein ingredients in finishing diets for rainbow trout (*Oncorhynchus mykiss*). University of Washington, Seattle, WA.

Opportunities and Challenges of Fish Meal and Fish Oil in Animal Food

Richard Sellers
American Feed Industry Association, Arlington, Virginia

Introduction

The American Feed Industry Association (AFIA) is the national feed trade association representing feed manufacturers, ingredient suppliers, equipment manufacturers, pet food manufacturers, animal health manufacturers and distributors, and other service suppliers to the feed industry. AFIA members manufacture 75% of the primary commercial feed in the United States, including aquafeed.

AFIA suggests examining the trend of removing the term "byproduct" and focusing on "co-product" or "protein product." This is the term better accepted by the pet food customer and one more palatable to the feed industry in general. The marine co-product industry should focus on the customers' needs, whether it be the feed or pet food industry. Also, the principal concerns of both industries are safety and consistency of product. Secondary to that is the "quality" of the product provided, as determined by the customer's criteria, such as protein balance, sanitation, and/or consistency.

U.S. feed industry overview

The U.S. feed industry is characterized by approximately 77 million metric tons of finished or ready-to-eat feed. This amount is difficult to determine due to the inherent problem of defining a feed given so much on-farm mixing. Cattle is the largest segment of this industry, and integrated poultry feed is the most tightly controlled. Aquafeed is by far the smallest with just over one million metric tons. Pet food has the highest net income and is not included in the figures for finished feed.

The industry utilizes 300-400 ingredients in manufacturing feed, and use of these is generally determined by "least cost" or "best cost" formulation (e.g., linear programming). The industry is experiencing considerable consolidation of firms, creating larger and larger, but fewer and fewer, feed companies.

There has been a considerable increase in regulatory oversight due to heightened sense of counter-terrorism policies. This has led to the hiring of 400 new U.S. Food and Drug Administration (FDA) investigators. An additional 1,000 are targeted in the fiscal year beginning October 1, 2002. This heightened regulatory environment has led to more controls on imported animal protein products for both animal health control and reduction of the counter-terrorism threat.

Benefits and uses of fish meal and oil

Twelve marine product ingredients have been approved for use in feed by the Association of American Feed Control Officials (www.aafco.org). These include crab meal, shrimp meal, condensed fish solubles, anchovy feed fat, menhaden feed fat, tuna feed fat, fish byproducts, dried fish protein digest, condensed fish protein digest, fish digest residue, menhaden fish meal, and Peruvian anchovy fish meal. Use of names other than these or ingredients other than these is prohibited by both state and federal laws.

The most common of these products relating to Alaskan fisheries co-products is likely fish byproducts, but the rendered product is fish meal. The fish byproducts legal definition is as follows:

> Fish by-products must consist of non-rendered, clean unde-composed portions of fish (such as, but not limited to heads, fins, tails, ends, skin, bone and viscera) which result from fish processing. If it bears a name descriptive of its kind, it must correspond thereto. Any single constituent used as such may be labeled according to the common or usual name of the particular portion used (such as fish heads, fish tails, etc.)

This definition is several decades old and reflects availability and use of the product at the time of definition development. It is possible and quite easy to change such a definition within the AAFCO process.

The definitions and process may be found in the annual publication of AAFCO, which for this year is *AAFCO 2002 Official Publication.* Many of the definitions are updated annually at the AAFCO Annual Meeting occurring each August. There is a yearlong process of reviewing petitions for new and amended ingredient definitions, which usually includes manufacturing, toxicology, analytical methods, and other data required in the process. FDA scientists do intensive reviews of new products to determine the safety and utility in feed. New chemical entities may require more formal "food additive petitions" to the agency, which may cost several million dollars and several years of regulatory review. Since many of the new entities may not have patent rights, exclusivity may be a problem, as the FDA petition process does not grant such exclusivity, except in the case of approved animal drugs.

Marine products are used primarily in the United States in aquaculture, poultry, and swine feed and pet food. They are excellent sources of balanced protein and B vitamins, and a limited source of essential fatty acids. Fish oil is an excellent source of essential fatty acids, but its high price owing to the human supplement demand makes for lower feed demand. However, it is an important attractant in aquafeeds.

Regulatory issues

As indicated earlier, only definitions approved by the U.S. Food and Drug Administration (FDA) and AAFCO may be used on feed labels, and similarly only those products may be used in feed. Blends of marine products may also be sold and can be called virtually anything (as long as the title is not false or misleading). However, a blend of marine products must have an ingredient listing for commerce. It should be noted that use of numbers in a product name are assumed by regulation to be the protein level unless otherwise designated.

The federal Food, Drug and Cosmetic Act requires each supplier to provide an unadulterated product. There are few tolerances established for marine products, and adulterants, when found, are normally dealt with on a case-by-case basis using established toxicology scientific literature for determination of adulteration.

Since there are no pesticides approved for use on fish, the tolerances allowed for pesticide residues in marine products is zero. High levels of detection and limited testing usually reveal few violations, if any, annually.

Heavy metals are fairly well documented in the scientific literature with respect to known tolerances, even though these are not spelled out in the regulations. The National Research Council published in 1981 *Mineral Tolerance of Domestic Animals* which is utilized by state and federal officials for determining violation of statutory feed adulteration. These levels are incorporated in the AAFCO publication for extrapolation from ingredient to finished feed.

Animal drug residues in marine tissues are limited to those approved for use in fish, which currently are very few. Discovery by the agency of a drug residue in fish tissue is de facto proof that the producer has adulterated the food fish and that the drug so used is also adulterated, as is the feed used as a vehicle for the drug. Serious civil and criminal sanctions can be applied for such a violation. However, such violations in fish and marine products are quite rare.

The major adulteration concern of marine products is *Salmonella* adulteration. FDA has for many years been sampling for *Salmonella* in feed and feed ingredients for information purposes. The agency has taken no regulatory actions when finding *Salmonella*. However, the firm so sampled is notified by an "information letter" from FDA when a positive result is found. FDA has a policy of "*Salmonella* negative" feed, but little enforcement has occurred.

The agency does find that marine products regularly contain *Salmonella* when sampled at the consignees. This is likely recontamination and usually can be traced to handling or transport vehicle.

In October 2002, the Centers for Disease Control and Prevention (CDC) published a scientific article in *Clinical Infectious Diseases* by three CDC scientists stating the FDA should enforce the *Salmonella* negative policy and implement mandatory HACCP (hazard analysis critical control points) for the feed industry. Unfortunately, the article was full of old references and is being disputed by reputable microbiologists working in feed microbiology. However, as a matter of public policy, the feed industry cannot ignore this CDC position.

AFIA endorses the Animal Protein Producers Industry (APPI) *Salmonella* Reduction and Eradication Program which has been operating for over 25 years. Details of this program can be found at www.animalprotein.org. Marine protein product producers should join such programs to reduce their potential for *Salmonella* contamination.

Quality issues

Feed mill managers are noticeably reticent to utilize marine products for several reasons, including potential for rancidity, handling difficulty, "fishy" odor, consistency, and others. Most large firms utilize "quality lists" that require ingredient suppliers to provide technical information for review, including site visits and manufacturing specifications. Once a firm is on a feed company's quality list, product specification deviations may result in the supplier's removal from such list and much difficulty in gaining access to the list again.

AFIA regularly reminds its members that there are "no good deals in agriculture" and that foreign products may be suspect if offered at a very low market price. This is because it may be difficult to visit the manufacturing site or view product specifications, including independent, third-party testing. An occasional foreign product testing may reveal violative results which can occur after finished feed product manufacturing and shipping.

Contamination issues

Some contamination issues mentioned earlier include *Salmonella*, heavy metals, violative drug residues, and product degradation. Other contamination issues may include persistent organic pollutants (POP), which includes dioxins, polychlorinated biphenyls (PCPs), and other dioxin-like compounds.

The U.S. Environmental Protection Agency is reviewing a dioxin reassessment in the United States, which may be released soon. This report will not establish tolerances, but raise the level of public awareness of dioxin contamination in animal food products. Also, the National Academy of Sciences and sister organizations are providing a government funded report on options to reduce dioxin in the food supply. More can be found at www.nas.edu.

Of great concern is the European Union's adoption on July 1, 2002, of maximum and action levels of dioxin in feed and feed ingredients. Finished feed maximum levels of 0.75 parts per trillion (pptr) have been established, but these do not include PCBs, which may be included at a later date. The U.S. does not support these levels and protested their establishment due to the paucity of data and sample results collected. FDA is unlikely to establish such tolerances due to the long process required to do so, but it continues to collect feed and feed ingredient samples to analyze for dioxins.

FDA's feed survey for 2000 included some 47 samples and found an average of 2+ pptr of dioxin (including PCBs) in Atlantic marine products. Lower levels were reported in Pacific products. These low levels are not of regulatory concern to the FDA. However, results of sample programs from 2001 and 2002 have yet to be released. A new program for 2003 has just been released. The purpose of these programs is to build a database of dioxin levels to gauge what future actions need to be taken to reduce dioxin in the food and feed supply.

As a general rule, AFIA recommends no samples be analyzed for contaminants unless a policy to deal with positive findings has been fully vetted with each firm's quality, management, and legal personnel. A positive adulterant result for products already shipped into interstate commerce could have potentially devastating regulatory, legal, public relations, and liability consequences. Since few tolerances have been established, such determination of adulteration rests solely on the FDA's shoulders.

Industry changes

Intense consolidation continues in the feed and feed ingredient industries, brought on by competition, market vulnerability, livestock and poultry supply shifts, and other variables. One cooperative is in bankruptcy and others may follow. The overall state of the feed industry is financially strong and feed is readily available throughout the United States at competitive prices. With the current glut of milk, beef, poultry, pork, and turkeys in the market, consolidation of the feed industry will likely continue for several years.

The issues of food safety and environmental handling are prime topics for the industry and present many formidable challenges.

Conclusion

The opportunities for selling marine products from Alaska into the U.S. feed industry abound. The feed industry regularly seeks out and embraces viable, good protein sources from reputable suppliers with a consistent, regular supply of product. Continued research is needed to bring new products to the marketplace and find new markets for old products. The feed industry looks for partners with safe, affordable, and efficient products for sale to continue to supply American producers with the best, safest, and cheapest feed products for a growing protein demand worldwide.

Developing Unique Feed Ingredients from Fish Byproduct Components

Peter J. Bechtel
University of Alaska Fairbanks, USDA Agricultural Research Service Laboratory, Fairbanks, Alaska

Abstract

Fish processing byproducts can be used to make fish meals and oils for use in the aquaculture, livestock, and poultry industries. The volume of fish processing byproducts available each year from fish harvested from Alaska waters is estimated to be over one million metric tons. Fish meals are made in Alaska from combined fish processing byproduct components; however, much is discarded. It is possible to collect and process the different fish byproduct components from automated high-speed fish processing lines. Individual fish processing byproduct components from pollock include heads, frames, viscera, and skin. Chemical and nutritional properties of the individual byproduct stream components have been determined. The amount of soluble protein varied, with viscera having the greatest percent of solubility at low temperatures and skin having the greatest percent soluble protein at higher temperatures. Estimated rat protein efficiency ratio (PER) values ranged from 2.1 for skin to 3.1 for viscera, and all byproducts had a greater than 90% protein digestibility using pepsin. Unique feed ingredients can be produced by combining byproduct components in different amounts to develop products with different protein, fat, and mineral contents, and different protein quality and solubility characteristics.

Introduction

Over 65% of the total wild fish harvested for human consumption in the United States comes from Alaskan waters. The three species from which the most byproducts are produced include Alaska pollock (*Theragra chalcogramma*), Pacific cod (*Gadus macrocephalus*), and pink salmon (*Oncorhynchus gorbuscha*). As shown in Table 1, the total harvest in

Table 1.   2000 seafood harvest in Alaska.

Species	Metric tons
Alaska pollock	1,067,738
Salmon	319,472
Pacific cod	226,709
Flatfish	141,530
Others	149,800
Total	1,905,249

2000 of these three species of fish was estimated at 1.6 million metric tons from a total catch of over 1.9 million metric tons (NMFS 2000, ADFG 2000). In Alaska waters, three times as much pollock is harvested as salmon or cod.

From the harvest of 1.9 million metric tons of fish over 1 million metric tons of processing byproducts were produced in the year 2000 (Table 2). The 1 million tons of byproducts consisted of 289,384 t of heads, 273,161 t of frames, 328,962 t of viscera, and 75,368 t of skin. More than half of all fish processing byproducts, especially those processed at sea, are not utilized. The yield from processing pollock was estimated to be 34% overall which results in 66% byproducts (Crapo et al. 1993). In 2000 the amount of pollock byproduct was estimated to be 707,707 t, consisting of 32% viscera, 26% heads, 33% frames, and 9% skin.

Fish harvesters and processors want to maximize profitability and fish value by developing markets for fish byproducts and also by reducing disposal costs. There are opportunities for utilizing fish processing byproducts to create specialty feeds for different aquaculture species (Hardy 2003), farm animals, and pet foods. There is a growing need for specialized aquaculture feed ingredients to reduce feed costs, decrease nitrogen and phosphorus discharges into the environment, improve feed efficiency, provide nutrients to maximize biological functions, etc. Differences have been reported in the physical and chemical properties of different byproducts (Kizevetter 1971, Bechtel 2003); however, many of the nutritional properties have not been systematically evaluated.

It is possible to collect individual pollock byproducts (heads, frames, viscera, and skins) and after the chemical and nutritional properties have been determined it is possible to develop unique feed ingredients from the individual components or by combining them in specified amounts.

Table 2. 2000 Alaska fish processing byproduct components.

Components	Metric tons
Heads	289,384
Frames	273,161
Viscera	328,962
Skins	75,368
Total	1,007,490

Total byproducts was calculated using separate equation from individual byproducts.

Methods

Whole pollock, fillets, heads, viscera, frames, and skins were obtained in February, with replicates obtained on separate days. Samples were first cut into smaller pieces and ground through a 12 mm plate three times, then ground through a 6 mm plate two additional times, and then stored in plastic bottles at –80°C until analyzed. Samples for ash content were placed in a muffle furnace at 550°C for 6 hours. Lipid content was determined in triplicate from freeze-dried samples using a Leco FA-100 analyzer. Nitrogen content was determined in triplicate using a Leco FB-2000 nitrogen analyzer. Protein content was calculated as percent nitrogen times 6.25.

Samples for mineral analysis were prepared by ashing 1 g of sample overnight in a 550°C muffle furnace, and the residue was subsequently digested overnight in an aqueous solution containing 10% (v/v) hydrochloric acid and 10% (v/v) nitric acid. A Perkin Elmer Optima 3000 Radial ICP-OES was used for the analysis. The amino acid content was determined by digesting freeze-dried samples in 6 N hydrochloric acid for 24 hours at 110°C. Precolumn derivitization (AccQ.Tag, Waters Corporation) was used and amino acid content was determined with a Waters Alliance HPLC system and Waters 474 scanning fluorescence detector. Cysteic acid content was determined after performic acid oxidation (AOAC 2000). Connective tissue content was calculated by determining the amount of hydroxyproline in a sample using the AOAC official method 990.26 (AOAC 2000). The estimated rat protein efficiency ratio (PER) was calculated using the equation ($Y = -0.02290 X + 3.1528$ where Y is the estimated PER and X is the collagen content) developed by Lee et al. (1978). Pepsin digestibility of samples was completed using the AOAC Official Method 917.09 (AOAC 2000). Fresh solutions of pepsin (ACROS Organics) were made in 0.075 M hydrochloric acid at 0.2 and 0.002% (w/w).

Percent protein solubility was determined by a modification of the protein dispersibility index before and after heating samples to 85°C. Samples were heated at 23 or 85°C in a water bath for 30 minutes, cooled, and then homogenized in 30°C distilled water for 1 minute, placed in a water bath at 30°C for 30 minutes, and then spun at 10,000 × G for 15 minutes at 20-25°C. An aliquot of the aqueous layer was removed and the protein content determined. The percent soluble protein was calculated from the ratio of total mg of soluble protein in the aqueous phase per total milligrams of protein and expressed as a percentage.

The ANOVA procedure was used with Statistica release 6 software with byproduct parts and dependent variables (minerals, pepsin digestibility, estimated rat PER, and protein solubility). Post hoc analysis used the Duncan test and the level of significance used was $P < 0.05$.

Results and discussion

The Alaska pollock fishery is one of the largest in the world. It is managed with the goal of maintaining a sustainable fishery. The fish are harvested predominantly from the Bering-Aleutian area and processed both onshore and at sea. Pollock products include large amounts of surimi, which is exported for further processing, and fillets, which are often further processed into fish sticks, fish sandwiches, and like products. The pollock fishery is actively managed by the National Marine Fisheries Service and fish are harvested several times during the year.

The common byproducts derived from pollock processing are heads, frames, viscera, and skin. With advances in processing technology more of the muscle tissue is being removed from the frames and heads; however, markets have not been established for most pollock byproducts with the notable exception of roe. Pollock byproducts from the large onshore processing plants are used to make fish meals and oils. Due to space and facility limitations, at-sea processors will dispose of much of their pollock byproducts while at sea.

Fish meal and oil made from pollock byproducts are commodities that often compete with other sources of whole fish meals and oil from Peru, Chile, Norway, and other countries. Pollock meals often have a low fat content, typical of white fish, and pollock meals are made from byproducts that have much of the skeletal muscle removed. In the past, many white fish meals have had high ash contents but with improved processing procedures it is now common to remove bone fragments and thus reduce the ash content. Studies by Babbitt et al. (1994) and Rathbone et al. (2001) have evaluated white fish meals with different protein and ash content. Another processing approach is to utilize the individual fish processing byproducts that are readily available directly off the processing line for making feed ingredients and other products.

An approximate composition of dried individual pollock byproducts is given in Table 3. These calculated values are expressed on a 92% dry weight basis to allow some comparison to other fish meal products that would contain 8% water in the final product. As can be seen from Table 3, the protein content was above 66% for all byproducts except viscera (41%). Protein was determined as percent nitrogen times 6.25 and although this value is commonly used (Sosulski and Imafidon 1990), it may overestimate the protein content of some byproducts. Also of interest was the low fat content of less than 6% for heads, frames, and skin. Viscera had a fat content greater that 47% which could be due to the high fat found in liver. The ash content ranged from 3 to 19% for all byproducts with heads containing the highest ash content of 19.4%. It should be pointed out that the byproducts were simply dried, so they will contain a number of components such as soluble protein which could be lost during conventional fish meal processing (heating and pressing steps).

The amino acid content of the byproducts on a percent micromolar basis are listed in Table 4. As expected, the content of glycine was highest in skin which contains large amounts of collagen, and lowest in viscera and fillets which contain smaller amounts of connective tissue. The content of lysine, threonine, and valine were lowest in skin. Several studies have reported amino acid analysis of individual byproducts including viscera (Freeman and Hoogland 1956, Olley et al. 1968, Dong et al. 1993), liquified frames (Ferreira and Hultin 1994), and Pacific whiting solid waste (Benjakul and Morrissey 1997). Gunasekera et al. (2002) reported the nutritional evaluation of selected byproducts from carp offal, fish frames, and trout offal. Onodenalore and Shahidi (1996) reported the amino acid composition of hydrolysates from shark. Mambrini and Kaushik (1995) published an interesting paper on predicting amino acid requirements for different fish species from their amino acid composition.

The content of selected minerals in pollock byproducts are listed in Table 5. As expected, the calcium and phosphorus content were highest for heads and frames which contain large amounts of bone. The ratio of calcium to phosphorus was approximately two for heads and frames. The content of the heavy metals such as cadmium and lead were low in all pollock byproducts. Zinc content was lowest for fillets and the iron content was highest for viscera.

Protein digestibility was estimated with the pepsin digestibility method. Using the standard 0.2% pepsin solution, digestibility was 92% or greater for all byproducts (Table 6). At one hundredth the pepsin concentration (0.002%) the digestibility was 93.8% or greater. Sugiura et al. (1998) determined the apparent protein digestibility of a number of fish meal and byproduct feed ingredients using trout and salmon. They reported apparent protein digestibility for herring, anchovy, and menhaden meals to range from 88 to 95%. Generally, fish processing byproducts and meals have a high degree of protein digestibility.

Table 3. Calculated pollock byproduct composition: adjusted 8% moisture.

	% water	% protein	% fat	% ash
Whole fish	8	65.7	15.5	10.8
Fillets	8	84.9	1.9	5.2
Heads	8	67.4	5.2	19.4
Frames	8	73.6	3.9	14.5
Viscera	8	41.2	47.2	3.6
Skin	8	87.5	1.7	2.8

Calculated from Bechtel 2003.

Table 4. Amino acid composition of pollock byproducts as percent micromoles.

	Whole fish	Heads	Frames	Viscera	Skin	Fillets
ALA	7.68	7.94	7.48	7.72	8.34	7.71
ASP	6.26	5.86	5.48	3.81	3.93	5.38
ARG	7.97	7.80	7.53	8.56	6.85	7.06
CYS	1.12	0.92	0.96	1.23	0.27	1.26
GLU	8.63	8.09	7.92	7.46	6.53	9.18
GLY	14.41	17.46	16.88	10.63	33.69	11.47
HIS	2.66	2.58	2.75	3.15	2.03	3.08
ILEU	4.39	3.87	3.95	4.80	2.19	4.55
LEU	7.70	6.88	6.99	8.35	3.98	7.83
LYS	5.50	5.72	6.31	5.59	3.40	6.38
MET	2.94	2.76	2.96	2.83	2.11	3.44
PHE	4.82	4.47	5.07	5.76	3.01	5.88
PRO	5.21	5.90	5.63	6.48	8.55	5.21
SER	7.03	7.01	6.57	6.78	6.58	5.53
THR	5.77	5.27	5.73	6.24	3.80	6.29
TYR	2.50	2.60	2.98	4.79	1.70	4.55
VAL	5.41	4.87	4.82	5.82	3.05	5.20

Number of replications per part is 3.
Unpublished data from P. Bechtel and R. Johnson.

Table 5.   Mineral content of pollock byproducts.

	Ca	Mg	Na	P	K	Cd	Fe	Pb	Zn
Whole fish	2.5[a]	0.15[a]	0.89[a]	1.9[a]	1.1[a]	1.1	43.4	<2	57.5[acd]
Heads	6.6[b]	0.22[b]	1.3[b]	3.6[b]	0.72[b]	<0.03	47.6	<2	68.6[bc]
Viscera	0.27[c]	0.07[c]	0.37[c]	0.78[c]	0.62[b]	0.62	74.4	<2	63.5[abc]
Frames	5.9[b]	0.2[b]	0.82[a]	3.5[b]	1.1[a]	<0.03	36.7	<2	61.4[abc]
Skin	0.63[c]	0.07[c]	0.34[c]	0.53[c]	0.14[c]	<0.03	26.4	<2	49.2[ad]
Fillet	0.13[c]	0.15[a]	0.67[a]	0.98[c]	1.6[d]	<0.03	10.2	<2	21.6[e]

[abcde]Superscripts with different letters within a column are different (*P* < 0.05).
All values are averages of 3 replications.
Ca, Mg, Na, P, K were calculated as % dry matter; all others were calculated as mg/kg dry matter.
Unpublished data from P. Bechtel and R. Johnson.

Table 6.   Pepsin digestibility of pollock byproducts.

Pepsin %	Whole fish	Heads	Viscera	Frames	Skin	Fillets
0.2	96.8[bc]	95.6[bc]	92.0[d]	98.3[ac]	*98.8[a]	*98.8[a]
0.002	96.0[a]	95.3[a]	93.8[a]	95.0[a]	*99.1[a]	*98.8[a]

All values are averages of 3 replications; * indicates 2 replications.
[abc]Superscripts with different letters within a row category are different (*P* < 0.05).
Unpublished data from P. Bechtel and R. Johnson.

    The connective tissue content of pollock byproducts was calculated from hydroxyproline content and ranged from a low of 1% in viscera to a high of 45% in skin (Table 7). Frames and heads contained substantial amounts of connective tissue in the range of 7% to 10%. Connective tissue content was used to calculate the predicted rat protein efficiency ratio (PER). Using this method of calculating rat PER, the values for heads, viscera, whole fish, frames, and fillet were all in the range of 2.9 to 3.1 while skin had a much lower value of 2.1 (Table 7).

    Protein solubility is of interest when devising diets used in aqueous systems. The protein solubility method used to generate the values in Table 8 was to first heat the byproducts to either 30 or 85°C and then homogenize the samples in water and determine percent soluble protein. At 30°C viscera had a high percent soluble protein (62%) and heads, frames, and fillets had values in the range of 15% to16% (Table 8). After heating samples to 85°C the percent soluble protein was reduced in all byproducts except skin. This would be expected if the heating step denatured and aggregated the protein. Skin with its high connective tissue content would be expected to have increased protein solubility with increased

Table 7.   Percent connective tissue and predicted PER value of pollock byproducts.

	Whole fish	Heads	Viscera	Frames	Skin	Fillets
%CT	6.82[a]	9.78[a]	0.97[b]	7.56[a]	45.50[c]	1.59[d]
Rat PER	3.00[a]	2.93[a]	3.13[b]	2.98[a]	2.11[c]	3.12[b]

All values are averages of 3 replications.
CT = connective tissue. Percent CT is expressed as percent of total protein.
Estimated rat PER (protein efficiency ratio) calculated by method of Lee et al. 1978.
[abcd]Superscripts with different letters within a row are different ($P < 0.05$).
Data from Bechtel 2003.

Table 8.   Effect of heating on percent protein solubility of pollock byproducts.

	Whole fish	Heads	Viscera	Frames	Skin	Fillets
30°C	18.2[a]	16.3[a]	61.6[b]	15.3[a]	3.7[c]	15.2[a]
85°C	11.0[a]	11.3[a]	18.3[b]	9.9[ac]	49.1[d]	5.2[c]

All values are averages of 3 replications.
[abcd]Superscripts with different letters within a row are different ($P < 0.05$).
Data from Bechtel 2003.

temperature due to the denaturation of collagen and formation of gelatins which are soluble at elevated temperatures.

Summary

The composition of dried whole fish, heads, viscera, frames, fillets, and skins were 41-88% for protein, 2-47% for fat, and 3-19% for ash. The calculated connective tissue content was 2 and 45% for fillets and skin, respectively. Pepsin digestibility of the protein in all byproducts was greater than 92%. Mineral content and amino acid analysis had differences between byproducts. Estimated rat PER values ranged from 2.9 to 3.1 for all pollock byproducts, except skin which was lower. Protein solubility ranged from 4 to 49% at 85°C for all byproducts.

It would appear possible to develop products to fit niche markets by using different byproducts or by combining selected byproducts. Examples include low ash products, and products with different degrees of protein solubility, fat, and protein content, etc. There are processing implications if individual byproducts are to be made into feed ingredients, including devising alternate drying and concentration procedures and the possibility of eliminating fat separation steps for some byproducts. If all byproducts cannot be saved or processed it may be possible to use selected byproducts to make higher valued products.

References

ADFG. 2000. 2000 Alaska Commercial Salmon Harvest, Alaska Department of Fish and Game. http://www.cf.adfg.state.ak.us/geninfo/finfish/salmon/catchval/blusheet/00exvesl.htm.

AOAC. 2000. Official methods of analysis of AOAC International, 17th edn. Association of Official Analytical Communities, Washington, D.C.

Babbitt, J.K., R.W. Hardy, K.D. Reppond, and T.M. Scott. 1994. Processing for improving the quality of white fish meal. J. Aquatic Food Prod. Tech. 3(3):59-68.

Bechtel, P.J. 2003. Properties of different fish processing byproducts from pollock, cod and salmon. J. Food Process. Preserv. 27:101-116.

Benjakul, S., and M.T. Morrissey. 1997. Protein hydrolysates from Pacific whiting solid wastes. J. Agric. Food Chem. 45:3424-3430.

Crapo, C., B. Paust, and J. Babbitt. 1993. Recoveries and yields from Pacific fish and shellfish. Alaska Sea Grant, University of Alaska Fairbanks, Fairbanks. 32 pp.

Dong, F.M., W.T. Fairgrieve, D.I. Skonberg, and B.A. Rasco, 1993. Preparation and nutrient analysis of lactic acid bacterial ensiled salmon viscera. Aquaculture 109:351-366.

Ferreira, N.G., and H.O. Hultin. 1994. Liquifying cod frames under acidic conditions with a fungal enzyme. J. Food Process. Preserv. 18:87-101.

Freeman, H.C., and P.L. Hoogland. 1956. Processing of cod and haddock viscera: 1. Laboratory experiments. J. Fish Res. Board Can. 13:869-877.

Gunasekera, R.M., N.J. Turoczy, S.S. De Silva, and G.J. Gooley. 2002. An evaluation of the suitability of selected waste products in feeds for three fish species. J. Aquatic Food Prod. Tech. 11(1):57-78.

Hardy, R. 2003. Marine byproducts for aquaculture use. In: P.J. Bechtel (ed.) 2003. Advances in seafood byproducts: 2002 conference proceedings. Alaska Sea Grant College Program, University of Alaska Fairbanks, Fairbanks. (This volume.)

Kizevetter, I.V. 1971. Chemistry and technology of Pacific fish. (Translated in 1973 by Israel Program for Scientific Translations Ltd.). U.S. Department of Commerce, Springfield, Virginia.

Lee, Y.B., J.G. Elliot, D.A. Rickansrud, and E.C. Hagberg. 1978. Predicting protein efficiency ratio by the chemical determination of connective tissue content in meat. J. Food Sci. 43:1359-1362.

Mambrini, M., and S.J. Kaushik. 1995. Indispensible amino acid requirements of fish: Correspondence between quantitative data and amino acid profiles of tissue proteins. J. Appl. Ichthyol. 11:240-247.

NMFS 2000. 2000 Gulf of Alaska groundfish quotas and preliminary catch in round metric tons. National Marine Fisheries Service. http://www.fakr.noaa.gov/2000/goa00b.txt.

Olley, J., J.E. Ford, and A.P. Williams. 1968. Nutritional value of fish visceral meals. J. Sci. Food Agric. 19:282-285.

Onodenalore, A.C., and F. Shahidi. 1996. Protein dispersions and hydrolysates from shark (*Isurus oxyrinchus*). J. Aquatic Food Prod. Tech. 5(4):43-59.

Rathbone, C.K., J.K. Babbitt, F.M. Dong, and R.W. Hardy. 2001. Performance of juvenile coho salmon *Oncorhynchus kisutch* fed diets containing meals from fish wastes, deboned fish wastes, or skin-and-bone by-products as the protein ingredient. J. World Aquaculture Soc. 32:21-29.

Sosulski, F.W., and G.I. Imafidon. 1990. Amino acid composition and nitrogen-to-protein conversion factors for animal and plant foods. J. Agric. Food Chem. 38:1351-1356.

Sugiura, S.H., F.M. Dong, C.K. Rathbone, and R.W. Hardy. 1998. Apparent protein digestibility and mineral availability in various feed ingredients for salmon feeds. Aquaculture 159:177-202.

Nutritional Quality of Alaska White Fish Meals Made with Different Levels of Hydrolyzed Stickwater for Pacific Threadfin (*Polydactylus sexfilis*)

I. Forster
The Oceanic Institute, Waimanalo, Hawaii

J.K. Babbitt
National Marine Fisheries Service Utilization Research Laboratory, Kodiak, Alaska

S. Smiley
University of Alaska Fairbanks, Fishery Industrial Technology Center, Kodiak, Alaska

Introduction

Production of fish meal from whole fish or seafood byproducts is accomplished with a process known as the "wet-reduction" method. In this process, raw material is cooked to release lipid from cells and denature protein, and then it is mechanically pressed to separate solid material from water and oil. The solid fraction is called press cake. The liquid fraction is separated into two fractions, oil and stickwater. The latter contains soluble proteins. Material balance studies suggest that the stickwater may contain as much as 30% of the protein in the starting material. The amino acid composition of the protein in the stickwater is different from that in the press cake, the result of the separation of soluble proteins from insoluble structural proteins such as the contractile proteins. In many conventional wet-rendering processes, the stickwater is concentrated by evaporation, and added back to the press cake for drying. Stickwater is often treated with proteolytic enzymes to reduce the protein size and thus create a product that is easier to concentrate. The resulting product is known as whole fish meal.

Many fish meal plants, especially those with limited processing space or high energy costs, such as shipboard plants or some plants in Alaska, choose instead to discard the stickwater fraction. Thus, these plants produce meal known as press cake meal, rather than whole meal. Because the stickwater fraction contains valuable protein, minerals, and other components, the consensus is that whole meal may be superior to press cake meal as a feed ingredient. Surprisingly little is known about the exact nutritional value of stickwater, especially in the context of fish nutrition. Virtually nothing is known about the amount of stickwater to add to press cake meal to optimize nutritional value of the product.

Given the development of new technologies to concentrate stickwater, opportunities exist to produce very high-value whole meal from seafood byproducts. This study was conducted to determine the optimum amount of stickwater to add to press cake to produce the highest value fish meal product, both nutritionally and economically.

Materials and methods

Test ingredients

To evaluate the effects of incorporating fish solubles (stickwater concentrated to 41% solids under commercial processing conditions) in fish meal, four fish meals were produced by the Fishery Industrial Technology Center (FITC) of the University of Alaska Fairbanks, School of Fisheries and Ocean Sciences. White fish (pollock and cod) press cake (dried using a rotary disc steam drier under commercial processing conditions) and solubles were obtained from a commercial supplier and transported immediately to FITC, where the dried press cake was divided into four lots. Fish solubles were mixed with the press cake to achieve replacement of the protein in the meal with 10%, 20%, and 40% protein from the solubles. Mixtures of the press cake and solubles were dried at 65°C in an Enviro-Pak (Series MP500) drier to a final moisture content of 7%. The dried meals were chemically analyzed and samples of each meal sent to the Oceanic Institute in Waimanalo, Hawaii, for nutritional characterization. A Norwegian fish meal (Norsk LT-94) was used as a control.

Fish

Pacific threadfin (*Polydactylus sexfilis*) were obtained from the Oceanic Institute hatchery. Prior to the trial, the fish were maintained on a commercial feed (Marine Grower Moore-Clark, Vancouver, Canada).

Diets

Five experimental diets formulated to contain 40% crude protein and 9% crude lipid, were manufactured at the Oceanic Institute for this trial (Table 1) as follows: all the major dry feed ingredients were mixed for

Table 1.   Formulation of feeds used in Pacific threadfin growth trial. All values shown are "as fed."

Ingredient	Control %	M 0 %	M 10 %	M 20 %	M 40 %
Fish meal—LT 94	45.00	0.00	0.00	0.00	0.00
Alaska 0[a]	0.00	47.90	0.00	0.00	0.00
Alaska 10[a]	0.00	0.00	46.70	0.00	0.00
Alaska 20[a]	0.00	0.00	0.00	45.90	0.00
Alaska 40[a]	0.00	0.00	0.00	0.00	45.60
Wheat, whole	35.86	35.26	35.76	36.66	37.26
Vital wheat gluten	4.00	4.00	4.00	4.00	4.00
Brewers yeast	5.00	5.00	5.00	5.00	5.00
Soy lecithin	2.00	2.00	2.00	2.00	2.00
Fish oil	5.30	3.00	3.70	3.60	3.30
Chromic oxide	0.50	0.50	0.50	0.50	0.50
Min Px	0.06	0.06	0.06	0.06	0.06
Vitamin Px	0.40	0.40	0.40	0.40	0.40
Choline chloride	0.12	0.12	0.12	0.12	0.12
Stay C-35 (35% AA potency)	0.08	0.08	0.08	0.08	0.08
Potassium phosphate, dibasic	0.56	0.56	0.56	0.56	0.56
Sodium phosphate, dibasic	0.56	0.56	0.56	0.56	0.56
Magnesium phosphate	0.56	0.56	0.56	0.56	0.56
Total	100.00	100.00	100.00	100.00	100.00

[a]Experimental fish meals made in this study.

15 minutes in a Hobart food mixer (Model D-300, Hobart Manufacturing Corporation, Troy, Ohio). A warm (approx. 60°C) aqueous solution of sodium phosphate, potassium phosphate, choline chloride, and trace element premix was then added to the dry ingredient mix, to bring the moisture content of the resulting mash to approximately 34-35%. The mash was then blended for a further 15 min. Half the supplemental oil and lecithin and all the cholesterol were blended in a KitchenAid mixer (Model K5SS, KitchenAid, St. Joseph, Michigan), added to the mash, and mixed for a further 15 min. The resulting mash was then passed through a California Pellet Mill (Model CL5, San Francisco) fitted with a 2.5 mm diameter die. No steam was used and the pellet temperature at the die was below 70°C. The resulting moist pellets were then dried overnight in a drying cabinet using an air blower at approximately 38°C until the moisture level was below 10%. The vitamin premix and vitamin C source were then emulsified with the remaining oil and lecithin in a KitchenAid

mixer and this mixture was added to the dry cooled pellets by top coating using a Hobart D300 food mixer with a whisk beater. The finished pellets were then stored in plastic bins at 19-20°C until used.

Husbandry

At the commencement of the trial, animals were stocked into 20 rectangular glass aquaria (76 × 31 × 31 cm dimensions and 52 liter water volume) at an initial density of 12 fish per aquarium. Individual weight at stocking was 3.0 g (s.d. = 0.3 g). Water (25.9-26.9°C) was supplied to each aquarium at the rate of 1 liter per minute. Analysis of variance procedures revealed no significant variation in stocking weight among dietary treatments. A 72 h stress replacement period after stocking was observed to avoid bias in the treatments.

Each diet was fed to fish in four randomly selected aquaria for 8 weeks. All diets were fed by hand to satiation twice daily at 0800 and 1600 h. Each feeding consisted of at least two passes. Feeding was terminated when feed pellets remained from the previous pass. Each tank was siphoned before the first feeding of the day to remove diet residues and fecal matter. Animals were weighed in bulk by tank every two weeks and individually at the end of the trial (8 weeks). At weeks 4 and 6, two fish from each tank were removed at random. The weight of these animals was recorded. There was negligible mortality throughout this trial.

Statistical analysis

Feed conversion ratio (FCR) was calculated as the weight of the feed consumed by all the fish in a tank divided by the total accumulated weight gain of the animals. The final body weight and FCR data were subjected to analysis of variance procedures to determine the significance of treatment effects (using $P < 0.05$). Where significant experimental effects were detected, Tukey's test was employed to ascertain the significance of differences between treatments within each trial.

Results and discussion

The growth and feed conversion ratio of fish in the treatments was affected by treatment (Table 2 and Figure 1), with the best performance exhibited by those in the control treatment ($P < 0.05$). Of the fish fed the diets containing different amounts of added solubles, the poorest performance was found for the meal with no added stickwater. Fish fed diets made from fish meal made with 20% inclusion exhibited an increased final weight that was 50% higher than those fed diets with meal that had no stickwater. Growth response of the fish was affected by both increased feed consumption and improved nutritional quality. FCR is calculated as the amount of feed consumed per unit of growth, and is an inverse measure of nutritional quality. The improvement of the FCR in the fish fed

Table 2.   Final weight and feed conversion ratio (FCR) of fish fed for eight weeks on diets made from one of five fish meals. Values within a column sharing a superscript are not significantly different ($P > 0.05$).

Fish feed	Final weight (g)	s.d.	FCR (g/g)	s.d.
Control (LT-94)	34.9[a]	2.8	1.05[d]	0.03
M 0	20.4[d]	1.7	1.64[a]	0.06
M 10	22.8[cd]	2.1	1.40[b]	0.06
M 20	30.4[ab]	2.5	1.29[c]	0.03
M 40	28.1[bc]	2.9	1.36[b]	0.05

s.d. = standard deviation.

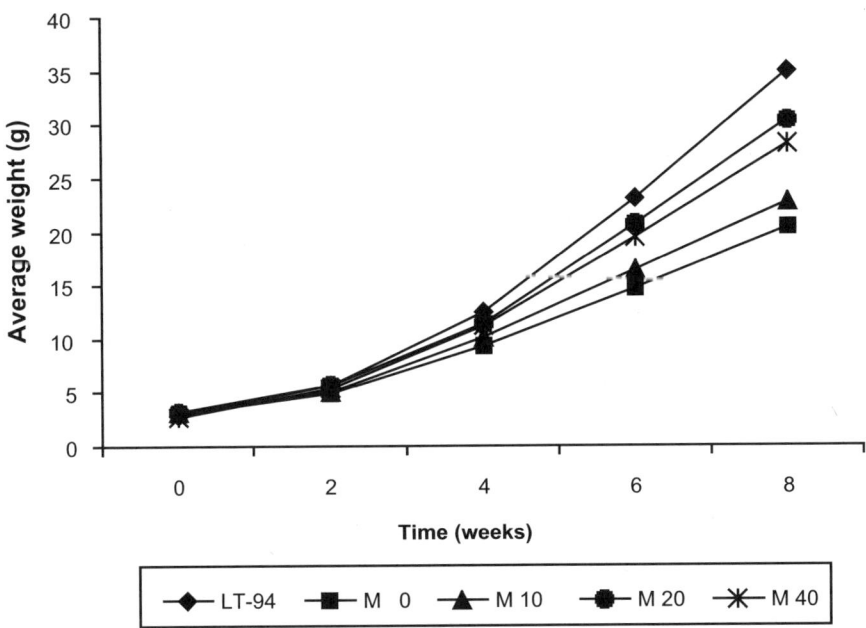

Figure 1. Average weight over time of Pacific threadfin fed diets containing a Norwegian fish meal (LT-94) or a fish meal with one of four levels of stickwater added (0, 10, 20, or 40%).

diets containing higher proportions of stickwater in the meal indicates that the nutritional quality of these meals was enhanced.

Similar results were found in a similar trial (unpublished data) conducted with Pacific white shrimp (*Litopenaeus vannamei*). In shrimp, the highest growth and feed utilization occurred in animals fed diets containing the highest level (40%) of stickwater and were the same as the animals fed the control diet (Norwegian fish meal).

In this study, addition of stickwater to press cake adds nutritional quality to the fish meal produced. This finding has implications for the fish processing, fish meal, and fish feed manufacturing industries. There is a cost to fish meal manufacturers for recovering stickwater and this cost needs to be offset by a higher economic value in the final product. In addition, the aquatic feed industry is interested in finding the cheapest, highest quality, and most reliable source of fish meal for inclusion in their products. Because of the higher nutritional quality of feeds manufactured with fish meals containing higher stickwater inclusion levels, these fish meals should command a higher price. This work has shown the relative importance of stickwater in improving the value of fish meal.

Acknowledgments

The authors appreciate the technical assistance of the staff members of the Aquatic Feeds and Nutrition program at the Oceanic Institute. This paper was prepared as part of the activities of the Alaska Fisheries ByProduct Utilization Program awarded to the University of Alaska Fairbanks by the U.S. Department of Agriculture, Agricultural Research Service, under agreement no. 59-5320-7-989. Mention of trade names or commercial products in this article is solely for providing specific information and does not imply recommendation or endorsement by the authors or their institutions.

Fish Oils: Properties and Processing

Subramaniam Sathivel
University of Alaska Fairbanks, Fishery Industrial Technology Center,
Kodiak, Alaska

Abstract

Natural fish oils contain omega-3 fatty acids which when part of a balanced diet are thought to have a number of positive effects on human health such as reducing heart diseases, increasing cardiovascular functions, and possible influences on brain growth during early infancy. Alaska fish processors annually produce only about 30,000 metric tons of unrefined fish oil from fish byproducts. Most of this oil is used for fuel and animal feed ingredients. Fish oil production in Alaska usually involves grinding, cooking of fish byproducts, decanting liquids from cooked solids, and separating the fish oil by centrifugation. Some smaller salmon processors have utilized a batch skimming operation for recovering oil from salmon heads. Unrefined fish oils contain non-triglycerides, such as free fatty acids and oxidized components that may reduce quality. These components need to be removed before use in the more lucrative markets. The longer these components remain in the oil, the greater their negative effect on final oil quality. Making high quality oils requires well-designed purification steps. Conventional fish oil refining is achieved through the following steps: degumming, neutralization, bleaching, and deodorizing. Adsorption technology is an alternative approach for effective removal of non-triglycerides and is gaining popularity as a cost effective non-thermal separation process.

Introduction

The world total production of fats and oils is in the range of 99 million metric tons per year; 20.5% is from soybeans, and marine oils from whole fish, fish liver, and other byproducts account for 1.88% (USDA 2000). Over one million metric tons of fish oils are used annually in foods or preparing foods (USDA 2000). Fish oils are usually priced competitively with other fats and oils available on the world market and have health benefits and a wide range of functional characteristics. One use of the partially hydrogenated fish oil is in the baking industry.

When the fish oil is winterized, a liquid fraction is obtained that can be blended with other soft oils such as soybean oil, and used as a salad oil. Fish oil can also be used for single-use shallow frying. Hydrogenated fish oil is used for deep-fat frying, either alone or as the major component of a blend with palm oil. Partially hydrogenated fish oil can be used for margarine and shortening production (FAO 1986) and in the production of food emulsifiers. Fish oil is used in a variety of products and processes, including soaps, fatty acids, fatty chemicals, leather tanning, protective coatings, lubricants and greases, pneumatic tool lubricants, rubber compounds, glazing compounds, gasket manufacture, core oils, tin-plating oils, rust proofing agents, refractory compounds, cutting oils, plasticizers, printing inks, linoleum, press-wood fiber boards, oiled fabrics, ceramics, ore flotation, fermentation substrates, illuminating oils, mushroom culture, fire retardant, polyurethane foams, and animal feeds. There is clearly a high demand for fish oil in the United States and other parts of the world because it can be used not only for food product development but also in other industries.

History of fish oil developments in the United States

In early 1640, the first marine-origin oil, whale oil, was produced in the United States. Whale oil is not a fish oil but a marine mammal oil. In 1811, menhaden oil was manufactured in the United States in Rhode Island as the first American fish oil. According to the literature, the menhaden fish were boiled in pots or kettles, then the fish were transferred to casks, and the oil and water pressed out by the pressing stone. Extracted oil was skimmed off the top and placed in barrels (Stansby 1978).

In 1860, a modern plant was established in Rhode Island, composed of steam cookers and mechanical screw presses. During the period 1873-1911, the U.S. production of fish oil remained stable at about 2 million gallons per year which was produced from 110,000 to 115,000 t of fish. In 1912, the production was increased to 6.6 million gallons from 323, 000 t of fish and remained at that level until mid-1940s (Fitzgibbon 1969). In 1940, the fish oil industry began to use centrifugation to separate oil from water and the recovered fish oil was used as an oil material in soap, paints, and linoleum. Due to the introduction of centrifugation, there was a rapid increase in production of menhaden oil in the United States, referred to as the "industrial revolution" of the menhaden Industry in the United States (Smith 1940). In 1953, more than 17 million gallons of oil was produced, and production peaked in 1956 with 22.5 million gallons of oil (Bimbo 1990). During the 1950s, researchers in the United

States began to study oils from other fish species, such as anchovy. It was reported that 931,000 gallons of anchovy and 920,000 gallons of tuna oil were manufactured in California in 1975. In 1976 the production of all American fish oil was 204 million pounds with the value of about 40 million dollars. Of this total production, 186 million pounds or nearly 90% was menhaden oil. Of the oil from other species, most of it was anchovy and tuna oils in about equal quantities (Stansby 1978). The U.S. National Marine Fisheries Service has reported that the total marine oil production in the United States was 129,000 t in 1985 and was composed of 98% (126,000 t) menhaden oil and 3,000 t of tuna, mackerel, anchovy, and other fish oils (NMFS 1985).

Effects of fish oil on diseases

Omega-3 fatty acids play a major role in human health (Kronhout et al. 1985). Natural fish oils have been claimed to help maintain heart and vascular health in humans (Haglund et al. 1998). Some reported beneficial effects of omega-3 fatty acids include heart diseases (O'Keefe and Harris 2000), cardiovascular functions (Simopoulos 1997), possible influences on brain growth during early infancy (Xiang et al. 2000), arthritis (Kremer and Jubiz 1987), glomerular nephritis (Thais and Stahl 1987, Robinson et al. 1987), lupus erythematosis (Accinni and Dixon 1979, Kelley et al. 1985), multiple sclerosis (Bates et al. 1989), strokes (Hirai et al. 1987, Budowski 1988), breast cancer (Karmeli 1987, Carter et al. 1987, Cohen 1987, Stampfer et al. 1987, Cave and Jurkowski 1987), colon cancer (Reddy 1987), certain skin disease (Kromann and Green 1980, Rhodes 1984), and blood pressure within the brain and retina (Neuringer et al. 1988).

Commercial processing of fish oil

Commercial processing of fish oil involves many steps (Bimbo 1990). After the fish are caught and transported to the factory, they are cooked with steam to denature the protein and to release bound water and fish oil (Bimbo 1989). The fish solids and liquid (fish oil and water) are separated by pressing. The pressed fish solids are called "press cake"; the liquid is called "press liquor." The press liquor contains particles of fish; therefore, the liquid is centrifuged and the particles removed and returned to the press cake (Bimbo 1989, 1990). The press liquor is heated and centrifuged to separate the fish oil from the stickwater (liquid and small suspended fish particles). At this stage, the oil is crude commercial fish oil. However, before the fish oil can be used in foods, it must undergo purification to produce a more pure and stable product. The steps used in further oil processing include degumming, neutralization, washing, drying, bleaching, filtration, deodorization, and stabilization. Degumming and neutralization are often done together and involve addition of sodium hydroxide solution to the heated oil. Free fatty acids and other components such as the phospholipids, trace metals, pigments, and oil-insoluble materials are removed (Young 1985a, 1985b; Bimbo 1987).

Rendering

Rendering is the extraction of fat or oil mainly from animal tissues by heat. Almost all the animal fats are recovered by rendering, whereas vegetable fats are obtained by crushing, solvent extraction, or both. In general, rendering can be wet or dry rendering. Wet rendering is carried out in the presence of large amounts of water. The fat cell walls are hydrolyzed by steam under pressure until they are partially liquefied and the released fat floats onto the surface of the water. The separated fat was traditionally removed by skimming, but now centrifugal methods are used.

In dry rendering, the tissues are dehydrated until brittle, and the fat cells break and release the fat. Both methods need a hydraulic or continuous screw press to complete the recovery. Generally, wet rendering is used for edible products where color, flavor, and keeping qualities are of prime importance. Dry rendering is preferred for inedible products where flavor and odor are secondary to production characteristics (Dormitzer 1956, Downing 1959).

In general, the wet rendering method is used to produce fish oil. The apparatus used in wet rendering is a vertical cylindrical steel autoclave or digester with a cone bottom designed for steam pressure of 40-60 psi and a corresponding high temperature. The vessel is filled with fatty material and a small amount of water. The usual digestion time is 4-6 hours. Under the high temperature employed, the fatty materials are cooked and the fat and oils released. There is a very efficient separation of the fat, which rises to the top of the vessel, leaving layers of solids and stickwater in the bottom. The fat is drawn off and purified from water and solid material by settling or by centrifuge (Norris 1982).

The optimum extraction was believed to take place at 100°C at atmospheric pressure. However, recent experiments have shown that the fat cells are ruptured before the temperature reaches 50°C. Therefore, theoretically it should be possible to separate liberated oil from solid materials at 50°C if separation of fat and oils from other components such as protein was complete. Furthermore, the coagulation of the fish protein was completed at about 75°C and the process is vary rapid. The investigation led to the conclusion that there was little to be gained by heating the material beyond 75°C or by using long heating times (FAO 1986).

The advantage of wet rendering is that an efficient recovery of fat is obtained with relatively simple equipment and that it is adaptable to a wide range of raw materials. In addition, there is little tendency for proteins and other substances to dissolve or disperse in the fat in the presence of water. On the other hand, wet rendering is slower and less efficient than dry rendering. Furthermore, some hydrolysis of fat occurs during steam rendering and the free fatty acids produced depend on the rendering time and temperature, storage temperature, and storage time of the fatty stock before it is processed (Norris 1982). Sathivel et al. (2003) reported a modified wet extraction method for fish oil. The extraction method was similar to wet rendering but the raw material was finely ground and heated with water at 70°C for 30 minutes. The oil was then separated from the protein and waste by centrifugation.

Pressing

The purpose of pressing is to squeeze out as much liquid as possible from the solid phase after the initial heating process. Pressing improves the oil yield. Both continuous single-and twin-screw presses are used in the fish meal industry today.

The screw in the single-screw press is designed with a taper and exerts an increasing pressure on the fish pulp by reducing the volume as it progresses through the press.

Crude oil

Oils obtained directly from rendering contain varying but relatively small amounts of naturally occurring non-glyceride materials that are removed through a series of processing steps.

For example, crude oils contain some free fatty acids, water, and protein that must be removed. Not all of the non-glyceride materials are undesirable elements. Tocopherols, for example, perform the important function of protecting the oils from oxidation and provide vitamin E. Processing is carried out in such a way as to control retention of these desirable substances. It is therefore necessary to determine certain key analytical values (Table 1) in order to modify processing conditions so as to obtain a product of satisfactory quality (Young 1985a).

Purification of fish oil

The general objective of processing fats and oils is the removal of impurities that cause the original product to have an unattractive color or taste or that cause harmful metabolic effect. Crude fats and oils intended for edible purposes are further processed to remove these substances while retaining desirable features. Recent emphasis is to preserve desirable natural components such as omega-3 fatty acids and tocopherols. Therefore, the process should be designed to preserve such components while reducing off-flavor, cholesterol, and other impurities (Takao 1986).

Degumming

Degumming is a treatment designed to remove the impurities such as phospholipids, free fatty acids, and trace metals with the least possible damage to the natural oil. Impurities consist of complex molecules and

Table 1. Crude fish oil quality assessment test and significance.

Test	Significance
Moisture (%)	For contractual and yield purposes
Insoluble impurities (%)	For contractual and yield purposes
Free fatty acids (%)	Contractual, yield, general quality purposes
Peroxide value (meq active oxygen/kg)	A measure of primary oxidation
Anisidine value	A measure of secondary oxidation
Iodine value	For fish oil type identification and as guideline for hydrogenation
Color	Quality indicator
Iron (μg/g)	Autoxidation catalyst
Copper (μg/g)	Autoxidation catalyst
Phosphorus (μg/g)	Refining treatment/catalyst poison
Sulfur (μg/g)	Refining treatment/catalyst poison
Soap (μg/g)	Presence of soap indicates adulteration with alkali-refined oil
Laboratory refining, laboratory hydrogenation, and laboratory bleachability	Quality assurance tests

Source: Young 1985b

are generally classified as "gum." The presence of gum in the oil affects the final oil color, flavor, foaming, and smoking stability, and are removed. Occasionally, the gum settles out in a storage tank and will cause high refining losses. The gum can be removed from oil with water or acids such as phosphoric acid or citric acid (Carr 1976, Cowan 1976, Norris 1982).

Degumming is not ordinarily carried out in fish oil processing, because fish oils are very low in phosphatides. In some refineries, however, an acid pretreatment, designed to hydrate gums and remove phosphorus and other trace metals, is applied to oil as it enters the alkali refining plant. The pretreatment of fish oils with phosphoric acid prior to caustic refining is a standard practice in Europe (Brekke 1980). Degumming has been shown to effectively reduce the mineral content of fish oil (Sathivel et al. 2002b) and lead, copper, arsenic, and zinc in menhaden oil were reduced as a result of the degumming process (Elson and Ackman 1978, Elson et al. 1981).

Neutralization

The term "refining" refers to any purifying treatments designed to remove non-glyceride impurities such as free fatty acids in the oil. Crude oils produced by rendering contain variable amounts of non-glyceride. Not all non-glyceride in crude oils are undesirable. However, most of the other impurities are objectionable and cause oil to foam, smoke, or precipitate in subsequent processing operations. The objective of refining is to remove the undesirable impurities from the oil with the least possible damage to glycerides, tocopherols, and other desirable impurities, and minimize the loss of oil. In general, the refining process can be done in two ways: alkali refining (neutralization) and physical refining.

Bleaching

Bleaching is designed to improve color, flavor, and oxidation stability of the oil by removal of compounds responsible for oil color and off-flavor. Many compounds in crude oil responsible for the color are broken down at high temperatures and the volatile products are removed under deodorization conditions. The bleaching step is also important to remove soap, trace metals, and sulfur compounds (Bimbo 1990). During bleaching, peroxides are broken down to aldehydes and ketones and these secondary oxidation products are adsorbed onto the "activated earth" surface such that the filtered oil after bleaching should have a low peroxide value (PV) compared with the oil before bleaching. The two types of commercial bleaching clays used in processing edible oils may be characterized as "natural bleaching earth" and "activated bleaching earth" (Richardson 1978).

Deodorization

Deodorization is the last major processing step in the refining of edible oils. Crude fish oils undergo rapid deterioration due to current harvesting and processing practices, and high concentrations of polyunsaturated fatty acids and other contaminants. Severe deterioration changes the sensory quality of fish oil. Off flavors in fish oil arise from metabolite contaminants, from fish oil protein spoilage, and from oxidation of the fish oil (Stansby 1971, 1973). For example, researchers have studied the volatile components of crude winterized menhaden oil by dynamic head space gas analysis and they found that many odor components are derived from lipid oxidation, including short chain saturated and unsaturated aldehydes, ketones, and carboxylic acids.

Undesirable compounds produced during handling and storing may affect the sensory quality of fish oil. Therefore, undesirable odors and volatile components are removed during refining and deodorization of food grade oils. Deodorization has been considered as a unit process that establishes the oil flavor and odor characteristics that are most readily recognized by the consumer (Zehnder 1976; Gavin 1977, 1978).

Fish oil production in Alaska

The Alaska fish processors annually produce over 30,000 t of unrefined fish oil from about 2,000,000 t of fish processed in Alaska. One good source of fish oil is the byproducts of salmon processing during the summer months. Unrefined fish oil is currently used as boiler fuel in the fish meal plants to reduce operating costs and is also sold to industrial users for use in animal feeds. Currently the value of the unrefined oils ranges from $0.75 to $1.50 per gallon. There are potential uses of this fish oil as a food additive or in the nutraceutical or cosmetics industry with a value as high as $7.50 or more per gallon (NMFS 2002). Salmon oils have significant amount of omega-3 fatty acids, which play an important role in human health.

The current fish oil production in Alaska involves grinding, cooking of fish processing byproducts, decanting liquids from cooked solids, and separating the fish oil by centrifugation. These unrefined oils contain non-triglycerides, such as free fatty acids and oxidized components that may reduce quality. These non-triglycerides need to be removed if the oils are to enter more lucrative markets. The longer these components remain in the oil, the greater their negative effect on final oil quality.

Finding more lucrative markets for this fish oil requires high quality products made using well-designed purification steps. Conventional fish oil refining is achieved through degumming, neutralization, bleaching, and deodorizing. The main disadvantages of conventional methods are high refining losses, oil oxidation, and high energy and/or processing costs. Studies have shown that an adsorption process used for edible oil purification removes non-triglycerides and is a cost effective process (Taylor and Ungermann 1984; Proctor and Palaniappan 1990; Palaniappan and Proctor 1990,1991; Toro-Vazquez and Rocha-Uribe 1993; Clark and Proctor 1993; Proctor and Toro-Vazquez 1996; Sathivel 2001). Use of chitosan and activated carbon as the adsorbents for adsorbing free fatty acids and oxidized components were investigated (Sathivel et al. 2002). The same researchers designed a fixed bed adsorption column and studied the selective adsorption kinetics of non-triglycerides from catfish oil (Sathivel et al. 2002, Sathivel and Prinyawiwatkul 2002). Compared to conventional oil purification, the main advantages of adapting adsorption technology for fish oil purification are lower refining losses, enhanced winterizing, less lipid oxidation, and less flavor reversion in the refined oil. The advantages are due to both low temperature and adsorption columns. Adsorption technology can potentially simplify current refining processing of the fish oil in Alaska.

References

Accinni, C., and F.J. Dixon. 1979. Degenerative vascular disease and myocardial infarction in mice with lupus-like syndrome. Am. J. Pathol. 96:477-492.

Bates, D., N. Carrtlidge, J.M. French, M.J. Jackson, S. Nightingale, D.A. Shaw, S. Smith, E. Woo, S.A. Hawkins, J.H.D. Millar, J. Belin, D.M. Conory, S.K. Gill, M. Sidey, A.D. Smith, R.H.H.S. Thompson, K. Zilka, M. Gale, and H.M. Sinclair. 1989. A double blind controlled trial of long chain n-3 polyunsaturated fatty acids in the treatment of multiple sclerosis. J. Neurol. Neurosurg. Psychiatry 52:18-22.

Bimbo, A.P. 1987. The emerging marine oil industry. J. Am. Oil Chem. Soc. 69: 706-715.

Bimbo, A.P. 1989. Recent advances in upgrading industrial fish to value added products. In: L.A. Johnson (ed.), New technologies for value added products from protein and co-products. American Oil Chemists Society, Champaign, Illinois.

Bimbo, A.P. 1990. Production of fish oil. In: M.E. Stansby (ed.), Fish oils in nutrition. Van Nostrand Reinhold, New York.

Brekke, O.L. 1980. Oil degumming and soybean lecithin. In: D.R. Erickson, E.H. Pryde, O.L. Brekke, T.L. Mounts, and R.A. Falb (eds.), Hand book of soy oil processing and utilization. American Soybean Association and American Oil Chemists Society, Champaign, Illinois.

Budowski, P. 1988. Omega-3 fatty acids in health and disease. World Rev. Nutr. Diet. 57:214-274.

Carr, R.A. 1976. Degumming and refining practices in the U.S. J. Am. Oil Chem. Soc. 53:347-352.

Carter, C.A., M.M. Ip, and C. Ip. 1987. Response of mammary carcinogenesis to dietary linoleate and fat levels and its modulation by prostaglandin synthesis inhibitors. In: W.E.M. Lands (ed.), Proceedings of the AOAC Short Course on Polyunsaturated Fatty Acids and Eicosanoids. American Oil Chemists Society, Champaign, Illinois, pp. 253-260.

Cave Jr., W.T., and J.J. Jurkowski. 1987. Comparative effects of omega-3 and omega-6 dietary lipids on rat mammary tumor development. In: W.E.M. Lands (ed.), Proceedings of the AOAC Short Course on Polyunsaturated Fatty Acids and Eicosanoids. American Oil Chemists Society, Champaign, Illinois, pp. 261-266.

Clark, P.K., and A. Proctor. 1993. Effects of lutein adsorption of adding polar solvents to silicic acid or to soy oil/hexane miscellas. J. Am. Oil Chem. Soc. 70: 1003-1007.

Cohen, L.A. 1987. Differing effects of high-fat diets rich in polyunsaturated, monounsaturated or medium chain saturated fatty acids on rat mammary tumor promotion. In: W.E.M. Lands (ed.), Proceedings of the AOAC Short Course on Polyunsaturated Fatty Acids and Eicosanoids. American Oil Chemists Society, Champaign, Illinois, pp. 241-247.

Cowan, J.C. 1976. Degumming, refining, bleaching, and deodorization theory. J. Am. Oil Chem. Soc. 53:344-346.

Dormitzer, H.C. 1956. Rendering. J. Am. Oil Chem. Soc. 33:471-473.

Downing, F.P. 1959. The production of meat and fat products through centrifugal rendering. J. Am. Oil Chem. Soc. 36:319-321.

Elson, C.M., and R.G. Ackman: 1978. Trace metal content of a herring oil at various stages of pilot-plant refining and partial hydrogenation. J. Am. Oil Chem. Soc. 55:616-618.

Elson, C.M., E.M. Bem, and R.G. Ackman. 1981. Determination of heavy metals in a menhaden oil after refining and hydrogenation using several analytical methods. J. Am. Oil Chem. Soc. 58:1024-1026.

FAO. 1986. The production of fish meal and oil. FAO Fish. Tech. Pap. 142, Revision 1.

Fitzgibbon, D.S. 1969. Historical statistics: Fish meal, oil and solubles. U.S. Fish and Wildlife Service, Bureau of Commercial Fisheries, Current Fisheries Statistics no. 5105. 30 pp.

Gavin, A.M. 1977. Edible oil deodorizing systems. J. Am. Oil Chem. Soc. 54:528-532.

Gavin, A.M. 1978. Edible oil deodorization. J. Am. Oil Chem. Soc. 55:783-791.

Haglund, O., R. Wallin, S. Wretling, B. Hultberg, and T. Saldeen. 1998. Effects of fish oil alone and combined with long chain (n-6) fatty acids on some coronary risk factors in male subjects. J. Nutr. Biochem. 9:629-635.

Hirai, A., J. Terano, H. Saito, Y. Tamura, and S. Yoshida. 1987. Clinical and epidemiological studies of eicosapentaenoic acid in Japan. In: W.E.M. Lands (ed.), Proceedings of the AOAC Short Course on Polyunsaturated Fatty Acids and Eicosanoids. American Oil Chemists Society, Champaign, Illinois, pp. 9-24.

Karmeli, R.A. 1987. Omega-3 fatty acids and cancer: A review. In: W.E.M. Lands (ed.), Proceedings of the AOAC Short Course on polyunsaturated Fatty Acids and Eicosanoids. American Oil Chemists Society, Champaign, Illinois, pp. 222-231.

Kelley, V.E., A. Ferritti, S. Izni, and T.B. Strom. 1985. A fish oil rich in eicosapentaenoic acid reduces cyclooxygenase metabolites, and suppresses lupus in MRL-lpr mice. J. Immunol. 134:1914-1919.

Kremer, J.M., and W. Jubiz. 1987. Fish oil supplementation in active rheumatized arthritis: A double-blinded, controlled crossover study. In: W.E.M. Lands (ed.), Proceedings of the AOAC Short Course on polyunsaturated Fatty Acids and Eicosanoids. American Oil Chemists Society, Champaign, Illinois.

Kromann, N., and A. Green. 1980. Epidemiological studies in the Upernavik district, Greenland. Acta Med. Scand. 208:401-406.

Kronhout, D., E.B. Bosschieter, and C. Coulander. 1985. The inverse relation between fish consumption and 20-year mortality from coronary heart disease. N. Engl. J. Med. 312:1205-1209.

Neuringer, M., G.J. Anderson, and W.E. Connor. 1988. The essentiality of n-3 fatty acids for the development and function of the retina and brain. Ann. Rev. Nutr. 1988:517-541.

NMFS. 1985. Fishery market news, weekly fish meal and oil prices. NOAA, NMFS, Silver Spring, Maryland.

NMFS. 1985. Current Fishery Statistics no. 8380. NOAA, NMFS, Silver Spring, Maryland, pp. 75-76.

NMFS. 2002. Fishery market news, weekly fish meal and oil prices. http://www.fakr.noaa.gov/sustainablefisheries/catchstats.htm. NOAA, NMFS, Silver Spring, Maryland.

Norris, F.A. 1982. Refining and bleaching. In: D. Swern (ed.), Bailey's industrial oil and fat products, vol. 2. 4th edn. John Wiley & Sons, New York, pp. 253-314.

O'Keefe, J.H., and W.S. Harris. 2000. Omega-3 fatty acids: Time for clinical implementation? Am. J. Cardiol. 85:1239-1241.

Palaniappan, S., and A. Proctor. 1990. Competitive adsorption of lutein from soy oil onto rice hull ash. J. Am. Oil Chem. Soc. 67:572-575.

Palaniappan, S., and A. Proctor. 1991. Evaluation of soy oil lutein isotherm obtained with selected adsorbents in hexane miscellas. J. Am. Oil Chem. Soc. 68:79-82.

Proctor, A., and S. Palaniappan. 1990. Adsorption of soy oil free fatty acids by rice hull ash. J. Am. Oil Chem. Soc. 67:15-17.

Proctor, A., and J.F. Toro-Vazquez. 1996. The Freundlich isotherm in studying adsorption in oil processing. J. Am. Oil Chem. Soc. 73:1627-1633.

Reddy, B.S. 1987. Dietary fat and colon cancer: Effect of fish oil. In: W.E.M. Lands (ed.), Proceedings of the AOAC Short Course on Polyunsaturated Fatty Acids and Eicosanoids, American Oil Chemists Society, Champaign, Illinois, pp. 233-237.

Rhodes, E.L. 1984. MAXEPA in the treatment of eczema. Br. J. Clin. Practice 38(Suppl. 31):115-116.

Richardson, L.L. 1978. Use of bleaching, clays in processing edible oils. J. Am. Oil Chem. Soc. 55:777-780.

Robinson, D.R., S. Tateno, B. Patel, and I. Hirai. 1987. The effect of dietary marine lipids on autoimmune disease. In: W.E.M. Lands (ed.), Proceedings of the AOAC Short Course on Polyunsaturated Fatty Acids and Eicosanoids. American Oil Chemists Society, Champaign, Illinois, pp. 139-147.

Sathivel, S. 2001. Production, process design and quality characterization of catfish oil. Ph.D. thesis, Louisiana State University, Baton Rouge.

Sathivel, S., and W. Prinyawiwatkul. 2002. A kinetic study: Selective adsorption of free fatty acids from crude catfish oil onto chitosan, activated carbon and/or activated earth. http://ift.confex.com/ift/2002/techprogram/session_1630.htm. Presented at the Institute of Food Technologists Annual Conference, Anaheim, California.

Sathivel, S., W. Prinyawiwatkul, and R.R. Bansode. 2002. A design for a fixed-bed adsorption column for removal of free fatty acids and oxidation compounds from crude visceral oil. http://ift.confex.com/ift/2002/techprogram/session_1626.htm. Presented at the Institute of Food Technologists Annual Conference, Anaheim, California.

Sathivel, S., W. Prinyawiwatkul, J.M. King, C.C. Grimm, and S. Lloyd. 2003. Oil production from catfish viscera. J. Am. Oil Chem. Soc. 80:377-382.

Simopoulos, A.P. 1997. Omega-3 fatty acids in the prevention management of cardiovascular disease. Can. J. Physiol. Pharmac. 75:234-239.

Smith, J.H. 1940. Advances in menhaden reduction. Chemical and Metallurgical Engineering 47:9-10.

Stampfer, M.J., W.C. Willett, G.A. Colditz, and M.D. Spezer. 1987. Intake of cholesterol, fish and specific types of fat in relation to breast cancer. In: W.E.M. Lands (ed.), Proceedings of the AOAC Short Course on Polyunsaturated Fatty Acids and Eicosanoids, American Oil Chemists Society, Champaign, Illinois, pp. 248-252.

Stansby, M.E. 1971. Flavors and odors of fish oils. J. Am. Oil Chem. Soc. 48:820-823.

Stansby, M.E. 1973. Problems discourage use of fish oil in American-manufactured shortening and margarine. J. Am. Oil Chem. Soc. 50:220A-225A.

Stansby, M.E. 1978. Development of fish oil industry in the United States. J. Am. Oil Chem. Soc. 55:238-243.

Takao, M. 1986. Refined fish oils and the process for production thereof. United States Patent no. 4,623,488.

Taylor, D.R., and C.B. Ungermann. 1984. The adsorption of fatty acids from vegetable oils with zeolites and bleaching clay/zeolite blends. J. Am. Oil Chem. Soc. 61:1372-1379.

Thais, F., and R.A.K. Stahl. 1987. Effect of dietary fish oil on renal function in immune mediated glomerular injury. In: W.E.M. Lands (ed.), Proceedings of the AOAC Short Course on Polyunsaturated Fatty Acids and Eicosanoids. American Oil Chemists Society, Champaign, Illinois, pp. 123-126.

Toro Vazquez, J.F., and A. Rocha-Uribe. 1993. Adsorption isotherms of sesame oil in a concentrated miscella system. J. Am. Oil Chem. Soc. 70:589-594.

USDA. 2000. Oilseeds and products. U.S. Department of Agriculture, Foreign Agriculture Service.

Xiang, M., G. Alfven, M. Blennow, M. Trygg, and R. Zetterstrom. 2000. Long-chain polyunsaturated fatty acids in human milk and brain growth during early infancy. Acta Paediatr. 89:142-147.

Young, F.V.K. 1985a. Interchangeability of fats and oils. J. Am. Oil Chem. Soc. 62:372-376.

Young, F.V.K. 1985b. The refining and hydrogenation of fish oil. Fish Oil Bulletin no. 17. International Association of Fish Meal Manufacturers, Herefordshire, U.K.

Zehnder, C.T. 1975. Deodorization 1975. J. Am. Oil Chem. Soc. 53:364-369.

Demonstrating the Use of Fish Oil as Fuel in a Large Stationary Diesel Engine

John A. Steigers
Steigers Corporation, Littleton, Colorado

Abstract

Approximately 8 million gallons of fish oil are produced annually as a byproduct of Alaskan seafood processing operations. Typically, about 2.8 million gallons of the total production volume is sold into domestic and international commodity markets, with the balance consumed on-site as boiler fuel. Inconsistent markets and difficult storage and transportation logistics often reduce the net value of the marketed oil to well below the cost of diesel fuel. The seafood processors and their associated communities generally are heavily dependent on diesel-fueled reciprocating engines for electric energy generation. The UniSea Fish Oil Demonstration Project is demonstrating the feasibility of using blends of fish oil and low-sulfur no. 2 diesel fuel in 2.3-megawatt, stationary medium-speed, two-cycle, engine-generator sets. The project entails assessments of the blended fuels' impacts on both engine exhaust emissions and engine operability and maintainability. Engine exhaust emissions resulting from the use of fuel blends ranging from 0 to 100% fish oil were measured at multiple engine loads. Results indicate up to 60% reduction in particulate matter, 33% reduction in carbon monoxide, and 78% reduction in sulfur dioxide emissions can be obtained using fish oil blends. These benefits are somewhat offset by an increase of up to 8% in nitrogen oxide emissions. Over a 10-month test period, the engines have operated normally in all respects utilizing a 50% fish oil fuel blend, to date consuming over 526,000 gallons of fish oil with no apparent adverse operational or maintenance impacts. Test program operations are anticipated to continue through October 2002.

Introduction

The goal of the UniSea Fish Oil Demonstration Project is to definitively evaluate the use of Alaska-produced fish oil as a practical supplemental fuel for a specific diesel-fueled engine-generator set in use for industrial energy production in rural Alaska. If the fish oil is found to be suitable as a supplemental fuel, it is anticipated that the practice may be adopted elsewhere in Alaska, which could result in substantial economic and environmental benefits both for the fish oil producers and for fuel consumers within the state.

UniSea, Inc., based in Redmond, Washington, owns and operates a large shore-based seafood processing facility located on Amaknak Island within the Unalaska/Dutch Harbor community in the Fox Island group of Alaska's Aleutian chain. Bering Sea pollock, one of several species processed by UniSea, yields a number of commercial products such as frozen fillets and surimi. The resulting processing wastes, e.g., the fish heads, skin, bones, and entrails, are directed to further on-site processing facilities to produce fish meal, bone meal, and fish oil. Typically, Alaskan shore-based processors recover 3 to 5% of the raw landed weight of pollock as fish oil. Statewide, Alaskan seafood processors, both shore-based and afloat, produce approximately 8 million gallons of fish oil annually (Steigers Corporation 2002). Of the approximately 2.8 million gallons of fish oil shipped from Alaska to external markets, 2.7 million gallons are produced by UniSea and three other major shore-based seafood processors in and near Unalaska/Dutch Harbor and the nearby island community of Akutan.

Well over half of the fish oil produced in the state is consumed at the producing facilities as a boiler fuel, with the balance sold to customers outside Alaska, primarily for use in animal feed and aquaculture but also as a human dietary supplement and for the manufacture of cosmetics and pharmaceuticals. However, due to Unalaska/Dutch Harbor's remote location, limited transportation options, and the sometimes-difficult logistical issues inherent in shipping fish oil, the economic challenges of marketing fish oil at times preclude even a break-even disposition of this commodity. In fact, the net market value of fish oil very often falls below that of diesel fuel. Accordingly, there is a desire on the part of UniSea and others to develop economically viable alternative uses for the fish oil. Use of fish oil as a locally consumed engine-generator fuel is a logical potential use that would likely benefit UniSea, both as a producer and as a consumer, as well as other fish oil producers and operators of diesel-fueled electrical generating units in the region.

The project's goal is to develop specific knowledge as to whether fish oil is an acceptable supplemental fuel, both with respect to the amount and type of air emissions generated and with respect to acceptable "durability" impacts on the engine. The project consisted primarily of a field demonstration conducted in two phases: engine emissions source testing while utilizing blended fuels and engine durability testing. Independent source-testing contractors conducted testing of engine emissions in October 2001 and July 2002. Multiple fuel blends, ranging from 100% low-sulfur diesel to 100% fish oil, were utilized in the test engine-generators at two or three operating loads. In addition to oxygen (O_2) and carbon dioxide (CO_2), the air pollutants nitrogen oxides (NO_x), carbon monoxide (CO), and particulate matter (PM) were evaluated. Sulfur dioxide (SO_2) is also a pollutant of interest, but those emissions were determined by mass balance calculations based on fuel sulfur content and fuel consumption rates and were not measured directly during source testing. The second portion of the project, durability testing, was intended to assess any impacts on engine operability and maintainability from long-term routine operation of the test engines utilizing a 50% diesel/fish oil fuel blend.

Fish oil characterization

The characteristics of fish oil are known to vary somewhat with the species of origin, the season, and the processor. The fish oil produced by UniSea is typical of that found in Alaska and is an amber to light orange–colored oil with a density of about 7.7 pounds per gallon, compared to no. 2 diesel fuel at about 7.1 pounds per gallon. A comparison of UniSea fish oil and no. 2 diesel conducted by Mr. Neil X. Blythe of the Fairbanks Morse Engine Division of BF Goodrich Company found that fish oil exhibits a substantially higher viscosity, is slightly more acidic, has a lower lubricity, and a higher flash point (Blythe 1996). Fish oil was reported to have a sulfur content of 0.004% by weight and a gross heat of combustion of 131,756 Btu per gallon. This may be compared to low-sulfur no. 2 diesel's 0.05% sulfur content and 137,000 Btu per gallon gross heat of combustion. Blythe's characterization of fish oil stated that it may be classed as a lipid, specifically a glyceride ester, and suggested that excessive hard deposits on exhaust gas path components and accelerated wear on components in contact with fuel may be a concern with sustained engine operation on fish oil blend fuels. Blythe's investigations, however, were not able to proceed for a sufficient period to be conclusive in this regard.

A recent analysis of a fish oil sample drawn from UniSea's July 2002 production run yielded a sulfur content of 0.0084% and a gross heat of combustion of 130,440 Btu per gallon (Intertek Testing Services 2002). The flash point of the fish oil sample was determined to be in excess of 230°F. This recent analysis is considered most representative of the fish oil utilized in this project.

Project description/methods

Test methods

UniSea owns and operates six Fairbanks Morse model 38TD8-1/8OP engines in its Dutch Harbor facility powerhouse. The UniSea engines were manufactured by the Fairbanks Morse Engine Division of BF Goodrich Company (Fairbanks Morse), shown in Fig. 1. The engines are 12-cylinder/opposed-piston, two-cycle, water-cooled, series turbo and blower air charging (no bypass) and are equipped with Woodward type EGB mechanical speed governors. The engines' primary fuel is low-sulfur (0.05% or less by weight) no. 2 diesel fuel. As directed by conditions of UniSea's air quality operating permit, all six engines' average fuel injection timing is set at no less than 41° high cam after inner dead center lower crank (H.C.A.I.D.C.L.C.). Three of the six engines operate at 720 rpm and are rated and permitted at 3,160 hp and 2,152 kW maximum. The remaining three engines operate at 900 rpm and, while rated at 3,960 hp and 2,826 kW, are permitted for and operationally limited to 3,223 hp and 2,300 kW. Note that, for consistency in the following discussions, the engines' respective permitted capacity, rather than rated capacity, is regarded as "100% load."

Fairbanks Morse engine-generator no. 6 (FM #6), a 720-rpm engine, was selected as the initial primary test unit for the October 2001 source testing and the October 2001 through October 2002 durability testing. The remaining five units were brought into the durability-testing program in June 2002 and, along with FM #6, will continue to operate on fish oil blend fuel through the end of the project testing in October 2002. FM #3 and FM #4, 900-rpm and 720-rpm, respectively, were source tested in July 2002.

Source testing

In October 2001, FM #6 was source tested at the following conditions: fuel blends (expressed in diesel to fish oil percent by volume) of 100/0, 75/25, 50/50, and 25/75 were each tested at engine load conditions of 100 and 67($\pm$5)%, with two 60-minute test runs at each condition. While useful information was obtained that is consistent with other tests, difficult test conditions attributable in large part to very adverse weather contributed to unusually variable results for the October testing. Accordingly, the October 2001 emission source test results are not further discussed in this report.

To gather a more conclusive emissions and operating data set and to support UniSea's anticipated air quality permitting initiatives, FM #3 and FM #4 were source tested in July 2002 at the following conditions: fuel blends of 100/0, 50/50, and 0/100 were each tested at engine load conditions of 100, 77, and 65($\pm$5)%, with three 60-minute test runs at each condition. Emission measurements included O_2, CO_2, NO_x, and CO for all

Figure 1. UniSea Powerhouse, Fairbanks Morse engine-generators nos. 4 through 1 (l-r).

runs. PM measurements were taken for all runs at all tested conditions except for the 50/50 and 0/100 fuel blends at 77 and 65% loads.

Durability testing

UniSea's Dutch Harbor powerhouse typically operates one to three FM engine-generators, depending on the level of the facility's seafood processing; each typically loaded at approximately 77% of full load (two smaller engine-generators located elsewhere in the facility provide a load-following function). For the period October 2001 through June 2002, FM #6 was operated 4,376 hours, consuming 221,400 gallons of fish oil blended 50/50 with low-sulfur no. 2 diesel fuel. Prior to start of testing in October 2001 and again in April 2002, technical representatives of Fairbanks Morse performed detailed inspections of FM #6 to first establish its baseline condition and then to assess any effects on the engine from operating with fish oil blend fuel. A number of less formal inspections were also conducted by UniSea operating staff when suitable opportunities arose. A final inspection of FM #6 will occur in November 2002 after the test program has concluded.

Starting in July 2002, the scope of the durability-testing program was expanded, with FM #1 through FM #5 being operated on the 50/50 fuel blend as well. Total fish oil consumption among the six powerhouse engines is currently 4,500 to 5,000 gallons per 24-hour day, seven days per week. Test operations are anticipated to continue through October 2002 by which time a total of 550,000 to 650,000 gallons of fish oil will have been consumed by the powerhouse engines in the course of project testing.

Fuel logistics and operations

Test fuels were blended on a batch basis in a dedicated 5,000-gallon mixing tank. Diesel fuel and fish oil were metered in from their respective storage tanks simultaneously in the appropriate volume ratios. Upon filling, the blend tank was recirculated with both a closed loop pump and a centrifugal fuel purifier to assure uniform mixing of the blended fuel prior to its use. The fuel purifier incorporates a fuel heater to raise the fuel blend temperature to about 90°F to increase the effectiveness of its oil/water/sediment separation. As-used fuel temperatures were not closely monitored by the project, and, while generally above 85°F, fuel blend temperature did drop as low as an estimated 40°F on occasion.

Results

The emission and engine operating data obtained from the July 2002 source testing of FM #3 and FM #4 are summarized in Tables 1 and 2, respectively (Alaska Source Testing 2002).

Engine operations

Both the 720-rpm and 900-rpm engines have operated routinely in all respects without apparent difficulty on all tested fuel blends. As anticipated due to the lower thermal energy content of fish oil relative to diesel, fuel consumption increased at higher fish oil content but remained comfortably within the governor and fuel system range of control. Fuel consumption rates for FM #3 and FM #4 as load and fish oil fuel content were varied are illustrated in Figs. 2a and 2b, respectively.

Starting the engines from either a warm or cold condition using the fish oil blend fuels was accomplished without difficulty. In fact, engine operators made the observation that the engines seemed to start slightly easier with fish oil than with diesel. Engine shutdowns (and subsequent restarts) were also accomplished without difficulty while fish oil blend fuels were utilized.

While detailed long-term trending has not yet been completed, unusual engine operating conditions have not been observed in the course of operations on blended fuels. An increase in engine fuel rack position due to the lower thermal content of the blended fuels was observed, as

Table 1. UniSea FM #3 (900-rpm) test data.

Runs		10-12	13-15	16-19	46-48	49-51	42-54	37-39	40-42	43-45
Fuel blend	(Diesel/fish oil)	100/0	100/0	100/0	50/50	50/50	50/50	0/100	0/100	0/100
Engine load	(kW)	2,274	1,778	1,493	2,266	1,791	1,502	2,267	1,789	1,475
Engine load	(% full load)	98.9%	77.3%	64.9%	98.5%	77.9%	65.3%	98.6%	77.8%	64.1%
Fuel use	(gph)	196.7	165.8	147.9	235.0	196.3	176.1	257.8	222.8	197.1
Stack flow	(acfm)	30,301	26,682	24,224	29,387	26,641	24,173	29,002	25,642	22,886
Exit velocity	(fps)	160.8	141.6	128.5	155.9	141.3	128.2	153.9	136.0	121.4
Stack temp	(°F)	572.3	550.0	526.7	582.0	559.7	537.7	570.0	545.0	530.0
O_2	(%volume)	15.1	15.5	15.8	14.9	15.3	15.6	15.0	15.3	15.7
CO_2	(%volume)	4.4	4.1	3.9	4.6	4.3	4.1	4.6	4.4	4.1
NO_x	(g/bhp-hr)	10.2	9.5	8.9	10.4	9.8	9.4	10.8	9.9	9.5
CO	(g/bhp-hr)	0.363	0.404	0.445	0.346	0.412	0.475	0.328	0.399	0.514
PM	(g/bhp-hr)	0.181	0.256	0.255	0.149			0.132		
NO_x[a]	(lbs/hr)	71.8	51.9	40.8	73.0	54.4	43.6	75.7	54.8	43.3
CO[a]	(lbs/hr)	2.55	2.22	2.05	2.42	2.28	2.21	2.30	2.21	2.34
PM[a]	(lbs/hr)	1.27	1.41	1.17	1.04			0.92		
NO_x	(lbs/gal)	0.365	0.313	0.276	0.311	0.277	0.248	0.294	0.246	0.220
CO	(lbs/gal)	0.013	0.013	0.014	0.010	0.012	0.013	0.009	0.010	0.012
PM	(lbs/gal)	0.0065	0.0085	0.0079	0.0044			0.0036		

[a]Normalized to 100% load basis.

Data expressed as average values from three 60-minute test runs.

acfm = actual cubic feet per minute; fps = feet per second; g/bhp-hr = grams per brake horsepower-hour.

Table 2.   UniSea FM #4 (720-rpm) test data.

Run		1-3	4-6	7-9	19-21	22-24	25-27	28-30	31-33	34-36
Fuel blend	(Diesel/fish oil)	100/0	100/0	100/0	50/50	50/50	50/50	0/100	0/100	0/100
Engine load	(kW)	2,218	1,740	1,471	2,211	1,766	1,481	2,269	1,748	1,470
Engine load	(% full load)	96.4%	75.7%	64.0%	96.1%	76.8%	64.4%	98.7%	76.0%	63.9%
Fuel use	(gph)	175.8	145.3	126.7	204.6	174.2	151.8	221.8	189.2	167.2
Stack flow	(acfm)	23,117	19,393	17,356	22,476	19,061	16,992	21,946	18,965	15,866
Exit velocity	(fps)	122.6	102.9	92.1	119.2	101.1	90.1	116.4	100.6	84.2
Stack temp	(°F)	647.7	621.3	597.3	646.7	624.7	598.0	650.3	618.3	581.0
O_2	(%volume)	15.1	15.5	15.8	14.9	15.3	15.6	15.0	15.3	15.7
CO_2	(%volume)	4.4	4.1	3.9	4.6	4.3	4.1	4.6	4.4	4.1
NO_x	(g/bhp-hr)	11.4	10.2	9.4	11.7	11.0	10.0	11.6	11.4	9.9
CO	(g/bhp-hr)	0.813	0.985	0.956	0.647	0.815	0.804	0.587	0.750	0.640
PM	(g/bhp-hr)	0.215	0.233	0.240	0.130			0.085		
NO_x [a]	(lbs/hr)	76.2	53.8	41.6	78.3	59.0	44.9	79.8	60.5	44.1
CO[a]	(lbs/hr)	5.45	5.19	4.25	4.33	4.35	3.60	4.03	3.96	2.85
PM[a]	(lbs/hr)	1.44	1.23	1.07	0.87			0.58		
NO_x	(lbs/gal)	0.434	0.370	0.329	0.383	0.339	0.296	0.360	0.320	0.264
CO	(lbs/gal)	0.031	0.036	0.034	0.021	0.025	0.024	0.018	0.021	0.017
PM	(lbs/gal)	0.0082	0.0084	0.0084	0.0043			0.0026		

[a]Normalized to 100% load basis.

Data expressed as average values for three 60-minute test runs.

g/bhp-hr = grams per brake horsepower-hour.

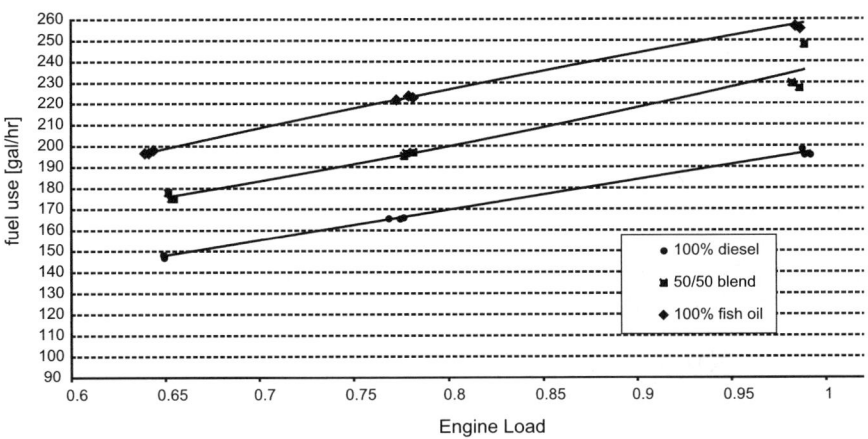

Figure 2a. Fuel use, UniSea FM #3 (900 rpm).

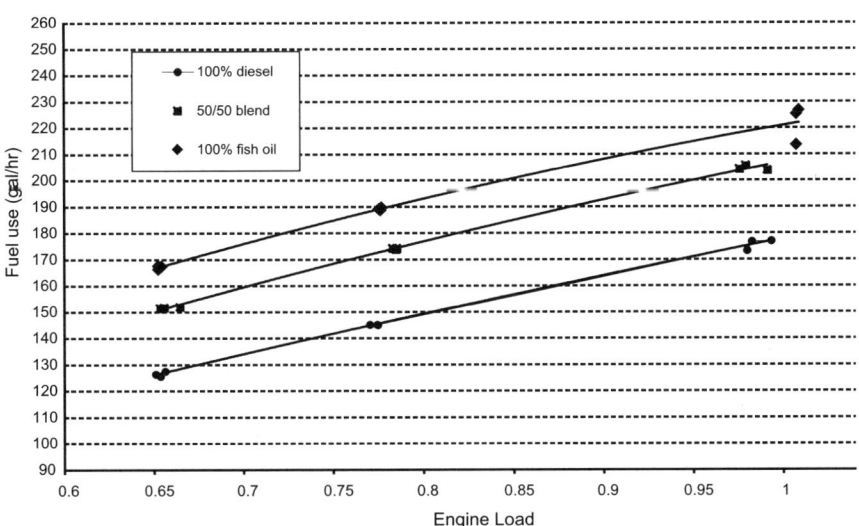

Figure 2b. Fuel use, UniSea FM #4 (720 rpm).

were higher engine-mounted fuel filter pressure differentials due to the higher viscosity of blended fuels.

FM #3 (900-rpm) emissions

As shown in Table 1, as fish oil content of the fuel blend increased, NO_x emissions, on a grams per brake horsepower-hour (g/bhp-hr) basis, increased on the order of 2 to 7% across the tested load range. Table 1 shows that FM #3 CO emissions did not vary significantly with fish oil fuel content. PM emissions decreased dramatically, however, with increased fish oil use. PM emissions dropped 40 to 60% (Table 1).

FM #4 (720-rpm) emissions

As shown in Table 2, as fish oil content of the fuel blend increased, NO_x emissions, on a g/bhp-hr basis, increased on the order of 2 to 8% across the tested load range. Table 2 shows that CO emissions decreased by 16 to 33% as fish oil fuel content was increased. PM emissions decreased 17 to 27% with increased fish oil content (Table 2).

SO_2 emissions

Fish oil produced by UniSea has sulfur content between 0.004% (Blythe 1996) and 0.0084% (Intertek Testing Services 2002) by weight. Based on the as-tested fuel consumption rates and on diesel fuel and fish oil sulfur contents of 0.05% and 0.0084%, respectively, reductions in engine SO_2 emissions of 30 to 78% were realized through the use of fish oil as engine fuel.

Fuel operations

The powerhouse was generally able to utilize fish oil as received from UniSea's on-site fish meal plant without further processing being necessary. On occasion, however, transient operating conditions in the final fish oil polishing stages of the fish meal plant resulted in brief incidents of a higher-than-normal content of suspended non-soluble proteins in the delivered fish oil. At those times, the powerhouse experienced difficulty with fuel purifiers and filters unable to handle the increased protein load. An additional purifier for the fish meal plant operation is currently being procured to address this issue. Most Alaskan fish oil producers have typically based their fish oil quality control standards and practices on the needs of their fish oil customers (largely animal feed and aquaculture operations) and of their boilers' fuel quality requirements. These standards and practices may not be adequate for fish oil intended for use as engine fuel. Accordingly, it would seem advisable that, for any large-scale use of fish oil as engine fuel, it is likely advisable for the consuming facility to have dedicated fish oil centrifugal fuel purifiers and/or suitable filtration equipment to ensure that any incidence of entrained water and insoluble protein or sediment in delivered fish oil does not create or contribute to adverse operating and maintenance conditions.

No problems were observed or experienced due to the higher viscosity of fish oil compared with diesel. At the UniSea powerhouse, fish oil is held in an uninsulated 25,000-gallon-capacity external storage tank and, thus, is subjected to cold winter temperatures. In practice, however, the fish oil is delivered from the fish meal plant somewhat warm, and turnover of the stored oil is such that fish oil never fell to temperatures that created adverse conditions. A heater-equipped centrifugal fuel purifier is available to recirculate the main fish oil storage tank if low fish oil temperature were to become an issue.

The batch blending of diesel and fish oil to achieve targeted blend ratios was found at times to be cumbersome and somewhat labor intensive, especially if achieving some degree of uniform mixing requires confirmation for operational or regulatory purposes. Potentially suitable commercially available adjustable-ratio in-line blenders are available and should be considered for use in this application.

Durability impacts

The assessment of any impacts on engine operability and maintainability will not be completed until after test operations conclude in October 2002. Results to date have been very encouraging, however, with no apparent adverse effects on the engines.

Through August 2002, the UniSea's six Fairbanks Morse engines have logged over 7,920 hours of routine operations on a 50/50 blend fuel, 4,850 hours on FM #6 alone, consuming over 526,000 gallons of fish oil while generating nearly 14,000 MWh. Inspections of fuel injectors and engine-mounted fuel pumps, likely candidates for accelerated wear, reveal no unusual wear patterns or rates. Visual inspections of exhaust gas path components such as piston ring seating grooves, exhaust ports, and exhaust turbine inlet rings, show no evidence of any unusual type or rate of hard deposits. The engine crankcase lubricating oil has been monitored closely and evaluated for lubricity at no less than 24-hour intervals, with no unusual conditions or consumption rates observed.

Conclusions

Based on the strong to-date project results, fish oil produced from wastes generated by the processing of Bering Sea pollock may be regarded as suitable as a displacement or supplementary fuel for the Fairbanks Morse model 38TD8-1/8OP and similar engines in a stationary electric-generation application when and where favorable economic, operating, and air quality permitting conditions exist.

Overall, dramatic decreases in CO, PM, and SO_2 are seen in exhaust gas emissions as fish oil content of the fuel blend increases, with an offsetting increase in NO_x. In terms of absolute magnitude, the changes in pollutant emissions are largely a wash with the tons of CO, PM, and SO_2 reduced approximately equal to the tons of NO_x increased. In identifying specific potential benefits of the use of fish oil as engine fuel, the pollutants of particular concern for the prospective consuming facility and its air quality environment would bear close consideration. As a case in point, the Unalaska/Dutch Harbor community within which UniSea is located has been identified as being of special concern for high ambient SO_2 levels by air quality regulatory authorities and a substantial reduction in SO_2 emissions may be deemed a desirable achievement even with an accompanying marginal increase in NO_x. Furthermore, in the case where a prospective consuming facility utilizes a higher sulfur diesel fuel than does UniSea (0.5% sulfur fuel is in common use in Alaska), the reductions realized in SO_2 (in excess of 95%) would far exceed the increase in NO_x emissions.

Any consideration of the use of fish oil as engine fuel should be done in the context of the prospective consuming facility's air quality permitting and overall regulatory environment. The Alaska Department of Environmental Conservation has air and water quality regulatory jurisdiction over UniSea's Unalaska/Dutch Harbor facility and has been supportive of the project's efforts. They have been increasingly supportive as higher-confidence emissions and operating data become available. Furthermore, fish oil, unlike diesel and other petroleum products, does not present the same adverse environmental impacts if spilled and is thus treated accordingly by regulatory authorities. This suggests a possible application for use as engine fuel in remote and environmentally sensitive locales, especially those with sub-optimal fuel storage and handling facilities.

Engines similar to UniSea's serving in stationary electric-generation applications are relatively rare in Alaska, and, thus, the general applicability of this project's results may be limited. In recognition of this consideration, the Alaska Energy Authority and Steigers Corporation are currently engaged in identifying potential Alaskan partners to conduct a fish oil demonstration project similar to that of this report but targeting engine types more commonly used in the state. Of particular interest are engines in wide use among electric generators in smaller rural Alaskan communities.

Acknowledgments

UniSea is conducting this project largely with its own resources, supplemented by loan funding from the Alaska Science and Technology Foundation. The Alaska Energy Authority and the U.S. Department of Energy's Regional Biomass Energy Program contributed additional grant funding and are providing technical support and guidance for the project. Steigers Corporation, on behalf of UniSea, developed the project concept, prepared proposals for and secured project funding and technical support, and continues to serve in the role of project manager. Further valued technical contributions were provided to the project by: Mr. Neil X. Blythe, manager of engine design/research and development for Fairbanks Morse Engine Division of BF Goodrich Company in Beloit, Wisconsin; Mr. Peter Crimp, development specialist for Alaska Energy Authority in Anchorage, Alaska; and Dr. Charles Peterson, professor of biological and agricultural engineering for the University of Idaho at Moscow, Idaho. The photograph is by the author.

References

Alaska Source Testing. 2002. Summary report: UniSea Dutch Harbor Seafood Processing Facility source emissions testing. Alaska Source Testing, LLC, Anchorage, Alaska.

Blythe, N.X. 1996. Fish oil as an alternative fuel for internal combustion engines. Fairbanks Morse Engine Division, BF Goodrich Company, Beloit, Wisconsin.

Intertek Testing Services. 2002. Report of analysis, fish oil sample (sample reference no. SF 02-18597). Intertek Testing Services/Caleb Brett, Valdez, Alaska.

Steigers Corporation. 2002. Alaska fish oil demonstration project: Fish oil resource report, Steigers Corporation/Alaska Energy Authority. Steigers Corporation, Littleton, Colorado.

Macro- and Micronutrient Composition of Fish Bone Derived from Alaskan Fish Meal Processing: Exploring Possible Uses for Fish Bone Meal

Ronald B. Johnson, Peter M. Nicklason, and Harold J. Barnett
National Marine Fisheries Service, Northwest Fisheries Science Center, Seattle, Washington

Abstract

Proximate, mineral, and amino acid analyses have been performed on fish bone collected from Alaskan pollock (*Theragra chalcogramma*) and Pacific cod (*Gadus macrocephalus*) fish meal processing waste from the 2001 and 2002 fishing seasons. Results show a significant amount of high quality protein (44.2 ± 2.7% dry weight) in the fish bone with a similar amino acid profile to that of white fish meal. Fish bone meal containing 90% solids is a good source of calcium and phosphorus for mammals and poultry with 15.33 ± 0.71% calcium and 7.69 ± 0.34% phosphorus. Concentrations of other minerals of nutritional interest in fish bone meal include magnesium (0.23 ± 0.23%), potassium (0.47 ± 0.09%), sodium (1.21 ± 0.12%), iron (50 ± 8 mg per g), manganese (34 ± 13 mg per g) and zinc (142 ± 31 mg per g). Plant available phosphorus was determined for three different grinds of fish bone meal resulting in a NPK (nitrogen-phosphorus-potassium) labeling of 6-9-0.5, 6-12-0.5, and 6-13-0.5 for grinds passing through –¼", U.S. #8, and U.S. #20 sieves, respectively. The use of fish bone meal for stream fertilization is discussed and results from an initial leaching experiment indicate fish bone may be a useful slow release phosphorus fertilizer, particularly in acidic streams.

Introduction

The landings of Alaskan pollock (*Theragra chalcogramma*) and Pacific cod (*Gadus macrocephalus*) in 2001 were 3.2 billion and 470 million pounds, respectively (NMFS 2002). Assuming 34 and 39% recovery (skin off, rib bone in), filleting operations produced almost 2.3 billion pounds of processing waste (Crapo et al. 1988). Much of this waste, along with undersized fish, is reduced into fish meal. Fish meal produced from these streams has high ash content (Nakprayon 1991) and is often screened to remove bone, primarily head and pin bones. There is a need to identify markets for fish bone and fish bone meal to recover production costs as well as to comply with federal regulations regarding the disposal of plant processing waste. Concomitantly, there is a need to explore additional refining techniques that would increase the value of fish bone to these markets or open new markets to this material.

Fish bone is currently being sold as both an animal feed ingredient and as a crude fertilizer. As an animal feed ingredient, it is attractive as both a phosphorus supplement and as a source of protein (A. Thulin, California Polytechnic University, San Luis Obispo, 2002, pers. comm.). Unlike mined phosphorus sources, which must be defluorinated by calcining or a similar process, bone typically does not contain high levels of fluorine. It is also a good source of calcium; however, there are several inexpensive sources of mineral calcium currently available to animal feed manufacturers (Cheeke 1999). Other benefits of utilizing fish bone meal as an animal feed ingredient are unclear. There is little information in the literature on the concentrations of other minerals of nutritional interest in fish bone meal. Hamada et al. (1995) examined the mineral composition of several fish of the North Pacific, but did not include Alaskan pollock or Pacific cod. There is also little information on typical proximate compositions of fish bone meal and the quality of the protein found in fish bone meal.

This paper presents proximate information and both mineral and amino acid profiles from fish bone meal produced from a conventional fish meal processing plant in Kodiak, Alaska, from a mix of Alaskan pollock and Pacific cod processing waste, in an effort to assist animal feed manufacturers when formulating feeds. In addition, samples were taken from the same plant at three different times—once during the 2001 fishing season and twice during the 2002 fishing season—to examine the consistency that can be expected from this material.

Crushed bone has been used for centuries as a phosphorus fertilizer (IFDC 1998). Its popularity in western civilizations diminished with the discovery and mining of phosphorus rock deposits in the late nineteenth century followed by the manufacture of various refined phosphorus fertilizers in the twentieth century. It is still being used in countries, China is an example, which do not yet have the capacity to either produce or import an adequate supply of modern phosphorus fertilizers (IFDC 1998).

In this regard, the market for bone as a crude fertilizer should diminish as fertilizer companies continue to expand their operations. However, bone meal is attractive to organic farming, which is continuing to grow as an alternative to modern farming practices, which are suspected to be non-sustainable and continually depleting the soil of trace minerals of nutritional interest (Batsell 2002). Unlike modern ammonium phosphate fertilizers, bone contains significant levels of several macro- and micro-nutrients for plant growth.

Fish bone generated from the fish meal plants in Alaska is separated from the meal after processing and requires further grinding to be used as either an animal feed ingredient or a fertilizer. The finer the grind, the more available is phosphorus and other minerals to plants and animals. This is especially of concern to fertilizer distributors. Fertilizer sold in the United States must list "available" phosphorus content in contrast to total phosphorus content (IFDC 1998). Available phosphorus is estimated from citrate soluble phosphorus determined by AOAC Method 960.03 (Horwitz 2000). The grind of bone has a significant effect on the amount of available phosphorus determined by this method. To further investigate this effect, we have performed particle size analyses and have determined the availability of phosphorus in three different grinds of fish bone meal.

A part of our research to introduce fish bone meal into new markets involved conducting leaching studies to determine if fish bone pellets could be a useful medium to return marine-derived nutrients to streams in the Pacific Northwest, where wild salmon (*Oncorhynchus* sp.) populations are diminished. The lack of salmon returning to streams in these watersheds has resulted in nutrient deficiencies, particularly in phosphorus and nitrogen. Food web processes are extremely limited in these streams and unable to sustain the historic levels of salmon fry once present. Returning nutrients to streams to stimulate periphyton growth and invertebrate populations has become an emerging issue in watershed restoration programs and was the subject of a special conference in Eugene, Oregon, in April 2001. Successful nutrient restoration efforts have been reported from placement of spawned hatchery salmon carcasses (Bilby et al. 1998) and direct stream fertilization with commercial fertilizer (Johnston et al. 1990, Ashley and Slaney 1997). In this study, leaching studies were performed on bone samples from three different species of fish to estimate the rate that phosphorus and other elements would be available to the stream ecosystem.

Developing new markets for fish bone meal likely will involve products for human consumption. Sada (1984) developed a refining process that produces a powdered fish calcium product from skipjack tuna bones (*Katsuwonus pelamis*). This product has been blended into entrees as a calcium supplement for school lunch programs in Japan. It also has been used as a supplement by the confectionary and candy industries. Another possible use for fish bone meal is an ingredient in baby food. Martinez et al. (1998) found significant increases in in vitro available calcium and phosphorus in fish-based infant weaning foods that were supplemented with either hake (*Merluccius* sp.) or sole (*Solea vulgaris*) bone. The potential use of fish bone in these markets should be considered in light of fish bone meal's nutritional qualities when it is used as an animal feed ingredient.

Materials and methods

Fish bone samples

Fish bone samples were collected from Kodiak Fishmeal Company once during the 2001 fishing season and twice during the 2002 fishing season. Kodiak Fishmeal Company is a conventional fish meal plant with a process similar to that described by FAO (1975). Fish bones were mechanically separated from the fish meal after final drying. The fish meal was produced from a mix of Alaskan pollock and Pacific cod processing waste. A commercial grind of fish bone meal, prepared by Kodiak Fishmeal Company, was reduced to finer grinds with a laboratory scale hammer-mill to investigate the effect of particle size on available phosphorus. Samples for proximate, mineral, and amino acid analyses were ground in the laboratory to pass a U.S. #20 sieve.

In addition to the commercial fish bone described above, fish bone pellets used in the leaching studies also included bone derived from a new modified silage process developed at our laboratory (Nicklason et al. 2003). Bone from this process originated from rock sole (*Pleuronectes bilineatus*) and pink salmon (*Oncorhynchus gorbuscha*).

Proximate analysis

Proximate analyses (ash, fat, protein, and moisture) were performed on ground fish bone samples from three different sampling occasions. Ash was determined by heating a 1 g sample in a 550°C muffle furnace for a minimum of eight hours. Fat was determined on a 1 g sample by super-critical fat extraction on a LECO FA-100 Fat Analyzer as described by Johnson and Barnett (2003). Protein was calculated by determining nitrogen concentration in a 0.25 g sample by Dumas combustion methodology on a LECO FP-2000 nitrogen analyzer and multiplying the result by 6.25 (Horwitz 2000). Moisture was determined on a 3 g sample by drying a minimum of eight hours in a 105°C oven.

Amino acid analysis

Amino acid analyses of ground fish bone samples were performed by the Experiment Station Chemical Laboratory (ESCL) at the University of Missouri. Analyses were performed by HPLC in accordance with AOAC Method 982.30 (Horwitz 2000). Twenty-three amino acids were quantified including all the amino acids considered essential for animal nutrition (Buttery and D'Mello 1994).

Mineral analysis

Mineral analyses were performed on ground fish bone samples from the 2001 fishing season. Samples were ashed as described above and digested overnight with an aqueous mixture of 10% hydrochloric and 10% nitric acid. Samples were analyzed for calcium, copper, iron, potassium, magnesium, manganese, sodium, phosphorus, strontium, and zinc by inductively coupled plasma–optical emission spectroscopy (ICP-OES) on a Perkin Elmer Optima 3000 Radial ICP-OES. Roberts et al. (1996) used this technology previously to determine mineral concentrations in human bone. Samples for mercury and lead analysis were lyophilized and digested with nitric acid in a closed vessel digestion microwave. Trace levels of mercury and lead were detected by hydride generation atomic absorption spectroscopy and graphite furnace atomic absorption spectroscopy, respectively, on a Perkin Elmer 5100 AA.

Available phosphorus

The particle size distributions of the various fish bone meal grinds were performed using U.S. standard sieves on a W.S. Tyler Ro-Tap shaker for 15 minutes. Available phosphorus was extracted from the grinds with water and ammonia citrate solutions as described in AOAC Methods 977.01 and 963.03, respectively (Horwitz 2000). Phosphorus levels in extracts were determined by ICP-OES. Available phosphorus was determined for each grind of fish bone meal in triplicate.

Leaching studies

Initial laboratory leaching studies were performed to approximate the rate that minerals would leach from fish bone placed in freshwater streams. Fish bone samples were ground to pass a U.S. #8 sieve and hand pressed into 5 g pellets. Four pellets were placed in 500 ml of deionized water and aliquots removed weekly and analyzed for phosphorus, calcium, and magnesium content by ICP-OES. Aliquots were large (over 100 ml) and the leaching flask was replenished by deionized water after removal of aliquot. Aliquots were filtered through a 0.45 micron syringe filter prior to analysis to remove any suspended solids or bacteria. To simulate worst-case conditions, flasks were kept in the dark, at 4°C, and minimally agitated when removing aliquots. Leaching studies lasted for 5 weeks. Leaching studies were performed for each bone species in triplicate.

Results

Proximate analyses of the fish bone sampled from the 2001 and 2002 fishing seasons are shown in Table 1. Results are listed on a dry weight basis

Table 1. Proximate analysis of fish bone generated from the reduction of Alaskan pollock and Pacific cod processing waste.

	KFC 1	KFC 2	KFC 3	Average ($n = 9$)
Fishing season	2001	2002	2002	
Ash	48.0 ± 0.1	50.1 ± 0.5	46.9 ± 0.5	48.3 ± 1.5
Protein	46.3 ± 0.1	40.6 ± 0.5	45.7 ± 0.4	44.2 ± 2.7
Fat	5.7 ± 0.1	9.3 ± 0.2	7.4 ± 0.2	7.5 ± 1.5

Numbers are mean (n = 3) percent dry weight, ± 1 s.d., error = 1 s.d. KFC = Kodiak Fishmeal Company.

due to large differences in moistures between samples (7.4-13.0%). Average moisture content was 9.9%. Ash content remained fairly consistent among the samples with more variation seen with protein contents.

Amino acid profiles and protein content of a sample from each of the three sampling times are shown in Table 2. Moisture content of these samples was approximately 10%. Amino acid profiles were fairly consistent among the samples and the total of the amino acids profiled accounted for over 94% of the protein present.

Mineral analyses of the fish bone samples are shown in Table 3. Results are expressed on a 90% solids basis. Phosphorus and potassium amounts are both elemental and expressed as the oxides used by the fertilizer industry (P_2O_5 and K_2O). Nitrogen concentrations are also included in Table 3 for comparison with other fertilizer compositions. Fish bone from the 2001 fishing season was analyzed for mercury and lead. Trace levels of both mercury and lead were detected in the fish bone; however, these concentrations are within the limits for its use as an animal feed ingredient (AAFCO 2002).

Available phosphorus was measured for the commercial grind of fish bone meal as well as for two other grinds prepared at our laboratory. Results are plotted in Fig. 1. The amount of available phosphorus increased with additional grinding. Particle size analysis of each grind was performed with a set of U.S. standard sieves and the profiles are listed in Table 4.

Leaching studies were performed on 5 g pellets of fish bone meal from three sources. The elements monitored were phosphorus, calcium, and magnesium. These elemental concentrations in the bone pellets are listed in Table 5 and weekly cumulative percentages of elements released into solution are listed in Table 6. The elemental concentrations of calcium and phosphorus in the ash of all three bone sources were similar. Magnesium levels were lower in the two fish bone meals derived from the modified silage process developed at our laboratory. The average pH

Table 2.   Complete amino acid profile of fish bone meal generated from the reduction of Alaskan pollock and Pacific cod processing waste.

Fishing season	KFC 1 2001	KFC 2 2002	KFC 3 2002	Average ($n = 3$)
Protein	38.7	38.1	41.5	
Alanine	2.66	2.67	2.75	2.69 ± 0.05
Arginine	2.56	2.59	2.94	2.70 ± 0.21
Aspartic acid	3.03	2.99	3.24	3.09 ± 0.13
Cysteine	0.27	0.27	0.32	0.29 ± 0.03
Glutamic acid	4.65	4.70	5.14	4.83 ± 0.27
Glycine	5.10	5.19	5.25	5.18 ± 0.08
Histidine	0.66	0.65	0.70	0.67 ± 0.03
Hydroxylysine	0.22	0.24	0.23	0.23 ± 0.01
Hydroxyproline	1.25	1.31	1.40	1.32 ± 0.08
Isoleucine	1.13	1.12	1.18	1.14 ± 0.03
Lanthionine	0.03	0.00	0.03	0.02 ± 0.02
Leucine	2.04	2.03	2.17	2.08 ± 0.08
Lysine	2.06	2.06	2.23	2.12 ± 0.10
Methionine	0.91	0.89	0.99	0.93 ± 0.05
Ornithine	0.07	0.06	0.06	0.06 ± 0.01
Phenylalanine	1.12	1.12	1.25	1.16 ± 0.08
Proline	2.51	2.55	2.66	2.57 ± 0.08
Serine	1.78	1.87	2.08	1.91 ± 0.15
Taurine	0.37	0.37	0.42	0.39 ± 0.03
Threonine	1.36	1.37	1.47	1.40 ± 0.06
Tryptophan	0.29	0.30	0.32	0.30 ± 0.02
Tyrosine	0.81	0.81	0.91	0.84 ± 0.06
Valine	1.50	1.44	1.51	1.48 ± 0.04
Total AA	36.38	36.63	39.25	

Numbers are percent by weight, averages ± s.d. Samples are approximately 90% dry matter. KFC = Kodiak Fishmeal Company.

**Table 3.   Mineral concentrations in fish bone meal generated from the re-
duction of Alaskan pollock and Pacific cod processing waste.**

	Percent by weight			
	KFC 1 ($n = 3$)	KFC 2 ($n = 3$)	KFC 3 ($n = 3$)	
Fishing season	2001	2002	2002	Average ($n = 9$)
Calcium	14.60 ± 0.11	16.16 ± 0.06	15.23 ± 0.43	15.33 ± 0.71
Phosphorus	7.35 ± 0.03	8.10 ± 0.03	7.62 ± 0.17	7.69 ± 0.34
Magnesium	0.26 ± 0.003	0.20 ± 0.01	0.23 ± 0.001	0.23 ± 0.02
Potassium	0.53 ± 0.005	0.52 ± 0.04	0.37 ± 0.04	0.47 ± 0.09
Sodium	1.14 ± 0.004	1.12 ± 0.04	1.37 ± 0.01	1.21 ± 0.12
Nitrogen Oxide	6.79 ± 0.04	5.85 ± 0.12	6.35 ± 0.12	6.33 ± 0.42
P_2O_5	16.8 ± 0.1	18.6 ± 0.1	17.5 ± 0.4	17.6 ± 0.8
K_2O	0.64 ± 0.1	0.62 ± 0.05	0.44 ± 0.04	0.56 ± 0.10
	Concentration (μg/g)			
Copper	2 ± 1	13 ± 3	ND	5 ± 6
Iron	380 ± 21[a]	46 ± 3	55 ± 10	50 ± 8[a]
Manganese	49 ± 1	18 ± 1	34 ± 1	34 ± 13
Strontium	1,047 ± 4	981 ± 9	953 ± 25	994 ± 44
Zinc	165 ± 8	114 ± 3	147 ± 40	142 ± 31
Lead	0.31			
Mercury	0.11			

[a]High iron concentration in KFC 1 is speculated to be due to residual rust on processing equipment and not
typical for these species of fish. Average concentration was calculated from KFC 2 and KFC 3 samples.
Concentrations expressed on a 90% dry matter basis, ± s.d. KFC = Kodiak Fishmeal Company.

of the leaching solutions varied among the three bone sources due to
different levels of residual acid left in the bone from the modified silage
process or, in the case of the bone from the commercial process, a complete absence of acid. Bone generated from the modified silage process
released phosphorus, calcium, and magnesium at a faster rate than bone
from the conventional process. The pH of one of the commercial fish bone
replicates was adjusted from 7.3 to 6.3 with 2 N hydrochloric acid in the
last week of the study. This resulted in the increased release of all three
measured elements. However, it was also observed that despite leaching
at a lower pH, calcium and phosphorus leached at a slower rate from
the rock sole bone than the pink salmon bone. The inverse was true for
magnesium which leached at a higher rate from the rock sole bone. All
elements leached at the lowest rate from the commercial fish bone.

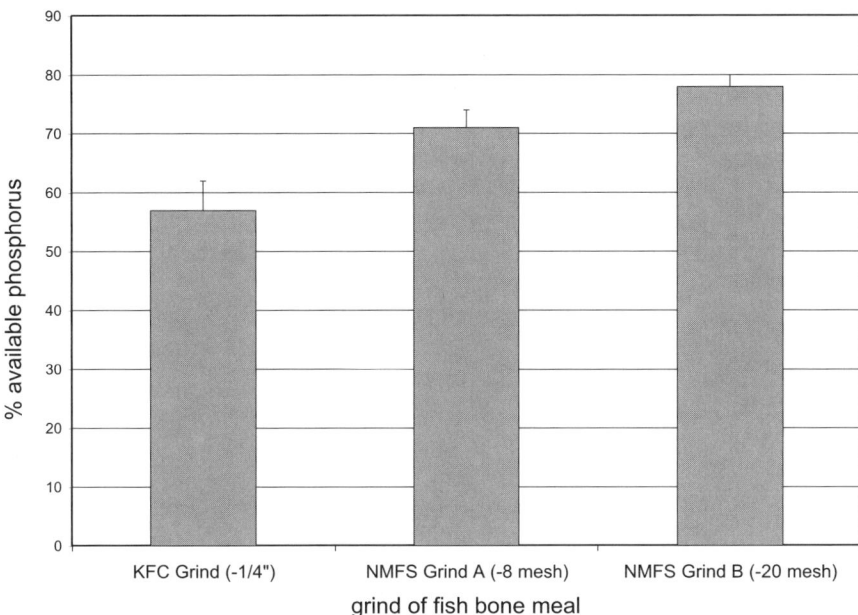

Figure 1. Increase in availability of fish bone meal phosphorus with additional grinding. Actual particle size analysis of these grinds is in Table 4.

Discussion

Evaluating fish bone meal as an animal feed ingredient

Proximate analyses, amino acid profiles, and minerals analyses of the fish bone meal from Alaskan pollock and Pacific cod processing waste demonstrate its potential as an animal feed ingredient. Like meat and bone meal, the fish bone meal sampled is a good source of protein with 44.2 ± 2.7% protein retained with the bone on a dry weight basis. Unlike meat and bone meal, fish bone meal is suitable for ruminant and non-ruminant feed formations due to the absence of transmissible spongiform encephalopathy diseases among marine fish (AAFCO 2000). Fish bone meal also contains over 7% phosphorus, which is highly available to mammalian animals and poultry (Hamilton 1996, Underwood and Suttle 1999). It is also a good source of calcium and contains other essential minerals including magnesium and zinc (Henry and Benz 1995, Baker and Ammerman 1995) as well as 7.5 ± 1.5% fish oil, which provides essential fatty acids and increases energy content.

Table 4. Particle size analysis of fish bone meal grinds and available phosphorus expressed as P_2O_5.

U.S. sieve #	Percent retained (cumulative)		
	KFC grind (–¼")	NMFS grind A (–8 mesh)	NMFS grind B (–20 mesh)
8	16.4	2.0	0.0
20	54.3	14.1	1.3
50	82.1	48.9	22.7
70	92.1	62.5	38.6
100	99.0	80.5	59.3
200	100.0	93.3	81.8
325	100.0	99.3	99.2
	Percent by weight ($n = 3$)		
Total P_2O_5	17.3 ± 0.5	17.6 ± 0.3	17.4 ± 0.1
Available P_2O_5	9.8 ± 0.6	12.5 ± 0.3	13.6 ± 0.2

KFC = Kodiak Fishmeal Company; NMFS = National Marine Fisheries Service.

Table 5. Composition of bone pellets used in leaching experiments.

Bone source	% Ash	% Nitrogen	Elemental composition in ash (% w/w)		
			Phosphorus	Calcium	Magnesium
Rock sole[a]	46.9 ± 0.4	6.7 ± 0.2	18.8 ± 0.6	31.8 ± 1.2	0.16 ± 0.01
Pink salmon[a]	30.3 ± 0.9	10.2 ± 0.1	17.0 ± 0.5	32.8 ± 1.5	0.15 ± 0.01
Pollock/cod[b]	47.8 ± 0.4	7.0 ± 0.1	16.5 ± 0.5	32.5 ± 1.1	0.63 ± 0.03

[a]Bone derived from NMFS modified silage processing.
[b]Bone derived from conventional fish meal processing.
Values are ± s.d., $n = 3$.

Table 6. **Cumulative amounts of elements released into solution during leaching trial.**

Bone source	Average pH	Average % of element leached from fish bone ($n = 3$)				
		Day 7	Day 14	Day 21	Day 28	Day 35
Phosphorus						
Rock sole[a]	5.7	5.5 ± 0.2	6.9 ± 0.2	7.6 ± 0.1	8.5 ± 0.2	9.1 ± 0.
Pink salmon[a]	6.2	5.8 ± 0.6	8.2 ± 0.5	9.6 ± 0.6	10.9 ± 0.7	12.0 ± 0.7
Pollock/cod[b]	7.5	0.6 ± 0.1	1.0 ± 0.1	1.2 ± 0.1	1.4 ± 0.1	1.6 ± 0.1
Pollock/cod[b]	6.3					1.9[c]
Calcium						
Rock sole[a]	5.7	5.0 ± 0.1	6.0 ± 0.2	6.6 ± 0.1	7.1 ± 0.1	7.4 ± 0.1
Pink salmon[a]	6.2	6.0 ± 0.7	8.3 ± 0.5	9.9 ± 0.7	11.5 ± 0.8	12.8 ± 0.9
Pollock/cod[b]	7.5	0.05 ± 0.002	0.10 ± 0.003	0.14 ± 0.003	0.18 ± 0.003	0.23 ±0.003
Pollock/cod[b]	6.3					1.1[c]
Magnesium						
Rock sole[a]	5.7	67.9 ± 1.2	74.4 ± 0.7	76.9 ± 0.3	79.5 ± 0.2	81.2 ± 0.2
Pink salmon[a]	6.2	37.9 ± 3.9	51.8 ± 1.8	56.9 ± 2.4	60.6 ± 2.5	63.3 ± 2.6
Pollock/cod[b]	7.5	4.0 ± 0.3	6.3 ± 0.3	8.0 ± 0.2	9.6 ± 0.3	11.2 ± 0.2
Pollock/cod[b]	6.3					29.9[c]

[a]Bone derived from NMFS modified silage processing.
[b]Bone derived from conventional fish meal processing.
[c]The pH of one pollock/cod replicate was adjusted to pH 6.3 on day 28.

The amino acid profiles of the fish bone meals sampled at the three sampling periods were similar. Distribution of the essential amino acids for animal growth in the fish bone meals closely resembled the distribution of amino acids in white fish meal (NRC 1998) with the exception of an elevated arginine level in the bone meal. The elevated level of arginine may be a benefit for inclusion in poultry feeds due to the absence of a urea cycle in those animals. The similarities in essential amino acid profiles combined with an average lysine content of 5.4 % (Table 7) suggest that, unlike other animal byproduct proteins which may contain high levels of connective tissue, the protein in fish bone meal is predominately from residual fish meal adhering to the bone. Fish meal is considered to be a superior protein source in animal feed formulations for its balanced amino acid profile, palatability, and digestibility (Cheeke 1999). Like other meat products, fish meal is a good source of undegradable or "bypass" ruminant protein. This bypass protein is especially important to growing heifers and steers and has been used to stimulate lactation in dairy cows and maintain milk protein levels (Akayezu et al. 1997, Cheeke 1999, Allison and Garnsworthy 2002). Since fish bone meal contains a significant amount of white fish meal, it may also be useful for dairy cattle feed with the additional benefit of supplying additional calcium needed for lactation.

Table 7. **Amino acid balance of fish bone meal protein compared to an "ideal" protein formulation for protein accretion in swine and other animal feed protein sources.**

	Fish bone meal	Ideal protein[a]	White-fish meal[b]	Soybean meal (dehulled)[b]	Meat and bone meal[b]	Feather meal[b]
	Percent by weight in protein					
Lysine	5.37		7.12	6.36	4.87	2.46
	Balance of essential amino acids to lysine					
Lysine	100	100	100	100	100	100
Arginine	127	48	90	115	137	270
Histidine	32	32	30	42	36	45
Isoleucine	54	54	58	72	53	186
Leucine	98	102	97	121	119	326
Methionine	44	27	39	22	27	29
Methionine + cysteine	57	55	54	47	47	228
Phenylalanine	55	60	51	79	65	193
Phenylalanine + tyrosine	95	93	96	139	107	309
Threonine	66	60	58	61	63	184
Tryptophan	14	18	15	22	11	26
Valine	70	68	68	75	81	283

[a]Data reprinted from NRC (1998).
[b]Data calculated from NRC (1998), Appendix 11 (Nutrient content of feeds).

The idea of an optimal balanced protein (ideal protein) is supported by the National Research Council (NRC) and others (Fuller 1989, Oldham 1994, Wilson 1994, NRC 1998). Table 7 shows the ratios of essential amino acids to lysine in several protein sources in animal feeds along with NRC's ideal protein for protein accretion in swine. Fish bone meal, fish meal, and soybean meal all have profiles close to that of the ideal protein. Cattle and poultry have similar amino acid requirements for protein accretion with the exception of additional essential amino acids (arginine, glycine, and proline) required for poultry (NRC 1994, 2001). Soybean meal is currently available for under $200 a ton (USDA 2002) and is the preferred protein source for many animal feeds. While soybean meal is useful in growing diets, there are several disadvantages to feeding soybean meal to young animals and feeds with milk protein and animal proteins are frequently used. Young pigs fed creep diets containing soy products may become sensitized to soy proteins and exhibit gastrointestinal distress after weaning due to the adsorption of soybean antigens (Cheeke 1999). Soybeans have also been shown to cause allergic reactions in young farm animals. (R. Kincaid, Washington State University, Pullman, 2002, pers. comm.).

In addition to needing a good balance of amino acids, young animals also have high requirements for calcium and phosphorus (NRC 1994, 1997, 1998; Underwood and Suttle 1999). Soybean meal contains low levels of calcium, and available phosphorus and supplementation with inorganic calcium and phosphorus sources are frequently needed (Soares 1995). Because of its balanced protein, high concentrations of available calcium and phosphorus, and the added benefit of containing an average of 7.5% fish oil, fish bone meal should prove to be a useful ingredient in young animal feeds. Depending on market conditions and transportation costs, fish bone meal also has the potential of being a less expensive source of protein and phosphorus in grower diets than soybean meal supplemented with dicalcium phosphate.

Evaluating fish bone meal as a fertilizer

Crushed bone has been used for centuries as a phosphorus fertilizer for crops. Currently, fish bone meal is being sold in the United States to organic farmers and individual gardeners. Blatt and McRae (1998) found cabbage and carrot yields from crops treated with organic fertilizers prepared with fish bone meal to be comparable to those treated with commercial fertilizers. Fertilizer labeling required by several states in the United States provides information on nitrogen, available phosphorus expressed as P_2O_5, and potassium expressed as K_2O (IFDC 1998). As fertilizers are sold on labeled content, we examined the increase in available phosphorus with additional grinding. The commercial grind of fish bone meal approximated a $-\frac{1}{4}$" grind. Fertilizer labeling for the original grind was 6-9-0.5 which increased to 6-12-0.5 and 6-13-0.5 after grinding with a laboratory hammer mill to pass U.S. #8 and #20 sieves, respectively. The grind produced by the hammer mill produced a large number of fines which is suspected to be driving the increase in availability. Grinding to the same mesh by other techniques which produce fewer fines, such as roll milling, may not achieve the same levels of availability. Conversely, material collected from any dust collecting equipment is suspected to have high availability and should be blended back into the product. Compared to phosphate rock, which is used by many organic farmers, the phosphorus in fish bone meal is more available, and requires less grinding to obtain a useful product. The grind currently available may be adequate for most organic farming applications; however, further grinding should be considered as a value added step.

In addition to phosphorus availability, mineral analyses of the fish bone meals showed significant levels of secondary macronutrients and micronutrients. Frequently commercial phosphorus fertilizers are deficient in these micronutrients, which eventually leads to low crop production after these minerals have been depleted from the soil (IFDC 2001). While this problem usually takes decades to appear, current commercial agriculture has been criticized as being "non-sustainable" with organic farming seen as a solution. Secondary macronutrients in the fish bone meal include 15.3% calcium and 0.2% magnesium and micronutrients in the fish bone meal include 50 mg per g iron, 34 mg per g manganese, and 142 mg per g zinc.

Evaluating fish bone meal for stream restoration

Our initial leaching studies of fish bone meal demonstrate its potential as a slow-release fertilizer for nutrient deficient streams. While fish bone meal may take longer to completely dissolve than commercial fertilizers, continual annual application will eventually result in consistent nutrient delivery. Early years can be supplemented with commercial fertilizers. From the results of the leaching trials, the length of time needed before supplementation can be discontinued appears to be related to the pH of the stream and the process the bone derived from. Bone generated from the modified silage process described by Nicklason et al. (2003) released phosphorus at a significantly greater rate than the bone from the conventional process. Also, lowering the pH of one pollock bone replicate from 7.4 to 6.3 in the last week of the study resulted in 2.8 times more phosphorus being leached into solution.

Figure 2 shows the weekly cumulative amounts of phosphorus leached into solution from the three different fish bone meals studied. After the second week, the leaching curves for all three fish bone meals became somewhat linear. The length of time estimated to reach near complete solvation of phosphorus from the fish bones into solution was estimated by extending this linear part of the curve. Phosphorus in the pink salmon bone is estimated to completely leach in 1.4 years at pH 6.2, phosphorus in the rock sole bone is estimated to completely leach in 2.4 years at pH 5.7, and phosphorus in the pollock/cod bone is estimated to completely leach in 11.2 years at pH 7.5. As mentioned previously, these rates reflect worst-case conditions. These initial leaching studies were performed in the dark at 4°C to minimize microbial activity. It is anticipated that leaching rates in freshwater streams will be significantly higher due to higher water temperatures and flows, UV radiation, and increased microbial activity.

It is unclear why the rock sole bone leached phosphorus at a slower rate than the salmon despite being at a lower pH. It is speculated that, during the acid digestion step in the modified silage process, apatite at the bone surface is partially converted to a more soluble calcium phosphate. Similar morphological changes occur in the industrial process of converting bone into a commercial dicalcium phosphate product as described by Choksi et al. (1980). If so, subtle changes in the modified silage process as well as bone particle size could affect the extent of any morphological

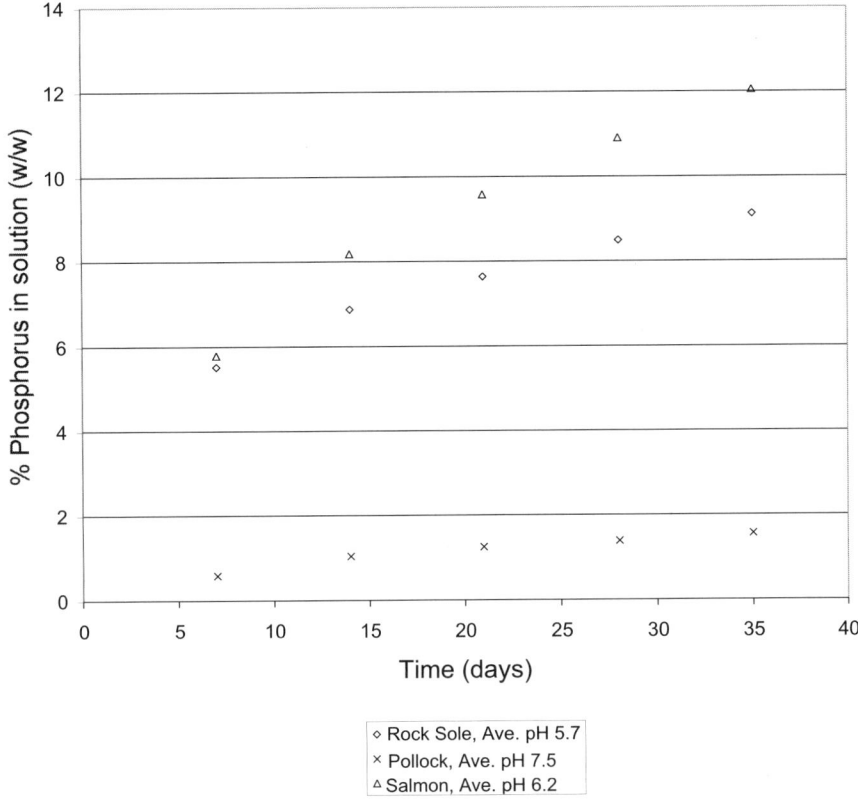

Figure 2. Cumulative percent of phosphorus leached from fish bone meal (w/w).

changes that would result in increased phosphorus solubility. Alternatively, the slower rate may be due to increased microbial activity in the salmon bone trials due to the higher protein content. Nitrogen levels in the salmon bone were significantly higher than in the rock sole (Table 5).

Nutrient restoration of salmon spawning streams by the addition of fish bone meal pellets has advantages over either salmon carcass placement or direct stream fertilization. It is a readily available, natural product that will return marine derived nutrients back to the watershed. The high concentration of phosphorus in fish bone makes it especially attractive for use in phosphorus deficient streams. Compared to adult salmon carcasses, which contain approximately 0.4% phosphorus by weight (Shearer et al. 1994), commercially available fish bone meal contains 7.7 % phosphorus by weight. This concentration of phosphorus would result in almost a twenty fold savings in delivery costs over whole carcass distribution. While not as concentrated as some commercial phosphorus fertilizers, fish bone meal has the similar advantage of salmon carcasses of supplying secondary macronutrients and micronutrients which are absent in many commercial fertilizers. Similar to its role as an organic fertilizer, fish bone meal would supply streams with additional nutrients essential to aquatic life.

The results from our initial laboratory leaching trials have spurred further outdoor studies to evaluate the overall phosphorus enrichment that can be expected in natural systems. Future work will be directed toward evaluating the response of periphyton and insect communities to the enrichment, as well as the formulation of a compressed bone pellet that would allow a single annual application.

Acknowledgment

The authors wish to thank Don Ernest, NMFS, Seattle, Washington, for providing technical expertise and providing the mercury and lead analyses.

Names of companies used in this paper are necessary to report factually on available data; however, the National Marine Fisheries Service (NMFS) neither guarantees nor warrants the standard of the product, and the use of the name by NMFS implies no approval of the products to the exclusion of others that may also be suitable.

References

AAFCO. 2000. Feed inspector's manual. Association of American Feed Control Officers, Oxford, Indiana. 290 pp.

AAFCO. 2002. Official publication. Association of American Feed Control Officers, Oxford, Indiana. 464 pp.

Akayezu, J.M., W.P. Hansen, D.E. Otterby, B.A. Crooker, and G.D. Marx. 1997. Yield response of lactating Holstein dairy cows to dietary fish meal or meat and bone meal. J. Dairy Sci. 80:2950-2963.

Allison, R.D., and P.C. Garnsworthy. 2002. Increasing the digestible undegraded protein intake of lactating dairy cows by feeding fishmeal or a rumen protected vegetable protein blend. Anim. Feed Sci. Technol. 96:69-81.

Ashley, K.I., and P.A. Slaney. 1997. Accelerating recovery of stream, river, and pond productivity by low-level nutrient replacement, chapter 13. In: P.A. Slaney and D. Zaldokas (eds.), Fish habitat rehabilitation procedures. Watershed Restoration Technical Circular No. 9. British Columbia Ministry of Environment, Lands and Parks and British Columbia Ministry of Forests, Vancouver, Canada, p. 341.

Baker, D.H., and C.B. Ammerman. 1995. Zinc bioavailabilty. In: C.B. Ammerman, D.H., Baker, and A.J. Lewis (eds.), Bioavailability of nutrients for animals. Academic Press, San Diego, pp. 367-398.

Batsell, J. 2002. Organic gains seal of approval. The Seattle Times, Seattle, Washington, Oct. 2, 2002, pp. E1-3.

Bilby, R.E., B.R. Fransen, P.A. Bisson, and J.K. Walter. 1998. Response of juvenile coho salmon (*Oncorhynchus kisutch*) and steelhead *O. mykiss*) to the addition of salmon carcasses to two streams in southwestern Washington, USA. Can. J. Fish. Aquat. Sci. 55:1909-1918.

Blatt, C.R., and K.B. McRae. 1998. Comparison of four organic amendments with a chemical fertilizer applied to three vegetables in rotation. Can. J. Plant Sci. 78:641-646.

Buttery, P.J., and J.P.F. D'Mello, 1994. Amino acid metabolism in farm animals: An overview. In: J.P.F. D'Mello, (ed.), Amino acids in farm animal nutrition. CAB International, Wallingford, U.K., pp. 1-10.

Cheeke, P.R. 1999. Applied animal nutrition. Prentice Hall, Upper Saddle River, New Jersey. 525 pp.

Choksi, A., A. Meeraus, and A. Stoutjesdijk. 1980. The planning of investment programs in the fertilizer industry. Johns Hopkins University Press, Baltimore. 333 pp.

Crapo, C., B. Paust, and J.K. Babbitt, 1988. Recoveries and yields from Pacific fish and shellfish. Alaska Sea Grant College Program, University of Alaska Fairbanks, Fairbanks, p. 50.

FAO. 1975. The production of fish meal and oil. FAO Fish. Tech. Pap., p. 54.

Fuller, M.F., R. McWilliam, T.C. Wang, and L.R. Giles. 1989. The optimal dietary amino acid pattern for growing pigs, 2. Requirements for maintenance and for tissue protein accretion. Br. J. Nutr. 62:255-267.

Hamanda, M., T. Nagai, N. Kai, Y. Tanoue, H. Mae, M. Hashimoto, K. Miyoshi, H Kumagai, and K. Saeki. 1995. Inorganic constituents of bone of fish. Fish. Sci. 61:517-520.

Hamilton, C.R. 1996. Meat and bone meal. Darling International, Irving, Texas, p. 11.

Henry, P.R., and S.A. Benz. 1995. Magnesium bioavailabilty. In: C.B. Ammerman, D.H. Baker, and A.J. Lewis (eds.), Bioavailabilty of nutrients for animals. Academic Press, San Diego, pp. 201-238.

Horwitz, W. (ed.). 2000. Official methods of analysis of AOAC International. AOAC (Association of Analytical Communities), Arlington, Virginia.

IFDC. 1998. Fertilizer manual. UNIDO, IFDC (United Nations Industrial Development Organization, International Fertilizer Development Center). Kluwer Academic Publishers, Dordrecht, Netherlands. 615 pp.

Johnson, R., and H. Barnett. 2003. Determination of fat content in fish feed by supercritical fluid extraction and subsequent lipid classification by thin layer chromatography–flame ionization detection. Aquaculture 216:263-282.

Johnston, N.T., C.J. Perrin, P.A. Slaney, and B.R.W. 1990. Increased juvenile growth by whole-river fertilization. Can. J. Fish. Aquat. Sci. 47:862-872.

Martinez, I., M. Santaella, G. Ros, and M.J. Periago. 1998. Content and in vitro availability of Fe, Zn, Mg, Ca, and P in homogenized fish-based weaning foods after bone addition. Food Chem. 63:299-305.

Nakprayon, S., T. Jamsai, and S. Jattupornpong. 1991. Nutrient composition of fish meal produced from fish waste product. In: 29th Kasetsart University Annual Conference: Animal Science, Veterinary Science, Aquaculture. Kasetsart University Press, Bangkok, Thailand, pp. 1-7. (Summary in English, text in Thai.)

Nicklason, P.M., H. Barnett, R. Johnson, M. Tagal, and B. Pfutzenreuter. 2003. Modified silage process for fish and fish processing waste. In: P.J. Bechtel (ed.), Advances in seafood byproducts: 2002 conference proceedings. Alaska Sea Grant College Program, University of Alaska Fairbanks, Fairbanks.

NMFS. 2002. Fisheries of the United States 2001. National Marine Fisheries Service, Silver Spring, Maryland. 126 pp.

NRC. 1994. Nutrient requirements of poultry. National Research Council. National Academy Press, Washington D.C. 155 pp.

NRC. 1998. Nutrient requirements of swine. National Research Council. National Academy Press, Washington D.C. 189 pp.

NRC. 2001. Nutrient requirements of dairy cattle. National Research Council. National Academy Press, Washington D.C. 381 pp.

Oldham, J.D. 1994. Amino acid nutrition of the dairy cow. In: J.P.F. D'Mello, (ed.), Amino acids in farm animal nutrition. CAB International, Wallingford, U.K., pp. 351-376.

Roberts, N.B., H.P.J. Walsh, L. Klenerman, S.A. Kelly, and T.R. Helliwell. 1996. Determination of elements in human femoral bone using inductively coupled plasma atomic emission spectroscopy and inductively coupled plasma mass spectrometry. J. Anal. At. Spectrom. 11:133-138.

Sada, M. 1984. Fish calcium. Infofish Marketing Digest 3:29-30.

Shearer, K.D., T. Asgard, G. Andorsdottir, and G.H. Aas. 1994. Whole body elemental and proximate composition of Atlantic salmon (*Salmo salar*) during the life cycle. J. Fish Biol. 44:785-797.

Soares, J.H. 1995. Phosphorus bioavailability. In: C.B. Ammerman, D.H. Baker, and A.J. Lewis (eds.), Bioavailability of nutrients for animals. Academic Press, San Diego, pp. 257-294.

Underwood, E.J., and N.F. Suttle. 1999. The mineral nutrition of livestock. CAB International Publishing, New York. 614 pp.

USDA. 2002. Current world market and trade report. USDA-FAS (U.S. Department of Agriculture, Foreign Agriculture Services), Oilseed Circular 09-02. USDA, Washington, D.C., Table 22.

Wilson, R.P. 1994. Amino acid requirements of finfish. In: J.P.F. D'Mello, (ed.), Amino acids in farm animal nutrition. CAB International, Wallingford, U.K., pp. 377-399.

An Overview of Fishery Byproduct Use in Russia for Culinary and Non-Food Purposes

Luidmila B. Mukhina
National Center of Quality and Safety of Fishery Products,
St. Petersburg, Russia

Introduction

The National Fish Quality Center was created in 2001 by the order of The State Committee for Fisheries of the Russian Federation. The center provides services to fish processing and catching enterprises in several spheres, from construction of processing plants to sales. Consulting is provided on equipment choice and placement, maintenance of hygiene standards, HACCP implementation, staff training, development of microbiological limits, shelf life determination, and the optimal usage of our bio-resources

The processing of fish and fishery products by land-based enterprises and vessels leads to the production of a large amount of byproducts. Accumulation and storage of spoiled byproducts can result in ecological and sanitary problems. On the other side, byproducts (heads, cartilage, milt, fins, bone, etc.) contain many vital substances.

The Russian government is very concerned about the human population's declining health and changes in nutritional intake that have taken place in the last ten years. As studies have shown there are shortages of proteins, fats, and vitamins in the diet of most Russians today. Compared with 1990, meat and meat product consumption has decreased by 32-40%, milk and dairy products by 33-37%, and vegetables by 35-40%. This results in the deficiency of proteins by 26-30%, of vitamins by 50-52%, and also some slight deficiencies of selenium (Se), calcium (Ca), iron (Fe), magnesium (Mg), and other micronutrients in the diet of the Russian population. The negative changes in food consumption have contributed to a decrease in health and lifetime expectancy. The Russian population has decreased at a rate of 1 million per year (2001 data) and lifetime expectancy has decreased to 65.6 years. In 1990 Russia was in seventh place in the world for per capita food consumption, and today it is seventy-first.

Recently, the Russian government has developed "the government policy concept of healthy nutrition of the Russian Federation population for the period till 2005." The concept includes a wide range of measures, and stresses the necessity of consuming high nutritionally valued protein, and supplying mineral and vitamin enrichment. To help meet the goals, fishery products should be considered as an important source of essential amino acids, omega-3 fatty acids, and micronutrients. Russia is a major fishing country and has many fisheries within its waters. Reduced world fishing resources is of concern to Russia as well as other countries. In Russia there is an intensive search for new raw material sources such as fishery byproducts and development of new products that are safe and of high nutritional value.

One of the most rational processing methods is fish mince production. Filleting reduces edible flesh on the carcass from 40-60% to 28-33% and the remaining tissue is available for mince production using more complex processing methods. Advantages of the more complex tissue recovery processing methods follow:

- Simultaneous processing of different kinds of fish and other seafoods.

- Use small size fish unfit for filleting.

- Provide material for new fishery products such as fish mince combined with new products and/or enriched with bioactive compounds.

- Potential source of tissues and components from non-traditional fishery materials.

Salmon byproducts

Specialists of the National Fish Quality Center in St. Petersburg, Russia, and the State University of Trade and Commerce Center in Russia have a joint scientific research program to develop methods to use salmon byproducts containing bone, such as heads, fins, and frames, to be added to pollock mince. Although fish flesh is rich in protein (Table 1), it lacks the minerals for a balanced diet, and fish flesh does not provide an optimal phosphorus and calcium ratio (Table 2). The goal is to make a natural component of salmon heads and backbones that is added to fish mince to improve its mineral content (Table 3).

The byproduct mince is a rich source of eicosanoic acids. The saturated to unsaturated fatty acid ratio is 1:2.9 for salmon byproduct mince and 1:2.5 for combined mince. By cooking the salmon byproduct mince the shelf life of the product can be prolonged. The ratio of calcium to phosphorus is 5.3:1 before cooking, and reaches 1:1.3 after cooking.

Table 1. Chemical composition (percent) of flesh and byproducts of pollock and salmon.

Parts	Moisture	Fat	Protein	Ash	Energy value (kcal)
Pollock					
Flesh	80.0	0.4	18.0	1.2	77.2
Bones	74.0	0.5	16.0	8.8	70.0
Fins	75.0	0.7	14.8	8.5	67.0
Chum salmon					
Head	70.8	10.2	14.8	3.4	156.0
Skin and flesh	71.3	5.3	22.0	1.5	139.0
Bones	62.4	11.0	18.2	90.0	176.0
Fins	66.4	6.6	16.9	9.8	130.7
Pink salmon					
Head	69.5	9.7	13.2	4.7	139.0
Skin and flesh	69.0	7.3	22.6	1.8	175.7
Bones	58.6	10.6	18.8	9.2	175.7
Fins	65.4	9.4	16.7	10.0	156.0

Table 2. Mineral composition (mg per 100 g) in pollock and salmon flesh.

Element	Pink salmon	Chum salmon	Pollock
Sodium (Na)	70.0	77.5	68.5
Potassium (K)	304.0	243.0	358.0
Calcium (Ca)	43.7	21.5	9.8
Magnesium (Mg)	41.5	23.4	30.8
Phosphorus (P)	227.0	232.0	268.0
Iron (Fe)	2.6	1.2	0.8

Table 3. Calcium, magnesium, and phosphorus content (mg per 100 g) in pollock mince, salmon byproduct mince, and compound mince before cooking.

Element	Salmon or pollock byproduct mince	Pollock flesh mince	Compound mince
Calcium (Ca)	4,555	39	3,100
Magnesium (Mg)	50	30	32
Phosphorus (P)	850	550	700

Compound mince is pollock flesh + salmon byproduct mince as 1:1.

Biofertilizer AUROS restores oil-polluted soils

Developing processing technologies for utilizing poor quality fish wastes undergoing decay and wastes from salted fish processing are a priority for Russia. These fish wastes can be utilized to produce fertilizers for agricultural uses. Many countries and areas have fish fertilizer such as the fish fertilizer available in Alaska.

Our group developed the method of composting and a processing line to produce a unique fertilizer called AUROS, which is made from fish processing wastes and pulp-and-paper and woodworking byproducts. It is the only biofertilizer on the domestic Russian market. Excess sludge and deposits from settling tanks can be utilized with fishery wastes and wood or paper fiber as a composting substrate. For composting it is recommended to add granulated or pilled dried byproducts from pulp-and-paper production, carton packing wastes from fish processing enterprises, or compound fillers such as carton and wood sawdust.

AUROS has been tested at the All-Russia Institute of Plant Protection and the State University of Agriculture. It contains a full range of nutrients necessary for plant growth and has higher values than other organic fertilizers. AUROS has two times more nitrogen, phosphorus, calcium, and magnesium than manure or compost. It is non-toxic to animals and humans. Properties are shown in Table 4.

AUROS is a good soil builder, reduces soil acidity, enhances water holding capacity at the level of roots, and stimulates soil bacteria activity including potential oil-destroyers. AUROS is both effective for open air planting and for greenhouse use. It is recommended to add the fertilizer just before planting and then 1-3 times during the growing season. The recommendations for open air use are 20-30 g per vegetable seedling (tomato, cucumber, vegetable marrow, peppers, eggplant, cabbage, and beet), and 80-100 g per m² for seeded vegetables (radish, carrot, parsley, and celery). The nutrient is base mixed with the soil. Experiments with AUROS on St. Petersburg vegetable farms demonstrated increases in tomato and cucumber harvest by 20-40%.

Table 4. Properties of AUROS bio-organic fertilizer.

Chemical composition		Physical characteristics	
Component	% dry matter	Parameter	Rating
Phosphorus (P$_2$O$_5$)	1.09	Nutrient solubility	Good
Potassium (K$_2$O)	0.12	Moisture	15-20%
Calcium (CaO)	1.83	Mechanical strength	Weak
Magnesium (MgO)	0.23	Blocking property	None
Iron (Fe$_2$O$_3$)	0.08	Apparent density	0.25-0.3 g/vol cm
pH = 7.5-8.0			

It is a well-known fact that unfavorable growth conditions lead to nitrate accumulation in plants. Experiments on lettuce, in which conditions were adjusted to heightened nitrate accumulation, showed that AUROS implementation was comparable to "Kemira" fertilizer (available in Finland) having similar nutrient value and 28% less nitrates. AUROS inhibits the growth of phytopathogenic fungus and reduces, by 3-4 times, the rotting effect leading to vegetable root decay. This is particularly important for vegetables grown in greenhouses. AUROS contains considerable quantity of "Trihodermin" which is used as a growth stimulator and root protector for plants in Russia.

The All-Russia State Research Institute has also demonstrated AUROS efficiency in reclamation of oil-polluted soils in combination with agricultural measures. AUROS has basic nutrient substances for plants and soil microflora, good solubility in water, high water-holding ability, hydraulic permeability, and is a good nutrient medium for oil-oxidizing microflora. These properties permit usage of this universal biofertilizer for restoration of oil-polluted soils.

The research on AUROS application has been carried out at the Research Institute of Geography of St. Petersburg State University since 1999. AUROS was used as a biostimulator of processes in the restoration of topsoil and herbage in oil-polluted areas. Results indicated the high efficiency of this biofertilizer in soil phytodetoxification of oil-polluted soil, with concentration of 1.0-5.0 liters per m^2. The restoration time of polluted areas depends both on level of pollution of soil and on conditions for activity of vegetation and oil-destroying microorganisms. Seeds of perennial grass planted in the oil-polluted soil with AUROS resulted in an increased plant survival of 45%. This effect allows acceleration of the process of vegetation restoration in oil-polluted areas.

The long-term monitoring of vegetation formation on polluted sites with oil concentrations of 4.0 liters per m^2 demonstrated the positive effect of the biofertilizer. One year after pollution and the subsequent treatment of the area with perennial grass, the site with AUROS was 60% covered with grass, while the area without AUROS application was only 35% covered.

Fish mince with additives consisting of vegetables and edible fish bones

Specialists at the Moscow Plekhanov Institute of the National Economy (N.I. Shapoval and others) used fish-bone tissue together with vegetable fillers (edible fibers) to prepare new culinary products from fish mince. The pectin substances of carrot and cabbage (up to 30% of the mince mass) results in an increase of water-retention and a reduction in moisture loss during thermal treatment of the mince.

The fish bones of hake were completely softened at a temperature of 95-98°C for 105 min at pH 5.0, or at 112-120°C for 60 min. The fish bones treated in this way had good water retention, due to the presence of gelatin. An investigation of various amounts of added fish-bone tissue on fish mince properties such as water retention, viscosity, and flavor of final products indicated that the optimal content of fish-bone tissue was 15% of the mince mass.

The amino acid composition of fried and steam cooked culinary products indicated a reduction of overall amino acid content; however, the balance between essential and nonessential amino acids remained at the level of control samples. Analysis of the mineral composition of prepared culinary products demonstrated an increase in the content of potassium, magnesium, iron, calcium, and phosphorus. The ratios of calcium-phosphorus and calcium-magnesium changed favorably as shown in Table 5. Studies on rats have shown (Table 6) that the vegetable fillers and the fish-bone tissue do not reduce the protein efficiency ratio of the developed products (increase in weight per gram of the consumed protein). Decrease in cholesterol and blood urea nitrogen content was seen in laboratory animals.

Mince made from heads, bones, fins, and scales of freshwater fish

Investigators from the St. Petersburg Trade-Economy Institute used fish bones, heads, fins, and scales of freshwater species (whitefish, bream, rudd) to manufacture culinary products such as fish cakes (S.A. Denisova, V.V. Shevchenko, and others). Organoleptic evaluations indicated the highest quality was achieved when 10% additives were used in Alaska pollock mince. However, quality of fish mince with additives of fish bones and fins (without thermal treatment) was lower than control samples. The

Table 5. **Ratio of calcium, phosphorus, and magnesium in culinary products made of fish mince with vegetable and fish-bone tissue additives.**

Name of product	Fried		Steam cooked	
	Ca:P	Ca:Mg	Ca:P	Ca:Mg
Natural fish "schnitzel"	1:5.21	1:0.9	1:5.14	1:0.9
Fish cakes with carrot additive	1:3.88	1:0.75	1:3.69	1:0.75
Fish cakes with cabbage additive	1:3.59	1:0.84	1:3.96	1:0.96
Fish cakes with fish-bone tissue additive	1:0.97	1:0.2	–	–
Optimal ratio	1:1	1:0.7		

N.I. Shapoval.

Table 6. **Biological values of fried culinary products made of fish mince with vegetable fillers and fish-bone tissue additives, fed to groups of animals.**

Parameters	Natural fish "schnitzel"	Fish cakes		
		With carrot additive	With cabbage additive	With fish-bone tissue additive
Increase in weight (g)	71.4	74.3	79.0	67.8
Protein consumption (g)	36.9	35.9	44.0	36.9
Protein efficiency ratio	1.93	2.07	1.80	1.84
Content in blood serum				
Overall protein (g%)	6.60	5.56	5.56	6.60
Urea nitrogen (mg)	16.0	14.0	15.0	10.0
Cholesterol (mg%)	93.5	92.0	94.6	90.0

N.I. Shapoval.

additives caused a grayish or brownish tint and a drier, slightly more fragile texture.

Fish mince with added scales was not inferior to the control sample and even surpassed it, due to decreased wateriness of the product. The scales gave the fish cakes a rich consistency and delicate harmonious taste, which was not present in others samples. The introduction of 10-25% scales in the culinary products made it possible to achieve an optimal ratio of calcium to phosphorus of 1:1 to 1:1.5 and also increase the content of magnesium 1.5 to 2.0 times, phosphorus 4 to 8 times, and calcium and iron 20 to 50 times. The consumption of 100 g of product met the human requirements for phosphorus, calcium, and iron.

Fermented fish mince with additives of lactic acid bacteria

One direction of the food development efforts has been in fermented fish products. A partial hydrolysis of proteins by proteolytic fermentation improves taste, aroma, and consistency. Another promising method developed for the meat industry uses an active lactic acid bacteria. The organic acids formed by the bacteria reduce the product pH, which creates optimal conditions for the action of proteolytic fermentation used in meat curing, and the organic acids also inhibit the growth of spoiling microorganisms thus prolonging the shelf life of the product. The lactic acid bacteria used require partially hydrolyzed products and carbohydrates for reproduction. The combination of proteolysis and lactic acid bacteria provides new opportunities for creation of better fish products.

Specialists (N.V. Velichkovskaya, V.D. Bogdanov) of the Far East State Technical Fishery University developed a technology for the manufacture of Alaska pollock fish mince with proteolytic fermentation preparations of crab viscera and mesophilic lactic acid streptococci. Production of lactate from bacteria is enhanced with the addition of a substrate such as sugar (2-3% by mass).

As a result of combined actions of proteolytic fermentation and lactic acid bacteria, hydrolysis of proteins in fermented fish mince achieved 30%. Water-retention after the treatment of the fish mince by fermentation preparations increased by 20-28%, and when lactic acid bacteria were present values were higher. The fermented fish mince with lactic acid bacteria loses the unpleasant fish mince smell and gets a noble smell of crab meat, so the taste of final products is improved.

Processing small pelagic and mesopelagic fishes for food purposes

An important feature when characterizing modern day fish raw materials is the growing number of small size fishes, which are traditionally used

Table 7.   Chemical content (percent) of samples of fish fillet and TIRM made of horse mackerel and hake.

Product	Moisture	Nitrogenous substances (× 6.25)	Fat	Ash
Horse mackerel fillet	77.30	20.60	1.30	0.75
Horse mackerel TIRM	76.97	20.36	1.27	1.30
Hake fillet	78.44	17.60	1.33	1.36
Hake TIRM	80.41	15.93	1.46	1.71

A.I. Mglinets and others.
TIRM = finely crushed fish mass.

Table 8.   Amino acid content of proteins in horse mackerel mince (g per 100 g product).

Amino acids	Fillet	TIRM	Fish skin	Fish bones
Isoleucine	0.33	0.69	0.58	0.46
Leucine	0.88	1.64	1.59	0.73
Lysine	0.93	1.75	1.82	0.76
Sulfur-containing	0.35	0.57	0.56	0.55
Aromatic	0.77	0.85	0.73	0.65
Threonine	0.56	0.98	1.15	0.62
Valine	0.48	0.83	0.91	0.46

A.I. Mglinets and others.
TIRM = finely crushed fish mass.

Table 9.   Content of macro- and microelements in the samples made of horse mackerel (mg per 100 g product).

Samples	Fe	Cr	P	Cu	Pb	Zn	Mg	Ca	K	Na
Fillet	100	4.0	24.0	0.18	4.0	0.3	1.8×10^3	5.0×10^2	2.3×10^4	8.9×10^3
TIRM	140	9.7	22.0	0.13	4.0	4.0	1.3×10^3	4.4×10^3	1.8×10^4	9.7×10^3
Fish bones	800	4.0	29.9	0.90	4.0	2.9	3.4×10^3	5.5×10^4	8.5×10^3	1.5×10^4

A.I. Mglinets and others.
TIRM = finely crushed fish mass.

for animal feed. Development of methods to use these raw materials for human food purposes is vital. A process was developed that makes mince using low value, small size fish which are prepared by cutting whole fish into small slices, scalding and washing them, followed by further processing (V.P. Skachkov).

In the 1980s the methods and equipment were developed that made it possible to produce a finely crushed fish mass (TIRM) from small size raw material. At the same time the useful properties of fish-bone tissue and fish skin were identified (A.I. Mglinets and others). Processing is based on a two-stage crushing of fish. The rough crushing of frozen fish briquettes is made by reducing pieces to 10-20 mm. Then after deep freezing to –25°C, the fish is again crushed in a special rotor crusher to particles of 100-200 microns. The small particle size reduces the taste of fish-bone tissue in the semi-finished product TIRM.

Proximate analysis (Table 7) of TIRM in comparison with fillets of the same fishes indicates the amount of mineral increase. All essential amino acids can be found in TIRM and their content was higher than in the fish fillet (Table 8). The fatty acid composition of lipids from TIRM and fillet were determined. The sum of the saturated fatty acids were the same, but TIRM had a greater content of monounsaturated fatty acids than fillets. Lipid from fillet had a higher content of polyunsaturated lipids than lipids from TIRM. TIRM in comparison with the fillet has a higher content of calcium, iron, and phosphorus (Table 9).

Specialists from the Kaliningrad State Technical University (O.Y. Mezenova, N.Y. Kochelaba, A.B. Odintsov) have developed methods for manufacture of weakly salted fish-protein pastes, which are prepared like preserves from whole small size fish (sprat, mesopelagic fishes).

Conclusion

Analyzing the available published materials on the development of products from non-traditional fish raw materials (fish wastes) indicate not all developments were conducted with the purpose of obtaining functional food products. Taking into account the shortage of fish raw material, some authors had goals such as a complete and integrated processing of raw materials, and increasing the availability of cheap fish products in the region, etc. Our studies investigate important aspects of manufacture of functional products such as:

- Providing choice of the nutrient source based on safety (heavy metals, radionuclides).

- Regulating the amounts of micronutrients in the enriched final product.

- Calculating the micronutrient amount after thermal treatment of the products.

- Estimating nutritional efficiency of enriched product in laboratory animals.

Bait for the Longlining Fishery of Snapper (*Pagrus auratus*)

Francisco Ignacio Blaha
Fisherman, Auckland, New Zealand

Abstract

Natural baits such as squid and pilchard have been the favored choice of fishermen for the longline fishing of snapper (*Pagrus auratus*) in New Zealand. Overexploitation of fish stocks together with an increase in demand for fish for human consumption has decreased the resources of natural baits at cost effective prices. This has led to attempts in the development of artificial or reconstituted baits. In this study, baits were prepared using differing proportions of industry surplus Greenshell™ Mussel (*Perna canaliculus*) and gurnard (*Chelidonichthys kumu*), both known to be part of the snapper diet. These species also have shown high amounts of glycine and alanine, which are known to stimulate feeding behavior in snapper.

Acceptability tests with captive snapper showed that formulations containing 70% of mussel, 30% of gurnard, and 5% of sodium alginate elicited a response that was greater than other combinations of the three ingredients. This preferred formulation was then used in sea trials following extrusion into fibrous casings and gelification by immersion in baths of calcium chloride solution. The hardness obtained after a 7.5% calcium chloride bath was judged the most suitable for longlining sea trials.

Captures per 1,000 hooks of test bait were compared to the same quantity of hooks baited with control squid bait. Control bait was 83.93% more effective than the test bait. Although the catches on the test bait were low, the selectivity toward snapper was 29.25% higher.

Reasons for the reduced total catch compared to the control may be due to less desirable texture and taste of the test bait, compared to natural foods of snapper. In terms of ease of handling by the fishermen, the form and consistency were well liked.

Introduction

Snapper

Snapper (*Pagrus auratus*) is a member of the Sparidae family of sea breams. This species ranges throughout most of the temperate to subtropical water masses of the western Pacific Ocean. *Pagrus major, Chrysophrys auratus,* and *Chrysophrys major* are synonyms for *Pagrus auratus* (Paulin 1990). Snapper are opportunistic carnivores and eat a wide variety of benthic and demersal organisms (Muller 1998).

Longlining

Longlining is a method of catching fish using a line and series of baited hooks, which is known as "the gear." It is defined as a passive fishing method, i.e., the gear is stationary and the encounter between gear and fish is a result of the fish moving to the gear, based on the attraction of fish to bait. The bait serves as the source of smell and taste stimuli to lure the fish to the gear and to ingest the baited hook. The success of capture is also dependent on the ability of the hook to catch and retain the fish until it is brought on board the fishing vessel.

Longlining is the favored method for the capture of snapper in New Zealand, resulting in a higher valued fish compared to that obtained by trawling, because the fish does not get damaged in the trail net and is generally alive when brought aboard. It has been regarded as one of the most efficient and environmentally sound methods of fish capture, on the basis of size selection, species selection, survival after escape, ghost fishing, fish quality, energy consumption/pollution, and impact on the seabed (Bjordal and Løkkeborg 1996).

Hooks

The main variables of a fishhook are size, shape, and coating, which give an indefinite number of possible combinations. Fishhook terminology is a somewhat confusing subject, since there is no strict international standardization of definitions.

Baits

Fishing with baited hooks is based on the target species' demand for food, and on the bait properties of stimulating the food-searching behavior of fish through odor release. Once the fish has been attracted to the longline, the physicochemical nature of the bait triggers the fish to attack and ingest the baited hook. A key factor in this process is the smell and taste stimuli generated by the bait, i.e., the concentration and composition of the chemical compounds released by the bait (Bjordal and Løkkeborg 1996).

Traditionally natural baits have always been the choice for longlining, and the selection of which type is based mainly on the fisherman's experience, the availability, and price of those baits. The traditional natural baits used in the snapper longline fishery are pilchard (*Sardinops neopilchardus*) and arrow squid (*Notodarus gouldi*). However, these high quality natural bait resources are becoming less available at affordable prices.

The knowledge that specific chemicals can evoke feeding behavior (Atema 1980; Goh and Tamura 1980a,b; Carr and Derby 1986) has lead to efforts in the development of artificial and/or reconstituted baits (Bjordal and Løkkeborg 1996). These baits could have the following advantages: improving selectivity in targeting specific species and thus avoiding by-catch and bait loss; uniformity in shape and size of baits, thereby making mechanized baiting cheaper and more efficient; and finally, a bait prepared from recycling fishing industry surplus products would increase the overall efficiency of fishing operations.

Although no published research in this area has been reported for any species in New Zealand, some attempts have been made in some major fishing nations (Norway, Canada, United States, Spain) to develop suitable bait for particular species. Most of the prior mentioned research is not public knowledge, due to the commercial interest in this topic. However, three areas of importance have been identified:

1. Feeding attractants and stimulants

The ability of certain L-amino acids to modify behavioral activity has been confirmed for different species of the genera *Chrysophrys = Pagrus* (*C. major = Pagrus auratus*) by Goh and Tamura, (1980a,b), Fuke et al. (1981), and Shimizu et al. (1990). These authors have identified L-glycine and L-alanine as potent stimulants inducing feeding behavior.

Hughes et al. (1980) and Pickston et al. (1982) have evaluated the levels of 18 different amino acids on behavioral activity of 12 commercially harvested marine species in New Zealand waters. Three of those species are known to be part of snapper diet: gurnard (*Chelodinichthys kumu*), scallop (*Pecten novaezealandiae*), and Greenshell™ Mussel (*Perna canaliculus*) (Berquist 1994, Muller 1998).

Industry byproducts from Greenshell mussel and gurnard are available, and were chosen as a basic mixture to be tested for acceptability.

2. Binders

Binders can help to mix all of the nutritional composition evenly and stick them together to enable hook stability in the bait, avoid loss of nutrients/feed in water, and minimize pollution of the environment. Alginates have been the preferred binder on research done by the group lead by J. Boyer (Centre technologique des produits aquatiques, Ministere de l'Agriculture, des Pecheries et de l'Alimentation, Quebec, Canada, 1999, pers. comm.). Alginate, commercially available as alginic acid, commonly called sodium alginate, is already used in bulk quantities in the local fishing industry were it is mixed in low concentration with water as part of the glazing for frozen snapper.

Calcium chloride has been chosen as a calcium donor in the gelation process, based on previous studies (J. Boyer, pers. comm.), commercial availability, and dilution properties. However, the potential for calcium to influence preference by fish has not been studied, as it exceeded the scope of this work.

3. Reinforcement material

Sausage casings have been cited by Bjordal and Løkkeborg (1996) as suitable reinforcement material. Four types of casings were commercially available and included those made from intestine, collagen, cellulose, and plastic. The cellulose casings were used in this work because of strength, commercial availability, and size to fit the hooks.

Materials and methods

Bait sausage production.

Raw materials used were frozen Greenshell mussel meat and frozen gurnard heads and filleted bodies. Frozen raw materials were minced in a Dimak mincer model DMK-3000 to form a paste, then further mixed and homogenized in different proportions (see below) with Pronova Frostgel SFPH sodium alginate (Norpro NZ Ltd., Auckland) in a food processor Kenwood model Chef. The alginate was added at 2.5% and 5% by weight.

A Kenwood Chef sausage-making nose model A-12 was fitted to make the bait sausages. Nojax clear/black size 29/70 fibrous casings (Oppenheimer New Zealand Ltd., Wellington) were used to form 50 cm long bait sausages of approximately 250 grams each; both ends were tied using 2 mm diameter hemp string. Two to four sausages were made per treatment.

The gelation baths were carried out in 30 liters polystyrene bins containing 10 liters of Lancaster calcium chloride (Bronson & Jacobs PTY., Auckland) in tap water solution at 5%, 7.5%, and 10% by weight at ambient temperature. Bait sausages were labeled and immediately placed under refrigeration at 1°C. Sausage composition is shown in Table 1.

Preference evaluation

To evaluate the snapper preference for the different types of mixture, a modification of the methodology described by Fuke et al. 1981 for the identification of feeding attractants was used.

Two-year fry (23 to 28 cm in body length) from Pah Farm Snapper Breeding Station, Kawau Island, New Zealand, were used for experimentation. Thirty fish were transferred to a commercial 1 m³ bin directly from the rearing ponds, with running seawater, temperature between 16 and 17°C during experimentation, and indoor artificial lighting.

Baits used in this experiment had not been submerged in the gelation baths. One of the two bait sausages, for each mixture ratio, was cut in half midway along the total length of the sausage, and a 2.5 cm length slice

Table 1. Composition of 10 experimental bait sausages.

Mixture	Gurnard %	Mussel %	Alginate %
1	0	100	5
2	30	70	5
3	50	50	5
4	70	30	5
5	100	0	5
6	0	100	2.5
7	30	70	2.5
8	50	50	2.5
9	70	30	2.5
10	100	0	2.5

was taken as a test sample. The sample was attached to the tip of a 1mm monofilament fishing line and suspended at 7 to 10 cm from the bottom of the tank, near the incoming seawater hose.

The responses of the 30 fish to the sample were measured as frequency of biting per minute. Each sample was tested three times, and the intervals between samples were of at least 5 minutes. A 2.5 cm length slice of a 2 cm diameter pale red and black striped garden hose was used as control during a minute before every test. Its presence in the water a minute before the test had the purpose of getting the fish accustomed to the presence of an unusual object. The samples were tested in numerical order during the morning and, using the second sausage, a duplicate assay at random sample order was completed during the afternoon using the same fish. Results were expressed as an average of biting per minute.

Bait sausage hardness

By direct visual and tactile observations during the acceptability tests, it was determined that bait hardness and casing adherence were inadequate to withstand sea and gear handling conditions. Consequently, increasing the hardness of the sausage bait became an objective of further research. The hardness of the bait sausage mixtures 7 and 2, containing 2.5% and 5% in weight of sodium alginate respectively, were measured after immersion in a gelation bath at increasing concentrations of calcium chloride.

The baths were carried out in 30 liter polystyrene bins containing 10 liters calcium chloride solution at the following concentration: 5.0%, 7.5%, and 10%.

All compression measurements were performed on a Texture Analyser (Stable Microsystems, Surrey, England), interfaced with Texture Expert Version 1.0 software program (Stable Microsystems, Surrey, England) on sample slices (3 cm diameter) placed longitudinally on a stainless steel flat plate (4 cm diameter) for uniaxial compression. The crosshead speed was maintained at 0.5 mm per second, with automatic force trigger sensor set at 0.2 N. Hardness (N) was defined as maximum force of peak at 40 seconds.

Sea trials

Longlining technique

The longlining equipment used for the sea trials was a Mustad Autoline gear. The spacing between hooks was fixed at 1 meter and for this experiment 550 hooks were used on an independent transferable rack. The trials were performed on board four vessels using the same longlining method and the same 550 hook rack. The trial was done once on each vessel during standard fishing operations. The hooks used were standard Tainawa Mustad # 17 (2.0 cm gap) on 1mm monofilament gangion, as they are the preferred type by the fisherman. Sliced arrow squid was used as control bait for the sea trials.

Baiting methodology

The method followed Løkkeborg (1990b, 1991) and Løkkeborg and Johannsessen (1992) for measuring the effectiveness of two different bait types by baiting the lines in alternate clusters of about 50 similarly baited hooks. Sausages were peeled, and cut into approximately 2.5 cm slices for each hook.

The following configuration was used to remove possible variations: (a) first 50 hooks of control bait, (b) 1-2 hook space and a color marker, (c) 50 hooks with test bait, (d) 1-2 hook space and a marker, and (e) 50 hooks with control bait until the line was set with 250 baited hooks for each type of bait.

The area selected to perform the trial is the east coast of the Coromandel Peninsula, a traditional longline snapper fishing ground. The depth of setting varied between 50 and 190 meters, as most of the gear was set following bottom contour. The data was collected during hauling of the line after an overnight set of approximately 9 hours, and the results transcribed to a preprinted form to be further analyzed once ashore.

Data analysis

Data collected from 20 clusters (1,000 hooks) lured with test bait were compared with the data obtained from the same number of clusters of hooks lured with control bait.

Table 2.   Snapper bites per minute.

Mixture	Control	Bites/min numeric order			Bites/min random order			Avg.
1	0	24	27	23	26	20	27	24.50
2	0	30	a	a	a	a	a	a
3	1[b]	a	29	23	25	28	30	27.00
4	0	23	24	26	22	24	25	24.00
5	0	20	21	21	21	22	19	20.67
6	0	23	26	24	25	26	24	24.67
7	0	30	a	a	29	a	a	a
8	0	25	28	26	25	27	28	26.50
9	0	24	23	24	22	24	21	23.00
10	0	21	17	23	19	18	20	19.67

[a]Counting was impossible due to the high frequency of biting.
[b]This represents the single bite on the plastic hose control.

Results

Mixture preference evaluation

The results of the preference evaluation are summarized in Table 2. Mixtures 2 (70% mussel:30% gurnard:5% alginate) and 7 (70% mussel:30% gurnard:2.5% alginate) were the most appealing to snapper, with such a high frequency of biting that counting was impossible.

The results suggest that the preference toward the bait diminished as the relative amount of mussel meat was reduced in the composition; however, mussel meat itself (mixtures 1 and 6) does not induce the same level of feeding behavior as the mixtures. Thus, mixtures containing 100% gurnard mince (samples 5 and 10), triggered the lowest level of attraction among those measured. These results could suggest some form of synergism between components of the mixture and warrant further investigation in the future.

The concentration of sodium alginate did not seem to alter significantly the biting per minute count. Yet, the results for counts on 5% alginate were slightly higher in comparison to those on 2.5%. Mixture samples 2 and 7 were chosen for further experimentation on increasing hardness.

Bait sausage hardness

Result for hardness analysis on triplicates of mixture samples 2 and 7 after 5%, 7.5%, and 10% calcium chloride bath showed an increase in hardness as a function of the increasing concentrations of the sodium alginate and calcium chloride in the bath. The highest hardness values are for mixture sample 2. The lowest hardness values were found in samples treated with 5% calcium chloride baths for both mixture samples. However, the figures do not show marked differences in hardness on sample mixtures after treatment in the 7.5% and 10% calcium chloride geling baths.

Based on these results, 7.5% calcium chloride solution was chosen as a geling bath for mixture 2 for further experimentation at sea trials, and mixture 7 was abandoned as being softer, thus potentially less practical for fishing purposes.

Sea trials

The results of the sea trials indicated that squid was 83.93% more effective as bait compared to the test bait. However, the test bait was 29.25% more selective toward snapper than the squid bait (statistical significance was not studied). In other words, squid caught on average 34.85 fish per cluster (50 hooks) of which 20.3 were the target species. In contrast, the test bait caught only 5.6 fish per cluster of which 4.9 were snapper. The distribution of snapper along the clusters of test baited hooks did not show preference for the hooks in the proximity of squid bait in comparison to those in the middle of the cluster.

Based on these results the commercial suitability of the test bait was seriously compromised, since the bait "does not catch enough fish" to be able to compete with squid as an economically viable alternative. Nevertheless, the high selectivity toward snapper was highly encouraging and has potential for further investigation.

As a further result, the firm consistency of the sausage baits in mixture 2, after the treatment in geling bath, enable the casings to be removed prior to baiting the hooks, therefore the reinforcement material is not as critical as presupposed for this type of bait sausage.

Discussion

The test bait represents a novel food item that the fish has probably never experienced before. This may cause a somewhat restrained response toward the bait and explain why there was a fairly low proportion of attracted fish that apparently swallowed the baited hooks. This proportion is probably lower for smaller fish, because they have a narrower diet range and therefore show a more restrained response toward novel food items.

Variability in the results between fishing vessels was not analyzed, because the main interest of the experience was in the total catches, and not in the particulars of each boat. The main reason behind using four boats was that no boat owner would accept setting up 2,000 hooks (a whole trip) without having a guarantee that the catches would cover the fixed costs of the trip. The fishing operations took place over a period of two weeks, and no major differences were expected on the seasonal and diurnal foraging cycles of snapper. The abundance was not visibly compromised by any special factor.

The competition toward the bait by non-target species was a primary objective of this study. Glycine and alanine are two of the most frequently cited feeding stimulants being reported, and were shown to affect 28 (80%) and 26 (74%) (respectively) of the 35 species of marine organisms studied by Carr et al. (1996). It is expected that other species that share the same habitat are attracted to the bait as well.

It is important at this point to state that the sausage form and consistency were well liked by all fishermen involved in the study, since it was practical to store, clean to handle, and easy to bait.

Conclusions

Under the present bait composition and form, the results of this study can be summarized as follows:

1. The test bait effectiveness does not suit industry needs.

2. The test bait selectivity toward snapper, under the limitations of this study, was higher.

3. The test bait form and consistency does suit industry needs.

4. Casing material is not as critical as presupposed.

5. Modifications on the taste and texture characteristics of the mixtures could lead to more promising results.

6. Development of baits based on seafood industry byproducts is a viable alternative for further study.

References

Atema J. 1980. Smelling and taste underwater. Oceanus 23:4-18.

Berquist, R.M. 1994. Patterns of activity and movement in New Zealand snapper, *Pagrus auratus*. M.S. thesis, The University of Auckland, Auckland, New Zealand. 125 pp.

Bjordal, A., and S. Løkkeborg. 1996. Longlining. Fishing News Books, Osney Mead, Oxford, England. 156 pp.

Carr, W.E.S., and C.D. Derby. 1986. Chemically stimulated feeding behaviour in marine animals and involvement of mixture interactions. J. Chem. Ecol. 12: 989-1011.

Carr, W.E.S., J.C. Netherton, R.A. Gleeson, and C.D. Derby. 1996. Stimulants of feeding behaviour in fish: Analysis of tissues of diverse marine organisms. Biol. Bull. 190:149-160.

Fuke, S., S. Konosu, and K. Ina. 1981. Identification of feeding stimulants for Red sea bream in the extract of marine worm (*Perineris brevicirrus*). Bull. Jap. Soc. Sci. Fish. 47(12):1631-1635.

Goh, Y., and T. Tamura. 1980a. Olfactory and gustatory responses to amino acids in two marine teleosts: Red sea bream and mullet. Comp. Biochem. Physiol. 66C:217-224.

Goh, Y., and T. Tamura. 1980b. Effects of amino acids in the feeding behaviour in Red sea bream. Comp. Biochem. Physiol. 66C:225-229.

Hughes, J.T., Z. Czochanska, L. Pickston, and E.L. Hove. 1980. The nutritional composition of some New Zealand marine fish and shellfish. N. Z. J. Sci. 23:43-51.

Løkkeborg, S. 1990. Reduced catch of under-sized cod (*Gadus morhua*) in longlining by using artificial bait. Can. J. Fish. Aquat. Sci. 47:1112-1115.

Løkkeborg, S. 1991. Fishing experiments with an alternative longline bait using surplus fish products. Fish. Res. 12:43-56.

Løkkeborg, S., and T. Johannsessen. 1992. The importance of chemical stimuli in bait fishing: Fishing trials with presoaked baits. Fish. Res. 14:21-29.

Muller, C.G. 1998. Can snapper (*Pagrus auratus*) (Pisces: Sparidae) feed visually at night? Master's thesis, The University of Auckland, Auckland, New Zealand. 98 pp.

Paulin, C.D. 1990. *Pagrus auratus*, a new combination for the species known as "snapper" in Australasian waters (Pisces: Sparidae). N. Z. J. Mar. Freshw. Res. 24:259-265.

Pickston, L., Z. Czochanska, and J.M. Smith. 1982. The nutritional composition of some New Zealand marine fish. N. Z. J. Sci. 25:19-26.

Shimizu, C., A. Ibrahim, T. Tokoro, and Y. Shirakawa. 1990. Feeding stimulation in sea bream, *Pagrus major*, feed diets supplemented with Antarctic krill meals. Aquaculture 89:43-53.

Shelf Life Extension of Mangosteen by Chitosan Coating

Attaya Kungsuwan, Bordin Ittipong, and Pawared Inthuserd
*Ministry of Agriculture and Cooperatives, Department of Fisheries,
Fishery Technological Development Division, Bangkok, Thailand*

Samrual Dokmaihom
*Ministry of Agriculture and Cooperatives, Department of Agricultural
Extension, Plant Protection Service Division, Bangkok, Thailand*

Abstract

Mangosteen is one of the economically important fruits of Thailand, which has a problem of short shelf life. This study investigates this problem through the application of chitosan to the fruit. Chitosan, a natural polysaccharide, is derived from crustacean shells and cephalopod pens. Chitosans applied in this study were obtained from crab shell (Cr, 89% degree of deacetylation, DD), shrimp shell (Sh, 89.2% DD), and squid pen (Sq, 85% DD). All chitosan types were dissolved in 1% acetic acid at the designated concentration (%, w/v) before spraying on the mangosteen. Six treatments were set up using 5 kg each of mangosteen. Those experiments were two controls (no spraying and spraying with 1% acetic acid) and four treatments (1% Sh, 2% Sq, 1% Cr, and 2% Cr). After drying, mangosteens were put into plastic bags and then kept at 10°C. The quality indices of mangosteen were hardening of the fruits, fruit color, and taste and juiciness of the flesh. Results showed that the most effective chitosan type and concentration was 2% Cr, which could extend the shelf life of mangosteens to more than 23 days with hardening of only 12%, compared to the control with a shorter shelf life with 22% hardening. This preliminary study has shown that chitosan can be applied as a tool for eradicating the post-harvest loss of mangosteen. Further experiments are needed on other chitosan types, packaging systems, and storage conditions for shelf-life extension.

Introduction

Mangosteen (*Garcinia mangostana* L., Clusiaceae) is one of the economically important fruits of the tropics. Its origin is in Southeast Asian countries and Indonesia. It can grow well in Thailand, Malaysia, Myanmar, Vietnam, and Cambodia. The mangosteen is round with a pericarp thickness of about 0.8-1.0 cm, and a dark purple color after ripening. Mangosteen is quite popular among Thai people; it is considered the "queen of fruit" there. Popularity has increased in foreign countries during the last decade. Other Asian countries, including Japan and Hong Kong, are big importers of Thailand's mangosteen. Table 1 shows production of mangosteen in Thailand between 1995 and 2000. Table 2 shows the export and value of mangosteen during 1995-2001.

The pericarp thickness of the mangosteen would indicate that it is a durable fruit that can withstand any impact. But in fact it is quite delicate. Research has indicated that the pericarp of mangosteen turns firm and hard after falling from the tree and hitting the ground. The higher the falling distance the quicker the pericarp hardens, due to lignin formation. This problem occurs in all mangosteens (Tongdee and Suwanakul 1989, Ketsa and Koopluksee 1993, Ketsa and Atantee 1998), and is a problem when mangosteens are exported. This study is to find a way to prolong the shelf life of mangosteen, in order to increase exports of mangosteens to Asian countries, especially Japan.

In Thailand the fishing industry is composed of harvesting, freezing, drying and salting, canning, and other processing. Table 3 shows the total catches of marine crustaceans and cephalopods during 1995-1999. For shrimp, especially black tiger shrimp, 90% is from aquaculture production, of which 90% is processed for export as frozen and canned products. For crabs 50% of total catches is exported as frozen or canned product.

Together with these industries are byproducts generated such as shrimp shells, crab shells, and squid pens. These precious byproducts are important raw materials for chitin and chitosan production. Both chitin and chitosan products have been utilized in many industries, even in medical science (Biagini et al. 1991, Miyazaki et al. 1996, Ventura 1996). Moreover, since agriculture is vital to Thailand, especially post-harvest technology upgrading, it is reasonable to carry out the research for such a goal.

This study was performed to evaluate different chitosans and different chitosan concentrations for prolonging the shelf life of mangosteen.

Materials and methods

Materials

Materials employed in this study were ripe mangosteens from Chanthaburi province (5 kg for each treatment, 10 fruits per kg), chitosan (crab shell

Table 1. Area and production of mangosteen in Thailand during 1995-2000.

Year	Area (rai)[a]	Production (tons)
1995	236,666	128,280
1996	265,568	143,311
1997	270,289	175,118
1998	292,604	155,044
1999	301,924	168,325
2000	352,301	176,763
Annual rate of increase (%)	9.5	8.9

[a]2.529 rai = 1 acre.
Source: Department of Agricultural Promotion, Thailand.

Table 2. Quantity and value of mangosteen exported during 1995-2001.

Year	Fresh mangosteen		Frozen mangosteen		Total	
	Quantity (ton)	Value (million Baht)	Quantity (ton)	Value (million Baht)	Quantity (ton)	Value (million Baht)
1995	3,117	65.721	704	46.121	3,821	111.842
1996	2,167	39.469	707	56.991	2,874	96.450
1997	2,812	62.376	436	37.005	3,248	99.381
1998	2,319	44.026	413	23.226	2,732	67.252
1999	5,001	104.832	281	25.895	5,282	130.727
2000	12,886	257.668	227	25.810	13,113	283.478
2001	18,388	408.430	329	21.167	18,717	429.597

Source: Department of Custom, Thailand.
US$1 = 41 Baht (exchange rate in February 2003).

Table 3.   Total Thailand crustacean and cephalopod catches, 1995-1999, in 1,000 t.

Year	Shrimp[a]	Crab	Squid and cuttlefish
1995	391.5	52.4	156.4
1996	373.9	52.9	173.2
1997	353.9	51.1	173.6
1998	349.3	58.0	188.1
1999	362.3	55.4	174.4

[a]90% was from aquaculture.
Source: Department of Fisheries (2002).

[*Portunus* spp.], 89% degree of deacetylation; shrimp shell [*Penaeus monodon*], 89.2% DD; and squid pen [*Loligo* spp.], 85% DD), and acetic acid (BDH Laboratory Supplies). All raw chitosan was obtained from local fish processing plants.

Methods

Chitosan solution preparation involved dissolving chitosan in 1% aqueous acetic acid. The concentrations of chitosan prepared were 1% shrimp chitosan (1% Sh, w/v), 2% squid pen chitosan (2% Sq), 1% crab shell chitosan (1% Cr), and 2% crab shell chitosan (2% Cr).

Each of the chitosan solutions was sprayed thoroughly on the pericarp of the mangosteens. Two controls were used—one with spraying with 1% acetic acid and another with no spraying. After spraying, the mangosteens were let dry before being packed in plastic bags and kept at 10°C. During refrigeration the appearances of mangosteens of each treatment were examined every day until being rejected.

The quality indices used were appearance, color, taste, and hardening of the pericarp, which is the biggest problem. Each week the color was checked. The flavor test was conducted by five panelists, and the scores were averaged.

Results and discussion

The most important index for consumer decision is appearance, including color, shininess, and firmness of pericarp.

The controls were rejected on 10 days (data not shown) due to a change to pale color and pericarp hardening. The four treatments: 1% Sh, 2% Sq, 1% Cr, and 2% Cr, prolonged the shelf life of mangosteen to 20 days. On the 23rd day the appearances and flavor for the 1% Sh, 2% Sq, and 1% Cr samples were undesirable and rejected; the 2% Cr was still acceptable.

Table 4. Evaluation of chitosan-coated mangosteen after 23 days of storage.

Code	Pericarp color	Pericarp hardening (%)	Remarks
Control	Pale purple	22	The pericarp changed from dark purple to pale purple and was not shiny. The calyx shrank with no green color left. The flesh was dull white or turned yellow with some spoilage seen as black or brown spots.
1% Ac	Pale purple	22	Pale pericarp. The calyx also turned pale green, dried, and shrank to about 20%. The flesh was somewhat whiter than the control but tasted sweet and sour.
1% Sh	Purple	3	Purple pericarp, with dry calyx but less than those of control and 1% Ac. The flesh was yellowish white, too moist, and less sweet which was nearly like the control.
2% Sq	Dark purple	26	The pericarp was dark purple but less shiny than those of 2% Cr. Fifty percent of the calyx was still green. Flesh was white and juicy with some spoiled fruits.
1% Cr	Purple	24	50% of the pericarp was pale. The rest was bright purple. Calyx dried up and shrank, with brown color. Flesh was stiff, not so white, and with some spoilage.
2% Cr	Dark purple	12	The pericarp was dark purple and still very shiny. Calyxes were mostly green with little shrinkage. Flesh was white, juicy, and stayed loose from the inner side of the pericarp (normal).

Ac = acetic acid; Sh = chitosan derived from shrimp shell; Sq = chitosan derived from squid pen; Cr = chitosan derived from crab shell.

In Table 4 are the results of sensory evaluation of chitosan-coated mangosteen after 23 days of storage. Crab chitosan of 2% was found to be the best concentration to preserve the mangosteens due to less hardening (12%), the best appearance, and desirable sweet and juicy taste. The hardening of 1% Sh was only 3% but the flesh was not acceptable. More studies are needed on properties of different chitosans.

Normally, we can keep mangosteens at room temperature for about 5-7 days or in the refrigerator for 7-10 days. In this study, chitosan prolonged the shelf life of mangosteen. The cost of chitosan coating is low and can help to increase shelf life. More studies should be conducted to optimize the effects of spraying chitosan on mangosteen to increase shelf life.

Acknowledgment

This study has been supported by the research fund of the Department of Fisheries, Ministry of Agriculture and Cooperatives, Thailand.

References

Biagini, G., R.A.A. Muzzarelli, R. Giardino, and C. Castaldini. 1991. Biological materials for wound healing. In: C.J. Brine, P.A. Sandford, and J.P. Zikakis (eds.), Advances in chitin and chitosan. Elsevier Applied Science, London, pp. 16-24.

Department of Fisheries. 2002. Fisheries statistics of Thailand 1999. Fisheries Economics Division, Department of Fisheries, Ministry of Agriculture and Cooperatives, No. 10/2002, Bangkok. 87 pp.

Ketsa, S., and S. Atantee. 1998. Phenolics, lignin, peroxidase activity and increased firmness of damaged pericarp of mangosteen fruit after impact. Post Harvest Biol. Technol. 14:117-124.

Ketsa, S., and M. Koopluksee. 1993. Some physical and biochemical characteristics of damaged pericarp of mangosteen fruit after impact. Post Harvest Biol. Technol. 2:209-215.

Miyazaki, S., N. Kawasaki, I. Serizawa, and K. Doi. 1996. Pharmaceutical application of chitosan: Drug release from oral mucosal adhesive films of chitosan and alginate. In: W.F. Stevens, M.S. Rao, and S. Chandrkrachang (eds.), Proceedings of 2nd Asia Pacific Symposium on Chitin and Chitosan: Environmental Friendly and Versatile Biomaterials. Asian Institute of Technology, Bangkok, pp. 291-294.

Tongdee, S.C., and A. Suwanagul. 1989. Post harvest mechanical damage in mangosteen. ASEAN Food J. 4(4):151-155.

Ventura, P. 1996. Lipid lowering activity of chitosan, a new dietary integrator. In: R.A.A. Muzzarelli (ed.), Chitin enzymology, vol. 2. Atec Edizioni, Ancona, Italy, pp. 55-67.

Nutraceuticals and Bioactives from Seafood Byproducts

Fereidoon Shahidi
Memorial University of Newfoundland, Department of Biochemistry,
St. John's, Newfoundland, Canada

Abstract

Seafood processing byproducts may account for up to 80% of the weight of the harvest and, depending on the species involved, include a variety of constituents with potential for use as nutraceuticals and bioactives. Nutraceuticals, defined as ingredients or extracts with clinically proven health promoting activity, including disease prevention and treatment, may be consumed in the medicinal form, as supplements or as functional food ingredients. Thus, a variety of seafood processing byproducts may render benefits above their nutritional value. As an example, highly unsaturated long-chain omega-3 fatty acids, derived from the liver of white lean fish, flesh of fatty fish, and blubber of marine mammals, exhibit important biological activities. They also serve as the building block fatty acids in the brain, retina, and other organs with electrical activity. Hence, inclusion of oils containing docosahexaenoic acid (DHA) in the diet of pregnant and lactating women as well as infants is encouraged. In addition, chitinous materials, carotenoids, and biopeptides, arising from proteins, may be recovered from processing byproducts of crustaceans, including shrimp, crab, and lobster. The health benefits of chitosan, chitosan oligomers, and glucosamine are related to the multifunctional role of these ingredients, including immunomodulatory activity. Meanwhile, antioxidative peptides with up to 16 amino acids in chain length have been isolated from skin of pollock and their use as nutraceuticals may prove beneficial. Hence, byproduct processed from seafoods may possess multifunctional roles and could serve as important value-added nutraceuticals and functional food ingredients.

Introduction

Seafoods have traditionally been used because of their variety of flavor, color, and texture. More recently, seafoods have been appreciated because of their role in health promotion arising primarily from constituent long-chain omega-3 fatty acids, among others. Nutraceuticals and bioactive components from marine resources and the potential application areas are listed in Table 1. Processing of the catch brings about a considerable amount of discard which may account for 10-80% of the total landed weight, depending on the species under consideration. The components of interest in seafood processing byproducts include lipids, proteins, flavorants, minerals, carotenoids, enzymes, and chitin, among others. The raw material from such resources may be isolated and used in different applications, including functional foods and as nutraceuticals. The importance of omega-3 fatty acids in reducing the incidence of heart disease, certain types of cancer, diabetes, autoimmune disorders, and arthritis has been well recognized. Such lipids are derived primarily from byproducts of the seafood processing industry and originate from the body of fatty fish, liver of white lean fish, and blubber of marine mammals. In addition, the residual protein in seafoods and their byproducts may be separated mechanically or via a hydrolysis process. The bioactive peptides so obtained may be used in a variety of food and non-food applications. The bioactives from marine resources and their application areas are diverse in general. A cursory account of nutraceuticals and bioactives from selected seafood processing byproducts is provided in this contribution.

Nutraceutical lipids
Marine oils

The long-chain omega-3 polyunsaturated fatty acids (PUFA) are of considerable interest because of their proven or perceived health benefits (Simopoulos 1991, Abeywardena and Head 2001, Shahidi and Kim 2002). These fatty acids are found almost exclusively in aquatic resources (algae, fish, marine mammals, etc.) and exist in varying amounts and ratios. While algal sources also provide minerals, such as iodine, as well as carotenoids and xanthophylls, fish body oil contains mainly triacylglycerols and fish liver oils serve as a source of vitamin A, among others. In addition, liver from other aquatic species, such as shark, contain squalene and other bioactives. Another source of long-chain omega-3 fatty acids is the blubber of marine mammals which contains eicosapentaenoic acid (EPA) and docosahexaenoic acid (DHA), similar to fish oils, as well as docosapentaenoic acid (DPA). It is worth noting that myristic acid is present in much smaller levels in the blubber oil from marine mammals than in algal or fish oils; this is a definite advantage when considering the atherogenic properties of myristic acid. In humans, DHA accumulates at a relatively high level

Table 1. **Nutraceutical and bioactive components from marine resources and their application area.**

Component (source)	Application area
Chitin, chitosan, glucosamine	Nutraceuticals, agriculture, food, water purification, juice clarification, etc.
Carotenoids/carotenoproteins /omega-3 fatty acids	Nutraceuticals, fish feed
	Nutraceuticals, foods, baby formula, etc.
Biopeptides	Nutraceuticals, immune-enhancing agents
Minerals/calcium	Food, nutraceuticals
Algae/omega-3, minerals, carotenoids	Nutraceuticals
Chondroitin sulfate	Arthritic pain relief
Squalene	Skin care
Specialty chemicals	Miscellaneous

in organs with electrical activity, such as retinal tissues of the eye and the neural system of the heart. While DHA and other long-chain omega-3 fatty acids may be formed from α-linolenic acid (ALA) (Fig. 1), the conversion efficiency for this transformation is very limited in healthy human adults and is approximately 3-5% (Emken et al. 1994). In infants and in adults with certain ailments, the conversion of ALA to DHA is less than 1% (Salem et al. 1996). Therefore, consumption of seafoods or marine oils, as such, is important. As shown in Fig 1, DHA may be retroconverted to DPA and EPA. Human feeding trials have indicated a retroconversion of DHA to EPA of about 10% (Coquer and Holub 1996).

The beneficial health effects of marine oils in reducing the incidences of coronary heart disease (CHD) have been attributed to their omega-3 fatty acid constituents (Simopoulos 1991). Omega-3 fatty acids are known to reduce the incidence of CHD by lowering the level of serum triacylglycerols and possibly cholesterol and also to lower the blood pressure in individuals with high blood pressure as well as to decrease the ventricular arrhythmias, among others. In addition, omega-3 fatty acids are known to relieve arthritic swelling and possibly pain, relieve type II diabetes, and to enhance body immunity. However, omega-3 fatty acids may increase fluidity of the blood and hence their consumption by patients on blood thinners such as coumadin and aspirin should be carefully considered in order to avoid any unnecessary complication due to vasodilation and possible rupture of capillaries. The omega-3 fatty acids, especially DHA, are known to dominate the fatty acid spectrum of brain and retina lipids, and play an essential role in the development of the fetus and infants as well as in the health status and body requirements of pregnant and lactating women.

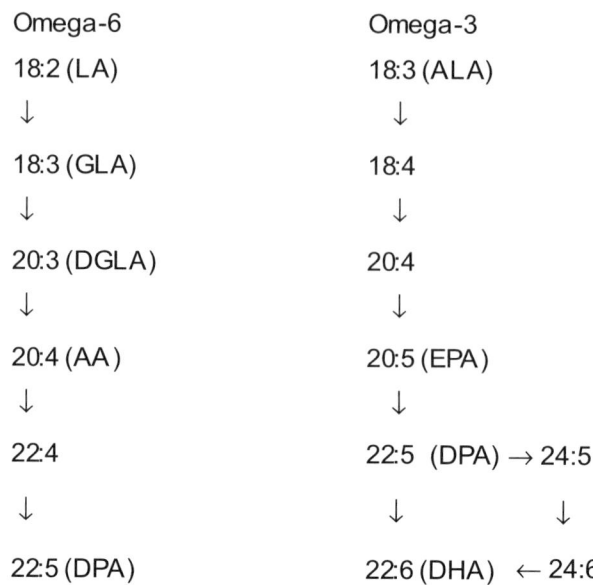

Figure 1. Essential fatty acids of the omega-6 and omega-3 families. Symbols are: LA = linoleic acid; GLA = γ-linolenic acid; DGLA = dihomo-γ-linolenic acid; AA = arachidonic acid; DPA = docosapentaenoic acid; ALA = α-linolenic acid; EPA = eicosapentaenoic acid; and DHA = docosahexaenoic acid.

Consideration of the three dimensional structures of unsaturated fatty acids demonstrates that bending of the molecules increases with an increase in the number of double bonds in their chemical structures, and this is further influenced by the position of the double bonds (i.e., omega-3 versus omega-6). These structural features in the triacylglycerol molecules as well as the location of the fatty acids in the glycerol molecule (i.e., sn-1, sn-2, and sn-3) may have a major effect on the bioavailability of fatty acids involved and their potential health benefits.

Two important sources of omega-3 fatty acids, namely menhaden oil (MO) and seal blubber oil (SBO) were considered in our work. Table 2 summarizes the fatty acids of MO, SBO, cod liver oil, and a commercial algal oil known as DHASCO (docosahexaenoic acid single cell oil). While omega-3 fatty acids, especially DHA, are primarily located in the sn-2 position in menhaden oil, they are mainly in the sn-1 and sn-3 positions of seal blubber oil (Table 3) (Wanasundara and Shahidi 1997a). These differences undoubtedly have a definite influence on their assimilation, absorption, and health benefits as well as reactions in which they are involved.

Table 2.   Major fatty acids of omega-3 rich marine and algal oils.

Fatty acid	Seal blubber	Cod liver	Menhaden	Algal (DHASCO)
14:0	3.73	3.33	8.32	14.9
16:0	5.58	11.01	17.4	9.05
16:1ω7	18.0	7.85	11.4	2.20
18:0	0.88	3.89	3.33	0.20
18:1ω9+ω11	26.0	21.2	12.1	18.9
20:1ω9	12.2	10.4	1.44	–
20:5ω3	6.41	11.2	13.2	–
22:1ω11	2.01	9.07	0.12	–
22:5ω3	4.66	1.14	2.40	0.51
22:6ω3	7.58	14.8	10.1	47.4

Units are weight percents of total fatty acids.
DHASCO = docosahexaenoic acid single cell oil.

Table 3.   Distribution of long-chain omega-3 fatty acids in menhaden and seal blubber oils.

Fatty acid	Seal blubber			Menhaden		
	sn-1	sn-2	sn-3	sn-1	sn-2	sn-3
EPA	8.36	1.60	11.2	3.12	17.5	16.3
DPA	3.99	0.79	8.21	1.12	3.11	2.31
DHA	10.5	2.27	17.9	4.11	17.2	6.12

Units are weight percents of total fatty acids.
EPA = eicosapentaenoic acid; DPA = docosapentaenoic acid; and DHA = docosahexaenoic acid.

Regardless of the source of long-chain omega-3 fatty acids, such oils must undergo appropriate processing. Therefore, refining, bleaching, deodorization, and addition of appropriate antioxidant stabilizers must be practiced in order to allow the use of these oils in food formulations. The type of food in which such omega-3 oils may be incorporated is listed in Table 4. These include foods that could be used within a short period of time and in products that do not develop off-flavors during their expected shelf life.

Table 4.   Food application of omega-3 oils.

Food	Country
Bread/hard bread	Australia, France, Germany, Ireland, Denmark
Cereals, crackers, noodles	France, Korea, Taiwan
Bars	U.S.A.
Pasta and cakes	France, U.K.
Infant formula	Australia, Brazil, Japan, New Zealand, Taiwan, U.K.
Milk, fortified	Argentina, Indonesia, Italy, Spain, U.K.
Juices, fortified	Brazil, Germany, Spain
Mayonnaise and salad dressings	Korea
Margarines and spreads	Ireland, Japan, U.K.
Eggs	U.S.A., U.K.
Canned tuna steak and seafood	Japan, U.S.A.
Tuna burger	U.S.A.

Omega-3 concentrates

For therapeutic purposes the natural sources of omega-3 fatty acids, as such, may not provide the necessary amounts of these fatty acids, and hence production and use of concentrates of omega-3 fatty acids may be required (Wanasundara et al. 2002). The omega-3 fatty acid concentrates may be produced in the free fatty acid, simple alkyl ester, and acylglycerol forms. To achieve this, physical, chemical, and enzymatic processes may be employed for concentrate production. The available methods suitable for large-scale production include low-temperature crystallization, fractional or molecular distillation, chromatography, supercritical fluid extraction, urea complexation, and enzymatic splitting (Wanasundara and Shahidi 1997b).

Among the simplest methods for concentrate production is fractional crystallization which takes advantage of the existing differences in the melting points of different fatty acids, as neat compounds or in different solvent systems. The more saturated fatty acids have higher melting points and may crystallize out of the mixtures and hence leave behind, in the liquid form, the more unsaturated fatty acids. Obviously, the free fatty acids and simple alkyl esters are more amenable to provide a higher concentration of omega-3 fatty acids than acylglycerols. This is because the latter mixtures consist of fatty acids with varying chain lengths and degrees of unsaturation in many different combinations in the triacylglycerol molecules.

Fractional distillation is another facile process for separation of mixtures of fatty acid esters under reduced pressure (0.1-1.0 mm Hg) (Brown and Kolb 1955). However, due to sensitivity of more highly unsaturated fatty acids to oxidation, one may use a spinning band column, which does not impose such limitations (Haraldsson 1984). While fractional distillation of menhaden oil ethyl esters increased the content of EPA from 15.9 to 28.4%, and DHA from 9.0 to 43.9%, molecular distillation afforded DHA with 90% parity (Max 1989).

Reverse phase chromatography has been used by Nakahara et al. (1996) to produce a DHA and DPA concentrate from marine microalgae. Teshima et al. (1978) used a silver nitrate–impregnated silica gel column to separate EPA and DHA from squid liver oil fatty acid methyl esters. The yield of the process for the fatty acids was 39% and 48%, respectively, with 85-96% EPA and 95-98% DHA purity. Similar studies on a variety of other oils have recently appeared in the literature using high performance liquid chromatography (Tekiwa et al. 1981, Adlof and Emken 1985, Hayashi and Kishimura 1993, Corley et al. 2000). More recently, centrifugal partition chromatography (CPC) has gained attention for production of omega-3 concentrates (Murayama et al. 1988, Goffic 1997). Wanasundara (2001) used a CPC technique to produce highly concentrated fatty acids such as EPA and DHA with a near quantitative yield.

Supercritical fluid extraction (SFE) is a relatively new process which is desirable for separation of PUFA. Since this method is based on separation of compounds based on their molecular weight and not their degree of unsaturation, a prior concentration step may be required in order to concentrate the omega-3 PUFA. Thus omega-3 fatty acids have been concentrated by SFE from fish oil and seaweed (Choi et al. 1987, Yamagouchi et al. 1986, Mishra et al. 1993). Fish oil esters were fractionated by SFE to obtain an oil with 60-65% DHA (Stout and Spinelli 1987).

Another possibility for concentration of omega-3 fatty acids is urea complexation. The natural acylglycerols are hydrolyzed to their fatty acid constituents in ethanol and the resultant components are allowed to crystallize in the presence of urea. The highly unsaturated fatty acids which deviate more and more from a near linear shape are not included in the urea crystals and remain in the liquid form, referred to as non-urea complexing fraction (NUCF). Meanwhile, saturated fatty acids and, to a lesser extent, mono- and diunsaturated fatty acids may be included in the urea to afford the urea complexing fraction (UCF). In this manner, depending on the variables involved, e.g., the amount of solvent, urea, and time and temperature, optimum conditions may be employed for the preparation of concentrates. If necessary, the urea complexation process may be repeated in order to enhance the concentration of certain fatty acids in the final products. We have used such techniques to prepare concentrates dominated by DHA, EPA, or DPA. The total omega-3 fatty acids in one such preparation from seal blubber oil was 88.2% and this was dominated by DHA (67 %) (Shahidi and Wanasundara 1999).

Finally, enzymatic procedures may be used to produce concentrates of omega-3 fatty acids. Depending on the type of enzyme, reaction time, temperature, and concentration of the reactants and enzyme, it is possible to produce concentrates in different forms, e.g., as free fatty acids or as acylglycerols. Thus, processes such as transesterification, acidolysis, alcoholysis, and hydrolysis as well as esterification of fatty acids with alcohols or glycerol may be employed.

Wanasundara and Shahidi (1998) have shown that enzymes might be used to selectively hydrolyze saturated and less unsaturated lipids from triacylglycerols, hence concentrating the omega-3 fatty acids in seal blubber and menhaden oils in the acylglycerol form. In this manner, the omega-3 PUFA content was nearly doubled. Furthermore, following urea completion, omega-3 concentrates obtained may be subjected to esterification with glycerol to produce concentrated acylglycerols. Upon glycerolysis of specialty alkyl esters from seal blubber oil, we found that monoacylglycerols (MAG), diacylglycerols (DAG), and triacylglycerols (TAG) were formed simultaneously. The amount of monoacylglycerols decreased continuously while that of triacylglycerols increased (He and Shahidi 1997). Depending on the structural characteristics of final products, the stability of acylglycerols was found to be better than that of their corresponding ethyl esters. Possible loss of natural antioxidants during processing may also affect the stability of products involved. Therefore, it is important to stabilize the modified oils using any of the recommended synthetic antioxidants or preferably natural stabilizers. Thus, TBHQ (tertiary butyl hydroquinone) at 200 ppm was able to inhibit oxidation of menhaden oil at 60°C over a 7-day storage period. Meanwhile, the inhibition effects were 32.5% for mixed tocopherols (500 ppm), 18.0% for α-tocopherol (500 ppm), 39.8% for mixed green tea catechins (200 ppm), 45.1% for EC (epicatechin), 48.2% for ECG (epicatechin gallate), 51.3% for EGC (epigallocatechin), and 50% for EGCG (epigallocatechin-3 gallate) (Shahidi and Wanasundara 2001). For seal blubber oil, the best protection of 56.3% was rendered by TBHQ at 200 ppm and 58.6% by ECG (200 ppm). Alpha-tocopherol inhibited the oxidation by only 14.2%.

Structured lipids

Structured lipids (SL) are triacylglycerols (TAG) containing combinations of short-chain fatty acids (SCFA), medium-chain fatty acids (MCFA), and long-chain fatty acids (LCFA) located in the same glycerol molecule and these may be produced by chemical or enzymatic means (Lee and Akoh 1998, Senanayake and Shahidi 2000). Structured lipids are developed to fully optimize the benefits of their fatty acid varieties in order to affect metabolic parameters such as immune function, nitrogen balance, and lipid clearance from the bloodstream. These specialty lipids may be produced via direct esterification, acidolysis, and hydrolysis or inter-esterification.

MCFA are those with 6-12 carbon atoms and are often used for production of structured lipids. As mentioned earlier, MCFA are highly susceptible to β-oxidation (Odle 1997). These fatty acids are not stored in the adipose tissues and are often used in the diet of patients with maldigestion and malabsorption (Willis et al. 1998). They have also been employed in total parenteral nutrition and formulas for preterm infants. Production of structured lipids via acidolysis of blubber oil with capric acid was recently reported (Senanayake and Shahidi 2002). Lipozyme-IM from *Mucor miehei* was used as a biocatalyst at an oil to fatty acid mole ratio of 1:3 in hexane, at 45ºC for 24 h and 1% (w/w) water (Senanayake and Shahidi 2002, 2003). Under these conditions, a structured lipid containing 2.3% EPA and 7.6% DHA at 27.1% capric acid (CA) was obtained. In this product, capric acid molecules were primarily located in the sn-1 and sn-3 positions (see Table 5), thus serving as a readily available source of energy to be released upon the action of pancreatic lipase. Incorporation of capric acid into fish oil TAG using immobilized lipase from *Rhizomucor miehei* (IM-60) was also reported (Jennings and Akoh 1999). After a 24 hour incubation in hexane, 43% capric acid was incorporated into fish oil while the content of EPA and DHA in the product was reduced to 27.8 and 23.5%, respectively. Similar results were obtained upon acidolysis of seal blubber with lauric acid (Senanayake and Shahidi 2003).

In an effort to produce specialty lipids containing both omega-3 PUFA and gamma-linolenic acid (GLA), preparation of such products under optimum conditions was reported (Senanayake and Shahidi 2000). GLA is found in relatively large amounts in borage oil (20-25%), evening primrose oil (8-10%), and black currant oil (15-18%). Using borage oil, the urea complexation process afforded a concentrate with 91% GLA under optimum reaction conditions.

Lipase-catalyzed acidolysis of seal blubber oil and menhaden oil with GLA concentrate (Spurvey and Shahidi 2000), under optimum conditions of GLA to TAG mole ratio of 3:1, reaction temperature of 40ºC over 24 h and 500 units enzyme per gram oil afforded products with 37.1 and 39.6% GLA incorporation, respectively. Of the two enzymes tested, lipase PS-30 from *Pseudomonas* sp. served better in the acidolysis process than *Mucor miehei* (Spurvey et al. 2001). Incorporation of GLA was in all positions and its content in the sn-2 position of both seal blubber oil and menhaden oil was 22.1 and 25.7%, respectively (Table 6). Thus, PS-30 served in a non-specific manner in the acidolysis process. The structured lipids containing GLA, EPA, and DHA so produced may have health benefits above those exerted by use of their physical mixtures.

Production of structured lipids containing GLA, EPA, and DHA may also be achieved using borage and evening primrose oils as sources of GLA and either EPA or DHA or their combinations (Senanayake and Shahidi

Table 5.   Enzymatic modification of seal blubber oil with capric acid.

Fatty acid	Unmodified	Modified	sn-1 & 3[a]
10:0	–	27.1	85.1
14:0	3.4	2.7	48.1
14:1	1.0	0.8	58.3
16:0	5.0	3.7	46.8
16:1ω7	15.1	11.9	55.5
18:1ω9+ω11	26.4	19.3	56.1
18:2ω6	1.3	1.7	66.7
20:ω9	15.0	9.1	72.5
20:5ω3	5.4	2.3	31.9
p22:1ω11	3.6	1.9	52.6
22:5ω3	4.9	3.0	76.7
22:6ω3	7.0	7.6	82.1

[a]Percent of modified fatty acid in sn-1 and 3 positions.
Units are percents of total fatty acids.
The enzyme used was lypozyme-IM from *Mucor miehei*.

Table 6.   Fatty acids of seal blubber oil (SBO), menhaden oil (MO) and their acidolysis products with γ-linolenic acid (GLA, 18:3ω6).

Fatty acid	SBO			MO		
	Unmodified	Modified	sn-1 &3[a]	Unmodified	Modified	sn-1 & 3[a]
14:0	3.36	2.40	58.3	8.18	4.55	53.3
16:0	5.14	3.04	51.1	19.89	8.78	53.5
18:1ω9	22.6	14.1	46.6	9.86	4.24	53.7
18:3ω6	0.59	37.1	77.9	0.43	39.6	74.3
20:1ω9	17.3	8.30	55.4	1.62	0.83	20.0
20:5ω3	5.40	3.80	84.6	12.9	11.0	65.9
22:5ω3	5.07	2.99	78.0	2.48	2.07	66.7
22:6ω3	7.73	4.36	79.2	10.0	6.56	77.4

[a]Percent of modified fatty acid in sn-1 and 3 positions.
Units are percents of total fatty acids.

1999a,b). The products so obtained, while similar to those produced by incorporation of GLA into marine oils, differ in the composition and distribution of fatty acids involved.

Bioactive peptides from marine resources

Cleavage of amide linkage in the protein chain leads to the formation of peptides with different numbers of amino acids as well as free amino acids. While enzymes with endopeptidase activity provide peptides with different chain lengths, exopeptidases liberate amino acids from the terminal positions of the protein molecules. Depending on reaction variables as well as the type of enzyme, the degree of hydrolysis of proteins may differ considerably. The peptides produced from the action of a specific enzyme may be subjected to further hydrolysis by other enzymes. Thus, an enzyme mixture or several enzymes in a sequential manner may be advantageous. The peptides so obtained may be subjected to chromatographic separation and then evaluated for their amino acid sequence as well as their antioxidant and other activities.

In a study on capelin protein hydrolysates, four peptide fractions were separated using Sephadex G-10. While one fraction exerted a strong antioxidant activity in a β-carotenelinoleate model system, two fractions possessed a weak antioxidant activity and the fourth one had a prooxidant effect. Two dimensional HPLC separation showed spots with both pro- and antioxidative effects (Amarowicz and Shahidi 1997). Meanwhile, protein hydrolysates prepared from seal meat were found to serve as phosphate alternatives in processed meat applications and reduced the cooking loss considerably (Shahidi and Synowiecki 1997). Furthermore, Alaska pollock skin hydrolysate was prepared using a multienzyme system in a sequential manner. The enzymes used were in the order of Alcalase, Pronase E, and collagenase. The fraction from the second step, which was hydrolyzed by Pronase E, was composed of peptides ranging from 1.5 to 4.5 kDa and showed a high antioxidant activity. Two peptides were isolated, using a combination of chromatographic procedures, and these were composed of 13 and 16 amino acid residues (Kim et al. 2001). The sequence of the peptides involved is given in Table 7 and compared with those of soy 75 protein hydrolysates (Chen et al. 1995). These peptides exert their antioxidant activity via free radical scavenging as well as chelation effects. Recently, proteases from shrimp processing discards were characterized (Heu et al. 2003) and application of salt-fermented shrimp byproduct sauce as a meat tenderizer was reported (J.-S. Kim, F. Shahidi, and M.S. Heu, unpublished results).

Table 7.   Antioxidative peptides from gelatin hydrolysate of Alaska pollock skin in comparison with that of soy 75 protein.

Peptide	Amino acid sequence
Alaska pollock skin	
P_1	Gly-Glu-Hyp-Gly-Pro-Hyp-Gly-Pro-Hyp-Gly-Pro-Hyp-Gly-Pro-Hyp-Gly
P_2	Gly-Pro-Hyp-Gly-Pro-Hyp-Gly-Pro-Hyp-Gly-Pro-Hyp-Gly
Soy 75 protein	
P_1	Val-Asn-Pro-His-Asp-His-Glu-Asn
P_2	Leu-Val-Asn-Pro-His-Asp-His-Glu-Asn
P_3	Leu-Leu-Pro-His-His
P_4	Leu-leu-Pro-His-His-Ala-Asp-Ala-Asp-Tyr
P_5	Val-Ile-Pro-Ala-Gly-Tyr-Pro
P_6	Leu-Glu-Ser-Gly-Asp-Ala-Leu-Arg-Val-Pro-Ser-Gly-Thr-Tyr-Tyr

Chitin, chitosan, and related compounds

Chitin is recovered from processing discards of shrimp, crab, lobster, and crayfish following deproteinization and demineralization (Shahidi and Synowiecki 1991, Shahidi et al. 1999). The chitin so obtained may then be deacetylated to afford chitosan (Shahidi and Synowiecki 1991). Depending on the duration of the deacetylation process, the chitosan produced may assume different viscosities and molecular weights. The chitosans produced are soluble in weak acid solutions, thus chitosan ascorbate, chitosan acetate, chitosan lactate, and chitosan malate, among others, may be obtained and these are all soluble in water. Chitosan has a variety of health benefits and may be employed in a number of nutraceutical and health-related applications. Chitosan derivatives may also be produced in order to obtain more effective products for certain applications. However, to have the products solubilized in water without the use of acids, enzymatic processes may be carried out to produce chitosan oligomers. Due to their solubility in water, chitosan oligomers serve best in rendering their benefits under normal physiological conditions and in foods with neutral pH. Furthermore, depending on the type of enzyme employed, chitosan oligomers with specific chain lengths may be produced for certain applications (Jeon et al. 2000).

Chitosans with different viscosities were prepared (Table 8) and used in an experiment designed to protect both raw and cooked fish against oxidation as well as microbial spoilage (Jeon et al. 2002, Kamil et al. 2002, Shahidi et al. 2002). The content of propanal, an indicator of oxidation of omega-3 fatty acids, was decreased when chitosan was used as an edible invisible film in herring. Furthermore, the effects were more pronounced as the molecular weight of the chitosan increased (Table 9). In addition, inhibitory effects of chitosan coatings in the total microbial counts for

Table 8. Characteristics of three different kinds of chitosans prepared from crab shell waste.[a]

Properties	Chitosan		
	I	II	III
Deacetylation time[b]	4 h	10 h	20 h
Moisture (%)	4.50 ± 0.30	3.95 ± 0.34	3.75 ± 0.21
Nitrogen (%)	7.55 ± 0.10	7.70 ± 0.19	7.63 ± 0.08
Ash (%)	0.30 ± 0.03	0.25 ± 0.02	0.30 ± 0.00
AV[c] (cps)[d]	360	57	14
DA[c] (%)	86.3 ± 2.1	91.3 ± 1.2	94.5 ± 1.3
Mv[c] (dalton)	1,816,732	963,884	695,122

[a]Results are expressed as mean ± standard deviation of three determinations.
[b]Deacetylation for chitosan I, II, and III was achieved using 50% NaOH at 100°C.
[c]Mv = viscosity molecular weight; AV = apparent viscosity; and DA = degree of deacetylation.
[d]cps = cycles per second.

Table 9. Content of propanal (milligrams per kilogram of dried fish) in headspace of chitosan-coated herring samples stored at 4°C.[a]

Chitosan	Storage period (days)					
	0	2	4	6	8	10
Uncoated	12.6 ± 3.4[a]	23.7 ± 4.2[b]	29.9 ± 4.2[c]	34.3 ± 1.9[c]	44.1 ± 4.0[c]	46.3 ± 2.4[c]
14 cps	13.8 ± 2.1[a]	18.3 ± 3.0[a]	24.6 ± 1.2[b]	30.9 ± 2.9[bc]	33.0 ± 0.8[b]	39.7 ± 0.9[b]
57 cps	12.6 ± 3.0[a]	15.5 ± 2.1[a]	19.7 ± 2.6[a]	24.9 ± 1.6[a]	22.8 ± 1.9[a]	24.2 ± 1.9[a]
360 cps	14.2 ± 2.4[a]	15.7 ± 2.6[a]	17.6 ± 2.2[a]	20.2 ± 1.4[a]	18.3 ± 2.4[a]	22.7 ± 1.3[a]

[a]Results are expressed as mean ± standard deviation of three determinations. Values with the same superscripts within each column are not significantly different ($P < 0.05$).

cod and herring showed an approximately 1.5 and 2.0 log cycles difference between coated and uncoated samples, respectively, after 10 days of refrigerated storage (results not shown). The monomer of chitin, N-acetylglucosamine (NAG), has been shown to possess anti-inflammatory properties. Meanwhile, glucosamine, the monomer of chitosan, prepared via HCl hydrolysis, is marketed as glucosamine sulfate. This formulation is prepared by addition of ferrous sulfate to the preparation. Glucosamine products may also be sold in formulation containing chondroitin 4- and chondroitin 6-sulfates. While glucosamine helps to form proteoglycans that sit within the space in the cartilage, chondroitin sulfate acts like a liquid magnet. Thus glucosamine and chondroitin work in a complementary manner to improve the health of the joint cartilage.

The byproducts in chitin extraction process from shellfish include carotenoids/carotenoproteins, and enzymes (Shahidi 1995, Shahidi et al. 1998, Shahidi and Kamil 2001). These components may also be isolated for further utilization in a variety of applications.

References

Abeywardena, M.Y., and R.J. Head. 2001. Long chain n-3 polyunsaturated fatty acids and blood vessel function. Cardiovascular Res. 52:361-371.

Adlof, R.O., and E.A. Emken. 1985. The isolation of omega-3 polyunsaturated fatty acids and methyl esters of fish oils by silver resin chromatography. J. Am. Oil Chem. Soc. 62:1592-1595.

Amarowicz, R., and F. Shahidi. 1997. Antioxidant activity of peptide fractions of capelin protein hydrolysates. Food Chem. 58:355-359.

Brown, J.B., and D.K. Kolb. 1955. Application of low temperature crystallization in the separation of polyunsaturated fatty acids and their compounds. Prog. Chem. Fats Other Lipids 3:57-94.

Chen, H-M., K. Muramoto, and F. Yamauchi. 1995. Structural analysis of antioxidative peptides from soybean β-conglycin. J. Agric. Food Chem. 43:574-578.

Choi, K.I., Z. Nakhost, V.I. Krukonis, and M. Karel. 1987. Supercritical fluid extraction and characterization of lipid from algal *Scenedesmus obliques*. Food Biotechnol. 1:263-271.

Coquer, J.A., and B.J. Holub. 1996. Supplementation with an algae source of docosahexaenoic acid increases (n-3) fatty acid status and alters selected risk factors for heart disease in vegetarian subjects. J. Nutr. 126:3032-3039.

Corley, D.G., S.G. Zeller, P. James, and K. Duffin. 2000. Process for separating a triglyceride comprising a docosahexaenoic acid residue from a mixture of triglycerides. International Patent PCT/US so /04166.

Emken, E.A., R.O. Adlof, and R.M. Gulley. 1994. Dietary linoleic acid influences desaturation and acylation of deuterium-labeled linoleic and linolenic acids in young adult males. Biochem. Biophys. Acta 1213:277-288.

Goffic, F.L. 1997. Countercurrent and centrifugal partition chromatography as new tools for preparative-scale lipid purification. Lipid Technol. 9:148-150.

Haraldsson, G.G. 1984. Separation of saturated/unsaturated fatty acids. J. Am. Oil Chem. Soc. 61:219-222.

Hayashi, K., and H. Kishimura. 1993. Preparation of n-3 PUFA ethyl ester concentrates from fish oil by column chromatography on silicic acid. Nippon Suisan Gakkaishi 59:1429-1435.

He, Y., and F. Shahidi. 1997. Enzymatic esterification of T-3 fatty acid concentrates from seal blubber oil with glycerol. J. Am. Oil Chem. Soc. 74:1133-1136.

Heu, M.S., J-S. Kim, F. Shahidi, Y. Jeong, and Y-J. Yeon. 2003. Characteristics of proteases from shrimp processing discards. J. Food Biochem. 27:221-236.

Jennings, B.H., and C.C. Akoh. 1999. Enzymatic modification of triacylglycerols of high eicosapentaenoic and docosahexaenoic acids content to produce structured lipids. J. Am. Oil Chem. Soc. 76:1133-1137.

Jeon, Y-J., F. Shahidi, and S.K. Kim. 2000. Preparation of chitin and chitosan oligomers and their application in physiological functional foods. Food Rev. Int. 16:159-176.

Jeon, Y-J., J.Y.V.A. Kamil, and F. Shahidi. 2002. Chitosan as an edible invisible film for quality preservation of herring and Atlantic cod. J. Agric. Food Chem. 50: 5167-5178.

Kamil, J.Y.V.A., Y-J. Jeon, and F. Shahidi. 2002. Antioxidative activity of different viscosity chitosans in cooked comminuted flesh of herring (*Clupea harengus*). Food Chem. 79:69-77.

Kim, S-K., Y-T. Kim, H-G. Byun, K-S. Nam, D-S. Joo, and F. Shahidi. 2001. Isolation and characterization of antioxidative peptides from gelatin hydrolysate of Alaska pollock skin. J. Agric. Food Chem. 49:1984-1989.

Lee, K., and C.C. Akoh. 1998. Structured lipids: Synthesis and applications. Food Rev. Int. 14:17-34.

Max, Z. 1989. High-concentration mixture of polyunsaturated fatty acids and their esters from animal and/or vegetable oils and their prophylactic or therapeutic uses. U.K. Patent GB2, 218, 984.

Mishra, V.K., F. Temelli, and B. Oraikul. 1993. Extraction and purification of T-3 fatty acids with an emphasis on supercritical fluid extraction: A film demonstration. Ber. Bunsen-Ges. Phys. Chem. 88:882-887.

Murayama, W., Y Kosuge, N. Nakaya, K. Nunogaki, J. Cazes, and H. Nunogaki. 1988. Preparative separation of unsaturated fatty acid esters by centrifugal partition chromatography. J. Liq. Chromatogr. 11:283-300.

Nakahara, T., T. Yokochi, T. Higashihara, S. Tanaka, T. Yaguchi, and D. Honda. 1996. Production of docosahexaenoic and docosapentaenoic acids by *Schizochytrium* sp. isolated from Yap Island. J. Am. Oil Chem. Soc. 73:1421-1426.

Odle, J. 1997. New insights into the utilization of medium-chain triglycerides by neonate: Observations from a piglet model. J. Nutr. 127:1061-1067.

Salem Jr., N., B. Wegher, P. Mena, and R. Uauy. 1996. Arachidonic and docosahexaenoic acids are biosynthesized from their 18-carbon precursors in human infants. Proc. Natl. Acad. Sci. U.S.A. 93:49-54.

Senanayake, S.P.J.N., and F. Shahidi. 1999a. Enzyme-assisted acidolysis of borage (*Borago officinalis* L.) and evening primrose (*Oenothera biennis* L.) oils: Incorporation of omega-3 polyunsaturated fatty acids. J. Agric. Food Chem. 47:3105-3112.

Senanayake, S.P.J.N., and F. Shahidi. 1999b. Enzymatic incorporation of docosahexaenoic acid into borage oil. J. Am. Oil Chem. Soc. 76:1009-1015.

Senanayake, S.P.J.N., and F. Shahidi. 2000. Structural lipids containing long chain omega-3 polyunsaturated fatty acids. In: F. Shahidi (ed.), Seafoods in health and nutrition: Transformation in fisheries and aquaculture: Global perspectives. ScienceTech Publishing Co., St. John's, Canada, pp. 29-44.

Senanayake, S.P.J.N., and F. Shahidi. 2002. Enzyme-catalyzed synthesis of structured lipids via acidolysis of seal (*Phoca groenlandica*) blubber oil with capric acid. Food Res. Int. 35:745-752.

Senanayake, S.P.J.N., and F. Shahidi. 2003. Enzymatic modification of seal (*Phoca groenlandica*) blubber oil: Incorporation of lauric acid. Food Chem. In press.

Shahidi, F. 1995. Role of chemistry and biotechnology in value-added utilization of shellfish processing discards. Can. Chem. Neur. 10:25-29.

Shahidi, F., and Y.V.J. Kamil. 2001. Enzymes from fish and aquatic invertebrates and their application in the food industry. Trends Food Sci. Technol. 12:435-464.

Shahidi, F., and S-K. Kim. 2002. Marine lipids as affected by processing and their quality preservation by natural antioxidants. In: T.C. Lee and C-T. Ho (eds.), Bioactive compound in foods: Effects of processing and storage. ACS Symposium Series 816, American Chemical Society, Washington, D.C., pp. 1-13.

Shahidi, F., and J. Synowiecki. 1991. Isolation and characterization of nutrients and value-added products from snow crab (*Chionoecetes opilio*) and shrimp (*Pandalus borealis*) processing discards. J. Agric. Food Chem. 39:1527-1532.

Shahidi, F., and J. Synowiecki. 1997. Protein hydrolysates from seal meat as phosphate alternatives in food processing applications. Food Chem. 60:29-32.

Shahidi, F., and U.N. Wanasundara. 1999. Concentration of omega-3 polyunsaturated fatty acids of seal blubber oil by urea complexation: Optimization of reaction conditions. Food Chem. 65:41-49.

Shahidi, F., and U.N. Wanasundara. 2001. Seal blubber oil and its nutraceutical products. In: F. Shahidi and J.W. Finley (eds.), Omega-3 fatty acids: Chemistry, nutrition, and health effects. ACS Symposium Series 788, American Chemical Society, Washington, D.C.

Shahidi, F., J.K.V. Arachchi, and Y-J. Jeon. 1999. Food application of chitin and chitosans. Trends Food Sci. Technol. 10:37-51.

Shahidi, F., Metusalach, and J.A. Brown. 1998. Carotenoid pigments in seafoods and aquaculture. Crit. Rev. Food Sci. Nutr. 38:1-67.

Shahidi, F., J. Kamil, Y-J. Jeon, and S-K. Kim. 2002. Antioxidant role of chitosan in a cooked cod (*Gadus morhua*) model systems. J. Food Lipids 9:57-64.

Simopoulos, A.P. 1991. Omega-3 fatty acids in health and disease and growth and development. Am. J. Clin. Nutr. 54:438-463.

Spurvey, S.A., and F. Shahidi. 2000. Concentration of gamma-linolenic acid (GLA) from borage oil by urea complexation: Optimization of reaction conditions. J. Food Lipids 7:163-174.

Spurvey, S.A., S.P.J.N. Senanayake, and F. Shahidi. 2001. Enzymatic-assisted acidolysis of menhaden and seal blubber oils with gamma-linolenic acid. J. Am. Oil Chem. Soc. 78:1105-1112.

Stout, V.F., and J. Spinelli. 1987. Polyunsaturated fatty acids from fish oils. U.S. Patent 4:675, 132.

Tekiwa, S., A. Kanazawa, and S. Teshima. 1981. Preparation of eicosapentaenoic and docosahexaenoic acids by reversed phase high performance liquid chromatography. Bull. Jpn. Soc. Sci. Fish. 47:675-678.

Teshima, S., A. Kanazawa, and S. Tokiwa. 1978. Separation of polyunsaturated fatty acids by column chromatography on silver nitrate–impregnated silica gel. Bull. Jpn. Soc. Sci. Fish. 44:927-929.

Wanasundara, U.N. 2001. Isolation and purification of bioactive compounds by centrifugal partition chromatography (CPC) and other technologies. Presented at the 2nd International Conference and Exhibition on Nutraceuticals and Functional Foods (Worldnutra 2001), November 26–December 2, Portland, Oregon.

Wanasundara, U.N., and F. Shahidi. 1997a. Positional distribution of fatty acids in triacylglycerols of seal blubber oil. J. Food Lipids 4:51-64.

Wanasundara, U.N., and F. Shahidi. 1997b. Structural characteristics of marine lipids and preparation of omega-3 concentrates. In: F. Shahidi and K.R. Cadwallader (eds.), Flavor and lipid chemistry of seafoods. ACS Symposium Series 674, American Chemical Society, Washington, D.C., pp. 240-254.

Wanasundara, U.N., and F. Shahidi. 1998. Lipase-assisted concentration of T-3 polyunsaturated fatty acids in acylglycerol forms from marine oils. J. Am. Oil Chem. Soc. 75:945-951.

Wanasundara, U.N., J. Wanasundara, and F. Shahidi. 2002. Omega-3 fatty acid concentrates: A review of production technologies. In: C. Alasalvar and T. Taylor (eds.), Seafoods: Quality, technology and nutraceutical applications. Springer, New York, pp. 157-174.

Willis, W.M., R.W. Lencki, and A.G. Marangoni. 1998. Lipid modification strategies in the production of nutritionally functional fats and oils. Crit. Rev. Food Sci. Nutr. 38:639-674.

Yamagouchi, K., M. Murakami, H. Nakano, S. Konosu, T. Kokura, H. Yamamoto, M. Kosaka, and K. Hata. 1986. Supercritical carbon dioxide extraction of oils from Antarctic krill. J. Agric. Food Chem. 34:904-907.

Omega-3 Fatty Acids in Health, Nutrition, and Disease: Future U.S. Market Considerations

Robert Katz
Omega-3 Research Institute, Inc., Bethesda, Maryland

Joyce Nettleton
Science Communications Consultant, Denver, Colorado

Abstract

Dyerberg and Bang's (1982) observation that the lack of cardiovascular disease in Greenland's western shore natives is associated with long-chain omega-3 polyunsaturated fatty acids in their diet, led to identification of fish oils as cardioprotective factors. As understanding of these cardioprotective effects increased, scientific groups recommended differing daily intake levels of EPA+DHA. We present a health-driven, new approach to estimating the market value of encapsulated food grade fish oil as well as crude fish oil at four different recommended daily intake levels of EPA+DHA. We introduced the concept of salmon equivalent of specific fish as the ratio of their average percent of EPA+DHA content by weight (grams of EPA+DHA per 100 grams fish) to the 1.2 average percent EPA+DHA content by weight of a 100 gram portion of Pacific salmon (*Oncorhynchus* sp.). Per capita consumption of EPA+DHA in Pacific salmon equivalents and the fraction of food grade fish oil consumed as encapsulated supplements were estimated at all four recommended levels. The ratio of the market values of food grade fish oil consumed as supplements (in the billion dollar range) to the wholesale value of the corresponding crude fish oil (in the tens of million dollar range) was found to be approximately 100:1. The implications of these results regarding the Alaska fish oil production are discussed.

Background

Long-chain omega-3 polyunsaturated fatty acids (omega-3 LC-PUFA) or fish oils are seafood byproducts. Their importance in human health and disease has been extensively studied in the last three decades. Thus, eicosapentaenoic acid (EPA) and docosahexaenoic acid (DHA), studied using fish oils, fatty fish, and purified omega-3 LC-PUFA ester derivatives, have been shown to be involved in coronary heart disease (GISSI 1999, Albert et al. 2002, Hu et al. 2002); cancer (Stillwell and Jensky 2002); metabolic syndrome (Simopoulos 1999); diabetes (Montori et al. 2000); pregnancy and infant development (Crawford 2000, Al et al. 2000); depression, bipolar disorder, and CNS in general (Stoll et al. 1999, Locke 2001, Katz et al. 2001); anger and hostility (Hibbeln 2001); and inflammatory and immune-based diseases including autoimmune diseases (Simopoulos 2002). In most of these disorders EPA+DHA insufficiency appears to displace the metabolic homeostasis of a healthy state, and supplementation appears therapeutic and preventive. Some of the physiological roles of the individual omega-3 LC-PUFA are unclear, but they are apparently involved in basic processes at different periods of life; for example, DHA is essential for normal brain and visual development of the fetus, newborn, and infant (Birch et al. 2000); and EPA is a potential antidepressant (Peet and Horrobin 2002) and regulator of endothelial responses (De Caterina et al. 2000). DHA is considered a "conditionally essential nutrient" for infant development, Carlson (2001), and it is likely that both EPA and DHA are required for optimum cardiovascular health and the prevention of chronic disease. Need for the precursor of EPA and DHA, alpha-linolenic acid, was recognized by the Institute of Medicine (IOM) in its most recent report (IOM 2002). It should be noted that while humans can convert alpha-linolenic acid to EPA and DHA, the process is highly inefficient and may be inadequate to provide optimum levels of EPA and DHA (Sprecher 1992, Cunnane et al. 1994). Intakes of EPA and DHA from fatty fish, the primary food source, are relatively low in the United States, (Kris-Etherton et al. 2000).

The U.S. Food and Drug Administration (FDA) approved the "generally recognized as safe" (GRAS) status of menhaden (*Brevoortia tyrannus*) oil, (FDA 1997) and a 1:1 mixture of DHA to arachidonic acid from microalgae, the latter to be added to infant formula (FDA 2001). In February 2002, the FDA approved a qualified claim for dietary supplements, linking omega-3 fatty acids (omega-3 FA) and coronary heart disease. The qualified claim states: "Consumption of omega-3 fatty acids may reduce the risk of coronary heart disease." The FDA evaluated the data and determined that, although there is scientific evidence supporting the claim, the evidence is not conclusive. It is possible that within the next 10 years there will be a recommended daily intake (RDI) for omega-3 LC-PUFA for the U.S. total population (TP) of 281,412,000 (U.S. Census 2000).

We estimate here future potential needs and market values for total EPA+DHA in food grade fish oil (FGFO) and crude fish oil (CFO). We also introduce a new concept for expressing omega-3 LC-PUFA levels or needs in terms of Pacific salmon (*Oncorhynchus* sp.) oil content in what we call salmon equivalent and discuss the present status of Alaskan fish oil production in light of these estimates.

Methodology and results

We propose a health-driven, new approach to estimating the market value of EPA+DHA in FGFO and CFO based on different levels of recommended daily intakes and on the per capita consumption of EPA+DHA from fish.

Four recommended daily intake (RDI) levels were used in calculations. Three of the four levels were recommended by experts (see 1-3 in Table 1), and the fourth was adopted from a clinical trial. The RDI values used in these calculations are shown in Table 1.

Estimation process

1. Initially, recommended yearly intakes (RYI) of pure EPA+DHA were estimated for the U.S. total population in metric tons per year (t per year) (Table 2). From these four values we estimated the total FGFO and the total CFO required to provide the RYIs of pure EPA+DHA for the U.S. total population (TP).

2. In the second phase we introduce the concept of salmon equivalent (SE). By definition, SE of a specific fish is the ratio of its average percent of EPA+DHA content by weight (g of EPA+DHA per 100 g fish) to the 1.2 average percent of EPA+DHA content by weight of a 100 g portion of *Oncorhynchus* sp. (Table 3). To illustrate these calculations, three fish were selected from the National Fisheries Institute's (NFI) list of the Seafood "Top Ten" U.S. Per-Capita Consumption by Species in Pounds (see legend to Table 3). Two of the selected fish have significant relevance to the Alaskan fishing industry. These are walleye pollock (*Theragra chalcogramma*), which is one of the main fish species harvested and processed in Alaska (harvested at 1,421,818 kg per year) and the various *Oncorhynchus* sp. (pink; sockeye or red; coho or silver; king or chinook; and chum or keta) (harvested at about 346,559 t per year not including the king salmon catch) (ADFG 2001). Tuna (*Thunnus* sp.) were added due to their high U.S. per capita consumption and their relatively high EPA+DHA content. The SE of all three fish species amounts to 2.28 kg per capita per year.

3. From the initial estimated RYI value of EPA+DHA for the U.S. (see Table 2) and the 1.2 average percent by weight content of EPA+DHA in salmon, we calculated an estimated value of EPA+DHA for the U.S. TP consumed as SEs (see Table 4). From this we estimated the amount of salmon each person in the United States would have to eat per year to reach the RYI at each of the four recommended levels. The difference

Table 1. Authorities recommending daily intake levels of omega-3 LC-PUFA and levels of the recommended daily intake (RDI).

No.	Recommending authority[a]	RDI of EPA+DHA (mg/day)
1	IOM, NAS[b]	160
2	AHA[c]	350
3	NIH 1999 workshop[d]	650
4	GISSI clinical trial[e]	850

[a]The first three recommendations are based on triglyceridic oils. The fourth is the amount of omega-3 LC-PUFA ethyl ester–based oil associated with a significant risk reduction in sudden cardiac death in a major secondary prevention trial (Marchioli et al. 2002).

[b]The Food and Nutrition Board of the Institute of Medicine (IOM), National Academy of Sciences (NAS) recently issued a report recommending a daily reference intake for omega-3 LC-PUFA of 160 mg per day for healthy people (IOM 2002). This is a very low recommended level and has drawn widespread criticism. Given that experts have reached consensus for much higher needs, we included some of the other recommendations reached by a scientific consensus development process (see below).

[c]The American Heart Association (AHA) recommends the consumption of two fatty fish meals, about 100 g portion size, per week (Kraus et al. 2001). Considering an average EPA+DHA content of 1.2 g in a 100 g portion of salmon (Kris-Etherton et al. 2000), the AHA RDI amounts to about 350 mg per day.

[d]Simopoulos et al. 2000.

[e]GISSI 1999, Marchioli et al. 2002.

Table 2. Estimated recommended yearly intake (RYI) of pure EPA+DHA for the U.S. total population (TP).[a]

No.	RDI of EPA+DHA (mg/day)	Estimated RYI of EPA+DHA for the TP (t/y)	Estimated RYI of EPA+DHA as FGFO[b] (t/y)	Estimated RYI of EPA+DHA as CFO[c] (t/y)
1	160	16,435	54,783	73,044
2	350	36,683	122,277	163,036
3	650	66,767[d]	222,556	296,741
4[e]	850	87,311	291,037	388,049

[a]The RYIs for the corresponding food grade fish oil (FGFO) and crude fish oil (CFO) for the U.S. total population of 281,412,000 were also estimated, (U.S. Census 2000).

[b]Menhaden (*Brevoortia tyrannus*) oil, used as a representative oil, contains about 30% EPA+DHA. *B. tyrannus* oil is a classical 1812 oil (180 mg EPA + 120 mg DHA in 1 gram of oil). Since RYI of EPA+DHA for TP constitutes only 30% of the weight of the selected fish oil, RYI of EPA+DHA as FGFO will be obtained by increasing it by 70%. Thus, $RYI_{FGFO} = RYI_{EPA+DHA} \div 0.3$.

[c]Up to 25% of CFO is lost during refining and purification (J. Crowther, Omega Proteins, Inc., Reedville, Virginia, 2002, pers. comm.). Thus, $RYI_{CFO} = RYI_{FGFO} \div 0.75$.

[d]Sample calculation using the RDI (recommended daily intake) of 650 mg per day. Sample calculation: $RYI_{650TP(CFO)} = [(RYI_{650TP}EPA+DHA) \div 0.3] \div 0.75$ and $RYI_{650TP(CFO)} = [(650 \times 10^{-9} \times 365 \times 281,412,000) \div 0.3] \div 0.75 = 296,741$ t/y.

[e]The number order in column 1, signifying the RDI levels used in calculations, is preserved throughout all tables.

Table 3. Calculated yearly per capita fish consumption of three fish species in salmon equivalents (SE).

Fish	Scientific name	EPA+DHA content of fish[a] (g/100 g fish) (average)		U.S. per person fish consumption[b] (kg/y)	Salmon equivalent[c] (SE)	Yearly per person fish consumption in SE (kg/y)
Pacific salmon	*Oncorhynchus* sp.[d]	1.0-1.4	1.2	0.63	1.00	0.63
Tuna (fresh)	*Thunnus* sp.[e]	0.5-1.6	1.05	1.54	0.87	1.34
Walleye pollock	*Theragra chalcogramma*	~0.5	0.5	0.75	0.42	0.31[f]

[a]Average values for *Oncorhynchus* sp. and *Thunnus* sp. were calculated from EPA+DHA content ranges provided in Kris-Etherton (2000) and the EPA+DHA content of *T. chalcogramma* was provided in Simopoulos and Robinson (1998). The EPA+DHA content varies by species, season, location of catch, packaging, cooking methods, etc.

[b]National Fisheries Institute's (NFI). http://www.nfi.org. Calculated by J.T. Assor, NFI, and converted to kg/y.

[c]Salmon equivalent (SE). The SE of a specific fish is the ratio of its average EPA+DHA content by weight (g) in a 100 g portion of that fish to the 1.2 g average EPA+DHA content of a 100 g portion of salmon. By definition SE of salmon = 1.0. Example: $SE_{tuna} = 1.05 \div 1.2 = 0.87$; tuna consumption SE = 1.54 × 0.87 = 1.34 kg/person/year.

[d]*Oncorhynchus* sp. = *O. gorbuscha*, *O. nerka* , *O. kisutch*, *O. tshawytscha*, and *O. keta* (Nettleton 1985).

[e]*Thunnus* sp. = *T. alalunga*, *T. obeses*, *T. thynnus*, *T. uthanus*, *T. palamis*, and *T. albakares* (Nettleton 1985).

[f]Total yearly per person fish consumption from all three species in SE equals 2.28 kg SE per person per year.

between the estimated (needed) per capita EPA+DHA consumption per year as SEs and total yearly per capita fish consumption as SEs (2.28 kg SE per capita per year) (Table 3) provides the amount of EPA+DHA that the population does not consume as fish. Although these figures are based only on three fish species, this limitation has only a small effect on an RDI of 350 mg per day or more.

4. Next we estimated the market value of CFO that corresponds to the fraction of FGFO consumed as supplements (Table 5) and the market value of the corresponding FGFO fraction as if it would be consumed solely as 1 g fish oil gelatin capsules (gel caps) (Table 6.) Assuming exclusive use of FGFO as gel caps seemed reasonable because the use of FGFO as an additive in foods is presently limited. The total number of gel caps multiplied by the cost of a standard capsule, priced at $0.075 (General Nutrition Corporation stores in the Washington, D.C., Metropolitan Area) provided the total retail value of the encapsulated FGFO. The two market values calculated are compared in the last two columns of Table 6.

Table 4. Estimated recommended yearly intake (RYI) of EPA+DHA for the U.S. total population (TP) expressed as salmon equivalents (SE) and the estimated fraction of EPA+DHA expressed as food grade fish oil (FGFO) supplements.

No.	Estimated RYI of EPA+DHA for TP (t/y)	Estimated RYI of EPA+DHA for TP, SE[a] (t/y)	Estimated fish consumption/person/y in SE[b] to meet EPA+DHA RYI (kg/y/person)	Total yearly per person fish consumption as SE[c] (kg/y/person)	Estimated fraction of pure EPA+DHA not met by fish consumption for TP[d] (t/y)	Estimated fraction FGFO not met by fish consumption for TP[e] (t/y)
1	16,435	1,370,000	4.87	2.28	8,740	29,133
2	36,683	3,064,000	10.86	2.28	28,982	96,607
3	66,767	5,564,000	19.77	2.28	59,067	196,890
4	87,311	7,276,000	25.85	2.28	79,610	265,367

[a]Estimated RYI of EPA+DHA for the TP expressed as SE is calculated based on the 1.2 average content of EPA+DHA provided in Table 3 for *Oncorhynchus* sp. Sample calculation (starting with data estimated in Table 2): 66,767 ÷ 0.012 = 5,564,000 t of SE/y. (If 100 g of salmon contain in average 1.2 g of EPA+DHA, 1.0 t of salmon will contain 0.012 t of EPA+DHA.) This is the amount of fish to eat in order to get estimate of RYI for TP assuming a 1.2% EPA+DHA fish tissue oil concentrate.

[b]Sample calculation: 5,564,000 × 1,000 ÷ 281,412,000 = 19.77 kg per person SE/y.

[c]Value derived from Table 3.

[d]Sample calculation: 66,767 × (19.77 − 2.28) ÷ 19.77 = 59,067 t/y. Pure EPA+DHA is not a commercially viable form for mass consumption. Therefore these values are converted to FGFO, which is then encapsulated or used as an additive in other foods. Both oil capsules and microencapsulated oils used as food additives are considered supplements.

[e]Sample calculation: 59,067 ÷ 0.3 = 196,890 t/y.

Table 5. **Estimated market value of crude fish oil (CFO) corresponding to the fraction of food grade fish oil (FGFO) consumed as supplements.**

No.[a]	Estimated amount of FGFO to be consumed as supplements by TP (t/y)	CFO corresponding to amount of estimated FGFO consumed as supplements[b] (t/y)	Estimated market value of CFO[c] corresponding to the fraction of FGFO consumed as supplements (million $/y)
1	29,133	38,844	21.5
2	96,607	128,809	72.1
3	196,890	262,520	147.0
4	265,367	353,822	198.1

[a]The numbering in column 1 (1-4) correlates with the four RDI levels of omega-3 LC-PUFA in Table 1.
[b]Sample calculation (estimates from Table 4): 196,890 t/y ÷ 0.75 = 262,520 t/y.
[c]The market value of CFO corresponding to the fraction of FGFO consumed as supplements was estimated using $560.00 net per t crude fish oil (Oil World 2002).

Table 6. **Estimated total retail value of encapsulated food grade fish oil (FGFO) consumed as supplements compared with the market value of corresponding crude fish oil (CFO).**

No.[a]	Estimated amount of FGFO to be consumed as supplements[b] t/y)	Corresponding no. of 1 g gelatin capsules (billion capsules/y)	Total approximate retail value of encapsulated FGFO consumed as supplements (billion $/y)[c]	Estimated market value of CFO corresponding to the fraction of FGFO consumed as supplements (million $/y)
1	29,133	29.1	2.18	21.5
2	96,607	96.6	7.24	72.1
3	196,890	196.9	14.77	147.0
4	265,367	265.4	19.90	198.1

[a]The numbering in column 1 (1-4) correlates with the four RDI levels of omega-3 LC-PUFA in Table 1.
[b]From Table 5.
[c]Assume $0.075 per capsule. This figure was based on average retail prices of 100-250 capsule bottles from Washington, D.C., area stores.

Discussion

There are three main marketable fish oil–containing raw materials: (i) fish and fish waste, (ii) crude fish oil (CFO) isolated mostly from fish, and (iii) food grade fish oil (FGFO) purified from the CFO.

By introducing the concept of salmon equivalent (SE) (see Methodology and results) we related the omega-3 LC-PUFA content of different fish (*Thunnus* sp. and *T. chalcogramma*) to the EPA+DHA content of the *Oncorhynchus* sp. Quantitatively, the concept expresses the total average amount of EPA+DHA consumed by one person during a year by eating the above three species of fish expressed as salmon in kg per year.

Thus, the SE could serve as a unifying standard for EPA+DHA content and facilitate their comparisons of different fish and products on the basis of their omega-3 LC-PUFA content, market value, and cost benefits. Although we used 1.2 g as the average EPA+DHA content for *Oncorhynchus* sp., use of 1.0 g as the average EPA+DHA content can be justified. This value would correspond to the latest recommendations of the American Heart Association regarding daily intake of EPA+DHA for individuals suffering from cardiovascular or coronary heart disease (Kris-Etherton et al. 2002).

Salmon equivalents allowed us to partition the total RYI of EPA+DHA at various RDI levels into two entities, one that is provided by fish and a second from supplements. By assuming complete encapsulation of the FGFO in 1.0 g gel cap supplements, the total retail value at the lowest RDI level (160 mg per day) is $2.18 billion. At the 650 mg per day RDI level, retail value reaches $14.77 billion. The total market value of CFO corresponding to the encapsulated FGFO is estimated at about $21.5 million and at about $147.0 million for the 160 and 650 mg per day levels, respectively. Although these calculations do not take production and other costs into consideration, the nearly 100-fold difference in market value in favor of the encapsulated FGFO supplement could serve as a powerful incentive for industry to develop new products using the FGFO.

It is instructive to compare these two market values for the approximately 30,000 ± 10,000 t per year of groundfish crude oils produced by Alaskan fisheries (S. Smiley, University of Alaska Fishery Industrial Technology Center, Kodiak, 2002, pers. comm.). The total wholesale market value of the 30,000 t per year CFO amounted to approximately $16.8 million, while the corresponding retail value of encapsulated supplements could reach approximately $1.69 billion. The incentive for increasing the processing of the *Oncorhynchus* sp. byproducts into salmon oil and further purifying the crude salmon oil into FGFO is high and suggests potentially significant increases in revenue if undertaken.

Summary

- Future recommended dietary reference intakes or recommended daily intakes of omega-3 LC-PUFA for the general population could average 650 mg per day per capita.

- For the total U.S. population of more than 281 million, the above recommendation would require about 222,556 t per year of FGFO, equivalent to about 296,741 t per year of CFO. For comparison, the production of Alaskan CFO from groundfish corresponds to about 30,000 t per year or about 10% of total estimated need.

- By expressing omega-3 LC-PUFA content of three varieties of fish as salmon equivalents (SE) we could estimate the corresponding amount of EPA+DHA consumed as SE (2.28 kg per capita per year) and the amount of FGFO that would need to be consumed as supplements (196,890 t per year, estimated at the 650 mg per day RDI level) to meet the RYI.

- Total retail value of the estimated fraction of FGFO corresponding to 196,890 t per year (consumed as supplements of 1.0 g capsules of an 1812 oil) would amount to approximately $14.77 billion. The corresponding value of a hypothetical encapsulated Alaskan FGFO could amount to approximately $1.69 billion. Both of these values are approximately 100 times the values of the corresponding CFOs.

- The significant escalation of market value of FGFO vs. CFO (100:1) could be a driving force for development of new products utilizing these health-relevant fish byproducts.

Acknowledgments

The authors thank the organizers of the 2nd International Seafood Byproduct Conference, November 10-13, 2002, Anchorage, Alaska, for inviting Robert Katz to present this paper.

Glossary of terms

1812 oil	fish oil that contains approximately 180 mg EPA and 120 mg DHA
CFO	crude fish oil
DHA	docosahexaenoic acid
EPA	eicosapentaenoic acid
FDA	U.S. Food and Drug Administration

FGFO	food grade fish oil
omega-3 LC-PUFA	long-chain omega-3 polyunsaturated fatty acids
RDI	recommended daily intake
RYI	recommended yearly intake
SE	salmon equivalent
TP	U.S. total population

Species

Brevoortia tyrannus	Atlantic menhaden
Oncorhynchus sp.	Pacific salmon
Theragra chalcogramma	Walleye pollock
Thunnus sp.	Tuna

References

ADFG. 2001. 2001 harvest statistics. Alaska Department of Fish and Game. http://www.cf.adfg.state.ak.us/geninfo/finfish/grndfish/catchval/01grndf.htm and http://www.cf.adfg.state.ak.us/GENINFO/finfish/salmon/CATCHVAL/BLUSHEET/01exvesl.htm.

Al, M.D.M., A.C. van Houwelingen, and G. Hornstra. 2000. Long-chain polyunsaturated fatty acids, pregnancy, and pregnancy outcome. Am. J. Clin. Nutr. 71(suppl):285S-291S.

Albert, C.M., H. Campos, M.J. Stampfer, P.M. Ridker, J.E. Manson, W.C. Willett, and J. Ma. 2002. Blood levels of long-chain omega-3 fatty acids and the risk of sudden death. New England J. Med. 346:1113-1118.

Birch, E.E., S. Garfield, D.R. Hoffman, R. Uauy, and D.G. Birch. 2000. A randomized controlled trial of early dietary supply of long-chain polyunsaturated fatty acids and mental development in term infants. Dev. Med. Child Neurol. 42:174-181.

Carlson, S.E. 2001. Docosahexaenoic acid and arachidonic acid in infant development. Seminars in Neonatology 6:437-449.

Crawford, M.A. 2000. Placental delivery of arachidonic and docosahexaenoic acids: Implications for the lipid nutrition of preterm infants. Am. J. Clin. Nutr. 71(suppl):275S-284S.

Cunnane, S.C., S.C. Williams, J.D. Bell, S. Brookes, K. Craig, R.A. Iles, and M.A. Crawford. 1994. Utilization of uniformly labeled 13C-polyunsaturated fatty acids in the synthesis of long chain fatty acids and cholesterol accumulating in the neonatal rat brain. J. Neurochem. 62:2429-2436.

De Caterina, R., J.K. Liao, and P. Libby. 2000. Fatty acid modulation of endothelial activation. Am. J. Clin. Nutr. 71(suppl):213S-223S.

Dyerberg, J., and H.O. Bang. 1982. A hypothesis on the development of acute myocardial infarction in Greenlanders. Scand. J. Clin. Lab. Investig. Suppl. 161:7-13.

FDA. 1997. Substances affirmed as generally recognized as safe: Menhaden oil. U.S. Food and Drug Administration. Federal Register, June 5, 1997, 62(108): 30751-30757. 21 CFR part 184. http://frwebgate.access.gpo.gov/cgi-bin/getdoc.cgi?dbname=1997_register&docid=fr05jn97-5.

FDA. 2001. Agency response letter. U.S. Food and Drug Administration. GRAS notice no. GRN 000041. http://www.cfsan.fda.gov/~rdb/opa-g041.html.

FDA. 2002. Letter responding to a request to consider the qualified claim for a dietary supplement health claim for omega-3 fatty acids and coronary heart disease. Office of Nutritional Products, Labeling, and Dietary Supplements, Center for Food Safety and Applied Nutrition, U.S. Food and Drug Administration. February 8, 2002. [docket no. 91N-0103]. http://www.cfsan.fda.gov/~dms/ds-ltr28-html.

GISSI. 1999. Dietary supplementation with n-3 polyunsaturated fatty acids and vitamin E after myocardial infarction: Results of the GISSI-Preventione trial. Gruppo Italiano per lo Studio della Sopravvivenza nell'Infarcto (GISSI) Miocardico. Lancet 354:447-455.

Hibbeln, J.R. 2001. Seafood consumption and homicide mortality. A cross-national ecological analysis. World Rev. Nutr. Diet 88:41-46.

Hu, F.B., L. Bronner, W.C. Willett, M.J. Stampfer, K.M. Rexrode, C.M. Albert, D. Hunter, and J.E. Mason. 2002. Fish and omega-3 fatty acid intake and risk of coronary heart disease in women. J.A.M.A. 287:1815-1821.

Innis, S.M. 1991. Essential fatty acids in growth and development. Prog. Lipid Res. 30:39-103.

IOM. 2002. Dietary reference intakes for energy and macronutrients. Institute of Medicine, National Academy of Sciences. National Academy Press, Washington, D.C.

Katz, R., J.A. Hamilton, A.A. Spector, S.A. Moore, H.W. Moser, M.J. Noetzel, and P.A. Watkins. 2001. Brain uptake and utilization of fatty acids: Recommendations for future research. J. Mol. Neurosci. 16(2,3):333-336.

Krauss, R.M, R.H. Eckel, B. Howard, L.J. Appel, S.R. Daniels, R.J. Deckelbaum, J.W. Erdman Jr., P. Kris-Etherton, I.J. Goldberg, T.A. Kotchen, A.H. Lichtenstein, W.E. Mitch, R. Mullis, K. Robinson, J. Wylie-Rosett, S. St Jeor, J. Suttie, D.L. Tribble, and T.L. Bazzarre. 2001. Revision 2000: A statement for healthcare professionals from the Nutrition Committee of the American Heart Association. J. Nutr. 131(1):132-146.

Kris-Etherton, P.M., W.S. Harris, L.J. Appel, et al. 2002. Fish consumption, fish oil, omega-3 fatty acids, and cardiovascular disease. Nutrition Committee, American Heart Association. Circulation 106:2747-2757.

Kris-Etherton, P.M., D. Taylor, S. Yu-Poth, P. Huth, K. Moriarty, V. Fishell, R.L. Hargrove, G. Zhao, and T.D. Etherton. 2000. Polyunsaturated fatty acids in the food chain in the United States. Am. J. Clin. Nutr. 71(suppl):179S-188S.

Locke, C.A., and A.L. Stoll. 2001. Omega-3 fatty acids in major depression. World Rev. Nutr. Diet 89:173-185.

Marchioli, R., F. Barzi, E. Bomba, et al. 2002. Early protection against sudden death by n-3 polyunsaturated fatty acids after myocardial infarction: Time-course analysis of the results of the GISSI-Prevenzione. Circulation 105:1897-1903.

Montori, V.M., P.C. Woolan, A. Farmer, and S.F. Dinneen. 2000. Fish oil supplementation in type 2 diabetes. A quantitative systematic review. Diabetes Care 23:1407-1415.

Nettleton, J. 1985. Seafood nutrition: Facts, issues, and marketing of nutrition in fish and shellfish. Osprey Books, Huntington, New York.

Oil World. 2002. Wholesale price of crude fish oil. October 10, 2002. Oil World (Rotterdam, The Netherlands) 45(41):493.

Peet, M., and D.F. Horrobin. 2002. A dose-ranging study of the effect of ethyl-eicosaenoate in patients with ongoing depression despite apparently adequate treatment with standard drugs. Arch. Gen. Psychiatry 59:913-917.

Simopoulos, A.P. 1999. Essential fatty acids in health and chronic disease. Am. J. Clin. Nutr. 70(3 suppl):560S-569S.

Simopoulos, A.P. 2002. Omega-3 fatty acids in inflammation and autoimmune diseases. J. Am. Coll. Nutr. 21(6):495-505.

Simopoulos, A.P., and J. Robinson, 1998. The omega plan. Harper Collins, New York, p. 351.

Simopoulos, A.P., A. Leaf, and N. Salem. 2000. Workshop statement on the essentiality of and recommended dietary intakes for omega-6 and omega-3 fatty acids. Prostaglandins Leukot. Essent. Fatty Acids 63(3):119-21.

Sprecher, H. 1992. Interconversions between 20- and 22-carbon n-3 and n-6 fatty acids via 4-desaturase independent pathways. In: A.J. Sinclar and R. Gibson (eds.), Essential fatty acids and eicosanoids: Invited papers from the Third International Congress. American Oil Chemists Society, Champaign, Illinois, pp. 18-22.

Stilwell, W., and L. Jensky. 2002. International workshop on cellular and molecular aspects of omega-3 fatty acids and cancer. J. Lipid Res. 43(9):1579-1580.

Stoll, A.L., C.A. Locke, L.B. Marangell, and W.E. Severus. 1999. Omega-3 fatty acids and bipolar disorder: A review. Prostaglandins Leukot. Essent. Fatty Acids 60(5-6):329-337.

U.S. Census Bureau. 2000. Census 2000 basics. U.S. Census Bureau. http://www.census.gov/mso/www/c2000basics/chapter2.html#results.

Functional Fish Protein Ingredients from Fish Species of Warm and Temperate Waters: Comparison of Acid- and Alkali-Aided Processing vs. Conventional Surimi Processing

Hordur G. Kristinsson and Necla Demir
Laboratory of Aquatic Food Biomolecular Research, Aquatic Food Products Program, Department of Food Science and Human Nutrition, University of Florida, Gainesville, Florida

Abstract

Two processes have recently been developed with the goal to manufacture functional protein isolates from fish muscle sources of low value. The processes are based on the solubilization of muscle proteins via electrostatic repulsion by acid pH (pH 2-3.5) or alkaline pH (pH 10.5-11.5) followed by their isoelectric precipitation. Many species of warm and temperate waters are underutilized in the sense that they or their byproducts are not used for human food consumption. The potential exists to utilize these sources and formulate high value food products by employing these new processes. Protein isolates were produced from select species (channel catfish, Spanish mackerel, croaker, and mullet) using the two processes. Results showed some variation between species, but in most cases higher protein recovery was obtained compared to conventional surimi processing. Lipid reduction was higher using the two processes compared to conventional surimi processing. The alkali process gave a protein isolate of improved color and substantially better oxidative stability than the acid process. Gelation properties of the alkali protein isolates was improved in most cases compared to those prepared using conventional surimi methods. The acid protein isolates had poor gelation properties, except for those made from croaker. The results suggest that alkali-aided processing may be a more successful means of producing functional protein ingredients from the four species studied in comparison with conventional surimi processing. The acid-aided process resulted in quality and functional problems not desirable for fish protein ingredients.

Introduction

Presently the demand for fish protein in the world is growing at a faster pace than the supply of traditional fish species (Hultin and Kelleher 2000). This has led to a world catch which is believed by many to be on the verge of or exceeding sustainable limits of our oceans with many common food fishes at the brink of endangerment (Kristinsson and Rasco 2000). The situation has directed attention to the abundant pelagic fish species and fish processing byproducts which are not primarily used for human food but directed toward animal feed production or fertilizer. Although the pelagic species are available in abundance they present the fish processor with numerous difficulties in part due to their seasonality, small size, abundance of oxidatively unstable lipids, large amounts of pro-oxidants (especially heme proteins), unstable muscle proteins of low functionality (due to low pH in muscle), and relatively high proteolytic activities (Okada 1980, Hultin 1994, Hultin and Kelleher 2000). The above factors hamper their direct consumption and greatly limit the possibilities to economically recover functional proteins from the pelagic species. Recovering proteins from byproducts presents a similar challenge.

Using conventional technologies to process fish and creating value-added fish products generally lead to limited utilization of the animal, and large amounts of protein-rich byproduct materials are lost and not recovered. Possibly more than 60% of fish tissue remaining after filleting (species dependent) is considered to be processing waste and not used as human food (Mackie 1982). This material is high in quality protein (≥10%) and other valuable compounds which could be utilized for human consumption. Producing value-added foods from fish such as surimi, which is, e.g., used to make imitation crabmeat, leads to further losses of protein. Surimi made from fillets will only yield ca. 50-70% of the original fillet protein (species dependent) because the process involves several washing steps to get rid of soluble compounds which may adversely affect the final product. In addition to processing waste it has been estimated that about 30% of the world's catch is transformed into fish meal (Rebeca et al. 1991) for animal feed and possibly about 30% of the biomass harvested will not find any utilization (Kristinsson and Rasco 2002). In light of an increasing world population, the danger of overfishing, and the limited utilization of our fish harvest, there obviously is a great need to utilize processing byproducts and underutilized species with more intelligence and foresight. Not only will increased utilization be environmentally sound but also at the same time it will create a larger dividend for the struggling seafood industry. However, for new processes aimed at increasing yield and expanding the use of unconventional fish sources to get acceptance from industry, they have to be more economically feasible than discarding the byproducts or using them for feed or fertilizer.

To address the problem outlined above a process was developed to economically produce functional protein isolates from fish sources of low value (Hultin and Kelleher 1999, Hultin et al. 2000). This process utilizes the pH-dependent solubility properties of fish muscle proteins for their separation and recovery from other components of muscle not desirable in a final product. The process involves subjecting a diluted (1:9) slurry of homogenized muscle tissue to a low (pH 2-3.5) or high (pH 10.5-11.5) pH. These pH values allow the muscle proteins to be solubilized, and disrupt the cellular membrane encasing the myofibrillar proteins. The acid and alkaline conditions used in the process are far enough from the muscle protein's isoelectric points (ca. pH 5-6) that the protein side chains gain a net positive (at acid pH) or negative (at alkaline pH) charge causing the proteins to repel each other and solubilize. The disruption of the muscle cell and solubilization of the proteins causes a great drop in solution viscosity, enough to enable cellular membranes to be separated from the soluble proteins by high force (ca. $10,000 \times g$) centrifugation, at the same time removing solids such as bones and scales and neutral fat. The essentially membrane- and lipid-free, soluble proteins are then recovered by isoelectric precipitation by adjusting the pH to ca. 5.5. The resulting isolate can be used for various purposes, e.g., as a functional food ingredient or directly used to produce value-added fish products such as surimi.

This process offers a new technology to recover and use protein for human consumption from the over 50 million tons of underutilized aquatic species and byproducts. The process creates a potential to use low-value, whole, dark-muscle fish species as a starting material to achieve a final product comprising a functional protein isolate of reduced lipid content, which greatly expands the potential use of the underutilized fish species of the world. The advantages of this process over other previously disclosed processes are many, including (1) the process leads to higher protein recovery since sarcoplasmic proteins are not removed, (2) the process is less labor consuming than conventional processes and allows better process control, (3) the process does not require the addition of salt to solubilize the proteins, and (4) the removal of membranes and lipids renders the final product more stable against oxidation and may remove fat-soluble toxins. The reduction in lipids is an important step in this process, since many of the dark muscle species are rich in triacylglycerols and membrane phospholipids. The membrane phospholipids are known to be the main substrate for oxidative reactions in fish muscle (Hultin 1994, Gandemer 1999) and their removal is expected to greatly enhance the stability of the final product. The increase in yield is also a key advantage in part since the raw material can be more responsively used.

The acid- or alkali-treated muscle proteins from several cold water species have been found to have good gelation properties compared to proteins from conventional processes (Undeland et al. 2002, Kristinsson and Hultin 2002b, Kelleher and Hultin 2000), making the process an attractive alternative to produce high value surimi-type products from low value sources. It remains to be investigated how these two processes compare in the production of protein products from fish species of warm and temperate waters and also how they compare to conventional surimi processing, which is the aim of the current study.

Materials and methods

Fish raw material

Four species were selected for this study. Skinless fillets from channel catfish were obtained fresh from a local supplier within 24 hours of harvest. Whole croaker, mullet, and Spanish mackerel were obtained packed in ice from Save-on-Seafood in St. Petersburg, Florida. The fish were 2-3 days old upon arrival to the laboratory and they were filleted, skinned, and processed immediately.

Production of surimi

Conventional laboratory-scale surimi processing was employed (Fig. 1a). First the fillets were ground in a Scoville grinder (Hamilton Beach) with 6 mm holes. The material was then gently mixed into three volumes of cold (4°C) water and stirred for 15 minutes following a 15 min period of settling. The next step involved dewatering the slurry by pouring it into a strainer lined with cheesecloth, then squeezing water out of the ground material by squeezing the cheesecloth. This process was repeated two times with the last wash including 0.2% NaCl to aid in dewatering. All steps were performed on ice. For the gelation tests the surimi was mixed with 5% sorbitol, 4% sucrose, and 0.3% sodium tripolyphosphate and frozen at –30°C.

Production of protein isolates from acid and alkali-aided processing

Protein isolates were produced using the acid- and alkali-aided processes as outlined in Fig. 1b. The processes involved grinding the fillets and homogenizing them for 30 seconds in nine volumes of water at 4°C in a Waring blender. The pH of the homogenate was then either lowered to pH 2.5 (for the acid process) or increased to pH 11 (for the alkaline process) by slowly adding 2M HCl or 2M NaOH, respectively, with slow but constant stirring. The solution was then transferred into centrifuge bottles and centrifuged at $10,000 \times g$ for 20 min at 4°C to separate the soluble proteins from the neutral lipids (i.e., storage fat), membranes (which contain

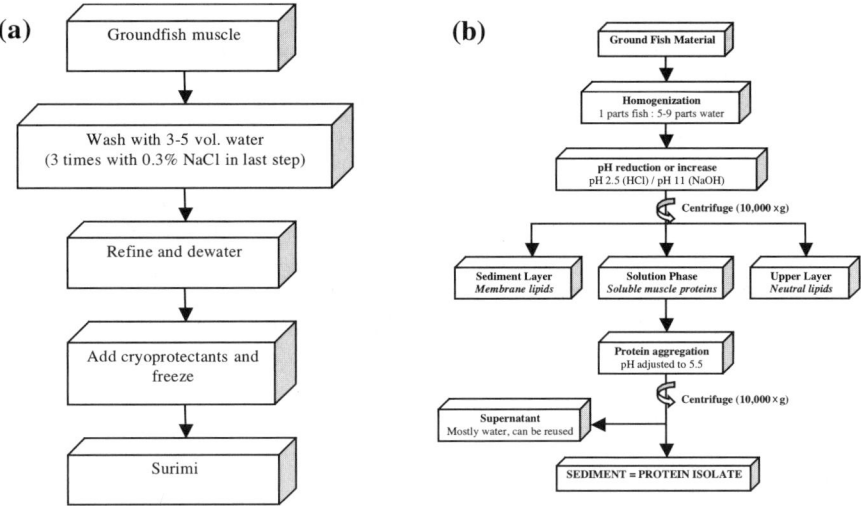

Figure 1. The processes employed in the study. (a) Conventional surimi processing. (b) The acid- and alkali-aided processes.

the membrane phospholipids), and solid material in the fish (mostly connective tissue and some bone residue). The middle phase containing the solubilized muscle proteins was then collected and its pH adjusted to 5.5 to aggregate the proteins followed by centrifugation at 10,000 × g for 20 min at 4°C to collect them. The sediment was the protein isolate (PI). For the gelation tests the PI was mixed with 5% sorbitol, 4% sucrose, and 0.3% sodium tripolyphosphate and frozen at −30°C.

Protein recovery and lipid reduction analysis

The raw material and its products (surimi and PI) were analyzed for protein content using the Biuret method (Torten and Whitaker 1964) and protein recoveries were calculated by dividing the total protein in the surimi or PI by the protein in the starting material. Lipids were extracted and quantified according to Bligh and Dyer (1959) and lipid reduction was calculated from the amount of lipids in the surimi and PI vs. that in the starting material.

Protein solubility and viscosity

Ground muscle was diluted in deionized water (1:9) and homogenized in a Waring blender for 30 seconds. The muscle protein homogenate was then separated into two fractions. One of the fractions was adjusted down to pH 1.5 while the other fraction was adjusted to pH 12. Samples were taken at 0.5 pH intervals for protein solubility and viscosity testing. For the protein solubility tests the protein samples were centrifuged at 10,000 × g for 20 min. The protein content of the middle layer was determined and total protein in the middle layer calculated. Percent soluble protein was determined by dividing the total protein content in the middle layer after centrifuging with the total protein content of the protein homogenate before centrifuging. For viscosity measurements samples were also taken at 0.5 pH intervals and viscosity measured using a single gap cylinder geometry in an AR2000 Advanced Rheometer (TA Instruments, New Castle, Delaware). Viscosity measurements were done at 5°C using an oscillatory time sweep with frequency set at 0.1 Hz and initial oscillatory stress at 0.1809 Pa.

Color analysis

Color of surimi and the protein isolate was determined by using a hand-held Minolta colorimeter. A minimum of three samples were taken on both uncooked and cooked surimi and protein isolate, and L, a, and b values were recorded and averaged. Whiteness was calculated according to Cortes-Ruiz et al. (2001). Cooked surimi and PI was made by heating samples in sausage casing at 80°C for 25 min and cooling on ice overnight.

Gelation properties

The surimi and PI were thawed at 4°C overnight and adjusted to 12% protein concentration with a solution containing the same cryoprotectant concentration as in the surimi and PI. The pH of the protein paste was adjusted to 7.0 with 1M NaOH or HCl and NaCl added to give 500 mM. The protein paste was then transferred onto a plate at 5°C on an AR2000 Advanced Rheometer. A parallel plate was then lowered on top of the protein paste to give a 1,000 micron gap. The paste was then heated from 5°C to 80°C and subsequently cooled to 5°C at 2°C per minute while the rheometer operated in an oscillation mode with the sample sheared at a constant frequency of 0.1 Hz with a maximum shear strain of 0.01. Viscoelastic changes on heating were studied by following changes in storage modulus (G'). The storage modulus characterizes the rigidity of the sample, where a higher G' translates to a stronger gel.

Lipid oxidation measurements

Samples of surimi and protein isolate were placed in plastic weighing boats and ziploc freezer bags and placed in cold storage set at 4°C. Samples (5 g) were collected every three days and frozen at –70°C to arrest oxidation. Samples were then analyzed for oxidation products (malondialdehyde) using the modified TBARS (thiobarbituric acid reactive substances) method described by Lemon (1975). Samples that represent day 0 were taken after ca. 2 hours from the start of the surimi and PI process (the time it took to process the material). Initial raw material was also measured for TBARS.

Microbial growth determination

Surimi and protein isolate were produced from aseptically handled raw material using sterile equipment for grinding, centrifuging, mixing, etc., and stored in plastic ziploc bags after production. Approximately 1 g samples were taken at different days for microbial growth determination and transferred to Butterfields buffer (pH 7.2) where they were mixed in a stomacher. Different dilutions were then placed on Petrifilm aerobic count plates (3M Microbiology Products) and incubated at 37°C for 48 hours before enumeration of colonies. Mean counts were reported as cfu per g.

Results and discussion

The four species selected for this study, channel catfish, Spanish mackerel, croaker, and mullet, are all species of commercial importance in the southern United States, from which products and byproducts may find use as substrates for the isolation of functional protein ingredients. The main focus of our study was to investigate how successful the acid and alkali-aided processes were compared to a conventional surimi process. A prerequisite for good protein recovery in the processes is good protein solubility at low and high pH. All species had protein solubilities between 70 and 90% at low and high pH. The solubility curve for catfish and croaker proteins is depicted in Fig. 2. The U-shaped curve is typical of muscle proteins that gain a net positive and negative charge at low and high pH, respectively, which leads to a great increase in solubility (Hamm 1994). There are, however, some differences in solubilities of the proteins from the different species, the reason for which is under investigation in our laboratory. Improvements in protein solubility at low and high pH will certainly be of importance as it would lead to potentially higher protein recoveries in the process.

A second important physical property of the homogenized muscle proteins at low and high pH is their viscosity. A significant drop in viscosity (below 50 mPa) is important to separate the soluble proteins from the cellular membranes and other insoluble matter in muscle (Hultin 2002). Taking catfish as an example it can be seen that there are major differences in viscosity at low and high pH, high pH showing a great increase in viscosity with a maximum at pH 9.5 (Fig. 3). This is most likely due to great electrostatic repulsion between the muscle proteins at that pH without sufficient solubilization. Viscosity increased also at low pH with two maxima, but being much lower than at high pH. To go below 50 mPa it was necessary to go below pH 3.5 and above pH 10.5, where solubility was also greatly increased. The pH values employed in our study were therefore within the limit needed for good separation of the membrane lipids from the proteins.

Protein recoveries were in almost all cases significantly higher when the acid and alkali processes were employed compared to the surimi

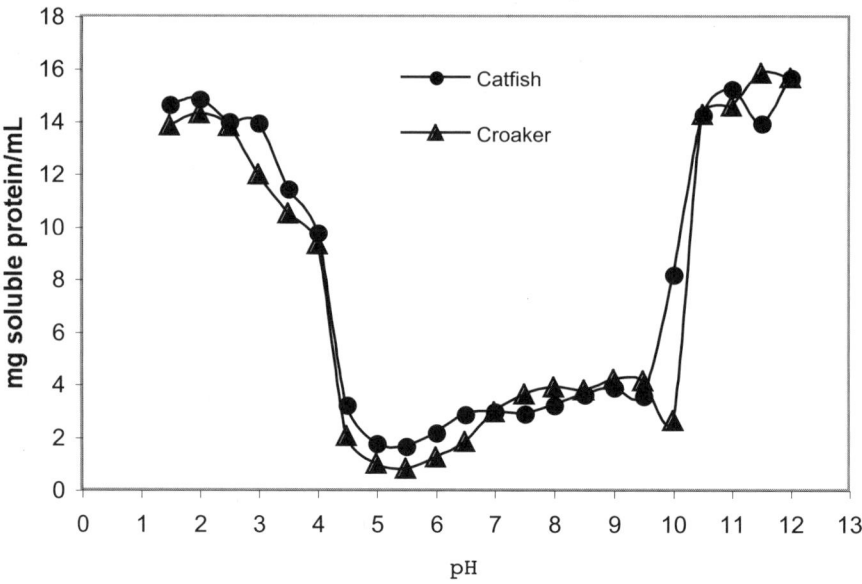

Figure 2. Solubility of fish muscle proteins as a function of pH. Muscle was homogenized in 9 volumes of water and pH adjusted with HCl and NaOH to decrease and increase the pH, respectively.

process (Table 1). The acid process yielded on average slightly higher recoveries compared to the alkali process. In the alkali process a sizable amount of heme and sarcoplasmic proteins remain soluble during the isoelectric precipitation step including some of the myofibrillar proteins, namely actin (data not shown). Undeland and coworkers (2002) recently reported similar findings for herring light muscle processed with the alkali process. This likely contributes to the lower protein recovery in the alkaline process. The lower recovery could also be due to lost protein in the upper lipid phase after the first centrifugation. The alkali treatment has been shown to make the muscle proteins more prone to interact with lipids and form emulsions (Kristinsson and Hultin 2002b) which would entrap some of the protein in the lipid phase. The upper lipid phase from the alkali process was in all cases significantly more voluminous than that generated in the acid process. The cause of the very high protein recovery obtained with mullet, using the acid-aided process, was that part of the sediment after the first centrifugation was collected with the soluble phase. This sediment was a very soft gel-like material that readily broke down when the soluble phase was collected. The much lower recovery for alkali processed mullet was in part because a large part of heme and sar-

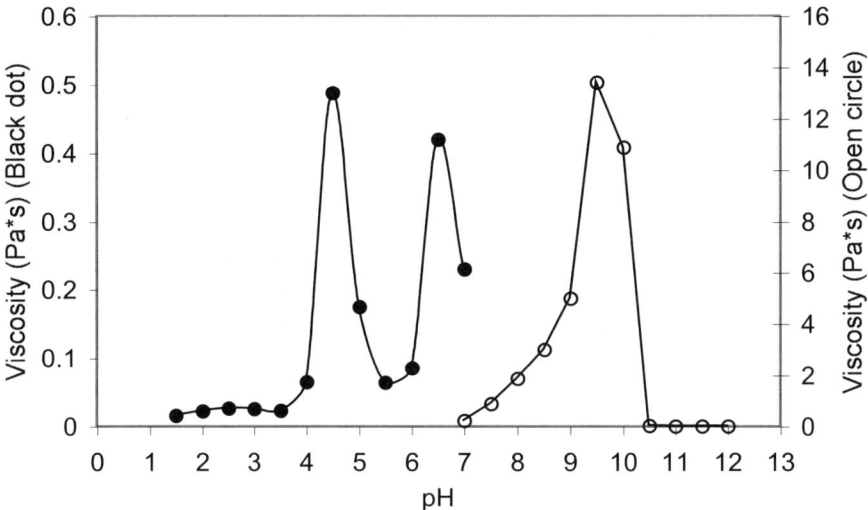

Figure 3. Viscosity of muscle proteins as a function of pH. Samples were prepared as described in Fig. 1. Note the two scales, one for each pH range.

Table 1. Protein recoveries from conventional surimi processing vs. acid and alkali-aided processing of fish muscle.

Product	Surimi	Acid PI	Alkali PI	Acid PI (skip 1st centrifugation)	Alkali PI (skip 1st centrifugation)
Catfish	62.3%[a]	71.5%[b]	70.3%[b]	85.8%[c]	82.1%[c]
Mackerel	54.1%[a]	73.6%[b]	69.3%[b]	N/A	N/A
Mullet	59.3%[a]	81.2%[b]	58.9%[a]	N/A	N/A
Croaker	57.7%[a]	78.7%[b]	65.0%[c]	N/A	N/A

Means within one species having different superscript letters are significantly different ($P < 0.05$).
PI = protein isolate.

coplasmic protein remained soluble during the isoelectric precipitation step (data not shown). The lower protein recovery of the surimi process compared to the other two processes has been reported to be due to washing away most of the sarcoplasmic proteins (Xiong 1997) and likely some of the proteins of the myofibrillar fraction (Lin and Park 1996).

To obtain an even higher protein recovery one can skip the first centrifugation step, going directly from low or high pH to 5.5 and recovering the protein isolate with centrifugation. This of course would not lead to a reduction in the membrane lipids. However, in species where lipid oxidation may not pose a problem (which we have found for catfish) this variation of the process is an excellent way to get increased protein recovery, as shown in Table 1 for catfish. Another potential means of increasing protein recovery is to re-suspend the bottom sediment after the first centrifugation at either low or high pH and then repeat the process. Cortes-Ruiz and coworkers (2001) increased protein recovery this way from 65.8 to 76% during acid-aided processing of brisling sardine muscle

The acid and alkali processes were also more successful than the conventional surimi process in reducing lipids present in the raw material (Table 2). In the acid and alkali process lipids are removed during the first centrifugation step in both the upper and sediment layer. Most of the storage fat and thus the more saturated fatty acids would be found in the upper phase while, if conditions are favorable, most of the unsaturated membrane phospholipids are expected to be sedimented during the first centrifugation (Hultin 2002). The greater emulsification ability of the alkali treated proteins may have contributed to the increased lipid removal, but at the cost of lower protein recovery. Lower lipid reduction for the acid processed mullet is likely due to the fact that more of the sediment was recovered in that process. The lower lipid reduction in the process where the first centrifugation step was skipped was expected since most all of the membrane lipids would be expected to be retained in the final protein isolate, including a sizable amount of the storage fat which may have co-precipitated with the proteins during the isoelectric precipitation.

Color of surimi and both protein isolates was generally good (Table 3). There were, however, some species differences. The darkest colors (i.e., the lowest L values) were found for mullet and croaker, which had the most dark muscle of the species tested. In most cases the color improved on cooking which is typical for surimi and surimi-like products. When the processes are compared in more detail it was generally found that the alkali process gave a product of higher L (and thus more whiteness) and lower b (less yellowness) value compared to the acid process. This is likely due to the greater removal of heme proteins in this process. The a values were less affected which is in agreement with results by Choi and Park (2002). The alkali process compared favorably with the conventional surimi process and produced products of whiter color for catfish and

Table 2. Lipid reduction using conventional surimi processing vs. acid and alkali-aided processing of fish muscle.

Product	Surimi	Acid PI	Alkali PI	Acid PI (skip 1st centrifugation)	Alkali PI (skip 1st centrifugation)
Catfish	58.3%[a]	85.4%[b]	88.6%[b]	45.4%[c]	61.2%[a]
Mackerel	72.1%[a]	76.9%[b]	79.1%[b]	N/A	N/A
Mullet	10.4%[a]	58.0%[b]	81.4%[c]	N/A	N/A
Croaker	16.65%[a]	38.1%[b]	68.4%[c]	N/A	N/A

Means within one species having different superscript letters are significantly different ($P < 0.05$).
PI = protein isolate.

croaker and less yellowness for catfish, mullet, and croaker. It is interesting to note that when the two variations of the process are compared, i.e., the one where the first centrifugation was skipped vs. the one with both centrifugation steps, it is evident that skipping the first centrifugation had a negative impact on color in terms of b values.

Gelation properties of the protein isolates were compared to those of the surimi (Table 4). The PI from the acid process had substantially poorer gelation characteristics than surimi and alkali PI for all species, except croaker. The alkali process, however, led to substantially stronger gels compared to conventional surimi except for the catfish. Gels produced using the modified alkali process where the first centrifugation step was omitted had similar gelling properties, compared to gels produced using the standard alkali process; however, gels produced in the standard acid process had lower values than in the modified acid process. The main difference between the standard PI and the PI from the modified process is that the latter contains most of the cellular membranes and also connective tissue. These two components may be the reason behind the better gelation properties. Kelleher found that when protein isolates were made from cod with and without membranes, the one with membranes developed a higher storage modulus on heating (Kelleher 2000). It is, however, worth noting that these gelation studies, which employ oscillatory tests, may yield different results than those employed on sausage type gels using torsion testing.

The question becomes: why do we see such a large difference between the three processes in terms of gelation? Surimi is largely composed of myofibrillar proteins since the sarcoplasmic proteins have been washed out. This is believed to increase the gelling ability since the sarcoplasmic proteins are believed by many to negatively affect gelation (Xiong 1997). However, other studies point to the opposite (Ko and Huang 1995, Morioka et al. 1997). Ongoing studies in our laboratory with catfish support a

Table 3.   Color values of surimi and protein isolates prior to and after gel formation.

Sample	Treatment	L	a	b	White-ness
Catfish surimi	Raw	70.4 ± 1.1	−0.9 ± 0.2	0.7 ± 0.4	70.4
	Gel	77.1 ± 0.1	−2.2 ± 0.1	3.1 ± 1.1	76.8
Catfish acid PI	Raw	73.8 ± 0.4	−3.6 ± 0.2	5.7 ± 0.3	72.9
	Gel	76.0 ± 1.1	−3.3 ± 0.1	6.5 ± 0.2	74.9
Catfish acid PI	Raw	75.1 ± 0.3	−2.3 ± 0.2	7.9 ± 0.5	73.8
(skip 1st centrifug.)	Gel	73.6 ± 0.5	−2.8 ± 0.4	8.6 ± 0.9	72.1
Catfish alkali PI	Raw	75.0 ± 0.7	−3.0 ± 0.2	0.2 ± 0.4	74.8
	Gel	80.0 ± 0.5	−3.2 ± 0.1	1.7 ± 0.2	79.7
Catfish alkali PI	Raw	78.4 ± 0.3	−2.2 ± 0.1	3.2 ± 0.0	78.1
(skip 1st centrifug.)	Gel	80.6 ± 0.6	−2.7 ± 0.1	4.4 ± 0.5	79.9
Mackerel surimi	Raw	78.4 ± 0.2	−1.5 ± 0.1	3.5 ± 0.1	78.1
	Gel	83.9 ± 0.9	−2.2 ± 0.2	5.7 ± 0.2	82.8
Mackerel acid PI	Raw	77.8 ± 0.6	−2.8 ± 0.1	8.0 ± 0.3	76.2
	Gel	75.5 ± 0.9	−2.3 ± 0.1	11.8 ± 0.4	72.7
Mackerel alkali PI	Raw	77.7 ± 1.0	−3.1 ± 0.2	4.5 ± 0.7	77.0
	Gel	77.2 ± 0.2	−3.0 ± 0.2	7.9 ± 0.7	75.7
Mullet surimi	Raw	73.3 ± 0.4	−0.2 ± 0.1	8.8 ± 0.2	71.9
	Gel	79.0 ± 0.8	−2.2 ± 0.2	9.7 ± 0.1	76.8
Mullet acid PI	Raw	68.3 ± 0.3	−1.3 ± 0.1	10.2 ± 0.2	66.7
	Gel	69.7 ± 0.6	−0.2 ± 0.3	14.7 ± 0.4	66.3
Mullet alkali PI	Raw	65.6 ± 0.6	−3.4 ± 0.1	3.2 ± 0.3	65.3
	Gel	72.5 ± 0.4	−2.8 ± 0.1	7.4 ± 1.4	71.4
Croaker surimi	Raw	64.7 ± 2.4	0.2 ± 0.1	6.4 ± 0.9	64.1
	Gel	73.0 ± 0.9	−1.9 ± 0.1	7.9 ± 0.4	71.8
Croaker acid PI	Raw	69.8 ± 0.7	−2.0 ± 0.3	7.8 ± 0.4	68.7
	Gel	73.3 ± 0.1	−2.0 ± 0.1	11.0 ± 0.2	71.1
Croaker alkali PI	Raw	67.6 ± 0.4	−2.1 ± 0.2	4.2 ± 0.4	67.3
	Gel	74.5 ± 0.1	−2.7 ± 0.1	7.8 ± 0.1	73.2

PI = protein isolate.

positive effect of the sarcoplasmic proteins on gel properties (Kristinsson and Crynen 2003). Both the acid and alkali PIs will have a sizable amount of the sarcoplasmic proteins present, but the majority of the protein is from the myofibrillar fractions. It is poorly understood how the myofibrillar and sarcoplasmic proteins are affected by the low and high pH used in the process for the species under investigation here. It is possible that the acid pH has a detrimental effect on protein conformation and thus the functional properties of these proteins, while the alkali treatment has a positive impact on the conformation and functional properties of

Table 4. **Storage modulus (G') of surimi and protein isolates after heating from 5 to 80°C and cooling to 5°C at 2°C per min.**

Product	Surimi	Acid PI	Alkali PI	Acid PI (skip 1st centrifugation)	Alkali PI (skip 1st centrifugation)
Catfish	1,704 Pa[a]	415 Pa[b]	1,821 Pa[a]	938 Pa[c]	1953 Pa[a]
Mackerel	879 Pa[a]	464 Pa[b]	1,642 Pa[c]	N/A	N/A
Mullet	1,629 Pa[a]	167 Pa[b]	2,059 Pa[c]	N/A	N/A
Croaker	1,349 Pa[a]	1,325 Pa[a]	2,298 Pa[b]	N/A	N/A

Means within one species having different superscript letters are significantly different ($P < 0.05$).
Protein concentration was 12% in 500 mM NaCl at pH 7.
PI = protein isolate.

the proteins. Kristinsson and Hultin (2002a) reported an increase in reactive –SH groups for cod myosin after alkali treatment, which could aid in the formation of a stronger gel than that of surimi. Stronger and more elastic gels were reported for alkali PI from rockfish compared to acid PI possibly due to more reactive sulfhydryl groups (Yongsawatdigul et al. 2001). Undeland and coworkers (2002) reported significantly harder gels for herring PIs made with the alkaline process when compared to the acid process.

What specific change is occurring with the proteins of the four fish species at acid and alkaline pH is unknown, and is under investigation in our laboratory. Previous studies with cod muscle proteins did show that gelation properties of myofibrillar proteins were largely unaffected by the acid and alkali treatments, while gelation properties of myosin were greatly improved after pH treatment relating to a partially denatured structure of the myosin head group (Kristinsson and Hultin 2002a,b). Similar results have been obtained with catfish myosin (Davenport and Kristinsson 2003) but other species are yet to be investigated. Protein hydrolysis may account for some of the lower gelation properties of the acid PI, as acidic conditions are known to activate muscle proteases which hydrolyze the muscle proteins and thus could lower their ability to form gels (An et al. 1994, Choi and Park 2002, Undeland et al. 2002). Undeland and coworkers (2002) found substantial hydrolysis of herring proteins at low pH on iced storage but no significant hydrolysis at alkaline pH. However, we tested for hydrolysis of catfish acid and alkali PIs and determined they had identical electrophoresis profiles indicating little hydrolysis had taken place (data not shown). This supports the previous idea that the acid treatment may have a negative influence on the conformation of these muscle proteins and their ability to interact to form gels on heating and cooling. It is under investigation to determine if the same holds true for the other three species.

An important quality aspect of any aquatic food product is lipid oxidation. Lipid oxidation can lead to undesirable appearance, odor, and taste problems, in addition to possibly being responsible for textural problems via interaction of oxidation products with proteins (Hultin 1994). In this study there were significant differences in oxidative stability of the samples from different treatments. In all cases the alkali PI has greater stability and thus less oxidation development than the acid protein isolate (PI) (Table 5). In all cases except for catfish substantial oxidation had already developed on day 0, which corresponds to values for the protein isolates after the process (ca. 2 hours from starting the procedures). This large difference between the acid and alkali PIs most likely stems from the fact that the acid PIs have a substantial amount of denatured heme proteins that co-precipitated with the muscle proteins at pH 5.5. At pH 2.5, used in this study, heme proteins denature rapidly and become very active catalysts of lipid peroxidation (Kristinsson 2002a,b). After pH readjustment from pH 2.5 to 5.5 the heme proteins only partly refold leaving an exposed hydrophobic surface and therefore they aggregate with the muscle proteins (Kristinsson 2002a). This leads to a higher concentration of heme proteins (most of which are denatured) in the acid PI vs. the alkali PI and thus more chance for lipid oxidation.

The alkali conditions used in this study on the other hand have very little impact on both conformation and solubility of heme proteins and it has been shown that pro-oxidative activity of heme protein is suppressed at high pH (Kristinsson 2002a). The alkali PI has thus not only substantially less heme proteins than the acid PI but the heme protein also would be expected to be in a native state after the process. This would explain the substantially better oxidative stability of the alkali PI. When the alkali PI is compared to surimi it has in most cases initially higher TBARS values, but after 6 days alkali PI and surimi have developed similar TBARS, except for mackerel where surimi has oxidized significantly more (Table 5). In Table 5 we compared the PI and surimi at their final preparation pH which is a disadvantage for PIs, because the PI is at pH 5.5 while the pH of the surimi ranged from pH 6 to 6.5. Oxidation develops faster at lower pH, especially when heme proteins are the main catalysts (Kristinsson 2002a). The higher initial values for the alkali PI vs. surimi are likely due to the low pH 5.5 protein aggregation step. The alkali PI has better oxidative stability compared to surimi after day 0 (Table 5). Future studies will be aimed at understanding the factors responsible for oxidation in the two processes and how these can be stabilized.

Microbial stability is another important property of protein products produced from fish muscle. Microbial spoilage would lead to odors, flavors, and appearances unacceptable to potential consumers. We were interested in following microbial growth with time for the protein isolates. As an example of this work we have included data for channel catfish in Table 6. From day 0 data it can be seen that the acid and alkali processes

Table 5. **Development of lipid oxidation in surimi and protein isolates as assessed by TBARS.**

Sample	Day	Surimi	Acid PI	Alkali PI	Acid PI (skip 1st centrifugation)	Alkali PI (skip 1st centrifugation)
Catfish	0	0.74 ± 0.54	0.78 ± 0.03	0.63 ± 0.05	0.49 ± 0.1	0.20 ± 0.02
	3	0.26 ± 0.20	1.06 ± 0.05	1.22 ± 0.27	5.87 ± 0.07	0.49 ± 0.03
	6	1.48 ± 1.30	3.88 ± 0.04	1.62 ± 0.09	5.08 ± 0.29	1.70 ± 0.03
Mackerel	0	5.65 ± 0.54	36.20 ± 0.10	12.93 ± 0.20		
	3	3.98 ± 0.18	29.34 ± 1.06	18.06 ± 0.16		
	6	33.93 ± 0.76	6.47 ± 0.08	21.32 ± 0.59		
Mullet	0	7.41 ± 0.50	43.23 ± 0.97	17.10 ± 1.77		
	3	8.10 ± 0.23	52.69 ± 0.24	18.22 ± 0.08		
	6	12.28 ± 0.42	38.53 ± 1.01	18.48 ± 0.78		
Croaker	0	3.03 ± 0.08	38.36 ± 0.29	10.08 ± 0.19		
	3	8.57 ± 0.084	30.50 ± 0.18	11.01 ± 0.04		
	6	7.63 ± 0.28	30.31 ± 0.12	10.22 ± 0.16		

Samples were kept at 4°C.
Day 0 is 2 h after the start of the surimi and PI process (the time it took to process the material).
PI = protein isolate.
TBARS = Thiobarbituric acid reactive substances.
Numbers are average ± s.d.

had reduced microbial content when compared to ground catfish. The alkali process led to the most reduction in microbial content. The reduction in microbial numbers was less for the regular surimi process. There are two potential ways bacteria can be reduced or killed during the new processes. First, the very low and high pH used may be enough to make microbial cells unviable or injured (Jay 1986). Second, the high centrifugal speed used in the first centrifugation may sediment the microbial cells along with the membrane lipids and insoluble material due to their high density. As both acid and alkali processes employ the same g-force during the first centrifugation it would be expected that about the same numbers of microorganisms would be centrifuged out. With this assumption in mind it may be deduced that the alkali conditions were more detrimental for the survival of these bacteria since the numbers were lowest for the alkali PI. Support for this comes from looking at microbial data after 10 days of storage where the alkali PI only has 3×10^2 organisms per g while the acid PI has 2.45×10^4 organisms per gram, suggesting less viable bacteria after the alkali treatment. The ground catfish and surimi had very high numbers and were highly spoiled after 10 days.

**Table 6.   Aerobic plate count of catfish su-
rimi and protein isolates.**

Sample	Day 0	Day 10
Catfish (ground)	6.32×10^3	7.71×10^5
Catfish surimi	2.38×10^3	1.55×10^6
Catfish acid	1.45×10^3	2.45×10^4
Catfish alkali	1.00×10^3	3.00×10^2

Numbers are average counts per gram tissue.

Conclusion

Comparing the two new processes with conventional surimi processing it was evident the alkali-aided process was overall most successful in preparing protein ingredients of low lipid content, good functionality, and stability. Even though protein recovery was higher for the acid process compared to both the alkali process and regular surimi processing, the alkali process led to more reduction in lipids, and removed more heme proteins giving better color and significantly better oxidative stability. In all cases but one the alkali process resulted in protein products with substantially improved gelling ability. To add to this the alkali process was most successful in reducing aerobic bacteria and had greatly improved stability against microbial spoilage compared to the other processes. It remains to be tested if the same holds true if the processes are scaled up to a commercial level.

Acknowledgments

This work was funded in part by grant 2002-34135-12467 of the USDA-CSREES Tropical/Subtropical Agriculture Research Program awarded to Dr. Kristinsson. The authors wish to extend their thanks to Ms. Bergós Ingadóttir for help with the solubility experiments, Ms. Ann E. Gore for help with preparation of the protein isolates and lipid analyses, Ms. Colleen Brown and Mr. Stefan Crynen for help with viscosity measurements, and Dr. Keith Schneider, Mr. Mike Madden, and Ms. Morgan Phares for help and advice with the microbial tests. We thank Save-on-Seafood for supplying us with fish for this study.

References

An, H., V. Weerasinghe, T.A. Seymore, and M.T. Morrissey. 1994. Cathepsin degradation of Pacific whiting surimi proteins. J. Food Sci. 59:1013-1017, 1033.

Bligh, E.G., and W.J. Dyer. 1959. A rapid method of total lipid extraction and purification. Can. J. Biochem. Physiol. 37:911-917.

Choi, Y.J., and J.W. Park. 2002. Acid-aided protein recovery from enzyme rich Pacific whiting. J. Food Sci. 67:2962-2967.

Cortes-Ruiz, J.A., R. Pacheco-Aguilar, G. Garcia-Sanches, and M.E. Lugo-Sanches. 2001. Functional characterization of protein concentrate from bristly sardine under acidic conditions. J. Aquat. Food Prod. Technol. 10:2-23.

Davenport, M.P., and H.G. Kristinsson. 2003. Low and high pH treatments induce a molten globular structure in myosin which improves its gelation properties. IFT (Institute of Food Technologists) Annual Meeting, Chicago, July 14. Abstract 42-9.

Gandemer, G. 1999. Lipids and meat quality: Lipolysis, oxidation, Maillard reaction and flavour. Sci. Aliment. 19:439-458.

Hamm, R. 1994. The influence of pH on the protein net charge in the myofibrillar system. Reciprocal Meat Conf. Proc. 47:5-9. American Meat Science Association, Savoy, Illinois.

Hultin, H.O. 1994. Oxidation of lipids in seafoods. In: F. Shahidi and J. R. Botta (eds.), Seafoods: Chemistry, processing technology and quality. Blackie Academic, Glasgow, pp. 49-74.

Hultin, H.O. 2002. Recent advances in surimi technology. In: M. Fingerman and R. Nagabhushanam (eds.), Recent advances in marine biotechnology, Vol. 7. Science Publishers, Inc., Enfield, New Hampshire, pp. 241-251.

Hultin, H.O., and S.D. Kelleher. 1999. Process for isolating a protein composition from a muscle source and protein composition. U.S. patent number 6,005,073.

Hultin, H.O., and S.D. Kelleher. 2000. Surimi processing from dark muscle fish. In: J.W. Park (ed.), Surimi and surimi seafood. Marcel Dekker, New York, pp. 59-77.

Hultin, H.O., S.D. Kelleher, Y. Feng, H.G. Kristinsson, M.P. Richards, I.A. Undeland, and S. Ke. 2000. High efficiency alkaline protein extraction. U.S. Patent Application No. 60/230,397.

Jay, J.M. 1986. Modern food microbiology. Van Nostrand Reinhold Company, New York. 642 pp.

Kelleher, S.D. 2000. Physical characteristics of muscle protein extracts prepared using low ionic strength acid solubilization/precipitation. Ph.D. dissertation, University of Massachusetts at Amherst.

Kelleher, S.D., and H.O. Hultin. 2000. Functional chicken muscle protein isolates prepared using low ionic strength, acid solubilization/precipitation. Reciprocal Meat Conf. Proc. 53:76-81. American Meat Science Association, Savoy, Illinois.

Ko, W.-C., and M.-S. Hwang. 1995. Contribution of milkfish sarcoplasmic protein to the thermal gelation of myofibrillar proteins. Fish. Sci. (Tokyo) 61:75-78.

Kristinsson, H.G. 2002a. Conformational and functional changes of hemoglobin and myosin induced by pH: Functional role in fish quality. Ph.D. dissertation, University of Massachusetts at Amherst.

Kristinsson, H.G. 2002b. Acid-induced unfolding of flounder hemoglobin: An evidence for a molten globular state with enhanced pro-oxidative activity. In press. J. Agric. Food Chem.

Kristinsson, H.G., and S. Crynen. 2003. The effect of acid and alkali treatment on the gelation properties of sarcoplasmic and myofibrillar proteins from channel catfish. IFT (Institute of Food Technologists) Annual Meeting, Chicago, July 15. Abstract 87-3.

Kristinsson, H.G., and H.O. Hultin. 2002a. Changes in conformation and subunit assembly of cod myosin at low and high pH and after subsequent refolding. Submitted to J. Agric. Food Chem.

Kristinsson, H.G., and H.O. Hultin 2002b. Effect of low and high pH treatment on the functional properties of cod muscle proteins. Submitted to J. Agric. Food Chem.

Kristinsson, H.G., and B.A. Rasco. 2000. Fish protein hydrolysates: Production, biochemical and functional properties. CRC Crit. Rev. Food Sci. Hum. Nutr. 40:43-81.

Kristinsson, H.G., and B.A. Rasco. 2002. Fish protein hydrolysates and their potential use in the food industry. In: M. Fingerman and R. Nagabhushanam (eds.), Recent advances in marine biotechnology, Vol. 7. Science Publishers, Inc., Enfield, New Hampshire, pp. 157-181.

Lemon, D.W. 1975. An improved TBA test for rancidity. In: A.D. Woyewoda, P.J. Ke, and B.G. Burns (eds.), New Series Circular. Department of Fisheries and Oceans, Halifax, Nova Scotia, pp. 65-72.

Lin, T.M., and J.W. Park. 1996. Extraction of proteins from Pacific whiting mince at various washing conditions. J. Food Sci. 61:432-438.

Mackie, I.M. 1982. Fish protein hydrolysates. Proc. Biochem. 17:26-32.

Morioka, K., T. Nishimura, A. Obatake, and Y. Shimizu. 1997. Relationship between the myofibrillar protein gel strengthening effect and the composition of sarcoplasmic proteins from Pacific mackerel. Fish. Sci. (Tokyo) 63:111-114.

Okada, M. 1980. Utilization of small pelagic species for food. In: R.E. Martin (ed.), Proceedings of the Third National Technical Seminar on Mechanical Recovery and Utilization of Fish Flesh. National Fisheries Institute, Washington, D.C. pp. 265-282.

Rebeca, B.D., M.T. Pena-Vera, and M. Diaz-Castaneda. 1991. Production of fish protein hydrolysates with bacterial proteases: Yield and nutritional value. J. Food Sci. 56:309-314.

Torten, J., and J.R Whitaker. 1964. Evaluation of the biuret and dye-binding methods for protein determination in meats. J. Food Sci. 29:168-174.

Undeland, I., S.D. Kelleher, and H.O. Hultin. 2002. Recovery of functional proteins from herring (*Clupea harengus*) light muscle by an acid or alkaline solubilization process. J. Agric. Food Chem. 50:7371-7379.

Xiong, Y.L. Structure-function relationships of muscle proteins. 1997. In: S. Damodaran and A. Paraf (eds.), Food proteins and their applications. Marcel Dekker, Inc., New York, pp. 341-392.

Yongsawatdigul, J., J.W. Park, and O. Esturk. 2001. Gelation characteristics of alkaline and acid solubilization of fish muscle proteins. Paper presented at the 2001 IFT Meeting (Institute of Food Technologists), New Orleans, June 23-27, 2001.

Bioactive Compounds from Sea Squirt Tunic Wastes

Sung-Hun Jeong and Sham-Hwan Ahn
Gyeongsang National University, Institute of Marine Industry, Tongyeong, Republic of Korea

Jose P. Peralta
University of the Philippines in the Visayas, Institute of Fish Processing Technology, Iloilo, Philippines

Seok-Joong Kang and Yeong-Joon Choi
Gyeongsang National University, Institute of Marine Industry, Tongyeong, Republic of Korea

Tae-Sung Jung
National University, College of Veterinary Medicine, Laboratory of Fish and Shellfish Diseases, Chinju, Republic of Korea

Byeong-Dae Choi
Gyeongsang National University, Institute of Marine Industry, Tongyeong, Republic of Korea

Abstract

Ascidians, commonly called sea squirts, are cultured or naturally grown species in Korea. The muscles of the sea squirt are popular seafood for the locals. But the tunic of the organism is normally considered waste. The wastes of the sea squirt tunic were crushed and pigments extracted with ethanol for 24 hours at room temperature. The carotenoid content of the tunic changed with season and region. Most analyzed carotenoids in the tunic showed a marked seasonal variation with a maximum in summer and a minimum in autumn. A total of 13 compounds were extracted from *Halocynthia roretzi*.

Glycosaminoglycans may be components of connective tissue or cell surface carbohydrates. Sulfated glycosaminoglycans can also be found in intracellular granules of a number of cells that participate in immune and inflammatory responses. The tunic was treated with 10 volumes of distilled water in a pressure cooker at 105°C for 3 hours. The extracted glycosaminoglycans were applied to a DEAE-cellulose column equilibrated with 0.5 M sodium acetate buffer (pH 6.0) and washed with 50 ml of the same buffer. Fractions were collected and assayed for hexoses and pentoses using phenol-sulfur acid; hexosamines and acetylhexosamines using metachromasia; and uronic acid using the carbazole reaction. This thermal extraction method was found effective for mass production of crude sulfated polysaccharides for industrial purposes.

Introduction

The major cultured species of sea squirt in Korea (*Halocynthia roretzi, Halocynthia aurantium, Styela clava, Styela plicata*) are found in the marine ecosystem. The larva of sea squirt is like a tadpole and swims for a specific period before settling on a suitable substrate and metamorphosing into the adult stage. The sea squirt has a cellulose-polysaccharide tissue known in the animal kingdom as a tunic (Tsuchiya and Suzuki 1962). The production of cultured sea squirt in Korea has increased every year after being reported to have anticancer properties (Watanabe et al. 1991).

The tunic has lots of carotenoids. Carotenoids are a group of fat-soluble pigments that contribute to the yellow, orange, and red colors found in all families of the plant and animal kingdoms, and are one of the most important natural marine pigment groups (Matsuno and Hirao 1989). Animals are unable to perform de novo synthesis of carotenoids but some bacteria, algae, yeasts, molds, and higher plants have this capacity (Simpson 1982, Liaaen-Jensen 1991). Carotenoids are responsible for the color of many important fish and shellfish products. Shrimp, lobster, crab, crayfish, trout, salmon, redfish, red snapper, and tuna have orange-red integument and/or flesh containing carotenoid pigments (Haard 1992). Salmon and trout raised in aquaculture are fed carotenoid pigments such as astaxanthin and canthaxanthin to color the muscle pink or red, and thus fulfill consumer expectations (No and Storebakken 1992). Commercially available salmon diets contain 40-75 mg astaxanthin or canthaxanthin per kg. Once the salmon reaches 300-400 g, supplementation with carotenoid pigments increases feed costs by 7-10% (Torrissen et al. 1989).

The glycosaminoglycans are a group of structurally related polysaccharides found as the carbohydrate moieties of proteoglycans and sometimes as free polysaccharides (Mathew 1975). Their functions are diverse. Glycosaminoglycans may all be components of connective tissue or cell surface carbohydrates involved in the cell's interaction (Rouslahti 1989). Sulfated glycosaminoglycans can also be found in intracellular granules of a number of cells which participate in immune and inflammatory responses (Okutani and Shigeta 1993). The tunic of an adult sea squirt contains mainly sulfated L-galactan, which is synthesized by the epidermal cells (Albano and

Mourão 1986, Mourão et al. 1997, Ruggiero et al. 1998), whereas at the larval stage, the main glycan found in the tunic is a complex sulfatedpolymer composed of D-glucose, L-fucose, and L-galactose (Albano et al. 1990). Pavão et al. (1995) described the presence of a unique dermatan sulfate-like glycosaminoglycan in the body of the ascidian *Ascidia nigra*, composed of $4-\alpha$-L-IdoA-$(2SO_4)$-$1\rightarrow3$-β-GalNAc-$(6SO_4)$ disaccharide units. Different from the mammalian counterpart the ascidian dermatan sulfate has no ability to potentiate heparin cofactor II inhibition of thrombin. Dietary fiber can be defined as those constituents of foods which are resistant to digestion by the secretions of the human gastrointestinal tract (Roberfroid 1993). Dietary fiber has been reported to have several physiological effects, such as preventing diabetes mellitus, lowering rates of hyper-cholesterolemia, reducing obesity, and delaying dumping syndrome, depending upon the physical and chemical properties of the individual fiber sources (Sung 1995).

This paper investigated carotenoid composition, omega-3 accumulation effect on eggs, physical characteristics of tunic glycosaminoglycans, and sensory characteristics of dietary fiber from the sea squirt tunic.

Carotenoid composition of the sea squirt tunic

Seasonal and regional variation of carotenoid content in the tunic of *Halocynthia roretzi*, cultured for three years at Tongyeong and Wolrae in Korea, was studied. The regional variation of carotenoids is consistent, but the carotenoid content of the tunic changed with seasonal variation (Table 1). The solvent used (acetone and acetone-methanol mixture) did not affect extraction of carotenoids from the sea squirt tunic. Most of the analyzed carotenoids in the tunic showed a marked seasonal variation with a maximum in summer (51.9 mg in August from Tongyeong, 54.5 mg in July from Wolrae) and a minimum in autumn (40.0 mg in September from Tongyeong, 38.9 mg in September from Wolrae).

Table 2 shows proportions of individual carotenoids in the ascidian tunic. Carotenoid content of tunic (49.2 mg per 100 g tunic) was much higher than that of muscles (2.35 mg per 100 g wet basis, data not shown). A total of 13 components were separated from the carotenoids extracted from *H. roretzi*. The carotenoids of *H. roretzi* accounted for are alloxanthin (31.3%), halocynthiaxanthin (15.5%), diatoxanthin (11.9%), diadinochrome (11.6%), mytiloxanthin (10.8%), and astaxanthin (7.8%).

Matsuno et al. (1985) investigated carotenoids in ten species of Tunicata. A new marine carotenoid, halocynthiaxanthin, was found to be present in most species of Tunicata. Alloxanthin was the principal carotenoid in six out of ten species of Tunicata. Most Tunicata examined were rich in the metabolic products of fucoxanthin (e.g., fucoxanthinol, halocynthiaxanthin, mythiloxanthin, mythiloxanthinone, amarouciaxanthin A and B) but scant in astaxanthin. In our studies most sea squirt tunics had different carotenoid composition and content.

Table 1.   Seasonal and regional variation of carotenoid content in the ascidian (*Halocynthia roretzi*) tunic.

	Tongyeong		Wolrae	
Month	A[a]	AM	A	AM
April	41.9±0.8[b]	43.1±0.5	52.4±0.3	46.6±0.5
May	47.6±0.6	45.0±0.9	55.4±0.3	51.0±0.7
June	49.2±0.6	47.3±0.7	53.3±0.9	52.7±0.6
July	50.4±0.4	45.1±0.7	54.5±0.4	51.3±0.3
August	51.9±0.6	46.6±0.8	46.8±0.9	47.3±0.4
September	41.0±0.4	40.0±0.5	38.8±0.7	39.3±0.5
Mean	46.9	44.5	50.2	48.0

[a]A = acetone; AM = acetone:methanol 1:1(v/v).
[b]All the data are presented as mean value of three determinations for each sample.
Units are mg/100 g wet basis.

Table 2.   Proportions of individual carotenoids in the *Halocynthia roretzi* tunic.

Carotenoids	Rf value[a]	Percent
Total concentration (mg/100 g tunic) = 49.2±0.9		
β-carotene	0.97	0.9±0.2
Lutein	0.92	0.3±0.1
Zeaxanthin	0.86	1.2±0.5
Pectenolone	0.70	0.8±0.1
Mytiloxanthinone	0.63	10.8±2.3
Astaxanthin	0.60	7.8±1.5
Alloxanthin[b]	0.52	31.3±6.4
Diatoxanthin	0.46	11.9±4.3
Diadinochrome	0.34	11.6±2.8
Halocynthiaxanthin	0.30	15.5±4.7
Mytiloxanthin	0.23	5.7±1.6
Fucoxanthin	0.14	1.4±0.6
Fucoxanthinol	0.11	0.8±0.2

[a]Thin layer plates on silica gel G. Developed in ethylacetate:dichloromethane 1:4 (v/v).
[b]Alloxanthin = cynthiaxanthin = pectenoxanthin.

Omega-3 enriched eggs

Diet plays an important role in maintaining health. Among the different products delivering essential nutrients to the body, the chicken egg has arguably a special place, being a rich and balanced source of essential amino and fatty acids as well as some minerals and vitamins. Recent developments suggest that in addition to well defined roles in energy metabolism and as constituents of biological membranes, the polyunsaturated fatty acids (PUFA) also have specific regulatory functions through the synthesis of different biologically active compounds, including eicosanoids (Surai and Sparks 2001). These PUFA have to be supplied by the diet (Nair et al. 1997). There are two classes of PUFA in the egg, which are metabolically and functionally distinct. The n-6 and n-3 PUFA absolute level, and the balance between n-6 and n-3 PUFA, in the diet are considered an important determinant of many metabolic functions in the human body (Simopoulos 2000).

Commercial table eggs contain a high proportion of n-6 PUFA (mainly 18:1 n-9) but are a poor source of n-3 fatty acids. Attempts were made to produce eggs high in n-6 and n-3 PUFA. The hen's diet is usually rich in flaxseed, linseed, or marine fish oil. As a result the egg's yolk is enriched with alpha-linoleic acid (ALA) and the level of DHA (docosahexaenoic acid) is also enhanced (Ferrier et al. 1995). The lipid content of the commercial DHA egg yolk is at a level of 270 mg DHA and 159 mg DPA (docosapentaenoic acid) per 100 g. The result of our two-week feeding trial research using sea squirt tunic carotenoids and marine algae (*Schizochytrium* sp.) mixed diets is shown in Fig. 1. The yolk lipid content increased from 143 to 429 mg of DHA and 116 to 348 mg DPA per 100 g during the two week feeding trial (Fig. 1). Although the lipid content is high, no fishy taste in the egg yolk was detected.

Physical characteristics of tunic glycosaminoglycans

Fat binding capacity, foaming, and emulsifying properties of glycosaminoglycans (GAGs) extracted from sea squirt tunic were summarized in Table 3. Except for the GAGs sample, all other samples did not show significant foam and/or emulsifying properties. The fat binding capacity ranged from 280% to 550%, with chitosan having the highest and GAGs having the lowest fat binding capacity. Knorr (1982) reported the fat binding capacity of chitin was greater than that of chitosan and microcrystalline chitin.

Foaming properties of proteins are important functional and quality attributes in ice cream, toppings, angel food cakes, and other foods. The foaming properties of protein are affected by intrinsic factors such as solubility, size, structural flexibility/stability, and concentration and

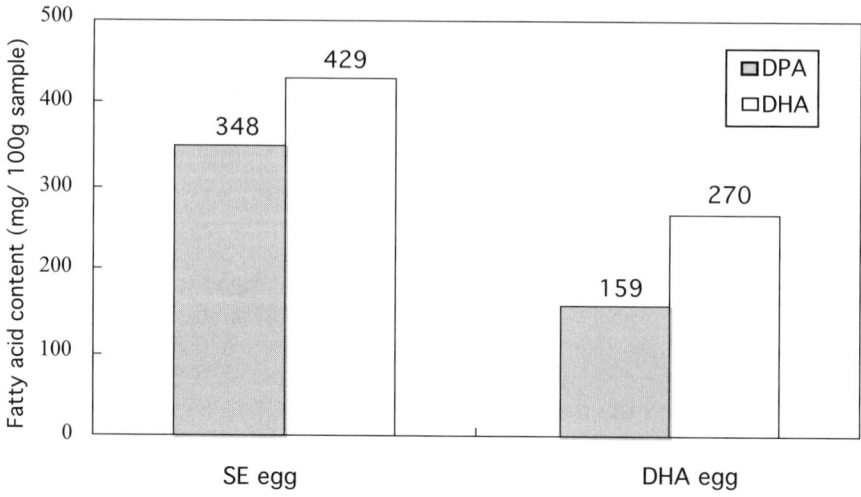

Figure 1. Docosahexaenoic acid (DHA), white) and docosapentaenoic acid (DPA, gray) content of egg yolk from hens fed with supplemented diet. SE egg = sea squirt tunic carotenoids in diet; DHA egg = commercial DHA egg.

Table 3.   Fat binding capacity, and foaming and emulsifying properties of sea squirt tunic, glycosaminoglycans (GAGs), chitin, and chitosan.

	Sea squirt tunic[a]	GAGs[b]	Chitin[c]	Chitosan[d]
Fat binding capacity[e]	320±71	280±75	530±106	550±81
Foam capacity[f]	0.05±0.01	0.70±0.03	n.s.[g]	n.s.
Foam stability[h]	0.03±0.00	0.50±0.05	n.s.	n.s.
Emulsion capacity[i]		49.2±7.3	n.s.	n.s.
Emulsion stability	n.s.	48.3±8.1	n.s.	n.s.

[a]Dried and ground sea squirt tunic.
[b]Glycosaminoglycans obtained by thermal extraction from tunic.
[c,d]Standard chitin and chitosan (Sigma Co., C-7170, C-3646).
[e]Fat binding capacity = final weight (g)/dried weight (g) × 100.
[f]Foam capacity = (total volume – drainage volume)/initial volume.
[g]n.s. = not significant.
[h]Foam stability = (total volume – drainage volume)/initial volume.
[i]Emulsion capacity = (emulsified volume/total volume) × 100.
Units are percents.

extrinsic factors such as pH, temperature, and salt concentration (Halling 1981). Values for foam stability of these samples were also measured (Table 3). GAGs formed strong foam as indicated by its high stability. But chitin and chitosan did not show significant foam development. Poole (1988) confirmed the potential value of low-viscosity chitosan as a foaming enhancer of acidic globular proteins. A limitation of chitosan as a foam enhancer is the reduction of the effective pH range.

Emulsifying properties of sea squirt tunic, GAGs, chitin, and chitosan were compared (Table 3). GAGs had good emulsifying properties. The emulsifying properties of sea squirt tunic, chitin, and chitosan were lower than GAGs. Knorr (1982) reported that chitosan and chitin did not produce emulsions, but microcrystalline chitin showed good emulsifying properties and was superior to microcrystalline cellulose. Increasing concentration of microcrystalline chitin (0.12-0.8 g per 100 ml water) had a positive effect on emulsion stability.

Characterization of GAGs from sea squirt tunic

Chondroitin sulfate was extracted using two treatments: enzyme and heat treatment. General compositions of the tunic components of sea squirt are similar between *Halocynthia roretzi* and *Halocynthia aurantium* (Table 4). Carbohydrate content was 47.7% to 48.0%, protein was 18.4% to 19.7%, and ash was as high 27.9% to 30.0%. Chondroitin sulfates were analyzed using the Yabe et al. (1987) method and sulfate was estimated using the Dodgson and Price (1962) method. Yields of chondroitin sulfates and sulfate in accordance with conditions of treatment showed 3.8 to 5.4% and 5.6 to 6.1% of yield in the autoclave treatment, respectively. In the case of enzyme treatment, the extraction yield of neutrase (Amore-Pacific Co., Seoul, Korea) was higher than autoclave treatment, but chondroitin sulfate content was low (S.H. Jeong, unpubl. data). Three volumes of 70% ethanol were effective to precipitate GAGs from the sea squirt tunic solution. Several sulfated polysaccharides have been isolated from the test cells of the ascidian *Styela plicata*. The most abundant polysaccharide is a highly sulfated heparin sulfate whose function has been evaluated by Cavalcante et al. 1999.

The HPLC chromatogram of the saccharides extracted from *H. roretzi* tunic is shown in Fig. 2. This acid hydrolyzed solution is composed of glucosamine, galactosamine, galactose, glucose, mannose, and small amount of fucose. The glucosamine peak was highest, and the second largest peak was galactosamine. The glycans extracted from the tunic of *H. roretzi* and *H. aurantium* were separated by anion exchange chromatography on a DEAE-cellulose column (10×1.5 cm) equilibrated with 0.5 M sodium acetate buffer (pH 6.0) and washed with 50 ml of the same buffer. Gel chromatographies of GAGs from *H. roretzi* (T-3, 4) and *H. aurantium* (K-3, 4) were performed on Sepharose CL-6B eluting with 0.5 M sodium

Table 4. Proximate composition and chondroitin sulfate content of the *Halocynthia roretzi* and *Halocynthia aurantium* tunic.

	H. roretzi	H. aurantium
Yield	5.4±0.8	3.8±0.3
Moisture	3.1±0.1	4.2±0.1
Ash	30.0±1.1	27.9±0.9
Protein	18.4±0.8	19.7±0.7
Lipid	0.5±0.1	0.5±0.1
Carbohydrate	48.0±0.7	47.7±0.9
Chondroitin sulfates[a]	57.1±3.1	48.7±3.2
Sulfates[b]	6.1±0.3	5.6±0.4

[a]Chondroitin sulfates were identified by Yabe et al. (1987) method.
[b]Sulfates were estimated by Dodgson and Price (1962) method.
Units are percents.

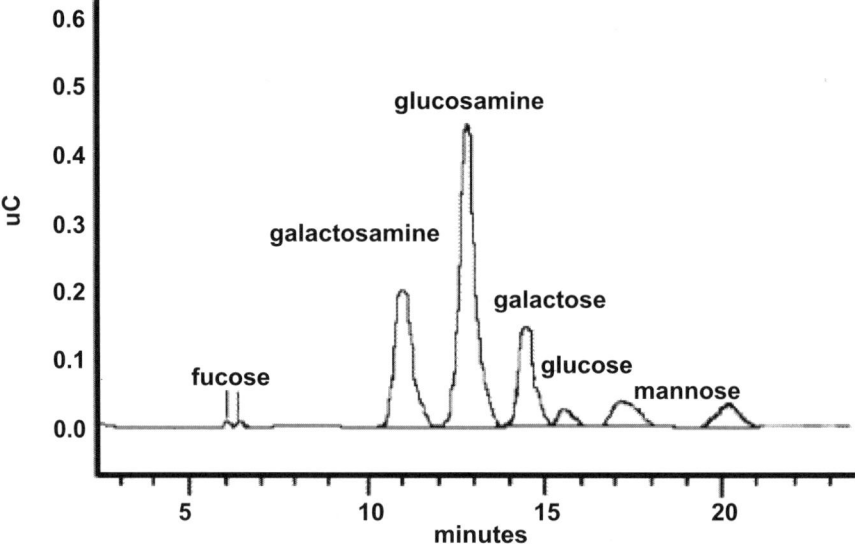

Figure 2. HPLC chromatogram of glycosaminoglycans obtained from *Halocynthia roretzi* tunic. Sample was dissolved in 100 microliters 16 mM NaOH and applied to a CarboPac PA1 column (4.5 × 250 mm) with Amino Trap cartridge (4.5 × 50 mm) and PAD2 with integrated amperometry detector.

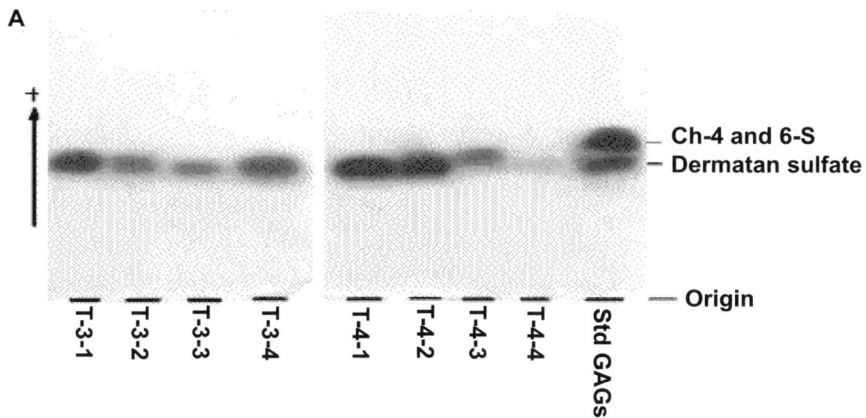

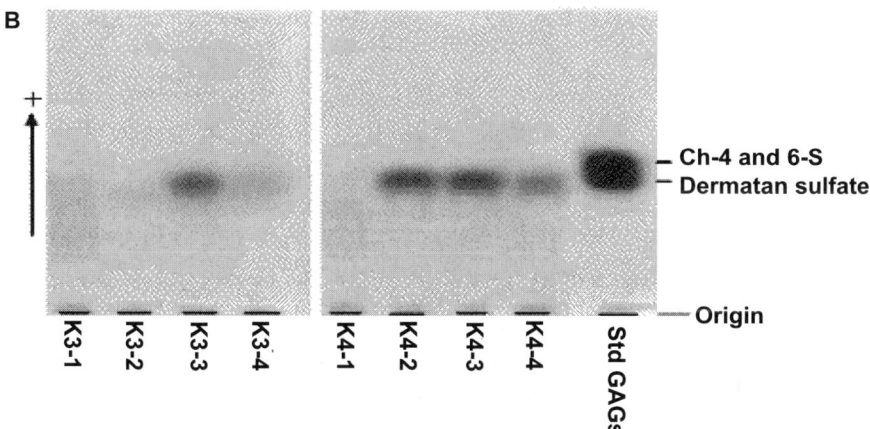

Figure 3. Agarose gel chromatographs of glycosaminoglycans from *Halocynthia roretzi* (A) (T-3-4), *Halocynthia aurantium* (B) (K3-4), and standards of chondroitin sulfate A (Ch-4-S) and C (Ch-6-S) and dermatan sulfate were separated on DEAE-cellulose and Sepharose CL-6B columns.

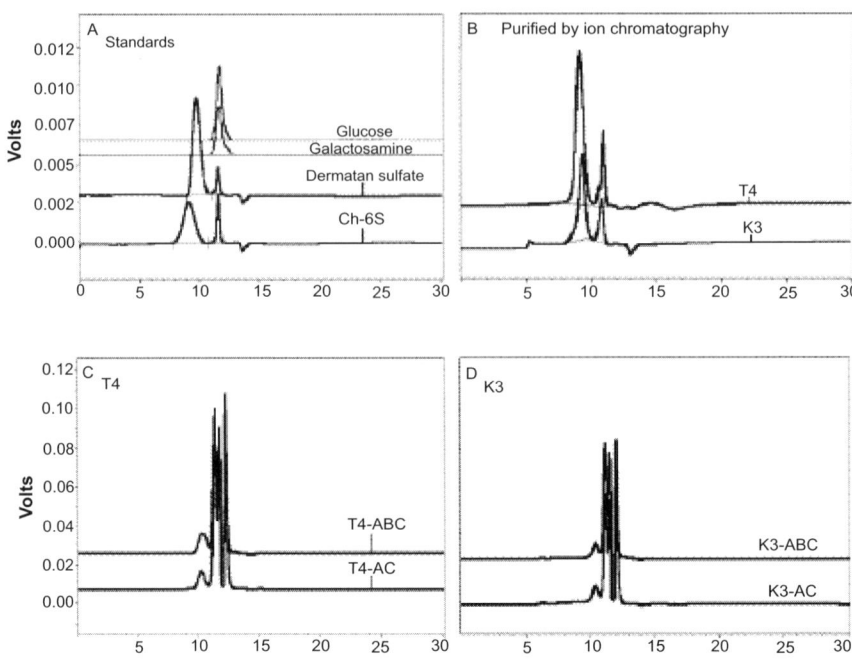

Retention time

Figure 4. HPLC chromatograms of purified glycosaminoglycans from sea
squirt tunic. A mixture of standard glucose, glucosamine, dermatan
sulfate, and chondroitin sulfate C (A) and the purified glycosaminoglycans from sea squirt tunic (B) and disaccharides formed by
chondroitin ABC and AC lyases (C, D) were applied to a 7.8 × 300
mm Phenomenex Biosep SEC-2000 column, linked to a Shimadzu
HPLC system. The column was eluted with a linear gradient of sodium phosphate buffer (pH 6.8) and monitored for UV absorbance
at 210 nm.

Table 5. Mean sensory scores for jellies prepared with refined dietary fiber from *Halocynthia roretzi* tunic.

	Control	5% fiber added	10% fiber added
Color	3.505[b]	4.005[a]	4.105[a]
Flavor	3.405[a]	3.505[a]	3.905[a]
Taste	3.705[a]	3.805[a]	3.905[a]
Texture	3.305[a]	3.605[a]	3.805[a]
Overall acceptability	3.205[a]	3.805[a]	4.105[a]

Scores in same row followed by different letter indicate significantly different at $P <$ 0.05.
From Byun et al. 2000.

acetate and each of them was separated into four fractions. The major polysaccharides purified from the sea squirt tunic migrate on agarose electrophoretic gel as a single and homogeneous metachromatic band with mobility between standards of dermatan sulfate and chondroitin sulfate C (Fig. 3). The disaccharides formed by exhaustive digestion of the T-4 and K-3 fraction (Fig. 4B) with chondroitin AC and ABC lyase were analyzed on HPLC (Fig. 4C and D). These results characterize the fraction as a dermatan sulfate–like glycosaminoglycan.

Sensory characteristics of dietary fiber from the sea squirt tunic

The main physiological effects of dietary fiber are primarily on gastric emptying and small intestinal transit time, resulting in an improved glucose tolerance and a decreased digestion of starch. The insoluble fibers that are not, or only marginally, fermented exert almost exclusively a bulking effect that shortens the transit time and increases the fecal mass (Roberfroid 1993). The ability of soluble and insoluble fiber to impart various functional properties in a food product is critical to their use in food quality and sensory attributes. Dietary fiber also can be modified to enhance various functional properties, including solubility, water-holding capacity (WHC), and adhesive, emulsifying, and foaming properties (Kinsella 1979). The dietary fibers from sea squirt tunic consisted of insoluble cellulose and hemicelluloses. Jellies were prepared using 5% and 10% dietary fiber isolated from *H. roretzi* tunic waste (Table 5). All the rheological parameters decreased in the enhanced jellies. Results indicated only color was different in 5% and 10% fiber enhanced samples.

References

Albano, R.M., and P.A.S. Mourão. 1986. Isolation, fraction, and preliminary characterization of a novel class of sulfated glycans from the tunic of *Styela plicata* (Chordata tunicate). J. Biol. Chem. 261:758-765.

Albano, R.M., M.S.G. Pavão, and P.A.S. Mourão. 1990. Structural studies of a sulfated L-galactan from *Styela plicata* (Tunicate): Analysis of the Smith-degraded polysaccharide. Carbohydr. Res. 208:163-174.

Byun, M.W., H.J. Ahn, H.S. Yook. J.W. Lee, and D.J. Kim. 2000. Quality evaluation of jellies prepared with refined dietary fiber from ascidian (*Halocynthia roretzi*) tunic. J. Korean. Soc. Food Sci. Nutr. 29:64-67. (In Korean.)

Cavalcante, M.C.M., P.A.S. Mourão, and M.S.G. Pavão. 1999. Isolation and characterization of a highly sulfated heparin sulfate from ascidian test cells. Biochim. Biophys. Acta 1428:77-87.

Dodgson, K.S., and R.G. Price. 1962. A note on the determination of the ester sulphate content of sulphated polysaccharides. Biochem. J. 84:106-110.

Ferrier, L.K., L.J. Caston, S. Leeson, J. Squires, B.J. Weaver, and B.J. Holub. 1995. Alpha-linoleic acid- and docosahexaenoic acid-enriched eggs from hens fed flaxseed: Influence on blood lipids and platelet phospholipids fatty acids in humans. Am. J. Clin. Nutr. 62:81-86.

Haard, N.F. 1992. Biochemistry and chemistry of color and color change in seafood. In: G.J. Flick and R.E. Martin (eds.), Advances in seafood biochemistry: Composition and quality. Technomic Publishing Co., Lancaster, p. 305.

Halling, P.J. 1981. Protein-stabilized foams and emulsions. CRC Crit. Rev. Food Sci. Nutr. 18:155-197.

Kinsella, J.E. 1979. Functional properties of soy proteins. J. Am. Oil Chem. Soc. 56:242-258.

Knorr, D. 1982. Functional properties of chitin and chitosan. J. Food Sci. 47:593-595.

Liaaen-Jensen, S. 1991. Marine carotenoids: Recent progress. Pure Appl. Chem. 63:1-12.

Mathew, M.B. 1975. Polyanionic proteoglycans. In: A. Kleinzeller, G.F. Springer, and H.G. Witmann (eds.), Connective tissue: Macromolecular structure and evolution. Springer-Verlag, Berlin, pp. 93-125.

Matsuno, T., and S. Hirao. 1989. Marine carotenoids. In: R.G. Ackman (ed.), Marine biogenic lipids, fats, and oils, Vol. I. CRC Press, Boca Raton, pp. 251-388.

Matsuno, T., M. Ookubo, and T. Komori. 1985. Carotenoids of tunicates. III. The structural elucidation of two new marine carotenoids, amarouciaxanthin A and B. J. Natur. Prod. 48:606-610.

Mourão, P.A.S., M.S.G. Pavão, B. Mulloy, and R. Wait. 1997. Chondroitin ABC lyase digestion of an ascidian dermatan sulfate. Occurrence of unusual 6-O-sulfo-2-acetomido-2-deoxy-3-O-(2-O-sulfo-α-L-idopyranosyluronic acid)-β-D-galactose units. Carbohydr. Res. 300:315-321.

Nair, S.S.D., J.W. Leitch, J. Falconer, and M.L. Grag. 1997. Prevention of cardiac arrhythmia by dietary (n-3) polyunsaturated fatty acids and their mechanism of action. J. Nutr. 127:383-393.

No, H.K, and T. Storebakken. 1992. Pigmentation of rainbow trout with astaxanthin and canthaxanthin in freshwater and saltwater. Aquaculture 101:123-134.

Okutani, K., and S. Shigeta. 1993. Inhibitory effect of sulfated derivatives of a marine bacterial polysaccharide on replication of human immunodeficiency virus in vitro. Nippon Suisan Gakkaishi 59:1433.

Pavão, M.S.G., P.A.S. Mourão, B. Mulloy, and D.M. Tollefsen. 1995. A unique dermatan sulfate-like glycosaminoglycan from ascidian. J. Biol. Chem. 270: 31027-31036.

Poole, S. 1988. The foam enhancing properties of low-viscosity chitosans. Proceeding from the International Conference on Chitin and Chitosan. Trondheim, Norway, pp. 523-531.

Roberfroid, M. 1993. Dietary fiber, inulin, and oligofructose: A review comparing their physiological effects. Crit. Rev. Food Sci. Nutr. 33:103-148.

Rouslahti, E. 1989. Proteoglycans in cell regulation. J. Biol. Chem. 264:13369-13372.

Ruggiero, J., M.A. Fossey, J.A. Santos, and P.A.S. Mourão. 1998. Charge distribution and calcium affinity of sulfated α-L-galactans from ascidians. Comparison between linear and highly branched polymers. Carbohydr. Res. 306:545-550.

Simopoulos, A.P. 2000. Human requirements for n-3 polyunsaturated fatty acids. 79:961-970.

Simpson, K.L. 1982. Carotenoid pigments in seafood. In: R.E. Martin, G.J. Flick, C.E. Hebard, and D.R. Ward (eds.), Chemistry and biochemistry of marine food products. AVI Publishing Company, Westport, Connecticut, pp. 115-137.

Sung, J.J. 1995. Physicochemical properties and utilization of dietary fiber. Food Sci. Ind. 28:2-23. (In Korean.)

Surai, P.F., and N.H.C. Sparks. 2001. Designer eggs: From improvement of egg composition to functional food. Trends Food Sci. Technol. 12:7-16.

Torrissen, O.J., W.H. Hardy, and K.D. Shearer. 1989. Pigmentation of salmonids: Carotenoid deposition and metabolism. Rev. Aquatic Sci. 1:209-227.

Tsuchiya, Y., and Y. Suzuki. 1962. Biochemical studies of the ascidian, *Cynthia roretzi* v. Drasche-VI. The presence of pseudokeratin in test. Bull. Jap. Soc. Sci. Fish. 28:222-226.

Watanabe, K., K. Nakamura, K. Yamaguchi, and S. Konosu. 1991. Biosynthesis of halocynine in the ascidian *Halocynthia roretzi*. Nippon Suisan Gakkaishi 57: 1587-1589.

Yabe, Y., T.H. Ninomiya, T. Kashiwaba, M. Tatsuno, and T. Okada. 1987. Determination of sodium chondroitin sulfate added in foods. J. Food Hygiene 28: 13-18.

Effect of Chitosan on Gelling Properties of Thai Catfish (*Pangasius sutchi*) Surimi

Attaya Kungsuwan, Bordin Ittipong, Sirirat Jongrittiporn, and Orawan Kongpan
Ministry of Agriculture and Cooperatives, Department of Fisheries, Fishery Technological Development Division, Bangkok, Thailand

Suthep Limsooksomboon and Chantip Limthongkun
Srinakarinthraviroth University, Faculty of Science, Department of Nutrition, Bangkok, Thailand

Abstract

Attempts have been made to study the feasibility of utilizing freshwater catfish, *Pangasius sutchi*, for surimi processing. Due to high fat content, a polysaccharide obtained from shrimp shell (chitosan) was applied as a gel-forming enhancer. Catfish was processed in surimi by conventional methods before adding the mixture of 2% chitosan in 2% acetic acid (w/v). Chitosan applied in this study was obtained from the bio-waste of cultured black tiger shrimp, *Penaeus monodon,* with 85% deacetylation. The experiments had five groups: one control, and four groups of catfish surimi that were further mixed with the above-mentioned chitosan solution at the volumes of 5, 10, 15, and 20 ml. All experiments were analyzed for quality of surimi using gel strength parameters and a folding test. The results showed that all treatments with the addition of chitosan had higher gel strength than that of the original or control (250.4 g·cm) which had the test score of B grade. The chitosan volume that provided the highest gel strength (490.1 g·cm) and folding test score of AA grade, was 15 ml (30 mg per 100 g surimi). The yield of surimi obtained from this study was 22% from whole, fresh catfish. This preliminary study has shown a potential of processing surimi from freshwater catfish, which is available in Thailand but is difficult to process and preserve due to its high fat content.

Introduction

Shrimp industries, especially frozen and canned products, are important
to the economy in Thailand. Most of the shrimp used in the industries are
normally obtained from culture, such as black tiger shrimp (*Penaeus monodon*). During the past decade, the black tiger shrimp culturing area has
increased, with the production as high as 275,544 tons in 1999 (Department of Fisheries 2002). Ninety percent of those cultured shrimp were
further processed for export. Wastes such as shells and heads generated
each day from these industries are approximately 30-40% of raw shrimp
weight. In general, these wastes will be used for animal feed, with a value
of 1-1.50 Baht per kilogram or about US$0.03. Adding value to the huge
amounts of industrial wastes has been our goal for decades.

Chitin and its derivative named chitosan are well known to be the
structural polysaccharides from shrimp shells and the integument of
other crustaceans (Muzzarelli 1977). Chitin has a molecular structure
similar in some respects to glucose. Moreover, while chitin is water insoluble, chitosan, a de-acetylated chitin, is water soluble in acidic pH. The
more acetyl groups eliminated from chitin, the greater the solubility of
chitosan. The percentage of the degree of deacetylation (%DD) is one of
the specifications required for chitosan trading. At present, these unique
naturally derived materials have high potential applications in agricultural industries in Thailand. There have been reports on the preservative
effect of chitosan on fishery products such as salted and dried fish and
fish balls (Kungsuwan et al. 1996, Ittipong et al. 2002).

Regarding freshwater fish culture in Thailand, striped catfish, *Pangasius sutchi*, is one of the kinds promoted in order to be in line with
the Self Sufficient System of His Majesty the King of Thailand. In such a
system, the fish left from household consumption would go to markets.
The production of striped catfish is around 7,000-10,000 tons in each
year (Department of Fisheries 2002) but the price of the raw fish is quite
low (18 baht per kilogram or about US$0.4). Moreover, fat content in
this kind of fish is quite high especially in the belly (24.2% fat) creating
rancidity problems which limits the utilization of this fish. Therefore,
the objectives of this study are to find ways to add value to fresh striped
catfish as surimi or other minced fish products through the addition of
chitosan, and also make it more stable. The surimi (minced fish) will be
able to achieve a better price and can be used for various food products
such as surimi-based products, fish noodle, and fermented fish.

In the process of making surimi, which is an intermediate product
for many types of products, the elimination of fat, color, and odor are the
keys to quality surimi along with the proper gel strength. Gel strength
formation in surimi is a phenomenon of systematic protein arrangement
into three-dimensional networks with water molecules in between (Lee
1984). To gain or achieve higher gel strength, the enzyme trans-glutaminase is known to be a key factor. But due to its high price, the application

is rather rare. Table 1 shows the Thai standard for surimi which is in accordance with the Codex Alimentarius (TISI 1990).

Objectives

The objectives of this study are to:

- Determine the gelling properties of striped catfish flesh.

- Evaluate the application of chitosan on gelling of striped catfish surimi.

Materials and methods

Materials

The materials used were striped catfish (Fig. 1), shrimp chitosan with 85% degree of deacetylation (Sea Fresh Co. Ltd.), sodium chloride, sucrose, sodium polyphosphate, sausage casing (25 mm diameter), and a rheometer (Rheo Tex SD-305).

Methods

1. Surimi process.

Fresh whole catfish were deheaded, gutted, skinned, and filleted, and the fillets were minced. The washing process to eliminate fat and other pigments was then performed using iced water (1:5 w/v) with standing time of 15 minutes, and then the mince was filtered through nylon screen. The washing process was repeated two additional times. To the washed minced meat, 3% sugar and 0.2% polyphosphate were added and mixed thoroughly to make fresh surimi for further studies. Figure 2 is the flow diagram for surimi processing (MFRD 1988).

2. Chitosan treatment.

Chitosan obtained from black tiger shrimp shells with 85% degree of deacetylation was used. The chitosan solution (2%) was dissolved in 2% aqueous acetic acid.

The surimi was separated into five samples of 1 kg each where one sample was the control. To the other four, chitosan solution was added with the volumes of 5, 10, 15, and 20 ml each and mixed well.

3. Surimi quality analyses

Specific properties of surimi from catfish were evaluated for gel strength, folding test, moisture content, and pH.

A. Gel strength.

The preparation for gel strength was done as follows. Each sample was mixed with 3% salt (w/w) and ice, in a silent cutter, to a moisture content of 65%. After mixing for 5-10 minutes, the surimi mixture was filled in the

Table 1.   Standard of surimi in Thailand.

Grade	Gel strength (g·cm)	Moisture (%)	pH
SA	>700	77-88	6.5-8.0
AA	450-700	77-88	6.5-8.0
A	350-450	77-88	6.5-8.0
B	<300	77-88	6.5-8.0

Source: Thai Industrial Standard Institute, Ministry of Industry, Thailand (1990).

Figure 1. Striped catfish *Pangasius sutchi* of marketable size.

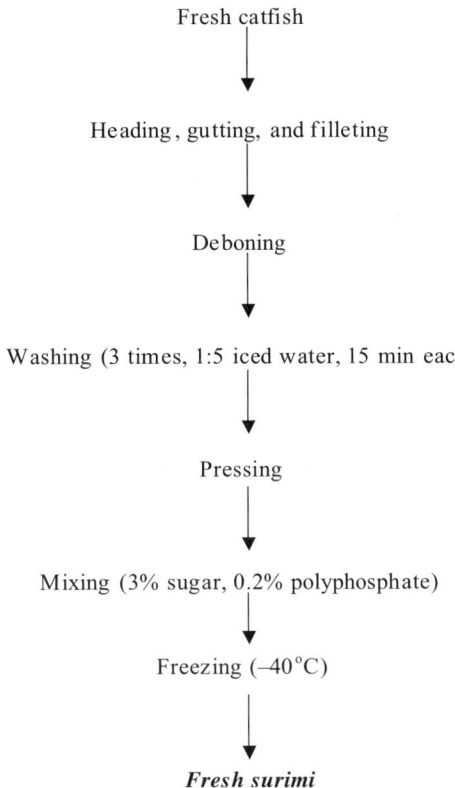

Fresh catfish

↓

Heading, gutting, and filleting

↓

Deboning

↓

Washing (3 times, 1:5 iced water, 15 min each)

↓

Pressing

↓

Mixing (3% sugar, 0.2% polyphosphate)

↓

Freezing (–40°C)

↓

Fresh surimi

Figure 2. Flow diagram of surimi processing.

sausage tube 15 cm in length and then both ends of the filled tube were tightly tied. The sausage tubes filled with surimi were set in a 40°C water bath for 20 minutes and were cooked in a 90°C water bath for another 20 minutes. Chilling was done in iced water after cooking until the product reached room temperature. These treated surimi tubes were then kept at 5°C overnight before testing.

To measure the gel strength, surimi tubes (at room temperature) were cut into pieces with a length of 2.5 cm and were taken for puncture test using a rheometer. The needle applied had a diameter of 5 mm. The force applied until the surimi piece was broken was recorded in grams. The distance from the start until the surimi piece was broken was recorded in centimeters. The gel strength was then calculated by multiplying the force by the distance. The unit of gel strength is g·cm. Figure 3 shows a diagram of gel preparation of each treatment (MFRD 1988).

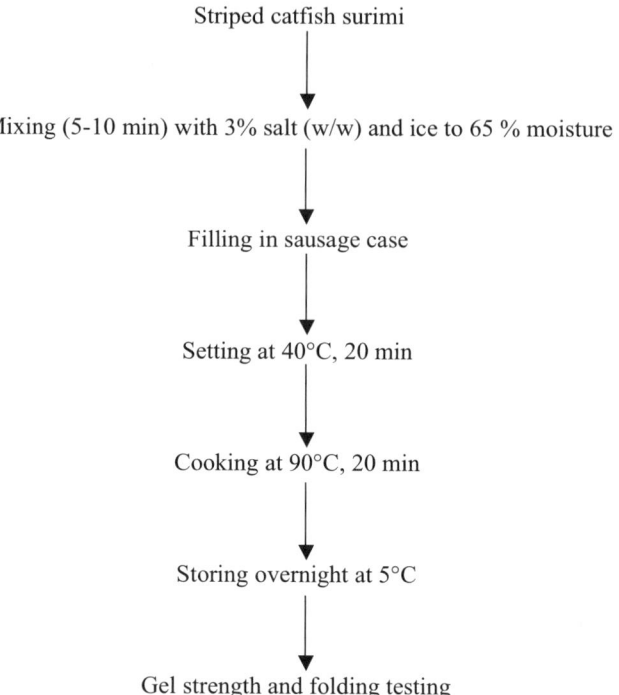

Striped catfish surimi

Mixing (5-10 min) with 3% salt (w/w) and ice to 65 % moisture

Filling in sausage case

Setting at 40°C, 20 min

Cooking at 90°C, 20 min

Storing overnight at 5°C

Gel strength and folding testing

Figure 3. Preparation of surimi samples for gel strength and folding tests.

B. Folding test.

The surimi prepared for the gel strength test was used. It was tested by slicing into thin pieces of 5 mm thickness and diameter of 25 mm (Kongpun et. al. 1998). The surimi pieces were then folded in half, and if there was no breakage they were folded again. Grading scores are shown in Table 2.

C. Moisture content.

Moisture content measure followed the AOAC method, section 7.003 (1984). The moisture of surimi should not exceed 80% (TISI 1990).

D. pH.

The pH of all samples was measured by mixing 5 g of surimi with 45 ml distilled water. The pH of good quality surimi is between 6.5 and 8.0 (TISI 1990).

Table 2. Folding test grading scores.

Grade	Description
AA	No break when folded the 2nd time.
A	Break a little when folded the 2nd time.
B	Break a little when folded 1st time.
C	Break (but does not separate) when folded the 1st time.
D	Break and separate from each other when folded the 1st time.

Table 3. Results of quality testing on all chitosan treated catfish surimi samples.

Treatment	Gel strength (g·cm)	Moisture (%)	pH	Grade (from gel strength)
1	247.98	62	7.6	B
2	377.23	65	7.4	A
3	437.05	65	7.4	A
4	491.35	65	7.4	AA
5	453.17	65	7.4	AA

Treatment 1 = control; 2, 3, 4, and 5 have chitosan added at 5,10,15, and 20 ml, respectively.

Results

The results of primary quality attributes gel strength, moisture, pH, and grade of surimi are shown in Table 3. It is clearly seen that adding chitosan increased gel strength. The highest gel strength was seen in treatment 4 where 15 ml were added, or 30 mg chitosan in 100 g of surimi.

The folding test indicated that all treatments had the grade of AA except the control, which was A grade (also see Method section for folding test). The high folding scores might be caused by the higher fat content than in other fish used for surimi, such as Alaska pollock. The high fat content could cause catfish surimi to be more tender and bendable.

From Table 3, the pH of all treatments were around 7.4-7.6 which fall within the standard level (Table 1). But in the case of moisture, the results showed that even the control had moisture content less than the minimum standard value of 77% (Table 1). These might be due to its high fat content (16% from original 22%). Moreover, the yield of fresh surimi from whole catfish was satisfactorily at 22% which is similar to conventional surimi production.

Discussion

This study has shown that there is potential to utilize striped catfish for surimi with the aid of chitosan, which is an important byproduct of shrimp industries in Thailand. Even though the gel strength obtained in this study was not as high as commercial surimi (> 700 g·cm), consumer tests (data not shown) indicated that catfish surimi was acceptable due to its softness and elasticity. More studies are needed on catfish surimi cost and shelf life, in order to answer the questions about commercial feasibility.

Moreover, considering the number of frozen shrimp factories (100) and daily production capacity of around 20 tons at each factory, there would be about 400-600 tons of shells per day as waste. This is a large amount of seafood industrial waste. Therefore if the byproducts of shrimp shells, namely chitin, chitosan, and their oligomers, are used in agriculture and other industries in Thailand, there would be benefits for all.

Acknowledgment

This research project was supported by the research fund of the Thailand Department of Fisheries together with the valuable contribution from Srinakarinthraviroth University.

References

AOAC. 1984. Official methods of analysis,14th edn. Association of Official Analytical Chemists, Washington, D.C.

Department of Fisheries. 2002. Fisheries statistics of Thailand, 1999. Fisheries Economics Division, Department of Fisheries, No. 10/2002, National Metal and Materials Technology Center (MTEC), Bangkok.

Ittipong, B., A. Kungsuwan, S. Jongrittiporn, and P. Inthuserdha. 2002. Preservative effect of chitosan in fishball. Proceedings of the 5th Asia Pacific Chitin-Chitosan Symposium and Exhibition, March 13-15, 2002, Bangkok.

Kongpun, O., P. Suwansakornkul, and J. Rungtong. 1998. Methods for measuring the properties of surimi. Fishery Technological Development Institute, Department of Fisheries, Ministry of Agriculture and Cooperatives, Special publication 1/1998, Bangkok.

Kungsuwan, A., B. Ittipong, and S. Chandrkrachang. 1996. Preservative effect of chitosan on fish products. In: W.F. Stevens, M.S. Rao, and S. Chandrkrachang (eds.), Proceedings of the 2nd Asia Pacific Symposium on Chitin and Chitosan: Environmental Friendly and Versatile Bio-materials. Asian Institute of Technology, Bangkok, pp. 193-199.

Lee, C. 1984. Surimi processing technology. Food Technol. 38(11):69.

MFRD. 1988. Hand-book on the processing of frozen surimi and fish jelly products in Southeast Asia. Marine Fisheries Development Department, SEAFDEC (Southeast Asian Fisheries Development Center), Singapore.

Muzzarelli, R.A.A. 1977. Chitin. Pergamon Press, New York.

TISI. 1990. Industrial standard of frozen surimi. Thai Industrial Standard Institute, Ministry of Industry, ISI935-2533, Bangkok.

Development of Pacific Whiting Fish Sauce: Market Potential and Manufacturing in the United States

S. Tungkawachara and J.W. Park
Oregon State University, Seafood Laboratory and Department of Food Science and Technology, Astoria, Oregon

Abstract

Manufacturing fish sauce from whole Pacific whiting (W) and a mixture (50/50) (WB) of surimi byproducts and whole fish was investigated to develop histamine-free fish sauce and to better utilize solid byproducts. This study also discusses the marketing potential and biochemical properties of the respective fish sauces.

Marketing potential for Pacific whiting fish sauce in the United States was evaluated through phone interview and consumer panels. Fish sauce import reached $16.6 million in the United States in 2000. Anchovy was the most popular raw material used (47%) in the fish sauce process and brand name was the most important factor in the purchase decision (52%). Consumer tests disclosed no significant difference in flavor acceptance and overall acceptance among fish sauce W, fish sauce WB, and commercial anchovy fish sauce ($P > 0.05$). As fermentation time increased, the degree of hydrolysis increased and higher total nitrogen and total amino acid contents were obtained; however, moisture content and pH decreased. Glutamic acid, alanine, leucine, lysine, and arginine were abundant after 9 months fermentation and accounted for 43.27% and 54.08% of total amino acids in fish sauce W and WB, respectively. Fish sauce W had a total amino acid concentration of 38.4 mg per ml, which was twice as much as WB fish sauce (16.6 mg per ml).

Based on positive consumer acceptance, good quality, and low cost, it was concluded that fish sauce from Pacific whiting had potential to replace imported anchovy fish sauce (which contains histamine).

Introduction

Fish sauce is a clear liquid condiment with an amber color and salty/mild cheesy flavors that is commonly used as a flavor enhancer or salt replacement for various food preparations in Southeast Asia. Fish sauce from anchovies is produced through natural fermentation by storing a mixture (1:3) of salt and minced fish at subtropical temperatures. In Thailand, where the majority of commercial fish sauce in the world is manufactured, fish sauce is processed through fermentation at ambient temperatures (30-35°C) for 12 to 18 months.

Fermentation is conducted by the combined function of proteolytic enzymes and halophilic microorganisms from the viscera (Sikorski et al. 1995). Endogenous fish enzymes are primarily responsible for degradation of muscle proteins during fermentation (Orejana and Liston 1982). The enzymes slowly hydrolyze the fish proteins during storage (Sikorski et al. 1995).

Pacific whiting (*Merluccius productus*), which is abundant off the Pacific Northwest, was excluded from commercial utilization until 1991 due to endogenous proteolytic enzymes that caused texture softening. An et al. (1994) reported that cathepsin B was the most active cysteine protease in Pacific whiting fish fillets. Serine proteases, cathepsin B–like enzymes, trypsin-like enzymes, and metalloproteases were almost equally responsible for protein hydrolysis at 35°C during fermentation of whole Pacific whiting (Lopetcharat 1999). In contrast, cysteine proteases were primarily responsible for the degradation of proteins from byproducts during fermentation at 35°C (Lopetcharat 1999).

Lopetcharat and Park (2002) were the first to study the feasibility of fish sauce production from Pacific whiting. However, a consumer acceptance profile for fish sauce from Pacific whiting was neither evaluated nor compared with the biochemical properties. Our objective, therefore, was to evaluate the marketing potential, in the United States, of fish sauce made from whole Pacific whiting or a mixture of surimi byproducts and whole fish based on consumer acceptance and the biochemical properties of fish sauce.

Sample preparation and analysis

Two sets of samples were ground: W for whole fish and WB for the mixture (1:1) of whole fish and surimi byproducts. Ground samples (W and WB) were homogeneously mixed with table salt at 3:1, placed in polyethylene bottles closed loosely with a lid, and kept at 35°C. Fermented fish mince was filtered using 8-fold cheesecloth followed by filter papers (Whatman no. 41 and no.1) with vacuum. The filtrate (liquid portion) was then subjected to biochemical analysis.

Information regarding imported fish sauce in the United States, including market size and the country of origin, was obtained from the

USITC Trade Database (USITC 2000). A list of importers of fishery products was obtained from the U.S. Department of Commerce (NMFS 1999). A marketing survey was performed using phone interviews to obtain various product information including the country of origin, package size, raw material, and other factors affecting the purchase decision. Seventy-five consumers were asked to evaluate the color, flavor, and overall acceptance of three fish sauce products: commercial anchovy fish sauce, Pacific whiting fish sauce made from whole fish (W), and fish sauce produced from a mixture (1:1) of whole fish and surimi byproducts (WB). Biochemical characteristics investigated at 0, 1, 3, and 9 months were pH, degree of hydrolysis (Hoyle and Merritt 1994), moisture content (AOAC 1995), total nitrogen (AOAC 1995), and amino acid concentration (Tungkawachara et al. 2003).

Results and discussion

Marketing survey

Market size and country of origin

In dollar value, fish sauce imported to the United States was $16.6 million in the year 2000 (USITC 2000). It increased by 3.7% when compared to imports from 1998. The majority of imports (72% of total value) were from Thailand. More than 60% of the total imports were brought into the United States through the West Coast ports and distributed to dealers and/or directly to retailers. The three major custom districts of entry for fish sauce imported to the United States in 2000 were Los Angeles (47%), San Francisco (17%), and New York (16.3%).

Dealers/importers interview

According to one of the largest fish sauce importers, the import price of fish sauce products was about $0.75 per kg. Brand name, customer's need, quality, and price were the factors affecting the purchasing decision for importers. The retail bottle (710 ml) made from either glass or plastic was the most frequently used container. Products were distributed with the original package without repackaging.

Importer A in Seattle (Washington) imports fish sauces from Thailand, Malaysia, the Philippines, Vietnam, Japan, China, and Korea. This company distributes the products to the Northwest region through wholesale brokers in Seattle as well as through direct sales to some Asian groceries in the Northwest. This importer carries both fish sauce and soy sauce products in equal proportions based on dollar value. Its customers make their purchasing decision based primarily on brand name and sweet flavor.

Dealer B, in Portland (Oregon) buys fish sauce from an importer in Seattle. Its products were primarily from Thailand and the Philippines. It distributes products to grocery stores and carries many kinds of Asian food products, and fish sauces represented 10 to 20% of total products based on dollar value. Customer's need and brand name were considered the most important purchasing factors.

Importer C in San Francisco (California) imports fish sauce only from Thailand. It works as an import agent for the biggest fish sauce company in Thailand. The three most famous brand names of fish sauce are directly distributed to groceries in the Northwest U.S. region. The yearly import quantity of fish sauce was about 5,000 metric tons, which is about 30% of the total fish sauce imported to the United States each year. All of the imported fish sauce was made from anchovies. The grade of fish sauce was considered the most important attribute in making a purchase decision followed by brand name and price. Color and sweet flavors were also reported to affect decision-making.

FDA has listed the tolerance level for histamine in fish products at 50 ppm in its HACCP guideline (hazard analysis critical control point) (Gingerich 1999). Ironically, the histamine content in fish sauce was not a concern for any of the fish sauce dealers/importers. Furthermore, they do not believe there is a potential problem of high histamine content in anchovy fish sauce. However, Putro (1993) reported that sardine fish sauce in Indonesia had a histamine content in the range of 140 to 230 ppm. In addition, Sanceda et al. (1996) found histamine in fish sauces from many countries, for example, 40 ppm in the Philippines fish sauce and 430 ppm in Thai fish sauce.

Retailer interviews

According to the retailer survey via phone interview, twenty-one grocers in Portland (Oregon) participated in this study. Information surveyed included country of origin, proportion of fish sauce products to entire food products sold (in dollars), package size, raw material used in fish sauce, and factors affecting the purchase decision. We found that most fish sauces available in grocery stores were made in Thailand (36%), followed by the Philippines (10%), Vietnam (10%), and Japan (10%). In more than one-third of grocery stores the proportion of fish sauce was less than 10% of the entire food products sold based on dollar value. Most fish sauce products (55%) were distributed in medium-size packages (401-750 ml) and 21% of the fish sauce products were distributed in large-size packages (751-1,000 ml). Raw materials commonly used for fish sauce include anchovy (47%), sardine (10%), crab (5%), krill/shrimp (10%), and 28% unknown. Brand name was the primary reason (52%) for grocery stores to make a purchasing decision involving fish sauces, followed by price (24%), quality attributes (sweetness, saltiness, and color) (14%), and customer's needs (10%).

Consumer acceptability

Tungkawachara et al. (2003) conducted a consumer acceptability test for Pacific whiting fish sauce fermented for 12 months. The majority of panelists (93.3%) were fish sauce users at home, consuming fish sauce at least once a month. As for ethnic groups, Asians composed 53.3% of the panelists and Caucasians composed 46.7%. The majority of panelists (89.3%) were between 20 and 59 years old and 68.0% of the panelists were female. There was no significant difference in overall acceptance (sensory) and flavor acceptance for all fish sauce samples ($P > 0.05$) (Table 1). Fish sauce WB, however, had lower color acceptance scores than the commercial anchovy fish sauce control ($P < 0.05$), whereas no significant difference in color acceptance was detected between fish sauce W and commercial anchovy fish sauce ($P > 0.05$).

Regarding consumer acceptance by ethnic group, Asians ($n = 40$) found all fish sauces similar for color, flavor, and overall acceptance ($P > 0.05$) (Table 1). Caucasians ($n = 35$), on the other hand, found all fish sauces the same for flavor and overall acceptance ($P > 0.05$), but not for color acceptance ($P < 0.05$) (Table 1). In terms of color, Caucasians found the commercial fish sauce and fish sauce W similar ($P > 0.05$), but found the commercial anchovy fish sauce different from fish sauce WB ($P < 0.05$). Regarding consumer acceptance by gender, there were no differences in terms of color, flavor, and overall acceptance ($P > 0.05$) (Table 1).

Biochemical properties

Fresh Pacific whiting flesh had neutral pH (6.93). The pH decreased gradually during fermentation reaching 5.4 after 9 months (Tungkawachara et al. 2003) (Table 2). The pH of fish sauce WB was lower than that of fish sauce W at all fermentation periods ($P < 0.05$). This was probably due to increased protein hydrolysis by the proteolytic activities, which results in more free hydrogen ions. Fermentation products containing organic acids such as lactic acid and acetic acid have been reported to lower the pH of fish sauce (Itoh et al. 1993, Funatsu et al. 2000, Michihata et al. 2000).

The degree of hydrolysis in fish sauce WB increased as fermentation time was extended. However, the degree of hydrolysis of fish sauce W increased up to 3 months, and then decreased (Tungkawachara et al. 2003). Comparing fish sauce W and WB, a significant difference in the degree of hydrolysis was found only at 9 months ($P < 0.05$). Degree of hydrolysis represents the extent of the hydrolytic degradation of protein (Adler-Nissen 1986), and Orejana and Liston (1982) concluded that endogenous enzymes are the major agents responsible for protein digestion during the fish sauce process.

For both W and WB moisture content decreased, while protein content increased as fermentation time increased (Table 2). This is likely due to protein hydrolysis and/or the possible moisture evaporation through

Table 1.　Consumer acceptance of fish sauce fermented for 12 months.

Attributes	Consumer group	Fish sauce samples		
		W	WB	C
Color acceptance	All consumers ($n = 75$)	6.40[ac]	6.12[bc]	6.92[a]
	Asian ($n = 40$)	6.58[a]	6.27[a]	6.80[a]
	Caucasian ($n = 35$)	6.20[ab]	5.94[b]	7.06[a]
	Female ($n = 51$)	6.46[a]	6.22[a]	7.00[a]
	Male ($n = 24$)	6.33[a]	5.92[a]	6.75[a]
Flavor acceptance	All consumers ($n = 75$)	6.07[a]	5.76[a]	6.31[a]
	Asian ($n = 40$)	5.93[a]	5.35[a]	6.30[a]
	Caucasian ($n = 35$)	6.23[a]	6.23[a]	6.31[a]
	Female ($n = 51$)	6.00[a]	5.61[a]	6.33[a]
	Male ($n = 24$)	6.33[a]	6.08[a]	6.25[a]
Overall acceptance	All consumers ($n = 75$)	6.09[a]	5.93[a]	6.37[a]
	Asian ($n = 40$)	5.90[a]	5.58[a]	6.35[a]
	Caucasian ($n = 35$)	6.31[a]	6.34[a]	6.40[a]
	Female ($n = 51$)	5.94[a]	5.82[a]	6.39[a]
	Male ($n = 24$)	6.54[a]	6.17[a]	6.33[a]

Results are expressed as the means of the acceptance of panels (n = number of panels).

Acceptance score: 9 = extremely like, 5 = neither like nor dislike, 1 = extremely dislike.

Means with the same letter in each row are not significantly different.

W = Fish sauce made from whole Pacific whiting.

WB = Fish sauce made from the mixture (1:1) of whole fish and surimi byproducts.

C = Commercial anchovy fish sauce.

Source: Adapted from Tungkawachara et al. (2003).

the loosely closed lid. During early fermentation, fish sauce WB had higher protein content than fish sauce W ($P < 0.05$). However, there was no significant difference at 9 months ($P > 0.05$). At 0 months of fermentation (referring to 1 day after fermentation), the moisture content of the two fish sauce samples were similar ($P > 0.05$). After the first month of fermentation, however, fish sauce WB had a lower moisture content than fish sauce W ($P < 0.05$). The lower moisture content in fish sauce WB was thought to be due to the greater protein hydrolysis by enzymes from intestines.

Total nitrogen in fish sauce is composed of protein nitrogen and non-protein nitrogen (NPN) compounds such as free amino acids, nucleotides, peptides, ammonia, urea, TMAO (trimethylamine oxide), etc. These compounds have been reported to contribute specific aroma and flavors (Finne 1992, Shahidi 1994). Total nitrogen content in fish sauce depends on fish species and the chemical composition of fish. About 80% of total

Table 2. Changes in pH, moisture content, and total nitrogen content of fish sauce during fermentation.

Fermen-tation time (months)	Sample	Moisture (%)	Total nitrogen (mg/100 g)	Salinity[a] (%)	pH
0	W	73.4±0.5	375±49	23.6	7.05±0.01
	WB	73.0±0.1	553±45	22.9	6.66±0.01
1	W	71.7±0.6	743±72	23	6.64±0.01
	WB	70.8±0.1	853±43	23.2	5.75±0.01
3	W	68.0±0.1	1,128±30	24.3	5.72±0.01
	WB	66.8±0.1	1,322±18	24.2	5.64±0.01
9	W	68.0±0.0	1,282±2	23.3	5.42±0.01
	WB	64.4±0.1	1,357±75	26.4	5.34±0.01

[a]Salinity was calculated by subtracting % protein, % moisture, and 0.75% fat from 100.

The mean and standard deviation were derived based on data obtained from triplicate runs.

W = Fish sauce made from whole Pacific whiting.

WB = Fish sauce made from the mixture (1:1) of whole fish and surimi byproducts.

Source: Adapted from Tungkawachara et al. (2003).

nitrogen in fish sauce remains in the form of amino acids (Dougan and Howard 1975).

Total nitrogen in fish sauce increased during fermentation (Table 2). At the early stages of fermentation (1 and 3 months), fish sauce WB had higher total nitrogen than fish sauce W ($P < 0.05$), possibly due to the greater degree of hydrolysis. After 9 months fermentation, fish sauce W and WB had a total nitrogen content of 1.28 and 1.36%, respectively. In comparison, Japanese fish sauces made from mackerel, sardine, and squid that were fermented for 12 months had a total nitrogen content of 1.89, 1.52, and 1.48%, respectively (Funatsu et al. 2000).

Amino acid composition

The free amino acid composition of Pacific whiting fish sauce fermented for 9 months is shown in Table 3 (Tungkawachara et al. 2003). Total free amino acids increased as fermentation continued. However, after 9 months of fermentation, total free amino acids in fish sauce W (38.37 mg per ml) were twice as much as fish sauce WB (16.56 mg per ml).

Glutamic acid, alanine, leucine, lysine, and arginine were abundant in the 9-month samples and accounted for 43.27 and 54.08% of total free amino acids in fish sauce W and WB, respectively. According to Benjakul and Morrissey (1997), in Pacific whiting muscle six major amino acids (glutamic acid, aspartic acid, lysine, leucine, arginine, and alanine) account for 56.54% of the total protein. Similar free amino acid profiles were found in skipjack tuna sauce (Cha and Cadwallader 1998) as well as in herring protein hydrolysate (Liceaga-Gesualdo and Li-Chan 1999). Ijong and Ohta (1996) found that the content of free alanine, isoleucine, glutamic acid, and lysine were high in a traditional Indonesian fish sauce.

Some differences in free amino acid profile between fish sauces W and WB were possibly due to the differences in protein composition. Fish sauce WB was made from raw materials containing a significant amount of fish skins, bones, heads, and other connective tissues, unlike fish sauce W, which was manufactured from whole fish.

Amino acids dictate the taste of seafood (Sikorski 1994). Taste values (Table 3) were calculated based on the amino acid concentration and taste threshold data according to the method described by Cha and Cadwallader (1998). Taste values of free amino acids from fish sauce W were about two times greater than those from fish sauce WB. Glutamic acid and aspartic acid showed the highest taste values and the lowest threshold value.

Various amino acids have their own taste. Lysine, alanine, glycine, serine, and threonine have a sweet taste, while arginine, leucine, valine, phenylalanine, histidine, and isoleucine give a bitter taste, and aspartic acid has a sour taste (Kato et al. 1989). A high content of glutamic acid in fish sauce might make an important contribution to the development of the umami taste (Komata 1990, Sanceda et al. 1990). Cha and Cadwallader (1998) stated that the specific free amino acids having sweet, sour, and bitter tastes may play a prominent role in the overall taste of fish sauce. Some volatile amino acids contribute to the aroma of fish sauce, as well. Glutamic acid gives a meaty aroma, whereas isoleucine and leucine give a sweet aroma. In addition, methionine gives a methyl sulfide–like aroma and phenylalanine gives a strong rose-like aroma (Saisithi et al. 1966).

Conclusion

Marketing information on fish sauce in the United States was reviewed, along with the biochemical properties of Pacific whiting fish sauce. Brand name was the primary reason for grocers to make a purchasing decision, followed by price and quality attributes (sweetness, saltiness, and color). The two fish sauce samples developed from Pacific whiting had similar biochemical properties during 9 months of fermentation. The consumer acceptance tests demonstrated that Pacific whiting fish sauces were similar in many aspects to imported anchovy fish sauce. Furthermore, solid byproducts from surimi processing can be utilized as a raw material for fish sauce.

Table 3. Free amino acid composition in fish sauce fermented for 9 months.

Amino acid	Concentration (mg/ml) W	WB	Taste threshold[a] (g/dl)	Taste value[b] W	WB
Arginine	3.074	1.516	0.05	61.5	30.3
Aspartic acid	1.698	0.891	0.003	566	297
Threonine	1.716	1.145	0.26	6.6	4.4
Serine	1.684	0.935	0.15	11.2	6.2
Glutamic acid	3.435	1.623	0.005	687	324.6
Proline	2.66	0	0.3	8.9	0
Glycine	1.036	0.482	0.13	8	3.7
Alanine	2.743	1.664	0.06	45.7	27.7
Valine	2.404	1.083	0.04	60.1	27.1
Methionine	1.366	0.217	0.03	45.5	7.2
Isoleucine	1.709	0.814	0.09	19	9
Leucine	3.528	1.873	0.19	18.6	9.9
Phenylalanine	1.31	0.625	0.09	14.6	6.9
Lysine	3.825	2.277	0.05	76.5	45.5
Histidine	0.199	0.065	0.02	10	3.3
Ammonia	0.2	0.286	NA	–	–
DL-allohydroxylysine	0.136	0.055	NA	–	–
Ornithine	0.031	0	NA	–	–
1-methylhistidine	0.27	0.064	NA	–	–
Taurine	0.529	0.335	NA	–	–
Urea	0	0.189	NA	–	–
Phosphoserine	0.104	0.037	NA	–	–
α-aminoisobutyric acid	0.026	0	NA	–	–
Hydroxyproline	3.656	0	NA	–	–
Tyrosine	0.525	0.263	NA	–	–
Cystathionine	0.201	0.115	NA	–	–
Total	38.374	16.555	–	–	–

[a]Taste threshold (Kato et al. 1989).
[b]Taste value calculated using Cha and Cadwallader (1998).
W = Fish sauce made from whole Pacific whiting.
WB = Fish sauce made from the mixture (1:1) of whole fish and surimi byproducts.
NA = Not available for taste threshold data.

Acknowledgment

This research was partially funded by the NOAA Office of Sea Grant, U.S. Dept. of Commerce, grant number NA76RG0476 (project number R/SF-19), and by appropriation made by the Oregon state legislature. The U.S. government is authorized to produce and distribute reprints for governmental purposes regardless of any copyright notation that may appear hereon. Additional support was also made by the National Fisheries Institute (NFI), Agricultural Research Foundation, Walter Jones Fisheries Development Award, Bill Wick Marine Fisheries Award, and American Institute of Fishery Research Biologists (AIFRB).

References

Adler-Nissen, J. 1986. Enzymic hydrolysis of food proteins. Elsevier Applied Science Publishers, London, pp. 90-93.

An, H., V. Weerasinghe, T.A. Seymour, and M.T. Morrissey. 1994. Cathepsin degradation of Pacific whiting surimi proteins. J. Food Sci. 59(5):1013-1017, 1033.

AOAC. 1995. Official methods of analysis, 16th edn. AOAC International, Arlington, Virginia.

Benjakul, S., and M.T. Morrisey. 1997. Protein hydrolysates from Pacific whiting solid wastes. J. Agric. Food Chem. 45:3423-3430.

Cha, Y.J., and K.R. Cadwallader. 1998. Aroma-active compounds in skipjack tuna sauce. J. Agric. Food Chem. 46:1123-1128.

Dougan, J., and G.E. Howard. 1975. Some flavoring constituents of fermented fish sauces. J. Sci. Food Agric. 26:887-894.

Finne, G. 1992. Non-protein nitrogen compounds in fish and shellfish. In: J.G. Flick and R.E. Martin (eds.), Advances in seafood biochemistry, composition and quality. Technomic Publishing, Lancaster, Pennsylvania, pp. 393-401.

Funatsu, Y., R. Sunago, S. Konagaya, T. Imai, K. Kawasaki, and F. Takeshima. 2000. A comparison of extractive components of a fish sauce prepared from frigate mackerel using soy sauce Koji with those of Japanese-made fish sauces and soy sauce. Nippon Suisan Gakkaishi 66(6):1036-1045.

Gingerich, T. 1999. New guidelines prevent excessive histamine product in bluefish and other scombroid fish. Commercial Fish and Shellfish Technology Fact Sheet, http://www.cfast.vt.edu/downloads/facts2.pdf. Virginia Cooperative Extension.

Hoyle, N.T., and J.H. Merritt. 1994. Quality of fish protein hydrolysates from herring (Clupea harengus). J. Food Sci. 59(1):76-79.

Ijong, F.G., and Y. Ohta. 1996. Physicochemical and microbiological changes associated with bakasang processing, a traditional Indonesian fermented fish sauce. J. Sci. Food Agric. 71:69-74.

Itoh, H., H. Tachi, and S. Kikuchi. 1993. Fish fermentation in Japan. In: C.H. Lee, K.H. Steinkraus, and P.J.A. Reilly (eds.), Fish fermentation technology. United Nations University Press, Tokyo, pp. 177-186.

Kato, H., M.R. Rhue, and T. Nishimura. 1989. Role of free amino acids and peptides in food taste. In: R. Teranishi, R.G. Buttery, and F. Shahidi (eds.), Flavor chemistry: Trends and developments. ACS Symposium Series 388. American Chemical Society, Washington, D.C., pp. 158-174.

Komata, Y. 1990. Umami taste of seafoods. Food Rev. Intl. 6(4):457-487.

Liceaga-Gesualdo, A.M., and E.C.Y. Li-Chan. 1999. Functional properties of fish protein hydrolysate from herring (*Clupea harengus*). J. Food Sci. 64(6):1000-1004.

Lopetcharat, K. 1999. Fish sauce: The alternative solution for Pacific whiting and its by products. M.S. thesis, Oregon State University, Corvallis. 114 pp.

Lopetcharat, K., and J.W. Park. 2002. Characteristics of fish sauce made from Pacific whiting and surimi byproducts during fermentation stage. J. Food Sci. 67(2): 511-516.

Michihata, T., Y. Sado, T. Yano, and T. Enomoto. 2000. Preparation of ishiru (fish sauce) by a quick ripening process and changes in the composition of amino acids, oligopeptides and organic acids during processing. J. Jpn. Soc. Food Sci. Technol. (Nippon Shokuhin Kogyo Gakkaishi) 47(5):369-377.

NMFS. 1999. NOAA National Marine Fisheries Service, Miami, Florida. http://www.nmfs.noaa.gov/trade/.

Orejana, F.M., and J. Liston. 1982. Agents of proteolysis and its inhibition in patis (fish sauce) fermentation. J. Food Sci. 47:198-203, 209.

Putro, S. 1993. Fish fermentation technology in Indonesia. In: C.H. Lee, K.H. Steinkraus, and P.J.A. Reilly (eds.), Fish fermentation technology. United Nations University Press, Tokyo, pp. 107-128.

Saisithi, P., B.O. Kasemsarn, J. Liston, and A.M. Dollar. 1966. Microbiology and chemistry of fermented fish. J. Food Sci. 31:105-110.

Sanceda, N., T. Kurata, and N. Arakawa. 1990. Overall quality and sensory acceptance of a lysine-fortified fish sauce. J. Food Sci. 55:983-988.

Sanceda, N., T. Kurata, and N. Arakawa. 1996. Accelerated fermentation process for the manufacture of fish sauce using histidine. J. Food Sci. 61:220-222.

Shahidi, F. 1994. Proteins from seafood processing discards. In: Z.E. Sikorski, B.S. Pan, and F. Shahidi (eds.), Seafood proteins. Chapman and Hall, New York, pp. 171-193.

Sikorski, Z.E. 1994. The contents of proteins and other nitrogenous compounds in marine animals. In: Z.E. Sikorski, B.S. Pan, and F. Shahidi (eds.), Seafood proteins. Chapman and Hall, New York, pp. 6-12.

Sikorski, Z.E., A. Gildberg, and A. Ruiter. 1995. Fish products. In: A. Ruiter (ed.), Fish and fishery products. CABI Publishing, Wallingford, U.K., pp. 315-346.

Tungkawachara, S., J.W. Park, and Y.J. Choi. 2003. Biochemical properties and consumer acceptance of Pacific whiting fish sauce. J. Food Sci. 68:855-860.

USITC. 2000. U.S. International Trade Commission-2000 tariff database, http://dataweb.usitc.gov.

Fish Sauce from Catfish (*Ictalurus punctatus*) Nuggets as Affected by Salt Content and Enzyme Addition

Somsamorn Gawborisut, Juan L. Silva, and Roberto S. Chamul
Mississippi State University, Department of Food Science and Technology, Mississippi State, Mississippi

Abstract

Exogenous enzymes in combination with different salt concentrations were added to make fermented sauce from catfish nuggets. The effects of bromelain at 0, 0.15, and 0.3% and NaCl at 11 and 17%, of diced catfish nuggets' weight, on sauce fermentation were studied. Samples were collected on days 1, 7, 14, and 21. Higher salt concentration resulted in lower formol nitrogen (FN) and higher soluble protein (SP), regardless of enzyme concentration or fermentation time. Adding exogenous enzyme, regardless of level, increased FN and SP. After 14 days, FN and SP did not increase further, regardless of treatment. At 11% salt, lactic acid (LA) content increased the first 7 days and decreased by 21 days. At 17% salt, LA was higher when treated with enzyme, and decreased at day 21. It is thought that some of the acid was used at the latter stages of fermentation, possibly by hetero-fermentative bacteria. At 11% salt, gas and a sour aroma were detected, whereas at 17% salt, no gas evolution was noted. A rancid aroma was noted on the samples treated with 17% salt. Fermentation could be accomplished at 11% salt and 0.15% bromelain, if fat is removed.

Introduction

During 2001, farm-raised catfish production in the United States totaled 260 million kilograms (USDA 2002). Mississippi, the largest catfish producer in the United States, produced 72% of live catfish (MSU Cares 2002). Catfish nuggets, obtained from the dressing/trimming process, average 7% of the live catfish weight (Silva and Dean 2001). The nuggets are low price food items because of their undesirable "mouth-feel" and high fat content. Value-added products produced from catfish nuggets may benefit the catfish industry. In southern Thailand, catfish is used for a traditional fermented fish (pla duk rah) (Trakullertsathien 2001), while drip obtained from the process can be used for fish sauce fermentation. Fish sauce is a condiment, which is part of the daily diet of more than 250 million people (Raksakulthai et al. 1986). Fish sauce is produced by mixing fish with salt at the ratio of 5:1 to 1:1 (Chaveesuk et al. 1993). The mixture is fermented for 6 to 12 months (Gildberg et al. 1984, Chaveesuk et al. 1993). To accelerate the fermentation processes, addition of exogenous enzymes such as fungal protease, pronase, trypsin, chymotrypsin, purified squid proteases, or crude squid proteases from hepatopancreas, have been added (Raksakulthai et al. 1986). Addition of papain in fish sauce during fermentation yielded a fairly good product (Beddows et al. 1976). Mixtures of trypsin and/or chymotrypsin (ratio 50:50 or 0:100) produced the most favorable results in terms of protein hydrolysis, while there was no difference among enzymes in terms of free amino acid composition (Chaveesuk et al. 1993). Pronase was used with some success to produce a fish sauce after four months of fermentation, but it was found to be expensive (Beddows et al. 1976).

Salt inhibits the activity of digestive proteases, particularly the pepsins, which are active under acid conditions (Gildberg et al. 1984). Pepsins are significantly inhibited by salt concentrations over 5%, but a low salt content allows for spoilage (Gildberg et al. 1984). No apparent spoilage occurred with salt concentrations of 10% or more (Beddows and Ardeshir 1979). Reduction of salt concentration did not increase the rate of hydrolysis and percent of protein conversion when bromelain was used (Beddows et al. 1976). The addition of acid at an ambient temperature promotes the activities of the proteolytic enzymes originating from fish (Beddows and Ardeshir 1979). Experiments with fish silage have shown that pH 4 gives the optimum speed of autolysis (Gildberg et al. 1984).

Use of low pH increased the rate and extent of proteolysis of ikan-bilis (*Stolephorus* sp.), with optimum conditions either pH 2.0 with 10% salt or pH 3.0 with 15% salt w/w (Beddows and Ardeshir 1979). Gildberg et al. (1984) concluded that acidification improves autolysis only at low salt concentration (5%, w/w). The yield of fish sauce decreased when salt concentration was increased. Autolysis was higher at pH 2-3, but the disadvantage of increasing the acidity is corrosion of equipment and utensils.

The objective of this project was to develop a fish sauce from undervalued catfish nuggets using reduced salt concentration and an exogenous enzyme.

Materials and methods
Catfish nugget sauce preparation
Catfish (*Ictalurus punctatus*) nuggets obtained from a local processing facility were iced and transferred to the Department of Food Science and

Technology, Mississippi State University. The nuggets were diced with a commercial food processor (Black and Decker APP FP 1300) for 15 seconds. Five hundred grams of diced nuggets were transferred to a glass jar. After that, bromelain enzyme (Sigma) (0, 0.15, and 0.3% of diced nuggets' weight) and NaCl (11 and 17% of diced nuggets' weight) were incorporated in the diced nuggets. One ml 0.006 M cysteine (Fisher) and glucose (Fisher) (3% of the nuggets' weight) were added to the mixtures to assist in the activity of bromelain and to enhance color development, respectively. The mixture was adjusted to pH 4 by adding 0.2 N HCl. The jars were incubated at 37°C (Precision Scientific). Samples of fish sauce were collected on days 1, 7, 14, and 21. The collected samples were filtered twice through Whatman filter paper no. 4 and Whatman filter no. 1, respectively.

Examination of fish sauce

Degree of hydrolysis (formol nitrogen, FN) was determined by modification of the method described by Chaveesuk et al. (1993). The result was expressed as mg of formol nitrogen per ml of fish sauce and calculated according to the following formula:

$$\text{mg formol nitrogen (FN)} = \text{ml NaOH (pH 7-8.5)} \times 0.1 \times 14$$

Soluble protein (SP) was determined by the Biuret method described by Cooper (1977) with bovine serum albumin (BSA) as a standard. Total nitrogen (TN) was determined using Kjeldahl Unit 21232+01 (Labconco Corporation) according to procedure (AOAC 1996). Results were expressed as percent nitrogen per ml of fish sauce compared to the blank and calculated according to the following formula:

$$\text{Nitrogen/ml (\%)} = (V_a - V_b) \times 1.4007 \times 0.1/\text{ml of sample}$$

where V_a = volume of 0.1 N HCl required for sample, and V_b = volume of 0.1 N HCl required for blank. Titratable acidity was determined by the AOAC method (1996). The result was expressed as percent lactic acid and calculated according to the following formula:

$$\text{TA (\% lactic acid)} = \text{ml NaOH} \times \text{Normality NaOH} \times 0.09 \times 100/\text{ml sample}$$

Experimental design

The experimental design was a factorial arrangement of treatment factors (concentration of salt × enzyme concentration × fermentation time) in a completely randomized design (CRD) with three replications. The hypothesis was tested at a significance level of 0.05. Least significant difference (LSD) of multiple comparisons of means at 95% confidence level were carried out as described by Petersen (1985) using Statistic Analysis System (SAS) version 8.2 software (SAS Institute, Cary, North Carolina).

Results and discussion

Salt content, enzyme concentration, and fermentation time did not interact significantly to affect degree of hydrolysis. Salt content and enzyme concentration independently affected degree of hydrolysis. Degree of hydrolysis (DH) significantly increased by 0.782 mg per ml when salt content was reduced from 17% to 11% (data not shown). The inhibitory effect of salt on the activity of digestive proteases was reported by Gildberg et al. (1984). On the other hand, Beddows et al. (1976) reported no inhibitory effect of salt on rate of hydrolysis when bromelain was used. The DH increased rapidly during the first week in all samples and decreased by day 14. The control (0% enzyme) produced a lower DH throughout the fermentation period. Addition of more enzyme (0.15% vs. 0.3% enzyme) did not affect DH (Fig. 1). Even though hydrolysis, DH, stopped after day 14 (Table 1), some substrate still remained. The decrease in hydrolysis may be attributed to a reduction in bromelain activity.

Salt content, enzyme concentration, and fermentation time interacted to affect soluble protein, SP. Salt content and enzyme had a significant interaction on soluble protein (SP). Soluble protein in catfish nugget sauce with 11% salt was higher than with 17% salt (Fig. 2). After day 14, the 17% salt treatment produced lower SP (Fig. 2). Therefore, salt concentration might suppress soluble protein. Fermentation time affected soluble protein production. The SP increased for the first 14 days, and remained constant thereafter (Table 1).

When 11% salt was used, after day 14, enzyme treated sauce had higher SP than the control (Fig. 3). When 17% salt was used, except for day 7, enzyme addition produced a higher SP compared to the control. Enzyme concentration did not affect SP, regardless of salt content (Fig. 3). Beddows et al. (1976) reported an increase in conversion of soluble solids to soluble protein (expressed as conversion of nitrogen) when 0.2% bromelain of mackerel weight was used. Addition of trypsin and chymotrypsin also increased soluble protein in herring sauce (Chaveesuk et al. 1993)

The interaction between salt content, enzyme concentration, and fermentation time did not have a significant effect on total nitrogen (TN). Salt content and enzyme concentration independently affected total nitrogen. Degree of hydrolysis significantly increased by 2.14%, when salt content was reduced from 17% to 11% (data not shown). Salt might suppress conversion of protein to nitrogen compounds during the fermentation process. Beddows et al. (1976) reported an increase in TN when salt was reduced from 20% to 10% of mackerel weight. Addition of enzyme, regardless of concentration, significantly increased TN in catfish nugget sauce (Fig. 4). Beddows et al. (1976) reported an increase in TN when 0.2% bromelain was used in mackerel. Addition of trypsin and chymotrypsin at 0.3% of herring weight increased TN in herring sauce (Chaveesuk et al. 1993). Total nitrogen varied due to fermentation time. The value rapidly increased

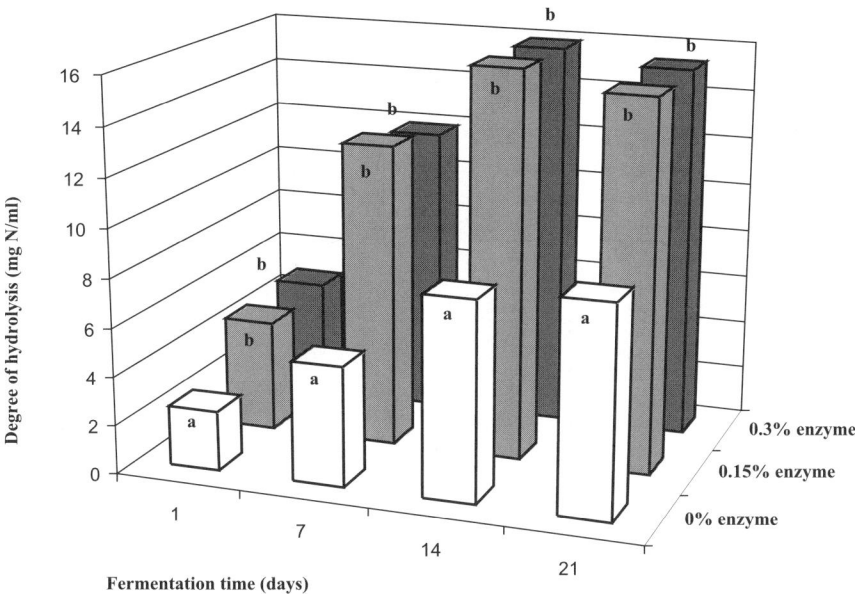

Figure 1. **Degree of hydrolysis of catfish nugget sauce as affected by enzyme concentration and fermentation time. Means within date not followed by the same letter differ (*P* ≤ 0.05).**

Table 1. **Protein and acidity of catfish nugget sauce as affected by fermentation time.**

Fermentation time (day)	Degree of hydrolysis (mg formol nitrogen)	Soluble protein (mg/ml)	Total nitrogen (%)	Titratable acidity (% lactic acid)
1	$3.90^a \pm 1.46$	$8.81^a \pm 1.78$	$8.77^a \pm 1.65$	$1.59^a \pm 0.47$
7	$9.73^b \pm 2.08$	$15.30^b \pm 3.25$	$15.59^b \pm 0.96$	$2.46^c \pm 0.22$
14	$13.31^c \pm 1.01$	$20.10^c \pm 2.97$	$19.29^c \pm 1.46$	$2.42^c \pm 0.21$
21	$13.31^c \pm 1.27$	$18.47^c \pm 3.22$	$20.26^c \pm 2.0$	$2.18^b \pm 0.45$

Mean ± standard error.
Means within a column not followed by the same letter differ (*P* ≤ 0.05).

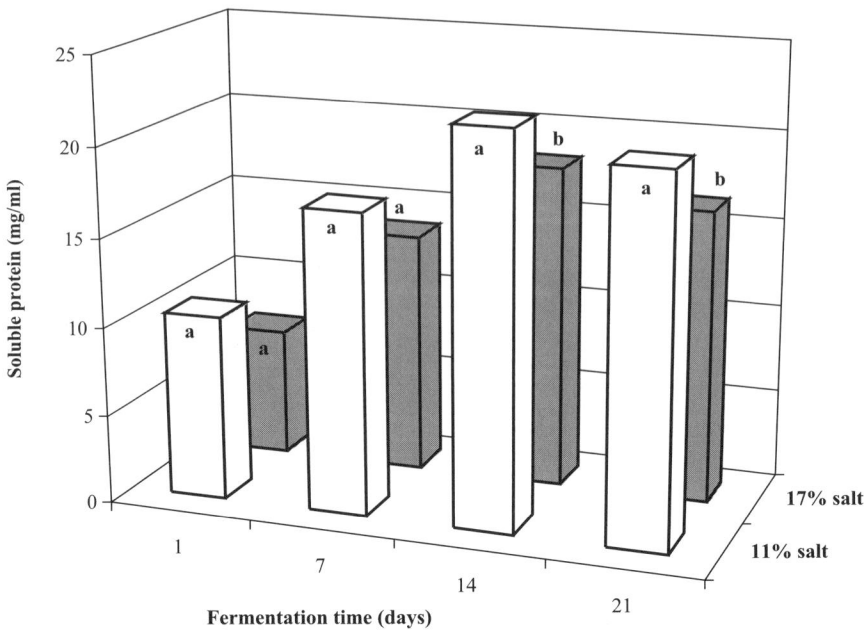

Figure 2. Soluble protein in catfish nugget sauce as affected by salt content for all enzyme concentrations and fermentation times. Means within date not followed by the same letter differ ($P \leq 0.05$).

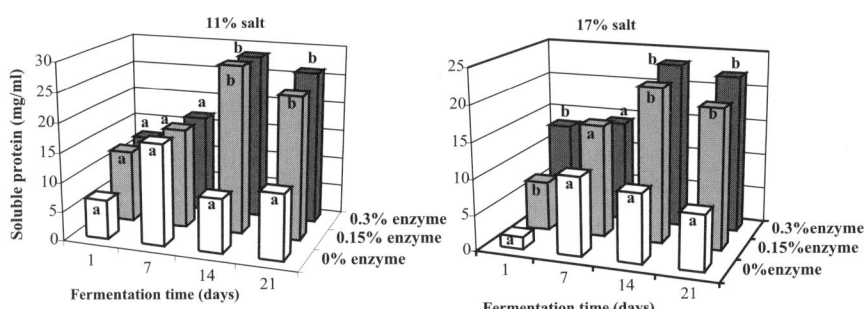

Figure 3. Soluble protein in catfish nugget sauce as affected by enzyme concentration and fermentation time. Means within date not followed by the same letter differ ($P \leq 0.05$).

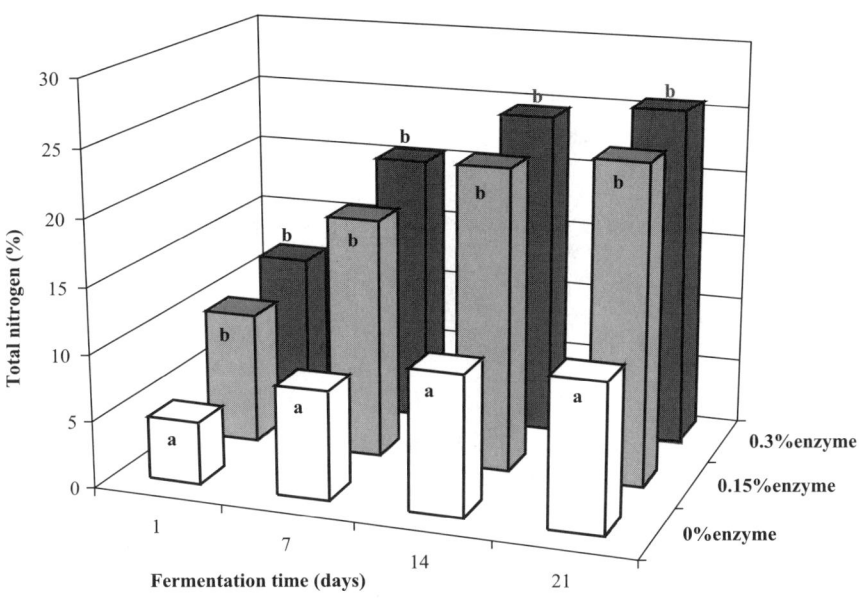

Figure 4. Total nitrogen of catfish nugget sauce as affected by enzyme concentration and fermentation time. Means within date not followed by the same letter differ ($P \leq 0.05$).

during the first week and slowly increased after day 14, and remained constant thereafter (Table 1).

Salt content, enzyme concentration, and fermentation time interacted significantly to affect titratable acidity (TA). Salt content and enzyme concentration interact significantly to affect TA. Initially, TA was lower for the 11% salt treatment (Fig. 5), but by day 7, TA was higher. After 14 days, TA was similar for both salt treatments. Enzyme addition produced significantly higher TA regardless of fermentation time (Fig. 6). The TA value increased at the beginning of fermentation but decreased by day 21 (Table 1).

Cheesy aroma, the aroma predominately found in oriental fish sauce, was very mild in the catfish nugget sauce irrespective of treatment (data not shown). A short fermentation period along with a high acid condition, which reduced growth of aroma-producing bacteria, might cause low flavor intensity. A sour aroma and gas bubbles were detected in the catfish nugget sauce with 11% salt. This may have been caused by bacteria using glucose as the carbon source. The pH adjustment achieved by adding 0.2 N HCl might provide additional water for the sauce, thus reducing salt concentration in the system leading to growth of some acid

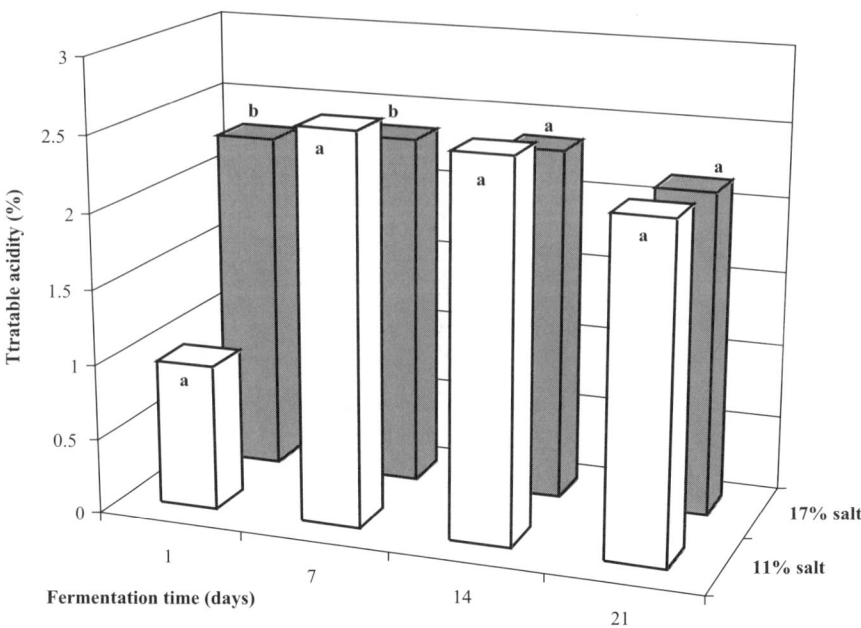

Figure 5. Titratable acidity (% lactic acid) of catfish nugget sauce as affected by salt concentration and fermentation time. Means within date not followed by the same letter differ (*P* ≤ 0.05).

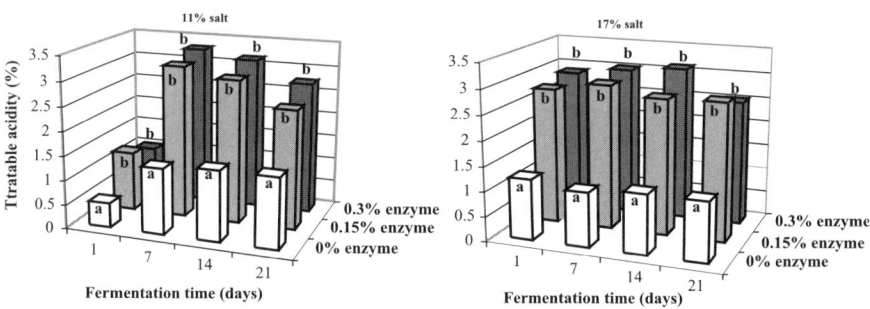

Figure 6. Titratable acidity (% lactic acid) of catfish nugget sauce as affected by enzyme concentration and fermentation time. Means within date not followed by the same letter differ (*P* ≤ 0.05).

and/or salt tolerant bacteria (Varnam 2002). Also, the hydrolyzed meat provided vitamins and amino acids for bacteria metabolism. At 17% salt, no gas bubbles or sour aroma occurred because salt concentration in the system might be high enough to inhibit acid or salt tolerant bacteria. The sour aroma occurring in fish sauce with 11% salt could alter the overall aroma of the sauce. Rancid odor was detected in the catfish nugget sauce because of oxidative rancidity, caused by high fat content in the nuggets. An amber color developed after a 4-month fermentation.

Conclusions

Based on chemical results, a catfish nugget sauce can be made in 14 days with 11% salt and a maximum 0.15% bromelain. Elimination of fat in catfish nuggets before fermentation, addition of antioxidants, or filling headspace of the jar with nitrogen, should be recommended to reduce rancid odor in catfish nugget sauce. To increase flavor intensity, flavorful components such as fresh ground shrimp heads as reported by Feng and Gate (2001) can be incorporated in the sauce. Replacing glucose with caramel color can be done to get amber color in the sauce. Determination of enzyme activity and sensorial studies are also warranted in further research.

Acknowledgments

The researchers would like to thank the Mississippi Agricultural and Fishery Experiment Station (MAFES), USDA-ARS Grant no. 320795-371072, and the Mississippi Catfish Processors.

References

AOAC. 1995. Official methods of analysis. Association of Official Analytical Chemists, 15th edn. AOAC International, Washington, D.C. 1,298 pp.

Beddows, C.G., and A.G. Ardeshir. 1979. The production of soluble fish protein solution for use in fish sauce manufacture, II. The use of acid at ambient temperature. J. Food Technol. 14:612-623.

Beddows, C.G., A.G. Ardeshir, and K.H. Steinkraus. 1976. The use of bromelain in the hydrolysis of mackerel and the investigation of fermentation of fish sauce. J. Food Technol. 11(4):379-388.

Chaveesuk, R., J.P. Smith, and B.K. Simpseon. 1993. Production of fish sauce and acceleration of sauce fermentation using proteolytic enzymes. J. Aquatic Food Product Technol. 2:59-77.

Copper, T.G. 1977. The tools of biochemistry. Wiley Interscience, New York. 678 pp.

Feng, J., and K.W. Gate. 2001. Adding fresh shrimp heads as a proteolytic enzyme source accelerated the fermentation process used to manufacture fish sauce. 2001 IFT Meeting abstract, New Orleans, Louisiana, June 23-27, 2001. Institute of Food Technologists, Chicago, Illinois.

Gildberg, A., J. Espejo-Hermes, and F. Magno-Orejana. 1984. Acceleration of autolysis during fish sauce fermentation by adding acid and reducing the salt concentration. J. Sci. Food Agric. 35:1363-1369.

MSU Cares. 2002. Aquaculture: Catfish. Current situation. Mississippi State University Extension Service, Coordinated Access to Research and Extension System (MSU Cares). www.//msucares.com/aquaculture/catfish.

Petersen, R.G. 1985. Design and analysis of experiments. Marcel Dekker, Inc., New York. 427 pp.

Raksakulthai, N., Y.Z. Lee, and N.F. Haard. 1986. Effect of enzyme supplements on the production of fish sauce from male capelin (*Mallotus villosus*). Can. Inst. Food Sci. Technol. 19:111-114.

Silva, J.L., and S. Dean. 2001. Processing channel catfish. USDA Southern Region Aquaculture Center, USDA-SRAC Pub. no. 183, Washington D.C. 6 pp.

Trakullertsathien, C. 2001. Fabulous fish. Bangkok Post, February 21, 2001.

USDA. 2002. Aquaculture Outlook. USDA Economic Research Service (ERS). http://usda.mannlib.cornell.edu/usda/reports.

Varnam, A. 2002. *Lactobacillus*: Occurrence and significance in non-dairy foods. Microbiol. Today 29(2):13-17.

Development of Food Products from Channel Catfish (*Ictalurus punctatus*) Byproducts

Juan L. Silva and T. Kim
Mississippi State University, Department of Food Science and Technology, Mississippi State, Mississippi

S. Danviriyakul
Rajabhat Institute Chandrakasem, Department of Agricultural Technology, Bangkok, Thailand

Abstract

About 25 million kg of catfish byproducts (5.5% nuggets and trimmings and the rest mince from frames) can be obtained annually. These tissues have adequate texture and quality, and form strong gels when comminuted with salt and other additives, and cooked. They can also be hydrolyzed into viscous slurries to be incorporated into products or whole muscle. This report presents some forms of food products that use these catfish byproducts. One product is heat-set patties, made from comminuted nuggets and minced product. Treatments including salt, whey protein concentrate (WPC), and sodium alginate were studied as the binders. Cooking yield and fat uptake were not affected by treatment. Panelists preferred a higher amount of whole muscle to provide texture to the patties, and patties with 50% nuggets and WPC. Patties made with only catfish mince were rated higher by panelists when mixed with 0.2% sodium tripolyphosphate (STPP), whether containing NaCl or KCl. Minced catfish was obtained by deboning frames and trimmings, followed by washing. The addition of antioxidants or additional washing steps did not increase the quality of the mince. A corn-dog type catfish product was also developed. The product was more tender when mixed for less than 5 minutes and made from fresh rather than frozen product.

Introduction

Catfish is the fifth most important seafood in the United States, with 1.16 pound per capita consumption (NMFS 2002). The industry processed over 270 million kg of catfish in 2001 (USDA 2002) and sales increased in 2002. The main products sold are the fillets and the nuggets (belly flap where peritoneal membrane is found), accounting for about 40-42% of the total weight of the product. The nuggets (~7% w/w) are not accepted well by consumers, selling at $1.00-$1.5 lower than fillets (McAlpin et al. 1994). This is because they contain the peritoneal membrane and the meat is tough. Frames, which are about 19% of the total weight of catfish, can provide meat (~25% mince recovery), 5% nuggets, and 0.5% trimmings (Silva and Dean 2001). If these byproducts are recovered, they will yield about 25 million kilograms of mince and intact muscle products. The mince can be produced for about $0.02 per kg (estimated from Dillard 1995). Some of these are currently sold to consumers (nuggets) at a low price, are converted into pet food (ground plus mince product), or are sent as offal to feed processing facilities, where they are worth about $0.05 per kg (McAlpin et al. 1994).

If the byproducts are converted into food, profit will be enhanced, waste will be reduced, consumers will be provided with convenience foods, and the industry will be more diverse. The food products developed could be targeted to institutional programs like schools, military, food service, and others. These products could be flavored, since catfish meat is very bland. The product is relatively easy to manipulate, and the taste and textural characteristics of catfish make it possible to develop a number of diverse products. The addition of seasonings, salt, breading, and other ingredients will increase product yield. There are many opportunities to develop different products that are convenient and nutritious. This report will briefly describe various products developed from catfish mince and whole muscle, some of these currently being commercialized.

Products made or tested

Mince and surimi

Catfish mince has been made from deboned meat and trimmings. Mince is usually made by washing the frames, draining them, and then deboning. The deboned mince slurry is washed and rinsed up to three times. The rinsed slurry is pressed through a rotary screen and dewatered using a screw press (Kim et al. 1996). Hoke et al. (2000) reported that washing mince was needed to obtain a better color with greater stability. This is accomplished by removing dark material, small bones, fat, and blood. One wash and drain cycle for the mince was sufficient to produce a high quality product. The washed mince results in lower fat and TBARs (oxidative rancidity) while increasing protein, iron, and Hunter "L" value—brightness (Table 1). Addition of antioxidants did not aid further in the storage stability of the washed mince.

Table 1. **Lipid, protein, and iron (dry basis) content, TBARs, and Hunter "L" values of unwashed and washed catfish mince.**

Variable	Unwashed	Washed
Lipid (%)	54.7[a]	18.20[b]
Protein (%)	43.6[a]	79.8[b]
Iron (%)	20.2[a]	29.7[b]
TBARs (mg/kg)	0.30[a]	0.20[b]
Hunter "L"	61.4[a]	66.4[b]

[a,b]Means, within a row for each variable, not followed by the same letter differ ($P < 0.05$).

TBARs = thiobarbituric acid reaction substances, mg malonaldehyde per kg mince.

Suvanich et al. (2000a) reported that unwashed mince had higher total volatile base nitrogen (TVBN) and lower salt-soluble protein (SSP) than twice-washed mince. They reported no change in pH or moisture content of the stored mince. In a parallel study Suvanich et al. (2000b) found that washed mince was lighter and less red and yellow than unwashed mince. They concluded that mince should not be stored longer than three days at 0 to 5°C or three months at –20°C (with cryoprotectants) for optimum quality attainment.

In our studies catfish mince of good quality and color were produced by washing the frames and the mince once, and value was increased by adding cryoprotectants. Production of surimi from catfish mince is feasible (Kim et al. 1996). Dillard (1995) concluded that the price of catfish surimi ranged from $0.27 to $1.33 per pound, when the cost of frames ranged from $0.0 to $0.15 per pound and the recovered mince was 15 to 20% (McAlpin et al. 1994).

Patties

Unfrozen catfish mince was obtained after being washed and kept at 1-2°C with ice. The mince was mixed with various ratios of NaCl to KCl totaling 1% (from 1% NaCl:0% KCl to 0% NaCl:1% KCl) and sodium tripolyphosphate (STPP) (0-0.3% w/w), and molded into patties. The patties were placed in high density polyethylene bags and frozen at –18°C. Patties were cooled and evaluated by a sensory panel and tested for yield stress (breaking stress) and total energy to break (Ghavimi et al. 1987). Breading and cooked yields ranged between 11-12% and 72-74% (0.1-0.3% STPP), respectively. Yield stress and break energy was higher for 1% NaCl:0% KCl patties (Table 2). Patties with 0.3% STPP had higher yield stress and breaking energy than low concentrations of STPP (Table 2). The functionality to bind water and extract salt-soluble proteins of NaCl is not greater than KCl. Panelists did not discern significant differences between the different NaCl or KCl treatments, but preferred products with 0.2-0.3% STPP. It was concluded that patties from minced catfish were best with 1% NaCl and 0.3% STPP. Partial substitution of NaCl by KCl may be feasible though there will be some textural losses.

Patties (2.5 cm) were also made by mixing catfish mince and ground nuggets at 30:70, 50:50, 70:30 (w/w) which also contained NaCl (1% w/w) and 0.3% STPP. Treatments were no binder (CTL), 0.3% whey protein concentrate (WPC), or 0.3% alginate (ALG). They were mixed for 4 minutes, made into patties, battered and breaded, and heat-set (pre-fried) at 177°C for 2 minutes. They were quickly cooled and stored frozen at –18°C until analyzed (Danviriyakul 1996). Compression force, shear force, and breaking energy were lower for AGL patties as compared to CTL or WPC patties (Silva et al. 1999). However, the mince to nugget ratio did not influence compression, shear force (Table 3), or breaking energy. Overall scores were highest for the patties with the highest nugget to mince ratio (Table 3). This is because the ground nuggets provided a sense of chewiness and mouthfeel of intact muscle as compared to mince. Baking yield averaged 96.5% for all patties. It is concluded that patties with at least 50% nugget meat, without additional binders, are acceptable to sensory panelists. The addition of WPC improved color (whiteness, brightness) without affecting texture.

Corn dog, smoked loaf, and other products

A restructured catfish product similar to a corn dog was developed by mixing diced catfish muscle (2.5 cm) with 1% NaCl and seasonings, stuffing into cellulose casings, smoking to 66°C internal temperature, and cooling overnight. After the casings were removed, the links were battered and fried at 185°C for 6-7 minutes (Abide et al. 1990). Product made with fresh fillet meat was more tender but had tougher skin than the product made with frozen fillets. Increased mixing time resulted in increased peak load (toughness) and chewiness, and decreased skin toughness (Table 4). This product could be made with fresh or frozen fish meat, and mixed with salt and seasonings for about five minutes.

A catfish loaf was made by dicing catfish muscle into 2.5 cm cubes, mixing it with 1% NaCl and seasonings, stuffing it into 10 cm casings, and smoking until done (82°C internal temperature). The resulting loaf was cooled and could be sliced and served cold or warm. The product had good gelling stability, and excellent flavor and texture.

Other products commercialized or tested have been canned mince (or mince added to chunks and canned), ground and/or diced meat for stews and similar products, and a suspension that can be incorporated back

Table 2.   Yield stress (lb$_f$) and break-ing energy (lb$_f$ × in.) val-ues for catfish patties as affected by salt type and amount of STPP added.

Treatment	Yield stress (lb$_f$)	Breaking energy (lb$_f$× in.)
NaCl (1%)	2.3[a]	1.0[a]
KCl (1%)	1.7[b]	0.6[b]
STPP (%)		
0.1	1.9[b]	0.7[b]
0.2	1.8[b]	0.6[b]
0.3	2.2[a]	1.0[a]

[ab]Means, within a column for each treatment, not fol-lowed by the same letter differ ($P < 0.05$).
STPP = Sodium tripolyphosphate.

Table 3.   Effect of nugget to washed mince ratio on sensory evaluation[1] and shear force of baked[2] reformed catfish patties.

Nugget :washed mince ratio	Color	Texture	Shear force (N)	Juiciness	Odor	Overall
70:30	7.2[a]	5.4[a]	323[a]	3.6[a]	3.6[a]	6.6[a]
50:50	7.2[a]	4.6[b]	308[a]	3.3[a]	3.3[a]	6.6[a]
30:70	7.0[a]	4.8[b]	273[a]	3.5[a]	3.5[a]	6.1[b]

[ab]Means, in the same column within treatment factor, not followed by the same letter differ ($P < 0.05$).
[1]Color and overall scores ranged from 1 = least desirable to 9 = most desirable.
Texture (chewiness); 1 = mushy, 5 = flaky-firm, 9 = most desirable.
Juiciness; 1 = soggy, 5 = moist, 9 = dry.
Odor; 1 = no fishy smell, 5 = mildly fish smell, 9 = too strong fish smell.
[2]Frozen heat-set patties were baked at 204°C for 25 min.

Table 4.   Peak load and sensory characteristics of a restructured catfish product as affected by mixing time.

Mixing time (min)	Peak load[1] (N)	Sensory attributes	
		Chewiness[2]	Skin toughness[3]
1	11.8[a]	7.9[a]	8.4[a]
5	13.7[b]	8[a]	7.4[ab]
9	15.6[c]	9.3[b]	6.5[b]

[1]Peak load after 50% deformation.
[2]Scores were 1.2 = soft and 13.8 = rubbery.
[3]Scores were 1.2 = tough and 13.8 = tender.
[abc]Means, within a column for each treatment, not followed by the same letter differ ($P < 0.05$).
Restructured catfish product was a corn dog–like product.

into whole muscle. The latter, called Suspentec[R] is a process whereby trimmings and other recovered meat is blended with other ingredients, passed through a reduction mill, reducing it to micron particles, and then the product is injected into whole muscle meat (Cozzini Group 1999).

Conclusions

Catfish solids can be recovered and made into edible products, reducing waste, increasing sales, providing low-cost and convenient products to the consumer, and increasing the diversity of available products. The key to commercialization of these products is the cost of the raw material and the quality of the final product. The technology is proven to work, yet the economics could be an obstacle to their commercialization. Some products are or have been in the market, yet no concerted effort has been put forth to bring these products to the masses.

Acknowledgments

The authors would like to thank the Mississippi Agricultural and Forestry Experiment Station (MAFES), USDA-ARS Grant No. 320795-371072, and the Mississippi Catfish Processors for their contributions.

References

Abide, G.P., J.O. Hearnsberger, and J.L. Silva. 1990. Initial fish state and mixing time effects on textural characteristics of a restructured catfish product. J. Food Sci. 55:1747-1748.

Cozzini Group. 1999. Suspentec[R]: A unique concept for added value. http://www.cozzini.com/

Danviriyakul, S. 1996. Physicochemical and sensory attributes of reformed breaded channel catfish patties as affected by raw materials and binding agents. M.S. thesis. Mississippi State University, Mississippi State.

Dillard, J.G. 1995. Economics of processing new value-added products from catfish. In: A.A. Hood and J.L. Silva (eds.), Proceedings of the Catfish Processors Workshop. MAFES Spec. Bull. 88-8. Mississippi Agricultural and Forestry Experiment Station, Mississippi State University.

Ghavimi, B., J.O. Hearnsberger, G.R. Ammerman, and L.R. Brown. 1987. Effects of salt and tripolyphosphate reduction on minced catfish patties. MAFES Research Report 12(6). Mississippi Agricultural and Forestry Experiment Station, Mississippi State University.

Hoke, M.E., M.L. Jahncke, J.L. Silva, J.O. Hearnsberger, R.S. Chamul, and D. Suriyaphan. 2000. Stability of washed frozen mince from channel catfish frames. J. Food Sci. 65:1083-1086.

Kim, J.M., C.H. Liu, J.B. Eun, J.W. Park, R. Oshimi, K. Hayashi, B. Oh, T. Aramaki, M. Sekine, Y. Horikita, K. Fujimoto, T. Aikawa, L. Welch, and R. Long. 1996. Surimi from fillet frames of channel catfish. J. Food Sci. 61:428-431, 438.

McAlpin II, C.R., J.G. Dillard, J.M. Kim, and J.L. Montanez. 1994. An economic analysis of producing surimi from catfish byproducts. MAFES Bull. 1013. Mississippi Agricultural and Forestry Experiment Station, Mississippi State University.

NMFS. 2002. U.S. seafood consumption. NOAA, NMFS, Washington, D.C. http://www.publicaffairs.noaa.gov/.

Silva, J.L., and S. Dean. 2001. Processed catfish: Product forms, packaging, yields, and product mix. USDA-SRAC Pub. 184. Southern Regional Aquaculture Center, Mississippi State University, Stoneville, Mississippi.

Silva, J.L., S. Danviriyakul, C. Handumrongkul, B. Wannappee, A. Wachalatone, and R.S. Chamul. 1999. Properties of branded, heat-set catfish patties depending on nugget mince ration and stabilizer. Abstracts, Institute of Food Technologists Annual Meeting, Chicago, p. 202.

Suvanich, V., M.L. Jahncke, and D.L. Marshall. 2000a. Changes in selected chemical quality characteristics of channel catfish frame mince during chill and frozen storage. J. Food Sci. 65:24-30.

Suvanich, V., D.C. Mardhall, and M.L. Jahncke. 2000b. Microbiological and color quality changes of channel catfish frame mince during chilled and frozen storage. J. Food Sci. 65:151-155.

USDA. 2002. Catfish processing. Agricultural Statistics Board, USDA-NASS (National Agricultural Statistics Service),Washington, D.C. http://USDA.mannlib.cornell.edu/.

Trends in the Utilization and Production of Seafood Byproducts

Hans Nissen
Atlas-Stord, Inc., Kansas City, Missouri

Abstract

Seafood companies in the North Pacific and Alaska generate a significant volume of seafood byproducts and could benefit environmentally and economically by utilizing these byproducts. Today many of the companies have integrated a fish meal plant to process seafood byproduct into valuable high quality fish meal and oil which is sold worldwide in competition with other protein meals. In order to justify the investment in a production facility to utilize the byproduct the annual volume must be considered. The trend has been to combine the byproduct from various plants in order to have enough volume to make the operation feasible.

In Alaska with many small processors at remote locations this can be a difficult task, and the reason much of the seafood byproduct today is still being dumped or disposed at landfills. We have considered a process where the byproduct is collected at the different sites and brought to a central location where it is hydrolyzed into silage and dried using a carrier liquid drying process (CLD) producing a stable product that can be marketed worldwide.

Introduction

As a fish meal equipment manufacturer we are in many cases the first in line to get the call when a fish processor, for various reasons, must consider how to utilize the fish byproduct from his process. The reasons can include new environmental regulations regarding disposal of byproduct, or an increase in disposal cost from the local landfill or from the person who hauls the byproduct away.

The questions are many, such as "What is included in a modern fish meal line to produce a high quality end product?" and "What about environmental impact?"

Trends in fish meal production
Material

Small oily fish like menhaden in the Gulf of Mexico, which is normally not suited for human consumption, is typically used as raw material for the production of fish meal and oil. Another source of raw material is the byproduct or offal from fish processing, such as heads, frames, liver, and viscera.

The quality of the raw material is typically measured by its freshness TVN (total volatile nitrogen). A low TVN indicates a fresh, less deteriorated product that can be used for premium quality fish meal.

Process

The fish meal and oil process is a separation process in which the content of water, oil, and solids of the fish are separated and the water removed by evaporation and drying.

The process has not been altered for a long period of time. Extensive development of the various equipment has been, and currently is, taking place in order to reduce energy consumption, increase product quality, and protect the environment. A modern fish meal line appears in the flow diagram in Fig. 1.

Step 1 is the cooking where the fish is heated and the protein coagulated in order to release the water and oil. Today the cooker is typically indirectly heated with steam and the flow controlled with respect to time and temperature.

Step 2 is mechanically dewatering the heated product using a strainer screw combined with a screw press. The press cake has a water content of approximately 50% and contains about 70% of the solids.

Step 3 is a separation of the oil from the press water using high speed decanters, where suspended solids are removed and now are called grax. This is followed by a centrifuge step where the oil is separated. The water is now called stickwater and contains 6-10% soluble protein.

Step 4 is the concentration of the stickwater from the centrifuge in an evaporator up to 40-50% solids. The evaporator is typically a "multi-effect" and can use steam, waste heat from the dryers, or electricity in the MVR (mechanical vapor recompression) evaporator.

Step 5 is the drying of the press cake mixed with grax from the decanter, and the soluble material from the evaporator, from approximately 55% to below 10% moisture. The drying process has the largest effect on the protein quality and several types of dryers are today being used. Direct fired rotary drum dryers are widely used in large volume plants and produce a fair average quality (FAQ) meal. Steam dryers are used in new installations and can be operated with minimum air intake securing optimal conditions for waste heat utilization and protection against pollution of the air. The steam dryer produces what is called steam-dried meal.

Fish byproduct

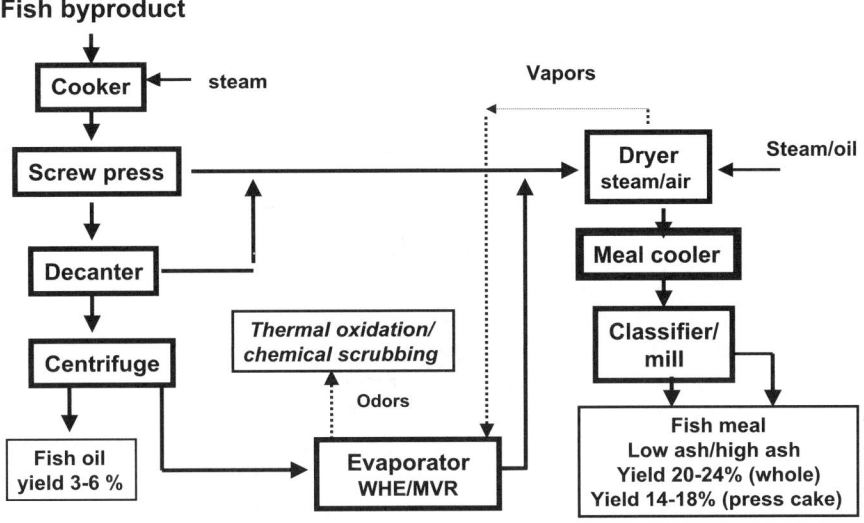

Figure 1. Flow diagram for a modern fish meal line. WHE/MVR = waste heat/ mechanical vapor recompression.

In order to produce specialty fish meal like LT (low temperature), Prime, etc., from fresh fish, the drying may take place in a vacuum steam dryer, keeping the product temperature below 70°C (158°F), or in an indirectly heated hot air dryer where the air is heated to a moderate temperature without getting in contact with the flue gases. Finally, a two-stage drying system using a steam dryer as the first step followed by a hot air dryer is being used to produce high quality fish meal, and at the same time making it possible to recover the waste heat in the evaporators.

Step 6 is the cooling, classification, and milling of the dried product to obtain the final product specification. Classification of the meal consists of a screening where the larger particles such as bones are separated in order to increase the protein level in the meal and produce a high and a low protein product.

Environmental protection

Normally the authorities, and more often the neighbors, require the fish meal plant to minimize emission of odors and polluted waste water. Newly designed fish meal plants are made air tight and installed with suction and/ or vacuum on all equipment. The non-condensable odors are treated in chemical scrubbers or incinerated in the combustion chamber of the steam boiler or the hot air dryer.

The water contained in the fish byproduct ends up as waste water. It can be discharged with the cooling seawater back to the ocean, or further treated in a waste water treatment plant in cases where zero discharge is required. In this case, the cooling water would be indirect, using a cooling tower, or alternatively using an air cooler for condensing.

Economy

Prices of fish meal and fish oil fluctuate to a great extent and depend on prices for soy protein and soy oil. Present prices are rather good which helps justify the investment in a fish meal plant.

A general trend, however, shows that production plants are getting larger and the production is concentrated in relatively few factories combining raw material from various sources or facilities. This is done in order to secure optimal use of labor, energy, etc., minimizing the production cost.

Smaller fish meal plants are being installed onboard factory trawlers. Figure 2 shows the feasibility of installing smaller fish meal plants depending on the length of the season. The calculation shows that a season of a minimum of 120 days per year and a capacity of greater than 2 tons per hour are required to make installation of a small fish meal plant feasible onboard factory trawlers.

Alternative processes

In Alaska many fish processors are located in remote areas where (1) it is not practical or possible to combine the byproducts, (2) there is a relatively small volume, and (3) the season is short. These processors will look for alternative processes that are less expensive or with the possibility to produce a special product that could be sold at a much higher level than fish meal and oil. Some of these alternative processes are

- Fish silage—liquid fish protein produced by using acid.

- Hydrolyzed fish products—liquid fish protein produced by using enzymes.

- Fish protein concentrate—liquid fish solubles not added back to the dryer.

Common for these processes is a liquid end product, which could be final dehydrated and thereby make it easier to get to the end user market due to lower shipping cost. A dehydrated product will also be of benefit because U.S. feed markets, among others, are based on handling dry products.

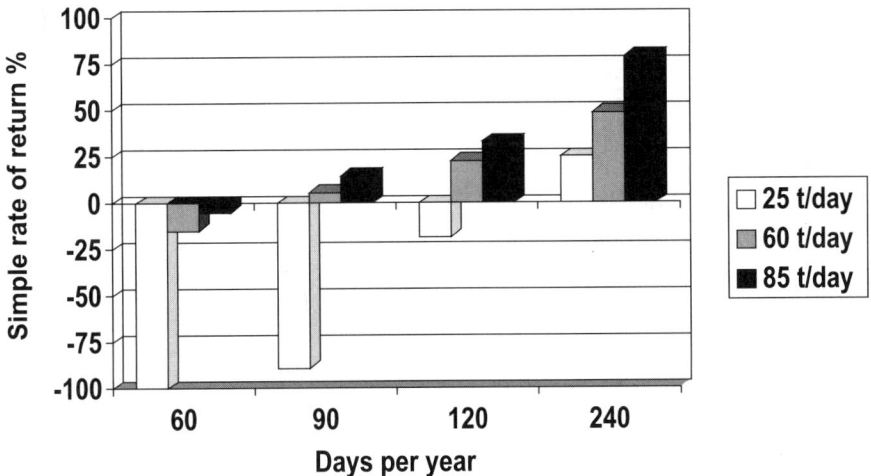

Figure 2. Feasibility of a small fish meal plant. Revenue is fish meal $500/t and fish oil $400/t. Cost is steam $20/t, electricity $0.05/kWh, and money 10% p.a. over 5 years.

Carrier liquid drying process

An alternative to existing drying processes for liquid such as spray dryers and others is the carrier liquid drying (CLD) process. Three applications for CLD are

- Separate dewatering of soluble fraction in conventional fish meal process.

- Final dewatering in fish silage process.

- Final dewatering in enzyme processes.

The CLD process appears on the flow diagram shown in Figs. 3 and 4, and the end product characteristics for stickwater during concentration are shown in Fig. 5. The basic processing conditions are

- Mixing ratio 1:6 to 1:20 (one part concentrate dry matter to 6 or 20 parts oil).

- Dry matter of concentrate not exceeding 35%.

- Evaporation temperature 280°F.

- Steam consumption 1.2:1 steam to evaporation.

- Electrical consumption 45kWh per ton evaporated.

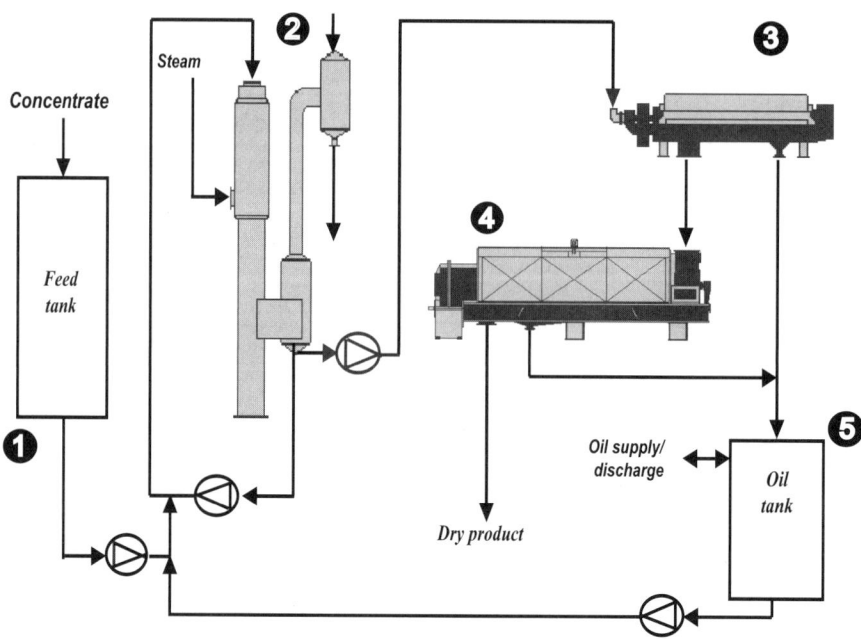

Figure 3. Flow diagram of carrier liquid dryer (CLD) process, with stainless steel components. (1) Feed tank; (2) Evaporator system; (3) Decanter; (4) Oil recirculation tank; (5) High pressure press, optional.

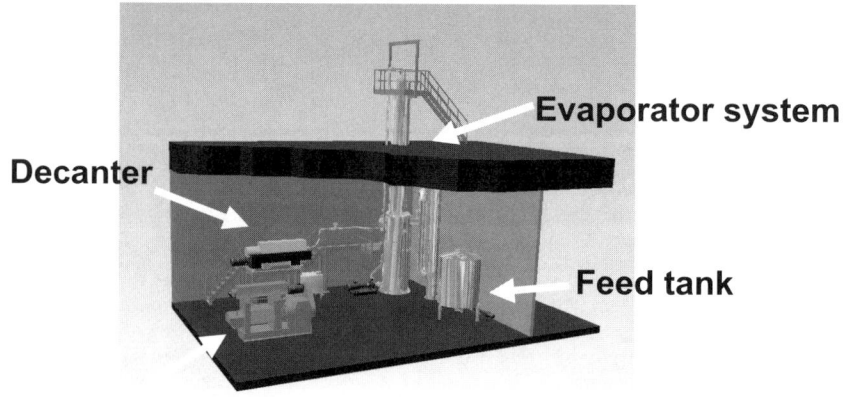

Figure 4. Diagram of carrier liquid dryer (CLD) system.

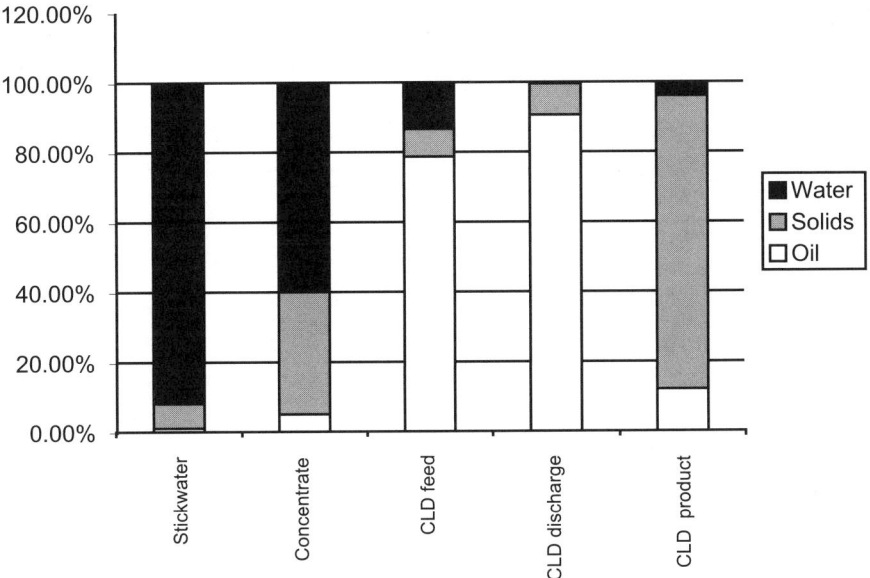

Figure 5. Composition progress in the concentration of stickwater. CLD is carrier liquid dryer.

Pilot testing has taken place on a variety of products including fish solubles and filleting waste. The final composition of the fish solubles was 87% nonfat solids, 10% oil, 3-5% moisture, and 70% protein. The final composition of the filleting waste was 78% nonfat solids, 18% oil, 2-4% moistures, and 67% protein.

In general, in the final product the residual oil content of 8-18% could be obtained without use of a high pressure press. The high pressure press could reduce the oil content in the final product up to 4%, and the high pressure press was not recommended for final products with less than 10% oil.

In summary the CLD system is an energy efficient and fast dewatering process that enables the production of a variety of products utilizing a compact and flexible design. The major disadvantage is that the oil used during processing must be reprocessed.

Economic impact

The above drying technique (CLD) could have an immediate economic and environmental impact on the existing fish meal plants in Alaska. Today most of the plants recover in their fish meal only 15-20% of the total dry solids from stickwater. The reasons are that adding more will increase

**Table 1.   Preliminary analysis of stickwater concentration econom-
ics from seven fish meal plants in the Aleutian Islands,
Alaska.**

	Percent	t/day
Capacity, each plant		600-1,200
Stickwater	68%	4,285[a]
Dry solids (protein), in stickwater	8%	343
Recovered as fish meal	18%	62
Discharged to sea		281

Annual revenue is $13.5 million ($400/t, 120 days)

Total cost is $3.0 million ($90/t, 120 days)

[a]1,132,000 gallons per day.

the salt content in meal, and end users have demand for a semi-press cake meal.

A preliminary calculation shown in Table 1 indicates the economic benefit of concentrating stickwater and incorporating it in the fish meal. For example, if the stickwater today being discharged to the sea was processed and dewatered and sold as secondary fish meal, earnings for all seven plants could be increased as much as US $10 million according to the above calculations.

Membrane Filtration of Stickwater

Leo D. Pedersen
Dantec Engineering Inc., Danville, California

Charles Crapo
University of Alaska Fairbanks, Fishery Industrial Technology Center, Kodiak, Alaska

Jerry Babbitt
National Marine Fisheries Service, Utilization Research Laboratory, Kodiak, Alaska

Scott Smiley
University of Alaska Fairbanks, Fishery Industrial Technology Center, Kodiak, Alaska

Abstract

In the Alaskan processing of seafood products, almost two thirds of the landings are discarded as byproducts. Much of the byproducts are converted to feed products, but a liquid fraction, stickwater, resulting from this processing is generally not recovered due to high water content and elevated salt content. The stickwater fraction contains from 20 to 40% of the initial byproduct solids. This project explored the potential for development of a membrane system capable of reducing the salt and increasing the solids content in stickwater. The membrane filtration system tested was capable of removing 85% of the water contained in the stickwater, resulting in approximately 25% solids in the concentrate. The salt concentration was reduced from 15%, without filtration, to 3.4% with membrane filtration. While the membrane filtration technology is capable of satisfactorily solving the high salt content problem, a higher solids content in the concentrate would be desirable to minimize the dryer load.[1]

---

[1] The use of enzymes for viscosity reduction of concentrated stickwater could make this possible.

Introduction

Alaskan fish meal processing operations differ from other countries in that the raw material is the byproduct from filleted fish rather than the processing of whole fish. The Alaskan raw material is of high quality due to fact that the fish has to meet human grade standards during catch and filleting, and therefore is "fresh" when received for processing into fish meal. However, the removal of the fillets results in a fish meal that has a significantly higher bone content, which results in a high ash content in the finished product. Further, due to climate conditions the caught fish is in some cases stored in seawater, which leads to higher than normal salt concentration in the fish meal.

The fact that the primary goal is the production of fish products for human consumption makes it more difficult to optimize the production schedule of the fish meal operation. Fish meal plants have little control over the volume to be processed and have to match the output from the "upstream" production. Further, the harvest volume of fish varies substantially during the year with a few months of very high volumes (pollock) followed by periods of much smaller volumes during the remainder of the year when the fishery for rockfish, salmon, cod, flatfish, etc., occurs. Design of efficient fish meal plants under these conditions is difficult. Plants that would be capable of meeting the peak requirements would be very capital intensive and would only be operated intermittently (and inefficiently) outside the peak season. The result has been that most plants are undersized and only utilize the press cake while dumping the press water (stickwater) into the ocean. However, changes in catch management have lowered the peak catches and extended the length of the season, which together with increased environmental concern has renewed the focus on increasing the efficiency of the fish meal operations to include the recovery of the stickwater fraction.

Fish meal processing

The major steps in a fish meal process are shown in Fig. 1. The cooked byproduct is dewatered in a press to minimize the drying load while the press water is dumped (most common) or can be concentrated. The stickwater concentration process is traditionally performed in thermal evaporators[2] but the development of high temperature–resistant membranes with low fouling tendencies has made this a promising new technology that could reduce capital, operational, and energy costs.

Stickwater is the aqueous phase produced in the large scale handling of fish waste. As shown, the fish processing byproducts are cooked, and

---

[2] Such evaporators have a considerable "hold-up" resulting in a lag time of 4 to 8 hours from the start of processing to the time when the product is actually ready to be added to the press cake. If the total production run in an off-peak situation is only 8-12 hours, very little of the stickwater could be added to presscake and dried.

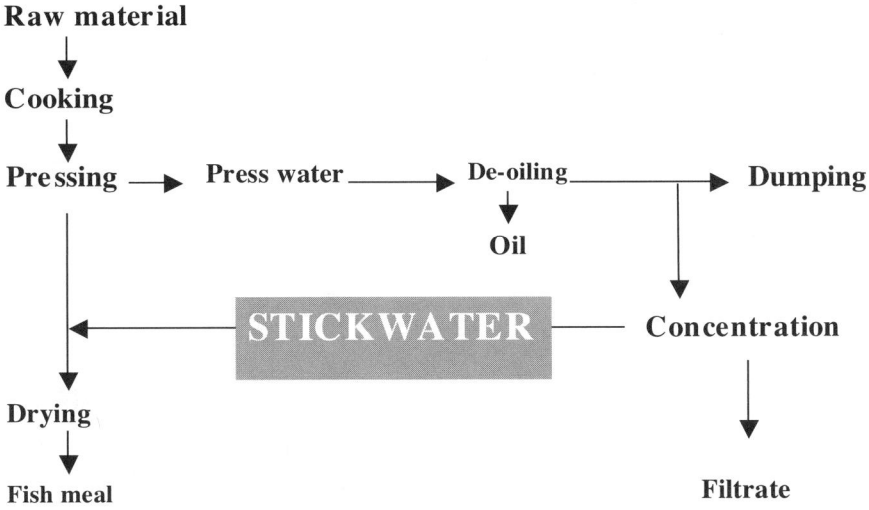

Figure 1. Fish meal production schematic.

the solids are separated from the liquid phase (presswater). The oil is then removed from the liquid phase, and stickwater is left. The stickwater can then be concentrated by evaporators, and may be sprayed back on the solids during drying. Stickwater has a very high concentration of water soluble organic molecules and is a problem for many processors because of its high salt levels. The high salt content limits the use of stickwater in animal and aquaculture feeds. In addition the high salt content causes severe corrosion of evaporators.

Byproducts from fillet processing contain a higher proportion of incompressible material (bones), which causes a proportionally higher amount of residual fish meat to be carried over in the stickwater fraction. In Alaska the stickwater contains 8 to 10% solids representing 33 to 45% of the total raw material solids, Fig. 2.[3]

Stickwater is a significant portion of the fish byproduct operation. For example, if there are 100 pounds of raw material, with a solid content of 19%, the amount of stickwater can range from 66 pounds at 4% solids to 79 pounds at 11% solids. Typically, the concentration of solids in the stickwater is about 7%, which means that the amount of stickwater would be approximately 71 pounds per 100 pounds of raw material.

---

[3] Processing of whole fish typically results in stickwater solids content of 5 to 6% representing only 16 to 22% of incoming raw material solids.

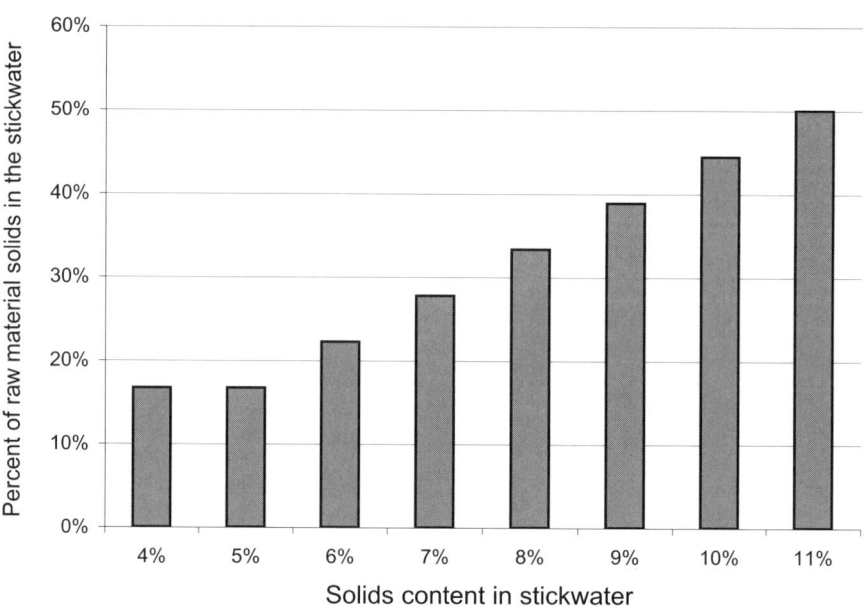

Figure 2. Solids carryover in stickwater.

Membrane filtration technology

Membrane filtration technology offers the advantage of lower capital and energy costs than for evaporators. In addition, membranes are polymeric or ceramic-based substrates that are inert to corrosion, solving a major problem inherent in current evaporation technology. In the past, the high salt concentration has resulted in evaporator failures after three to five years of operation due to corrosion.

Membrane filtration separates water and salts from larger molecules, such as proteins and solids, based on molecular size rather than evaporation. Membranes can selectively recover large protein molecules, while salts such as sodium chloride, calcium, and smaller nitrogen-based molecules will pass through the membranes with the water (Fig. 3). This process can be energy efficient and cost effective, even for smaller volumes.

Membrane filtration process

Several factors must be considered in the development of membrane filtration systems for a particular application. The key factors are membrane type, system design, membrane substrate, membrane design, and cleaning procedures. A diagram indicating the critical design components in a membrane filtration system is shown in Fig. 4.

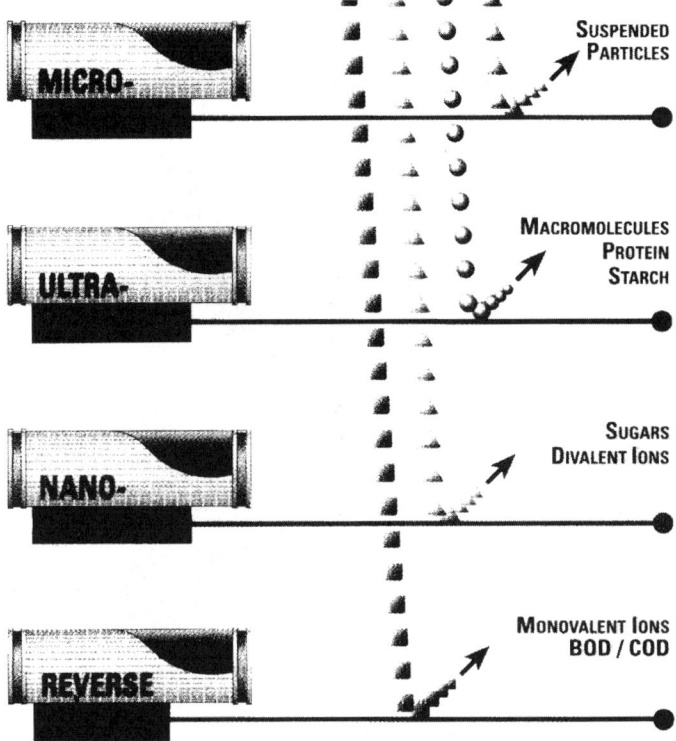

SUSPENDED
PARTICLES

MACROMOLECULES
PROTEIN
STARCH

SUGARS
DIVALENT IONS

MONOVALENT IONS
BOD / COD

Figure 3. Membrane filtration spectrum.

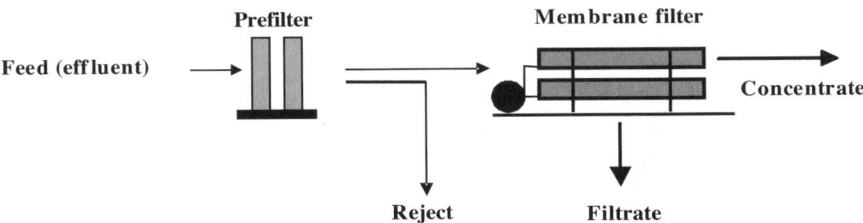

Prefilter

Feed (effluent)

Membrane filter

Concentrate

Reject

Filtrate

Figure 4. Membrane filtration process schematic.

Figure 5. Membrane filtration test unit.

Test procedure

The pilot system used for the testing is shown in Fig. 5. The system contained a pre-filter followed by two industrial standard filtration element housings. The pilot system was equipped with a pump configuration for allowing operation in batch and in continuous modes.

Prefiltration

Prefiltration was required to avoid product plugging the flow channels between membrane layers during operation. In this application fouling was a major problem in that the protein in the process waters had very high fouling tendencies. In addition, a filter can exert shear forces sufficient to denature protein and add to the fouling problem. Prefiltration with 100 micron pore size filters has been shown to reduce these problems (Pedersen et al. 1989). During the testing different prefiltration filters were used; however, only stainless steel screens with pore sizes above 70 microns performed satisfactorily. Smaller screen sizes and woven or pleated filters resulted in excessive pressure drops and ultimately in failure. In Fig. 6 the material build-up on a 70 micron screen after eight hours of operation is shown. Even though significant fouling occurred,

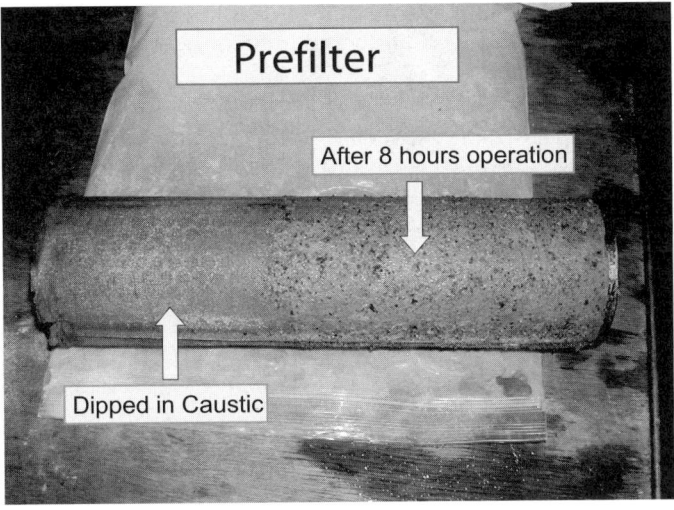

Figure 6. Prefilter, 70 micron stainless steel.

the pressure drop over the filter remained below 15 psi, and the fouling layers were easily removed by a caustic cleaning as shown in Fig. 6.

Membrane filtration module

Membrane filters can be configured as spiral wound, hollow fiber, and tubular. Tubular filters generally have superior performance on highly viscous products, such as liquid phases containing fish protein. For industrial applications, however, the spiral wound module is generally preferred for cost reasons.

The test configuration was based on the spiral wound module. A critical design parameter is the narrow flow channels in the spiral wound membrane, which requires a tight pre-filtration for removal of suspended solids in order to avoid plugging. Three different feed channel spacer sizes were tested: 0.75, 1.20, and 2.30 mm (30, 47, and 90 mil). The 1.2 mm spacer was found to perform the best. The 0.75 mm spacer resulted in excessive pressure drop over the module at higher concentrations due to the increased viscosity. The 2.30 mm spacer did not show any performance improvement over the 1.20 mm spacer.

Membrane pore size selection

The filtration system can be optimized to separate molecules of different sizes, such as oil, protein, and ash components in the fish processing byproduct. Two different pore sizes were tested (Table 1). A membrane

Table 1.   Analytical composition of stickwater, concentrate, and permeate processed by 50,000 and 10,000 dalton membranes.

	Stick-water %	Conc. 50,000 daltons %	Conc. 10,000 daltons %	Perm. 50,000 daltons %	Perm. 10,000 daltons %
Moisture	91.03	85.23	85.48	97	97.83
Solids	8.97	14.77	14.52	3	2.17
Ash	0.86	0.84	0.84	0.83	0.8
Protein	6.36	10.06	10.22	2.38	1.39
Oil	2.28	4.51	4.26	ND	ND
Total	100.53	100.64	100.8	100.16	99.98

ND = not detectable.

with a molecular weight cutoff of 50,000 daltons performed well but had a slightly lower retention than the 10,000 dalton membrane.

Membrane substrate

The membrane substrate should have low affinity to the product being processed to minimize membrane fouling. Protein, in particular, has a tendency to become attached to the membrane surface causing rapid fouling and a poor filtration rate. Membrane substrates with high hydrophobicity will be required for this application (Nicolaisen 2001).

Past test results showed that the membrane polymer PVDF performed better than most other substrates (Pedersen 2001). A recent development based on a polyacrylic-nitrile (PAN) membrane has resulted in a membrane with superior performance on oily substances and protein-rich fluids (Table 2). The PAN membrane substrate was selected for testing in this study.

Test results

Results of a typical PAN membrane test run is shown in Fig. 7. The initial solids in the stickwater were 10.6%, which were concentrated to 24% with the membrane system. This resulted in approximately 84% of the water removal needed for fish meal production.

The analytical composition of the stickwater processed by a 50,000 and a 10,000 molecular weight cut-off membrane is shown in Table 1. A further analysis of the protein contained in the permeate fractions show that these were short-chain compounds.

Table 2.   Hydrophilicity of different membrane substrates.

Membrane material	PTFE	Poly-propylene	PVDF	Unmodified PAN	Hydro-philic polysulfone	Ceramic	M-series Ultra-philic[R]
Contact angle[a]	112°	108°	66°	46°	44°	>30°	4°

→→→→→→→→→→→→Increasing hydrophilicity→→→→→→→→→→→→→→→→→

Oleophilic:	Oleophobic:
Repels water	Repels oils
Absorbs oil	Absorbs water
Fouls with free oils	Not fouled by free oils
Lower flux per foot	Higher flux per foot
Difficult to clean	Easier to clean

[a]Angle of edge of water droplet on the membrane.

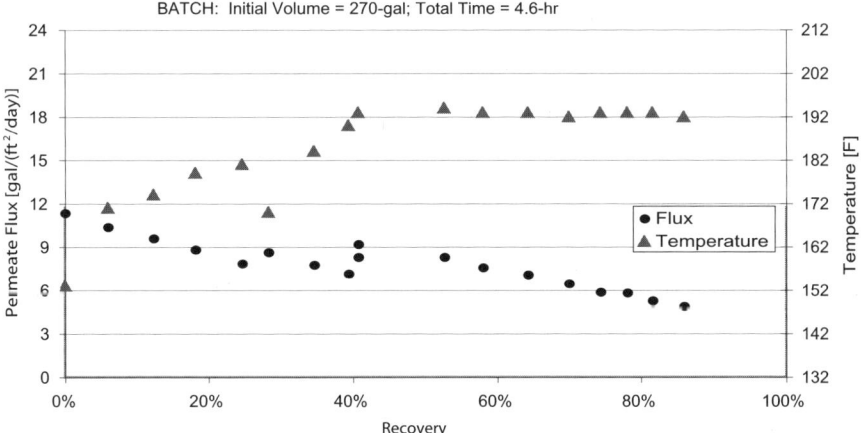

Figure 7.   PAN membrane performance as a function of concentration. Initial volume is 270 gallons; total time is 4.6 hours.

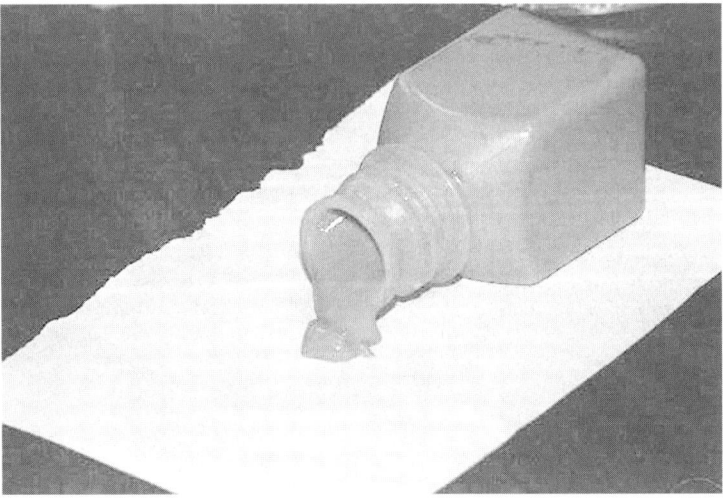

Figure 8. Concentrated stickwater.

Cleaning procedures

During operation, products gradually coat the membrane, reducing the filtration rate over time. The membrane system will require periodic cleaning. The cleaning typically consists of a caustic rinsing during which residual protein and other organic residues are solubilized and removed from the membrane surface. Depending on the nature of the foulant, however, oxidizers, surfactants, enzymatic treatments, and acidic rinses might also be needed. Given the nature of the feed stream, the development of a tailored cleaning protocol will likely be required.

Several cleaning regimes were tested. Application of a hot-water flush followed by a caustic cleaning (10 minutes at pH 11) effectively removed the organic fouling layer. Periodically an acid wash (10 minutes at pH 2) was also applied for removal of potential mineral build-up. A temperature above 60°C (140°F) was critical for achieving an efficient cleaning.

Summary

Concentration of stickwater by membrane filtration can remove 85% of the water. Higher concentrations might be achievable by use of enzymes to lower the viscosity. The salt content in the stickwater is lowered in a similar ratio to the water content, and the residual salt content is acceptable for inclusion of stickwater in the produced fish meal (Fig. 8). The concentrated stickwater has a light gray color without any burnt odor as sometimes is the case with evaporator concentrate. It should be well suited for blending back to press cake.

The membrane filtration system required cleaning after 6-10 hours of processing. The flux was easily restored by a hot-water flush or a 10 minute caustic cleaning.

The filtration rate ranged from 5 to 12 gallons per square foot per day (9 to 20 liters per square meter per day), which is acceptable for most commercial applications.

References

Nicolaisen, B. 2001. Membrane filtration of oil and gas field produced water. Osmonics Inc., Minnetonka, Minnesota.

Pedersen, L.D. 2001. Membrane technology for stickwater recovery in fishmeal processing. University of Alaska contract UAF 02-0026.

Pedersen, L.D., W.W. Rose, M.F. Deniston, and C.A. Merlo. 1989. Hyperfiltration technology for the recovery and utilization of protein materials in surimi process wash waters. NOAA, NMFS, project no. NA 86AA-H-SK140.

Utilization of Seafood Byproducts: Why Fish Meal?

Jette Kristensen
Alfa Laval Protein Technology, Inc., Soborg, Denmark

Current trends in Alaska fish byproduct processing

Although it has been mentioned several times in the past two days that marine protein is an excellent source of protein, the question remains, "How do we utilize it in the best way?"

In my opinion, none of the large fish meal manufacturers in Europe or South America have as fresh raw material as you find here in Alaska. In the last few years, there has been a lot of focus on trying to increase yield of the main Alaskan products, lox and surimi. Jerry Babbitt and others have concentrated on increasing the overall yields, and many plans have been developed to produce a second-grade surimi by taking the frames from recovery lines and pressing out most of the meat. This increases the overall yield for many customers up to 3-4% and means there is less protein in the byproduct fraction. We have actually started to cut more into the bones and in the future may try to find solutions that will allow us to cut even closer.

Three processing methods

With the raw material available from the main process, we have to produce an end product that is stable, i.e., we have to remove the oil. This can be achieved through three processes. The latter two processes are discussed in more detail in the section "Acid and enzyme processing methods" below.

- Coagulating the proteins with heat, allowing the fat and water to be removed from the product, which can then be dried and turned into fish meal.

- The use of acid.

- The use of enzymes or bacteria to break down the product.

Worldwide fish meal production

Scandinavia is producing fish meal with 70-72% protein and 10% ash, while South American–produced fish meal contains 65-67% protein and 17% ash. Some areas around the world are trying to make a high protein fish meal by removing ash during the process and basically drying the meat or protein. It is possible to reach up to 77-78% protein in the final product.

Looking at the fish meal produced in this region (Table 1, column 1), most of the protein is taken out in filleting blocks or surimi, which leaves much more ash in the final product. I would estimate that the figure for protein could therefore be lower than the 60% indicated for this region.

Fish meal protein quality

Normally, when one talks about fish meal applications, one talks about protein content, digestibility, and functionality.

True digestibility is difficult to measure, however. If you look at the first four types of protein content on the chart of different fish meals produced around the world (Table 2), they all fall in a range of 66-69% protein. If you look at true digestibility as measured by the displacing method, however, more variation among the products becomes evident. The biological value of the protein shows an even larger differentiation at the same level of protein quality.

The big difference in the various fish meal types is in the next protein utilization. Although some of the difference is attributable to the raw material, the process is also greatly responsible. Cooking and drying systems used (Table 2) are: FAQ (fair average quality) is produced in a steam dryer, sandeel is produced in a vacuum dryer, mackerel waste is produced in a hot-air dryer, and herring waste is produced in a vacuum dryer.

At the bottom of the table are some numbers for deboned pollock meal that were produced by Jerry Babbitt in a mid-1990s study. The figures demonstrate that when we remove the bones, we get a very high protein content and a significant increase in the net protein value. There is very high digestibility of the protein.

Niche production in Europe

Given the proper quick cooking and separation process, a small plant can produce a very high-quality fish oil. If the material is fresh and processed immediately, it is possible to produce a product that is so good that it does not need to be refined.

"Super prime quality" (total volatile nitrogen [TVN] < 10) is made from mackerel waste in the U.K., where the waste stream has been processed immediately. We have obtained fish meal with very low free fatty acids (< 0.1%), peroxide (< 1.0 meq per kg), and anisidin (< 1.0%).

Table 1.   Fish meal content.

	Fish meal (seafood processor)	FAQ meal (South America)	LT meal (Northern Europe)	Ultra flash meal (Northern Europe)
Protein	60%	65%	72%	77%
Ash	22%	17%	10%	5%
Water	8%	8%	8%	8%
Oil	10%	10%	10%	10%

FAQ = fair average quality. LT = low temperature.

Table 2.   Fish meal protein quality.

Meal type	Protein content %	Digest-ibility %	Biological value %	NPU %	Usable protein %
FAQ meal	67	90.9	89.0	80.9	54.2
Sandeel LT	69	92.2	89.9	82.9	57.2
Mackerel waste	66	96.0	90.7	87.1	57.5
Herring waste	69	93.2	90.5	84.3	58.2
Deboned white	83	99.6	96.3	95.7	79.4

FAQ = fair average quality. LT = low temperature.

"Prime quality" (TVN < 25) describes a seafood byproduct with waste mixed in, made in a small plant. We take out the fish oil and still get high quality fish meal: free fatty acids < 0.5%, peroxide < 2.0 meq per kg, and anisidin < 2.0%.

Small production schedules where you focus on removing one of the streams are possible. Mackerel waste is of interest due to its high oil content. European salmon also has a high oil content, as much as 25-30%, so that is another area where it makes sense to try to remove that oil in good quality.

The question remains as to what customers are doing with the protein and stickwater. At one plant in the Faeroe Islands, they remove high-quality oil from hatchery-raised salmon and then the stickwater and protein are mixed and frozen into blocks. The blocks are then shipped to Denmark and sold.

Another example is a mackerel factory where they are concentrating on removing the oil. Two other elements are also mixed in before freezing and, in this case, the end product goes to the mink and pig industries in Denmark.

In Alaska, we are not lucky enough to have a lot of poultry and pigs and the fish here do not have much oil, so we have to focus on protein.

Acid and enzyme processing methods

Let us examine in more depth the two processing methods that do not use heat. First, there is the acid method, also called "silage." In this process, we finely grind the product and then add acid. The acid and the bacterial enzymes break down and change the protein structure, resulting in a soluble, liquefied product. This process is difficult to control, so the numbers in Table 3 may not be completely correct. The protein will break down to polypeptides, free amino acids, and unfortunately some ammonia and other unwanted compounds.

Adding enzymes to quickly break down the product allows us to control the process. We can add enzymes that produce polypeptides or enzymes that produce more free amino acids (Figure 1).

The oil product aspect of these two processes is also important. If you are using acid to break down protein, it will have a big impact on the oil in the oil phase. Free fatty acids (FFA) will increase dramatically. The normal FFA range is between 8% and 10%. The peroxide value will also be high.

Enzymes have the benefit of attacking the oil fraction in a more controlled way. The normal FFA levels with the enzyme process will be between 2% and 3%, but because the process is quick, there will be little oxidation in the oil phase.

Table 4 compares advantages and disadvantages of the silage and enzyme processing methods.

- When we use silage and acid, it is very difficult to control the protein and oil quality. If we use enzymes, the quality of both those fractions can be controlled.

- The silage process takes a long time; enzymes speed up the process.

- Both processes can be batched, but normally you will need continuous process when working with enzymes because you work at critical temperatures. At temperatures around 50°C (115°F), bacteria growth is possible, so you must be careful.

- While the silage form usually must be used as a feed component, working with enzymes enables the possibility of food-grade end product, like extracts and marinades. Of course, enzyme product can also go to the pet food industry.

Table 3. Silage and fish liquid protein concentrate processing methods—content.

	Silage	Liquid protein concentrate[a]
Protein phase		
Polypeptide	33%	50%
Free amino acids	33%	50%
Ammonia and insoluble nitrogen	33%	1%
Oil phase		
Free fatty acids	8-10%	2-3%

[a]Faster processing using the addition of proteolytic enzymes.

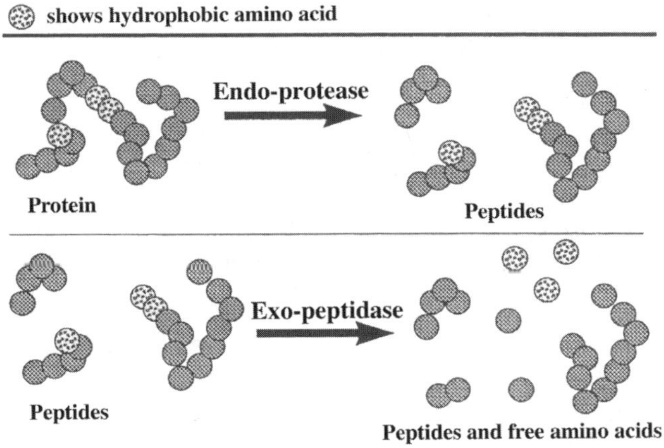

Figure 1. Enzyme hydrolysis of protein with endo-protease and exo-peptidase.

Table 4. Silage and fish liquid protein concentrate processing methods—advantages and disadvantages.

	Silage	Liquid protein concentrate[a]
Quality of protein and oils	Poor	Improved
Process time	Long	Short
Process	Batch	Continuous
Grade	Feed product	Pet food or better (possibly)
Pretreatment process	Simple	Complex

[a]Faster processing using the addition of proteolytic enzymes.

- Silage is a simple process, requiring just grinding and acid at the remote location and then relocation of that product to a central location. Working with enzymes is more complicated because of the need for more processing on-site. But you gain control over the quality of the end products.

- Depending on what type of enzymes you add, you can control the cleavage of the protein chains (Fig. 1). This allows you to decide if you want peptides or free amino acids. Peptides are the long chains that give you functionality. Free amino acids, on the other hand, are where you turn if you are looking more for flavor.

- When we process protein, bones, and oil into fish meal, we get a product that has its own protein content. The digestibility of the protein will vary depending on the process used. The silage process cannot control what the end protein fraction will be; however, the enzyme process does allow control over the degree and type of cleavage of the end product. This allows us to decide whether we want products with more flavor or functional properties.

Conclusion

Looking to the future, we will see more vegetable protein coming into agriculture. As we look at the future of marine protein, we need to decide whether we need digestibility, flavor, or functionality. The answer will determine the process we use. If we keep the products and their processes separate, i.e., if we have one process for protein and another for bone which results, for example, in bone meal or gelatin, then we will get high added value out of our byproducts.

In order to get a high-quality, high-value fish oil, the oil needs to be as fresh as possible when processed. Therefore, in the future we need to decide whether we should produce commodity fish meal or separate the material and optimize the process for each of the fractions.

Modified Silage Process for Fish and Fish Processing Waste

Peter Nicklason, Harold Barnett, Ron Johnson, and Mark Tagal
National Marine Fisheries Service, Northwest Fisheries Science Center, Seattle, Washington

Bob Pfutzenreuter
University of Alaska Fairbanks, Fishery Industrial Technology Center, Kodiak, Alaska

Abstract

Dried fish silage was made from fresh and frozen whole fish, and fresh and frozen viscera and processing waste. The Modified Silage Process (MSP) uses less acid and lower temperature than common fish silage processes and recovers over 90% of the fish solids and oil in a stable dry meal. Feeds made from different fish and processing wastes using the MSP performed as well as a commercial feed. Feed conversion ratios (FCR) for MSP feeds were 0.86 with juvenile salmon. An economic evaluation of the MSP using the Net Present Value model illustrates where the MSP may contribute to the processing of fishery byproducts in Alaska or other remote and seasonal situations.

Introduction

Fish meal, oil, and bone from whole fish or processing waste are valuable commodities that are currently in high demand. Limited supplies of fish raw material, used in the fish meal industry, and increasing consumption by a growing aquaculture industry point to a stable future market for these products (CEC 2002). There is a plethora of fish waste currently being discarded at a cost to processors. In addition to being lost potential byproduct income, this waste is an environmental liability that is threatening the operation of many processing facilities. However, fish waste could be turned into profits if processing economics could be improved. Nearly all fish meal, oil, and bone are produced by a conventional cooking and pressing process (Bimbo 2000). The design and economics of the conventional process dictate continuous operation and high volume throughput. This eliminates the processing of fish wastes in situations where there is sporadic operation or overall volume cannot support the minimum equipment costs.

An alternative process is the production of fish silage (Johnson et al. 1982). The fish waste is ground and mixed with acid. Acid inhibits bacterial spoilage and also provides the conditions that increase the activity of digestive enzymes in the fish. Fish waste liquifies over time, releasing the oil and bones which can then be separated. The remaining liquid is high in protein and can be neutralized by base, concentrated by evaporation, or used directly in the liquid form as a feed supplement or plant fertilizer. The advantage of the silage process is simple equipment for manufacture and stability against spoilage that allows for bulk storage in tanks for a year or more. The main limitation for the silage process is high moisture content, which greatly increases transportation costs of this material to market. Ensiling can be uneconomical for facilities that are remotely located and/or generate more waste than can be held in tanks at the site. Examples are seasonal fish processing facilities that have uneven daily production and are not located near a silage end-user. The United States has many plants, such as Alaska salmon canneries, that could benefit from a more efficient silage or waste process.

Three problems need to be solved to expand the silage process and make it available to more fish processing facilities.

1. Increased value for silage products.

2. Efficient water removal for transportation.

3. Simple process design.

Preliminary work on this problem initially focused on combining the oil with dried silage to produce high value nutrient dense feeds for aquaculture. Observations during this early work suggested that the silage process could be modified by using different amounts of acid and lower temperatures to produce high value products on a continuous or batch basis. The potential for increased revenue from dried high value silage feeds justified a closer examination of the overall process, including drying, to produce this material.

Material and methods

Fish waste

Numerous frozen and fresh fish materials were used for the study. Representative samples included whole frozen herring (*Clupea pallasii*), fresh Pacific whiting viscera, whole fresh Pacific sardine (*Sardinops sagax*), whole frozen Pacific whiting (*Merluccius productus*), whole frozen arrowtooth flounder (*Atheresthes stomias*), a variety of fresh and frozen waste viscera, and fresh and frozen egg stripped captive chinook salmon (*Oncorhynchus tshawytscha*). The chinook salmon were used in the later stages of this study and provided the most information in the refining of the modified silage process (MSP). Figure 1 shows a general material flow schematic for the modified silage process based on mature roe-stripped chinook salmon. The basis of the modified silage process is the use of less acid and low temperature. Mixing of the silage and mechanical separation of undigested material accelerates the process at low temperature.

Preparation of modified silages for feeding studies

Trial #1

Fresh whole Pacific sardines from Astoria, Oregon, and fresh Pacific whiting processing waste were chopped in a Hobart bowl cutter. Each was acidified with 25% sulfuric acid to a final concentration of 1.4% (w/w) and stored in buckets on ice for transport to the Northwest Fisheries Science Center in Seattle. The antioxidant ethoxyquin dispersed in water with TWEEN 80 (Sigma Chemical Co.) was added to samples at a level of 200 ppm. The temperature of the samples did not exceed 13°C during initial processing and the samples were stored at 3°C until final processing. In the following weeks the samples were processed following the general outline in Fig. 1. One percent NaOH was used to neutralize these samples before tray drying in forced convection at 75°C. Dried material was screened through 8 mesh sieve and stored at –10°C.

Trial #2

Eviscerated sablefish (*Anoplopoma fimbria*) hatchery mortalities and whole frozen arrowtooth flounder were ground and acidified with 1.4% sulfuric acid w/w and antioxidant added. Whole frozen Pacific whiting was thawed, ground through a 1.0 cm plate, and acidified with 1.1% w/w formic acid and antioxidant as above in trial #1. Frozen mature chinook salmon with eggs removed were ground and acidified with 1.1% formic acid. Further processing followed the general process shown in Fig. 1. The sablefish and arrowtooth flounder samples were tray dried as per samples in trial #1 above and blended together to form a high fat feed designated as BCAT. The Pacific whiting and salmon samples were drum dried using 80 pounds per square inch gage pressure steam.

Trial #3

The MSP feed for the third feeding trial described in this paper is the salmon that is defined above.

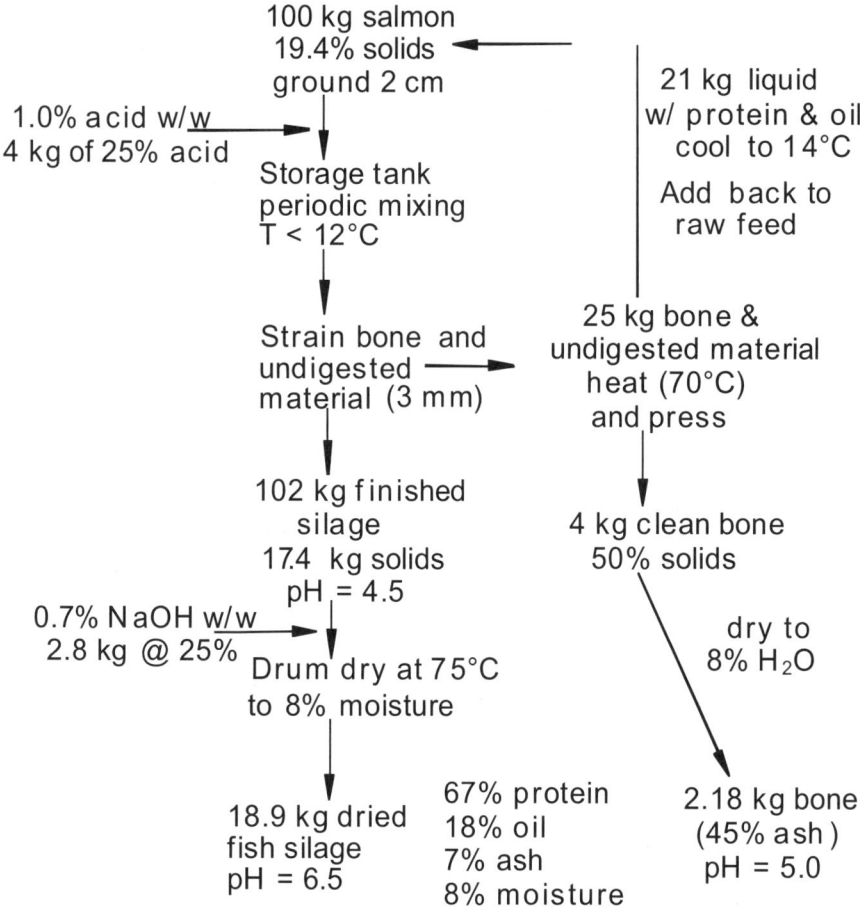

Figure 1. General material flow for modified silage process.

Feeding trials and feed formulations
Trial #1

In this feeding study sixty sablefish averaging 225 g were divided evenly into six tanks. Two tanks of fish were fed an experimental feed made from fresh Pacific sardine and two tanks of fish were fed feed made from Pacific whiting waste. Experimental feeds were made by blending the dry ingredients with 8% to 14% water and extruding at ambient temperature through 6 mm dies. Table 1 shows the ratios of dry ingredients used for all modified silage feed formulations. A third group of twenty fish were fed a commercial feed as a control. Fish were fed 3% of body weight five

Table 1. Experimental silage feed ingredients.

Dried silage material	977	g
Vitamin C	1.0	g
Choline (70% solution)	6.0	g
TM salt[a]	1.0	g
Vitamin mix[b]	15.0	g
Total	1,000	g

[a]USFWS, No. 3 (in Hardy 1989).
[b]USFWS, Abernathy (in Hardy 1989).

days a week. Weights were taken weekly over a five week period and feed adjusted weekly.

Trial #2

In this feeding trial sablefish of 100 g average size were divided among ten tanks with ten fish per tank. The four experimental feeds described previously and listed in Table 2 were used to feed two tanks of fish each. A control feed was used as a reference. The experimental feeds were made as described in trial #1 except a 4 mm die was used. Fish weights and growth were measured over a 10 week period. Fish rations were 3% of fish weights adjusted weekly.

Trial #3

Juvenile coho salmon (*Oncorhynchus kisutch*) averaging about 3 g were stocked in four tanks with 75 fish in each tank. Two tanks were controls fed a 2 mm commercial feed and two tanks were fed the experimental salmon feed left over from trial #2. The salmon feed from trial #2 was ground in a hammer mill and re-pelleted in a California pellet mill to 1.5 mm. Fish were fed to satiation five days a week. Fish weights and growth were measured over a 12 week period.

Chemical analysis

Proximate (moisture, protein, fat, and ash) and amino acid analysis were performed on all feeds used in trial #2 and trial #3. Moisture and ash were determined according to standard methods. (AOAC 1980) Fat was determined on a 1 g sample by supercritical fluid extraction using a LECO FA-100 Fat Analyzer. Protein was measured by determining nitrogen concentration in a 0.25 g sample by Dumas combustion methodology with a LECO FP-2000 Nitrogen Analyzer and multiplying the result by 6.25.

The Experiment Station Chemical Laboratory (ESCL) at the University of Missouri-Columbia performed amino acid analysis on the MSP salmon carcass. Amino acid analyses for the other MSP materials was determined by protein hydrolysis, pre-column derivatization with o-pthaldialdehyde (OPA) and 9-fluorenylmethyl chlorofomate (FMOC), and separation by HPLC using the Amino Quant method (HP, San Fernando, California). Use of trade names or companies in this publication does not imply endorsement of commercial products by the National Marine Fisheries Service.

Results and discussion

The course of this project evolved from the idea of producing silage feeds with added oil to re-defining the silage process and new methods to produce different products. The goal was to find product value and efficiencies that would make this waste handling process more available to the fish processing industry.

The object of the feeding trials was to establish a value for dried whole silage feeds. In the initial trial #1 the question was "Would the fish eat the feed and grow?" The fish weight gain for the feeds is shown in Fig. 2. From this data it can be seen that the experimental feeds had little or no negative effects on the fish.

Trial #2 was conducted to establish a feed conversion ratio to indicate potential value of MSP fish feed. The cumulative feed conversion ratio (FCR) is shown in Fig. 3. The FCR is the dry weight ingredients fed to fish divided by fish weight gain. Sablefish is a new species under study for commercial aquaculture development. This was one of the first formal feeding studies involving this fish. It can be seen in Fig. 3 that three of the MSP feeds performed equal or better than the commercial feed. The sardine feed had developed rancidity during eight months of storage and was not as acceptable to the fish. During this study there was a water quality problem that affected feed intake of all fish prior to the final weighing. Table 2 contains the proximate composition of the five feeds used in trial #2 and the whiting waste MSP feed from trial #1. The amino acid profiles for the dried MSP materials are in Table. 3.

Trial #3 focused on the best performing MSP salmon feed from trial #2. Figure 4 shows the dry weight feed consumption for the salmon MSP feed and the commercial control.

From observation in the feeding studies the control of the feeding environment, lighting, feed size, feed density, and other factors, also affected the feeding behavior of the fish. The main difference in MSP feeds was the density which determined the sink rate. All MSP feeds except the whole Pacific whiting feed from trial #2 sank quickly and the fish ate mostly on the bottom. The whole Pacific whiting MSP feed, made from poorly frozen fish, tended to float and subsequently sank slowly. Most of this feed was eaten on the surface or before it hit the bottom of the tank. Commercial feed had a slow sink rate and was eaten throughout the water column. In trial #3 the two tanks fed MSP salmon were near an overhead

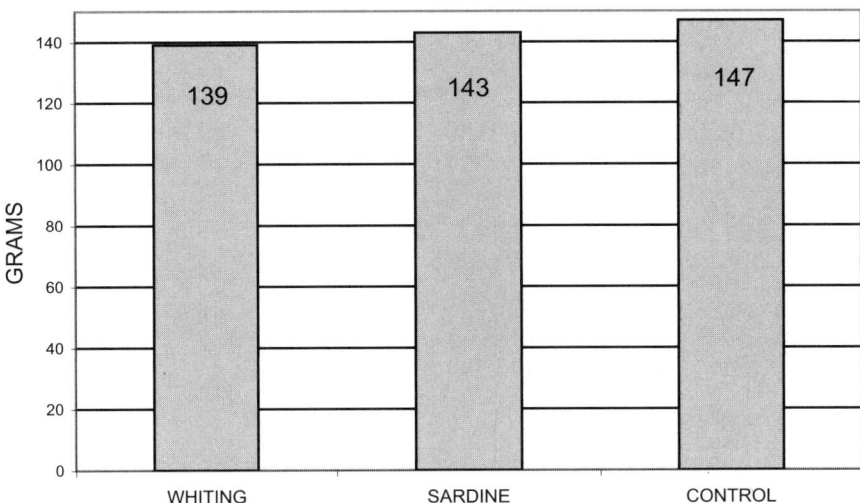

Figure 2. Average weight gain of sablefish in feeding trial #1.

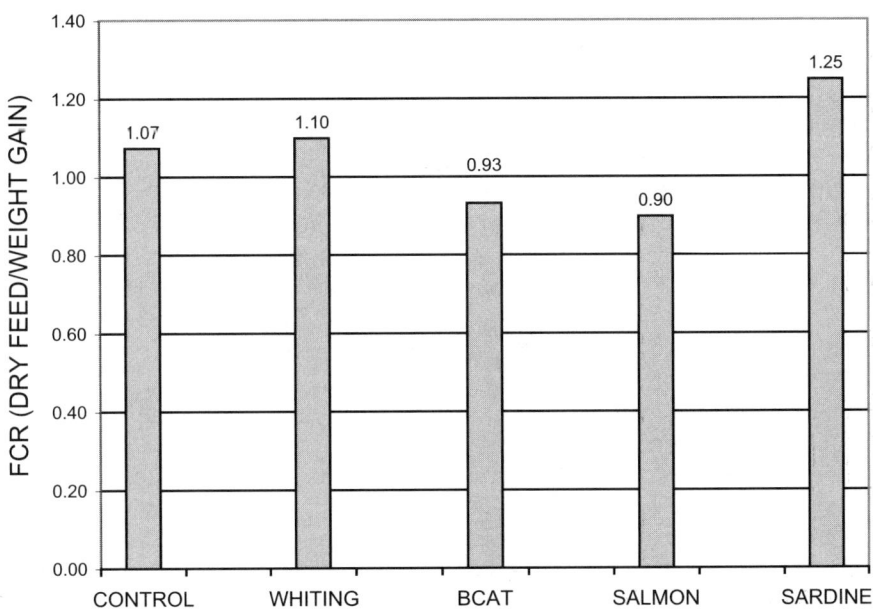

Figure 3. Feed conversion ratio for sablefish in feeding trial #2.

Table 2. Proximate composition of feeds used in trials #1, #2, #3.

Feed	Trial used in	Moisture content	Protein[c]	Fat	Ash	Carbo-hydrate[d]
Commercial control[a]	#1	22.0	59.0	19.2	12.8	9.0
Commercial control[b]	#2	20.7	59.5	18.8	14.2	7.5
Commercial control[b]	#3	20.5	58.9	18.5	13.8	8.8
Pacific whiting process waste	#1	16.2	56.8	30.3	12.9	
Whole frozen Pacific whiting	#2	14.7	62.3	25.2	12.5	
Whole fresh Pacific sardine	#1, #2	12.2	45.0	41.5	13.5	
Sablefish plus arrowtooth flounder (BCAT)	#2	8.3	46.9	40.8	12.3	
Spawned salmon carcass	#2, #3	16.2	63.8	20.3	15.9	

[a]From manufacturer.
[b]By analysis.
[c]Protein, fat, ash, and carbohydrate shown on dry weight basis.
[d]Carbohydrate by difference. No starch or binders added to MSP feeds.

light that eliminated shade when lids were raised for feeding. A board was placed over half the tank during feeding to give the fish a place (shadow) to hide. The MSP salmon feed sank quickly and much of it was eaten from the bottom of the tank. This was a problem later in the trial when the larger fish produced more waste which was periodically removed. The fish feeding on the bottom would stir up the waste and feed would be mixed or lost in the waste. The fish fed the commercial feed produced more visible waste that was attributed to undigested starch binders. This did not interfere with the commercial feed consumption and loss of feed in the bottom waste. The slow and consistent sink rate allowed the fish enough time to find and eat the commercial feed in the water column.

The basis of the MSP is low temperature and minimum amount of acid. The general rule for acid use developed during this study is based on one part concentrated (95%) acid for twenty parts dry weight solids. For example 100 kg waste containing 20% solids would be processed with a total of 1 kg concentrated acid. With mixing, this low acid was sufficient to solubilize all muscle and most soft tissue into a homogeneous paste at low temperature. The paste could then be strained out leaving bone with attached connective tissue, fins, some viscera, and heavy structures such as gill plates and snouts. This material contains a high percentage of connective tissue, protein, and some oil which upon heating melts and separates from the bone. Recovered protein and oil liquid can be added back to the main product stream (Fig. 1.)

Table 3.   Amino acid composition of MSP used in feeding trials.

	Salmon	Sardine	Whole whiting	Whiting waste	Arrowtooth flounder
ALA	6.78	5.35	5.89	5.97	5.37
ASP	9.09	6.62	10.87	9.96	10.43
CYS	0.97				
GLU	13.63	11.10	13.35	13.8	13.65
GLY	9.44	5.31	6.32	6.96	5.00
PRO	5.25	3.68	4.33	4.52	3.44
SER	3.62	3.44	4.71	5.09	3.93
TYR	2.89	2.87	3.09	3.97	3.00
Essential amino acids					
ARG	6.72	6.62	7.65	8.60	5.67
HIS	2.38	4.00	2.08	2.23	1.86
ILEU	4.08	4.18	3.09	4.54	4.07
LEU	7.02	7.22	7.00	8.26	7.13
LYS	7.49	6.51	6.97	7.32	7.43
MET	2.83				
PHE	3.83	3.95	4.05	4.44	3.79
THR	4.10	4.05	4.52	5.22	4.03
VAL	4.84	4.94	3.94	5.24	4.33
TRP	1.01				

Units are weight % protein.

The use of low temperature is based on the environment of the waste fish species and the melting point of their connective tissue (Sikorski et al. 1984). Fish such as cod, pollock, and whiting, have a connective tissue melting point around 12°C. Warm water species such as tuna and carp have a connective tissue melting point up to 27°C. When fish processed by MSP are held at or below their connective tissue melting point the protein breaks down more slowly. This keeps much of the oil distributed in the membranes and minimizes pooling and droplet formation. At lower temperatures proteolysis is reduced (Stone and Hardy 1986). The result is a high yield of high quality protein and oil. The oil is finely distributed in the digested tissue, so that when dried with the protein free flowing powder is formed. The end result of the MSP for fish waste is a nutrient-dense feed with low ash content.

From the feeding study results the MSP feeds performed similarly to the commercial control. In trial #3 the final FCR for the control was 0.83 and the final FCR for the MSP salmon was 0.86. Based on the composition and feed efficiency, the MSP feeds compare in quality to high grade

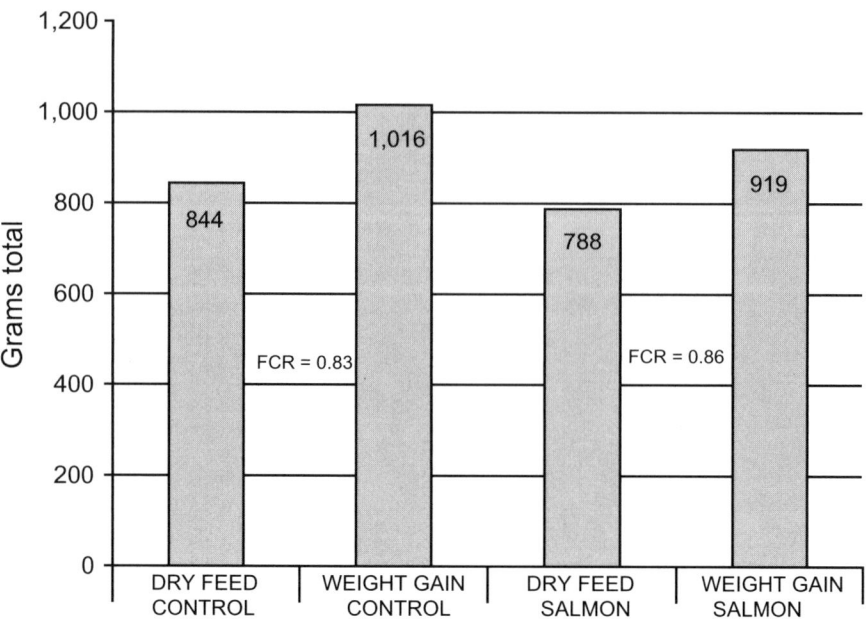

Figure 4. Total dry feed used and weight gain for juvenile coho salmon in feeding trial #3.

fish meal and could be used as the main ingredient in nutrient dense fish feeds. The value could range from $0.60 per kg for fish meal up to $1.40 per kg if manufactured into feed. Based on this information the following economic analysis was developed to examine the potential of the MSP for seafood wastes.

Economic analysis

Using the potential value of MSP material established above forms the basis for an economic analysis developed using the net present value (NPV) model (Valle-Riestra 1983). In this analysis all cash flows are discounted to time zero when the initial capital investment for the MSP is made. A desired return on investment is used to establish the risk of the project. If the NPV of all cash flows over the life of the project is positive then investment in the project is supported. Other variables are inflation, depreciation, and taxes. Cash flows include capital investment, cash expenses for operation, and sales revenue. The NPV model was applied to a hypothetical situation analogous to a seasonal salmon cannery. The daily capacity would be 30 t of raw waste material. Over a seventy day operating period a total of 1,500 t of raw material at 20% solids would be

Table 4. Net present value input for baseline estimate.

NPV	ROR	Capital investment	Cash expense	Revenue	Inflation	Depre- ciation	Tax rate
$5,200	0.08	$500,000	$50,000	$186,000	0.02	$100,000	0.4

processed into 300 t of MSP and 25 t of bone meal. All products would have 8% moisture. With sales of $600.00 per ton for MSP and $250.00 per ton for bone, the revenue is $186,250. Shown in Table 4 are the values used to establish a reference NPV for this project. For this case the NPV is positive. This project would meet the 8% return risk for investment.

In Fig. 5 the limit for capital investment is approximately $525,000. This is an analysis where the cash flow variable is changed with all other values held constant. For an existing fish processing facility that has available space, utilities, and support, the capital investment could be reduced below $525,000. This would make the investment more attractive and risk averse.

In Fig. 6 the annual limit for cash expenses is approximately $55,000. This cost could be reduced by operating with an existing facility that could provide labor, utilities, buildings, administrative, and sales support at lower cost.

The revenue of $186,000 from sales of MSP material as fish and bone meal meets the hypothetical NPV investment. This is shown in Fig. 7 where the revenue intersects the NPV value of zero. The revenue could be increased in two ways. Additional investment for pelleting MSP material for the feed market could double the value. The other revenue potential is charging for waste disposal. Both of these revenue enhancements could be used in a seasonal Alaska salmon cannery.

High quality feed could be sold to fish hatcheries. Currently all hatchery feed must be shipped to Alaska adding transportation and handling expense. Alaska fish hatcheries have excess fish above the amount required to provide eggs for enhancement programs. Selling eggs from these excess fish would help the finances of many hatcheries; however, it is unlawful to waste salmon flesh in Alaska. Producing high value MSP products could help solve the problem of disposal of carcasses at many hatcheries. Disposal costs would support an MSP business and also help hatcheries utilize excess salmon eggs and carcasses. In addition to the value or cost of fish waste as a disposal service there are the situations where waste is a cost of doing business. For fish processors where grinding and pumping waste back to the sea is being restricted, the rate of return on investment for adding an MSP operation may be acceptable at 0% if the company is not shut down or fined for waste discharge violations. A

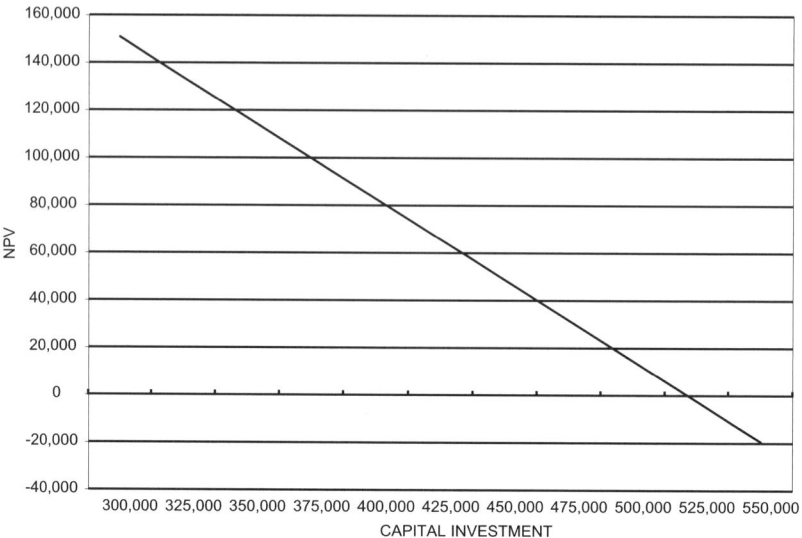

Figure 5. Net present value (NPV) versus capital cost variable. The cash flow variable changed while all other values are held constant.

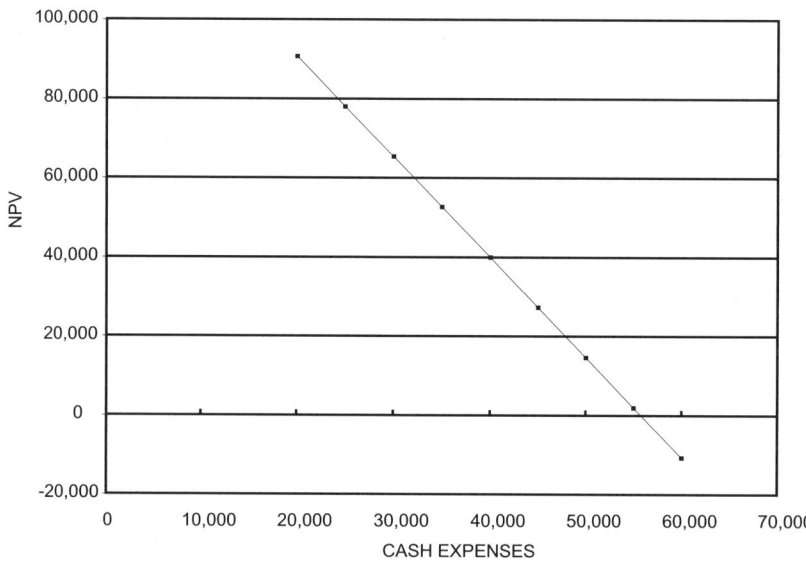

Figure 6. Net present value (NPV) versus cash expenses.

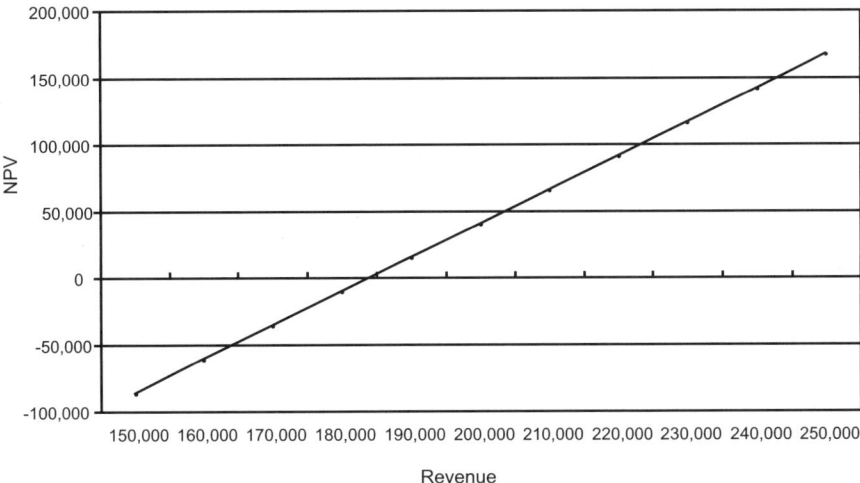

Figure 7. Net present value (NPV) versus revenue.

similar case is processing plants that must transport waste to a disposal facility at a cost, pay for disposal, or both.

A final advantage to the MSP is 100% recovery of fish material to products. There is no continuous discharge of streams with high BOD (biological oxygen demand) or COD (chemical oxygen demand). The acid eliminates much of the odor associated with the spoilage and processing of fish offal. In a well operated MSP plant the only discharge would be water vapor from drying of products. The MSP could possibly be used in populated areas where the odor and discharge are problems.

Conclusion

The three problems for improving the silage process stated in the intro-duction were addressed in this study.

1. Increased revenue for silage products. Overall revenue is enhanced by creating high value products and also total recovery of all solids entering the process. The ability to keep all oil and protein solids in the MSP feed stream is an advantage over other processes that separate oil and protein streams. The high feed conversion ratio and NPV project evaluation indicate that the modified silage process has the potential to economically solve fish waste problems for more processors.

2. Efficient water removal was demonstrated by the ease of drum dry-ing MSP material. Short-time flash drying is important in maintaining the high nutritional value of the protein and oil.

3. The simple process design of MSP minimizes capital investment and cash expenses. Acid and base use is reduced by up to 60% over other silage processes. Cooling for the process can come from local seawater or a small refrigeration package. With acid stabilization the process can be run as batch or continuous depending on time of season and material flow.

Proper formulation of feeds made with MSP material needs more work. Although the MSP feeds performed well there could be further per-formance gain with proper density, size, color, and flavor attractants.

References

AOAC. 1980. Official methods of analysis. 13th edn. Association of Official Analyti-cal Chemists, Inc., Arlington, Virginia.

Bimbo, A.P. 2000. Fish meal and oil. In: R.E. Martin, S.P. Carter, G.J. Flick, and L. Davis (eds.), Marine and freshwater handbook. Technomic Publishing Co., Inc., Lancaster, Pennsylvania, pp. 540-581.

CEC. 2002. A strategy for the sustainable development of European aquaculture. Commission of the European Communities, Communication from the Commis-sion to the Council and the European Parliament, 19.9.2002. Brussels.

Hardy, R.W. 1989. Diet preparation. In: J.E. Halver (ed.), Fish nutrition. Academic Press, San Diego, pp. 475-548.

Johnson, E.L., R.H. VanHaagen, and R. Stokes. 1982. Utilization of fish wastes as fish silage. U.S. Dept. of Commerce contract 81-ABH-00044.

Sikorski, Z.E., D.N. Scott, and D. Buisson. 1984. The role of collagen in the quality and processing of fish. Crit. Rev. Food Sci. Nutr. 20:301-344.

Stone, F.E., and R. Hardy. 1986. Nutritional value of acid stabilized silage and liqui-fied fish protein. J. Sci. Food Agric. 37:797-803.

Valle-Riestra, J.F. 1983. The present worth of cash flows. In: Project evaluation in the chemical process industries. McGraw-Hill Book Company, New York, pp. 87-90.

Process Accounting (PA) Applications to Milkfish Processing

Jose P. Peralta
University of the Philippines in the Visayas, Institute of Fish Processing Technology, Iloilo, Philippines

Abstract

"Process accounting" is a method that define the maximum process streams suitable for the conversion of raw materials to products, with the end result of increasing utility of the raw materials, minimizing losses as wastes, and implicitly mitigating and protecting the environment. The converted products are further clustered into upstream products and downstream products, defining its appropriate utility values. Upstream products are the main products, and have high utility values, while downstream products are those with relatively low utility values. Process accounting, as a management tool, is being advocated as a protocol in the design of seafood processing operations.

This paper discusses process accounting concepts and its application to milkfish (*Chanos chanos*) processing. The method defines the utility of the fish and its parts. Whatever residue the conversion process makes, utilities of the residue are also defined. Since water is also involved in the process, wastewater utility is defined. It is interesting to note that if all of the defined products/utilities are technically feasible and economically viable, waste materials or throwaway materials are minimized. Even the end of line, solid and liquid wastes from processing plants can be utilized.

Introduction

"Process accounting" (PA) are activities that define processes suitable for the conversion of raw materials to products with the end result of

- Increasing utility of the raw materials.

• Minimizing losses as wastes.

• Implicitly mitigating and protecting the environment.

PA should be defined before a processing plant is established. This method was introduced by Peralta in 1999 in the lecture "Process accounting: An alternative paradigm," an R.S. Benedicto Professorial Chair Lecture on Post Harvest Fisheries, as an alternative management tool in the design of fish processing operations.

Normally, fish processing plants are designed to process a single fish species, with one major product. Tuna canning plants are designed to produce canned tuna. Some of the byproducts of such processes are often overlooked, sometimes even ignored. The PA intent is to provide as many product and process streams as possible that could be made viable. This paper discusses application of PA initiatives in defining product and process streams in milkfish (*Chanos chanos*) processing that could be worth considering.

World milkfish production

FAO Fisheries Statistics (2001) have shown that milkfish is one of the important cultured fish species. World milkfish production started low in 1985 at 311,879 metric tons valued at US $351,675,000 (Fig. 1). Production improved the next year but went down the following year. In 1990 it peaked at 434,090 t, valued at US $626,748,000, but went down again the following year, and progressively dropped to 342,321 t, valued at US $604,203,000 in 1992. Production climbed the following year and continued to climb in 1994 at 400,845 t valued at US $727,211,000. Then it went down again the following year. In 1997, from 367,286 t valued at US $660,464,000, it continued to climb to an all time high of 461,857 t valued at US $715,091,000 in 2000.

Philippine milkfish production

In the Philippines, milkfish is also one of the most important cultured fish species. In 1994, milkfish production was at 135,682 t valued at Philippine peso P 7.1 million (US $142 million). It progressively climbed to 170,677 t valued at P 13,654 million (US $273.08 million) in 1999 (Fig. 2).

Milkfish are cultured in brackish fishponds, fish pens, and cages in lakes and coastal areas. There are about 222,907 ha of brackish fishponds throughout the Philippines.

Philippine export trade of milkfish

Total Philippine milkfish exports: the aggregate of frozen, fresh/chilled, whole and smoked milkfish in 1994 was at 122 t (US $582,000). It reached a peak of 254 t (US $888,000) in 1995 but went down to 163 t (US $374,000) the following year. Exports were steady at 204 t (US $518,000) in 1998,

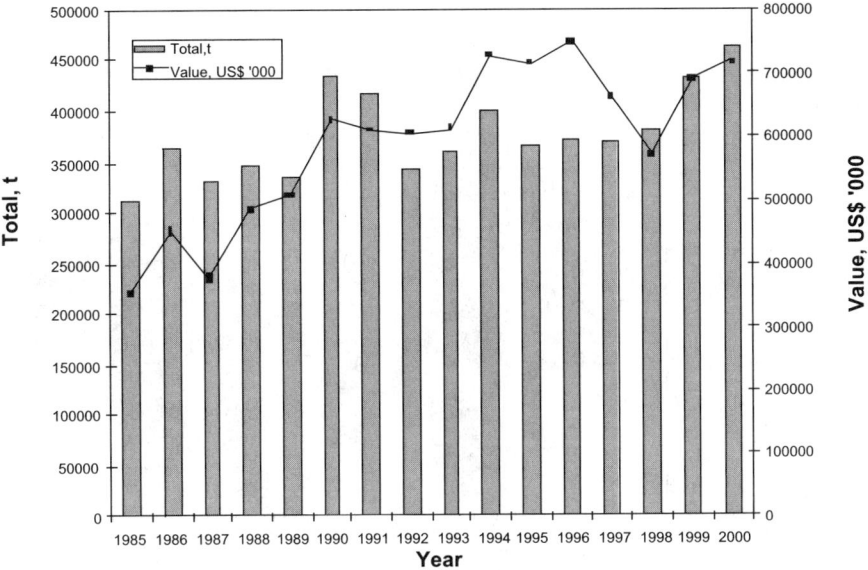

Figure 1. World milkfish production, 1985-2000. Source: FAO 1996, 2001.

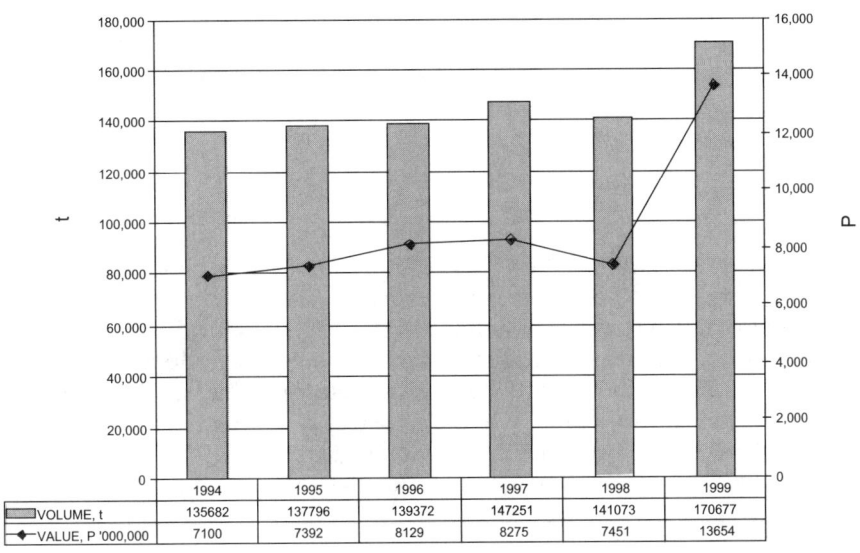

	1994	1995	1996	1997	1998	1999
VOLUME, t	135682	137796	139372	147251	141073	170677
VALUE, P '000,000	7100	7392	8129	8275	7451	13654

Figure 2. Philippine milkfish production, 1994-1999. P = Philippine peso, PSO ~ US $1. Source: Univ. of Asia and the Pacific 2000.

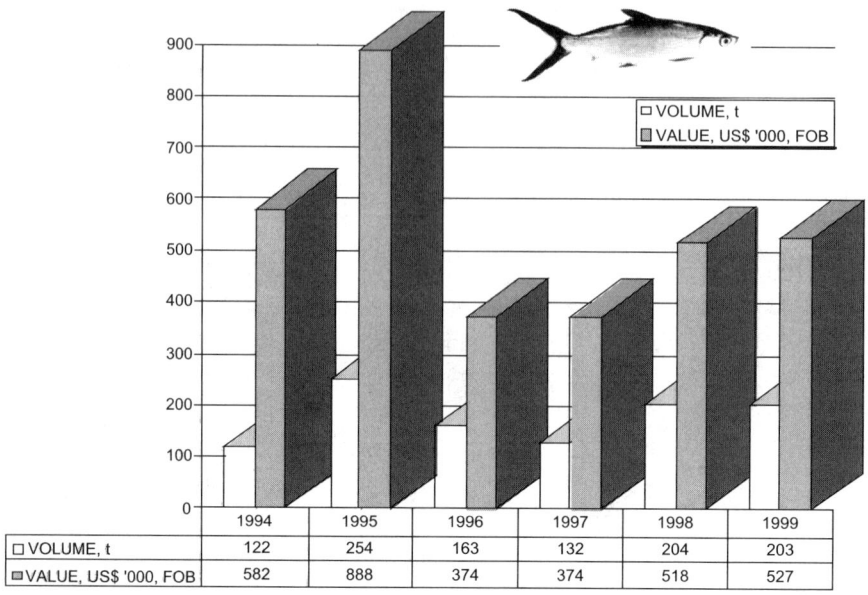

	1994	1995	1996	1997	1998	1999
□ VOLUME, t	122	254	163	132	204	203
▣ VALUE, US$ '000, FOB	582	888	374	374	518	527

Figure 3. Milkfish exports: frozen, fresh/chilled, whole, and smoked, 1994-1999. Source: Univ. of Asia and the Pacific 2000.

and 203 t (US $527,000) in 1999 (Fig. 3). Major destinations for frozen milkfish included Canada, Bahrain, United States (mainland), Hawaii, and Guam (Fig. 4).

Methods

The concept

Process accounting is a method that defines as many product conversion streams as possible in a production process, thereby minimizing the accumulation of wastes. It also defines the utility value of the products by clustering it as upstream products and downstream products (Fig. 5).

Upstream products are normally the main products, and products that have high utility value, while downstream products are those with relatively low utility value. Figure 6 emphasizes the limits of the concept's application. The product stream will be pursued only if profit potential is acceptable.

Applications to processing of milkfish

Milkfish is normally sold as fresh fish. At times, it is minimally processed: deboned and marinated. Some small entrepreneurs went further and

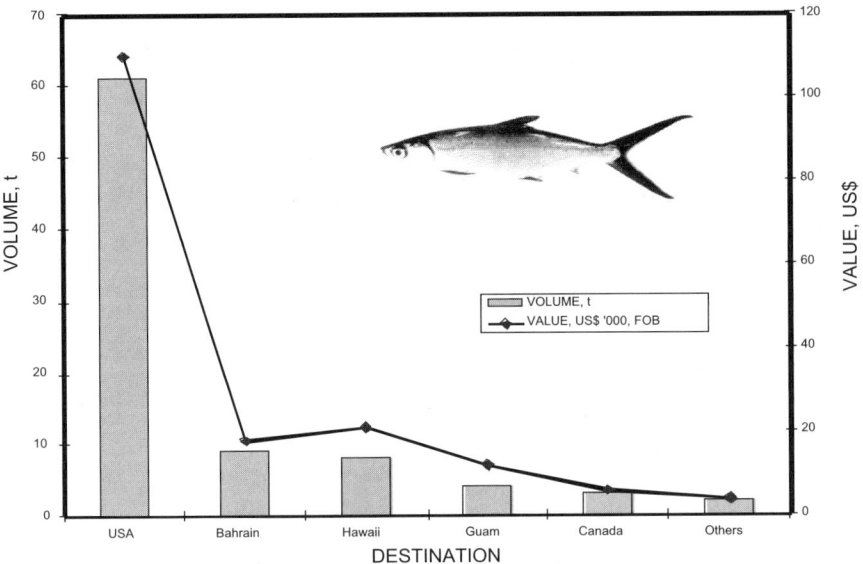

Figure 4. Frozen milkfish exported to country of destination. Source: Univ. of Asia and the Pacific 2000.

converted and processed the fish into smoked, dried products. Others converted the fish into other value-added products such as rolls, skin crackers, cured as ham, etc. Some large and small companies also processed milkfish into bottled and canned products. Figure 7 illustrates some of the value-added products developed.

The milkfish is not perceived by processors as raw material with functional parts, similar to chicken. With this change in paradigm, the whole milkfish may be transformed into several functional parts, with different utilities.

The fish head is chopped off and considered one of the unwanted parts in most industrialized countries. In some Asian countries such as the Philippines, the head is one of the favorite fish parts. If the head is chopped off, with some meat below the collar, and several heads placed in a styrofoam tray and then shrink wrapped, the packages could be sold chilled or frozen for fish soup. In the Philippines, sinigang, a fish soup with tamarind or lemon and vegetables, is a local delicacy. There are prepared vegetable packages (minimally processed) in the wet market or grocery store for this.

The fillets could be deboned and sold as such. Sometimes they are placed in styrofoam packages and shrink wrapped. An alternative is to dip the fillets in batter, then fry and pack frozen. Milkfish is a very spiny

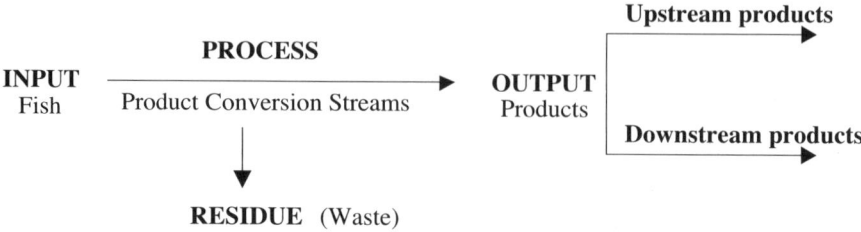

Figure 5. Fish processing flow. Source: Peralta 1999.

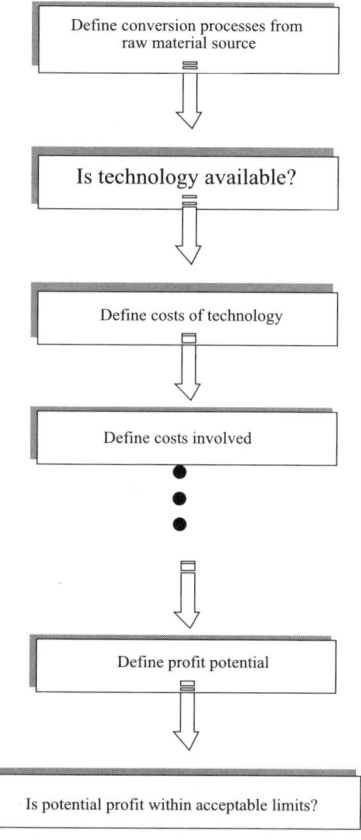

Figure 6. Limits of application of the Process Accounting concept. Source: Peralta 1999.

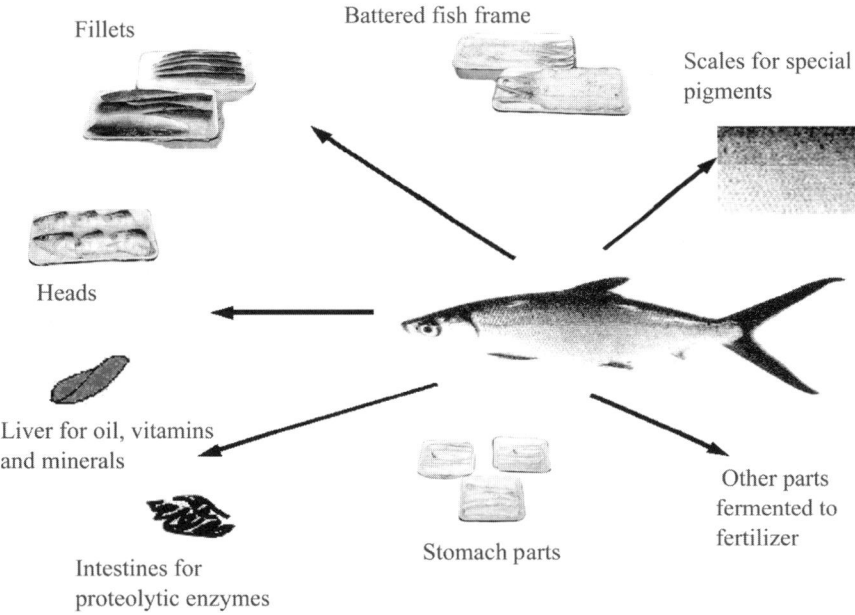

Fillets

Battered fish frame

Scales for special pigments

Heads

Liver for oil, vitamins and minerals

Intestines for proteolytic enzymes

Stomach parts

Other parts fermented to fertilizer

Figure 7. Utility of milkfish parts (*Chanus chanus*).

fish, and it takes a lot of eating skills to consume without having the bones removed first.

The belly flaps could be separated and sold as special parts. This would cost more than the fillets because in the Philippines, belly portions are preferred over fillets. Belly flaps could also be placed in styrofoam packages and shrink wrapped.

The skeletal frames, with some meat intact, are normally fed to cats, or simply thrown away. However, if these are pressure cooked, then dipped in batter and deep fried, it becomes a crispy treat. This product is sold as fish flakes in some Philippine restaurants. As a consumer product, the pre-battered, ready to fry products are placed in styrofoam packages, sealed in polyethylene pouches and frozen.

The liver portion is also segregated. If you consider 20 tons of milkfish at a time, the liver portion becomes considerable. Fish oil could be extracted from the liver. The oil is rich in omega-3 and -6 fatty acids. Also, it contains vitamins and minerals. The meal residue, which is high in protein, could still be used as an ingredient in pet food formulation.

The intestines could be processed for proteolytic enzymes. These proteolytic enzymes have industrial applications, e.g., in the manufacture of detergents, tanning of leather, etc. The other parts—gills, internal organs, and others—could be processed into fish meal. Considerable protein and calcium are in this meal.

The fish scales could become an ingredient in pearl-like paints. Some multinational paint companies have found a special chemical carrier for these fine pigments. Application of these special paints includes plastic, paper, metal, and wood products. Solid parts that could not be converted into fish meal could be bio-digested into organic fertilizer. Figure 7 shows utilization of fish parts.

How about the washings or the waste water? Waste water has a high organic load, which is mostly soluble protein. It is for this reason that most government regulatory bodies prohibit dumping of effluent wastes from processing plants into nearby bodies of water.

Initial work has focused on converting liquid wastes into organic liquid fertilizer. The goal of the initiative is to make the conversion into fertilizer as cheap as possible. There is considerable demand for organic liquid fertilizer by the orchid and cut-flower industry. The fertilizers, in both liquid and solid forms, are considered as downstream products.

In the culture of milkfish in the Philippines, inorganic fertilizer is used in the pond to produce an algal bloom. The local Philippine term for it is lablab. This zooplankton/phytoplankton–rich bloom serves as food for the milkfish fingerlings. The solid and liquid fertilizer made from milkfish processing wastes could be used back to the ponds to produce the algal fish foods for the culture of milkfish. Ergo, the cycle continues.

Conclusion

Process accounting is a management tool. It provides the process designer a broader perspective on alternative processes that could be implemented. Furthermore, it provides the process designer an opportunity to cluster upstream processes (main processes with products of high utility value) as differentiated from downstream processes (products of low utility value). Thus, emphasis and focus could be made on products of high utility.

The ultimate goal of PA is zero waste. The method directly addresses other environmental issues and concerns relative to fish processing operations. PA is being advocated as a protocol in the design of fish processing operations.

As applied to milkfish processing, the method defines the utility of the fish and its parts. Whatever residue the conversion process makes, utilities of the residue are also defined. Even utility of end of line waste water is defined.

It is interesting to note that if all of the defined products/utilities are technically feasible and economically viable, then waste materials are minimized. Even the end of line, solid and liquid wastes can be used to make fertilizers, which can be used back for milkfish aquaculture.

References

Dept. of Agriculture. 1999. 1999 Philippine fisheries profile. Bureau of Fisheries and Aquatic Resources, Dept. of Agriculture, Manila, Philippines.

FAO. 1996. Aquaculture production statistics. FAO Fish. Circ. 815, Rev. 8, 1985-94.

FAO. 2001. Fisheries statistics, FAO. http://www.fao.org/fi/statist/fisoft/FISHPLUS.asp.

Peralta, J. 1999. Process accounting (PA): An alternative paradigm. A Roberto S. Benedicto Professorial Chair Lecture on Post Harvest Fisheries, U.P. Visayas, Miag-ao, Iloilo, Philippines. Unpublished.

Univ. of Asia and the Pacific. 2000. Food and yearbook and directory, M.Ed. Agribusiness, Univ. of Asia and the Pacific, Manila, Philippines.

Obtaining Human Food from Whole Underutilized Fish

A. Gelman
Kimron Veterinary Institute, Bet Dagan, Israel

U. Cogan and S. Mokady
Technion-Israel Institute of Technology, Haifa, Israel

V. Drabkin and L. Glatman
Kimron Veterinary Institute, Bet Dagan, Israel

Abstract

The annual worldwide fish catch still falls short of supplying the animal protein necessary for feeding the growing human population. Nevertheless, many ecological niches are filled by small underutilized fish species that weigh between 5 and 30 g, which cannot be processed economically with the present-day technologies; therefore large amounts of protein are not exploited. We have developed to the pilot-plant level the technology of a process for preparing mince out of small fish: Kinneret bleak (*Mitrogrex terra-sanctae*), a freshwater fish native to Israel, and sardinella (*Sardinella aurita*) and European sardine (*Sardina pilchardus*), both of which are typical pelagic fish.

The procedure is based on a physico-chemical process that enables mince to be obtained from small fish without previous preparation. In tests the yield varied between 60 and 65% of wet material, and the minces contained 16-19% protein. The minces obtained could be kept frozen, provided that antioxidants and vacuum packing were used. From the minces were made diverse traditional products (frankfurters and fish balls) as well as nontraditional products (fish chips, noodles, and ravioli). The sensory grade for these products ranged between 5.3 and 9.6 on a 10-point scale.

Introduction

Nowadays fish supplies for human consumption exceed 100 million metric tons annually; these fish are mostly caught in the world oceans and in internal water bodies, or obtained by aquaculture (FAO 2000). Nevertheless, many ecological niches are filled by small fish species weighing between 5 and 30 g, which cannot be processed economically with the present day technologies; they are therefore underutilized and thus large amounts of fish protein are not exploited. Meantime, there is an urgent need to increase the consumption of aquatic protein in order to alleviate the problem of undernutrition in developing countries (Bello 1989, Bertullo 1989, Yang 1999). The world's consumption of fish could be more than doubled if presently underutilized or unused resources were brought into the human food chain (Grantham 1981).

Only a small part of the small-fish catch is processed into canned or smoked products, or is sold at local market places. The remaining small fish are mainly used in manufacturing products for animal feeding, such as fish protein concentrate, fish meal, and fish hydrolysates.

Present day technologies for the production of fish mince usually include the steps of cutting, removal of the head, and gutting. The remainder of the body is subjected to deboning, i.e., mechanical separation of the meat from the bones. This process is economical only for relatively big fish, above 100 g; it is rather impracticable for small fish, because of the excessive manual labor inherently involved. The inherent instability of the underutilized small fish, which is due to the relatively high rate of autolysis and rapid bacterial spoilage, and the high catch rates achieved, necessitate fast, mechanized systems for optimal utilization. Mince technology is seen by many workers to offer the best potential for increased exploitation. Major species that have been studied for their potential for mince production from whole fish include the mackerels (Lee and Toledo 1976, Bussman 1977), herring (Stodolnik 1979), sardines (Ishikawa 1978), the oily, non-pelagic capelin (Strom 1980), etc. In addition to mechanical deboning systems, several chemical and biochemical techniques have been developed (Rasekh and Metz 1973). Less work has been done on preparing minces from tropical and sub-tropical species. However, of all the fish resources considered for mince production, the small pelagics seem to present the most intractable problems. These problems arise mainly from the high, but variable, levels of polyunsaturated fats, the effects of fat degradation products on taste and texture, and the contamination of minces by the highly proteolytic gut contents (Grantham 1981).

Preliminary studies by Gelman et al. (1985, 1989) showed that production of mince from whole small pelagic fish without preliminary preparation might be feasible. The process was based on exploiting the endogenous enzymes of fish to facilitate subsequent separation of the meat from undesirable tissues. However, these studies addressed only a single fish species, and no scale-up experiments were carried out.

The basic objectives of the present study were development and scaling up of the technology for mince production from whole small fish without preliminary dressing, to study the storage characteristics of the frozen mince, and to obtain nutritious, inexpensive food products for human consumption, based on the mince.

Materials and methods
Fish
Kinneret bleak (*Mitrogrex terra-sanctae*), of the family Cyprinidae, is a small, sardine-shaped freshwater fish, native to Lake Kinneret (Sea of Galilee), and is the most common species in the lake. It is a schooling, pelagic fish, generally up to about 14 cm in total length, silvery colored and with a strong skeleton. The fish were caught with a purse seine, with weight 10-25 g. The fish were transferred to the laboratory in sealed plastic bags in a Styrofoam box containing sliced ice, and were stored at 0-2°C pending processing. The elapsed time from fish packing to arrival at the laboratory was 2 h. Sardinella (*Sardinella aurita*), weight 15-40 g, were harvested in Haifa Bay and were transferred to the laboratory as described for Kinneret bleak. Sardine (*Sardina pilchardus*), weight 25-90 g, were harvested off the northern coast of Mauritania, and were frozen and stored in blocks of 15 kg each, at −18°C for 2 months.

Analyses
1. Proximate composition
Analyses of proximate composition (total protein, oil, ash, and moisture) were carried out as described in the Official Methods of the AOAC (Cunniff 1995). All samples were analyzed in duplicate.

2. Salt solubility measurements
A modification of Dyer's method was used for the determination of salt-soluble nitrogen (SSG). A 2-g sample of fish flesh or mince was homogenized in 48 ml of a 5% NaCl solution (pH 7.0-7.5) at 0°C for three 1-min. periods, with 2-min. intervals. The homogenate was centrifuged at 12,780 g for 20 min. at 0°C in a Sorvall Superspeed RC2-B refrigerated centrifuge, and the protein content of the supernatant liquid was assessed according to the Kjeldahl method.

3. Thiobarbituric acid (TBA) value
Malonaldehyde was determined according to Vyncke (1970). Absorbance at 538 mm was read on a Beckman spectrophotometer.

4. Total volatile basic nitrogen (TVB-N)
TVB-N was determined with a Kjeltec 1026 (Sweden). Ground samples (2 g) to which magnesium oxide (2 g) was added were placed in the distillation

tube of the Kjeltec distilling unit (Prabin and Co. 1987). The tube contents were boiled and distilled in 25 ml of boric acid in the presence of indicators (a mixture of methyl red and bromcresol green). The distillate was collected in a receiver flack, filtered with 0.01 M sulfuric acid, and expressed as grams of nitrogen per kilogram of fish.

5. Organoleptic analysis

Organoleptic analysis of the fish mince–based products was performed by evaluating four characteristics (taste, smell, texture, and appearance) utilizing a 10-point scale. The higher scores denoted better quality. A panel comprising 10 trained and experienced tasters performed the tests; the members comprised women and men, 25-50 years old. The allocated scores were averaged over all tasters.

6. Bacteriological analyses

Total aerobic bacterial counts (mesophilic and psychrophilic bacteria) and coliform bacteria were determined. The mince samples, weighing 10.0 g, were homogenized in a Stomacker 400 (Seward Laboratory, UAC House, U.K.; model BA 6021) for 1 min. in 90 ml of sterile peptone (0.1% w/v). Tenfold serial dilutions of the same diluting medium were plated on plate count agar and MacConkey agar (Difco, Detroit, MI). The plates were incubated at 30°C for 2 days for counting mesophilic bacteria and at 10°C for 7 days for counting psychrophilic bacteria.

Results

1. Fish mince preparation

Preparation of fish mince was carried out according to the scheme shown in Fig. 1, with some variations according to the species involved. Pilot-plant runs were conducted with 20-30 kg fish in a single operation. The fish were partially thawed and cut into slices (1-3 cm thick) which were rinsed in cold tap water, so as to remove most of the viscera. Then the fish were placed in the reactor with an equal weight of water, and the pH of the mince was adjusted to the desired level (4 for Kinneret bleak and 4.5 for the other species) by acetic acid addition. The incubation temperatures, suitable for cathepsines activation, were 40, 35, and 30°C for Kinneret bleak, sardine and sardinella, respectively. The incubation was continued for 40 min., and the fish mass obtained was then suitable for the grinding and separation steps. The finisher (Langsenkamp, Model 185-SL) separated the meat from the bones, scales, and fins. Sodium bicarbonate was added to the fish mass to adjust the pH to 6.5, and the mass was placed in the cheesecloth bags for water removal. Then the mince was whitened by adding an aqueous solution of 0.6% hydrogen peroxide (H_2O_2) for 10 min. at 30°C. The quantity of H_2O_2, and the time, temperature, and pH used for mince whitening had been determined in

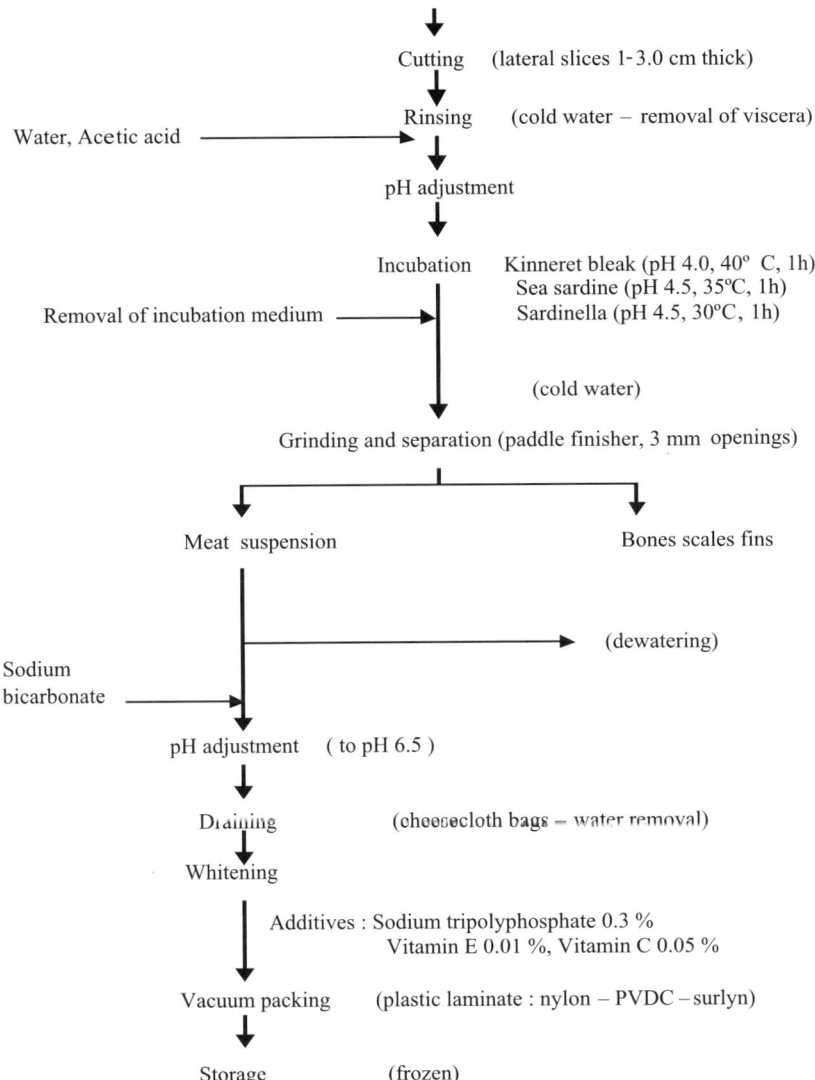

Figure 1. Flow sheet for pilot-scale production of mince.

preliminary experiments, which showed that the addition of 0.6% H_2O_2 for 10 min. at 30°C resulted in 25% whitening, independently of the pH (Fig. 2). The degree of whitening was measured in a HunterLab Model D25A-2 Colorimeter (HunterLab, Reston, VA, USA). Residual H_2O_2 was determined as described in the Official Methods of the AOAC (Cunniff 1995).

Some food additives were incorporated into the mince before packaging: sodium tripolyphosphate (0.3%), and vitamins E (0.01%) and C (0.05%) which acted as antioxidants in the lipid and aqueous phases, respectively. The resulting mince was vacuum packed in a plastic laminate comprising three layers of nylon, polyvinylidene chloride, and surlyn. The resulting product was a light gray homogenous paste, of neutral odor and taste. The mince was stored at –18°C. The Kinneret bleak mince was used for preparing fish balls, fish balls with sesame, and noodles; the sardinella mince was used for preparing fish chips, frankfurters, and ravioli.

2. Yield

The yields of fish minces obtained and their chemical compositions are given in Table 1. The results are expressed as percentages of the whole fish. The pilot-plant operation resulted in rather high yields: 60-65% of the wet weight of the whole undressed fish. The protein contents of the minces were very similar to the initial contents in all fish species. The fat contents of all the minces were similar to one another, irrespective of the initial fat contents. The ash contents in the minces decreased 1.5-2 fold.

3. Storage stability

The stored mince was sampled and assayed for weight loss, soluble protein content, and chemical and microbial parameters at given time intervals.

The weight loss in all three fish species was 30-35% at the beginning and 40-45% in the end of 200 days of storage (Fig. 3). The soluble protein decreases during storage were also similar in all species, though the levels at the beginning of storage differed (Fig. 4).

Results of TVB-N measurements are presented in Fig. 5: the initial levels of TVB-N in all kinds of mince were low (0.8-1.1 g per kg) and hardly changed by the end of storage (0.9-1.2 g per kg). The fat rancidity developed differently in all three fish species. The TBA values hardly changed in Kinneret bleak mince during storage (around 0.2 mg per kg), and changed insignificantly (from 2 to 2.5 mg per kg) in sardinella mince. The initial TBA level of sardine mince was 6 mg per kg, and it rose to a significant 10.4 ± 1.0 mg per kg at the end of storage (Fig. 6). No residual H_2O_2 was detected in the mince after treatment.

The minces of the three fish species studied had relatively low initial bacterial total counts, comprising psychrophils and coliforms. Total counts of aerobic mesophilic bacteria (Fig. 7) and psychrophils hardly changed during storage. The populations of coliform bacteria decreased

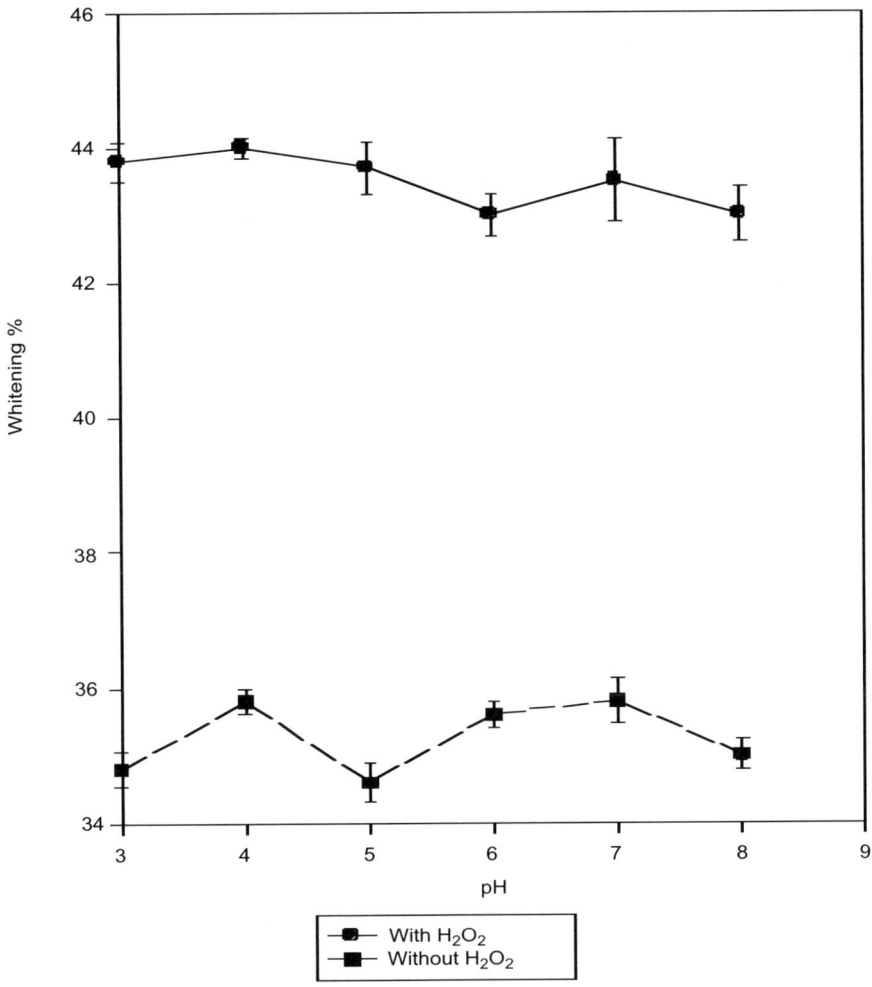

Figure 2. Effect of pH on whitening of mince.

Table 1. **Yield and proximate composition of raw and minced fish.**

	Dry matter %	Protein %	Fat %	Ash %	Yield %
Kinneret bleak					
Whole	26	16.6	5.0	3.5	100
Mince	24	17	4	2	65
Sardine					
Whole	37	16.6	16	4.2	100
Mince	23	16	5	2	60
Sardinella					
Whole	29.1	19.5	5.1	4.55	100
Mince	23.8	19.0	3.6	1.2	62

to 10-15% of the initial value (10^2 CFU per gram) in all kinds of mince, at the end of storage.

4. Products

Diverse products were prepared from the mince: fish chips, frankfurters, fish balls, fish balls with sesame, noodles with 20% fish mince added, and ravioli filling (Fig. 8). The results of organoleptic evaluation are shown in Table 2. The highest scores were awarded to noodles with 20% fish mince added: 7.1-9.6 according to the parameters used. The cooked product was characterized by piquant taste. The ravioli filling consisted of 80% mince, and the organoleptic properties of the ravioli were evaluated rather high (scores of 7.8-9.1). Chips containing 48% of mince, and fish balls containing 44-60% of mince also received rather high scores (5.3-8.5 and 6.5-8.7, respectively). Frankfurters and fish balls with sesame, both containing 55-60% mince, were evaluated lower, with scores of 6.5-7.2 and 6.3-7.9, respectively.

Discussion

Present-day technologies for obtaining fish mince require that the fish be headed, gutted, and split, with the main portion of the backbone removed, prior to mechanical separation. The need for preliminary dressing is the main reason for the underutilization of small pelagic fish, and the result is that large amounts of fish protein remain unexploited. Through the application of appropriate mince technology, it should be possible to

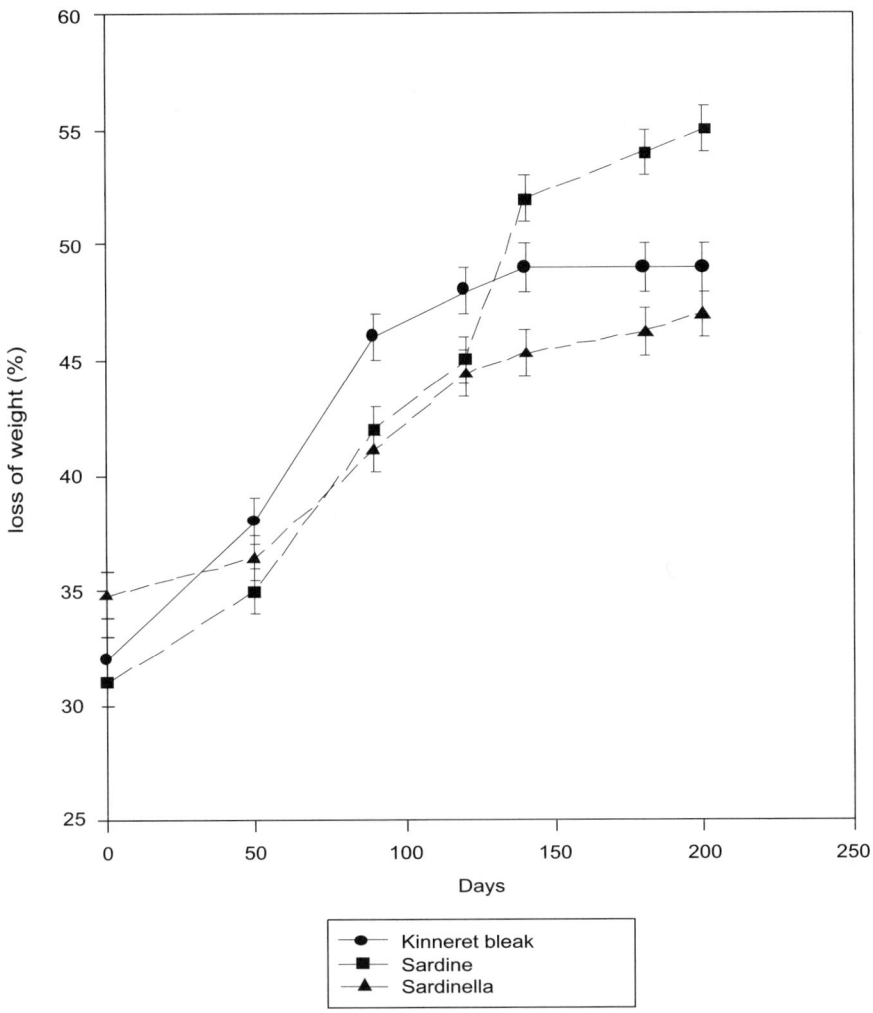

Figure 3. Weight lost from the minces during storage.

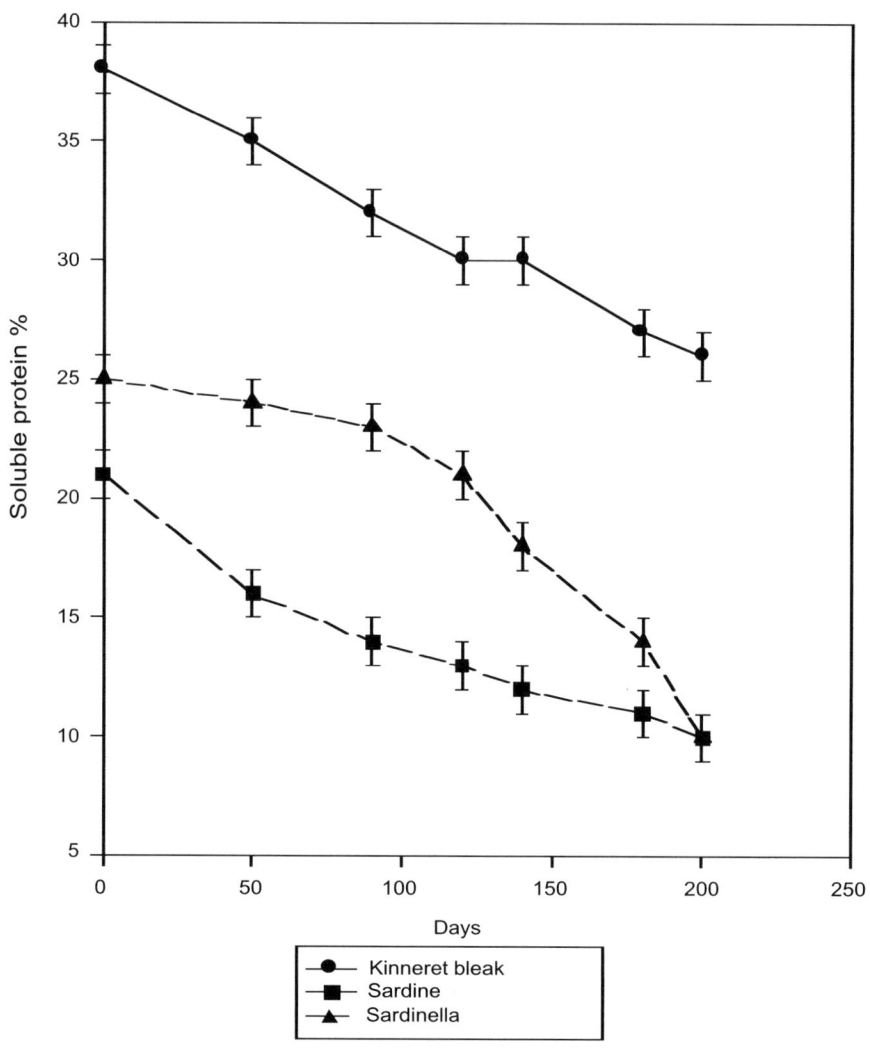

Figure 4. Changes in the soluble protein contents during storage.

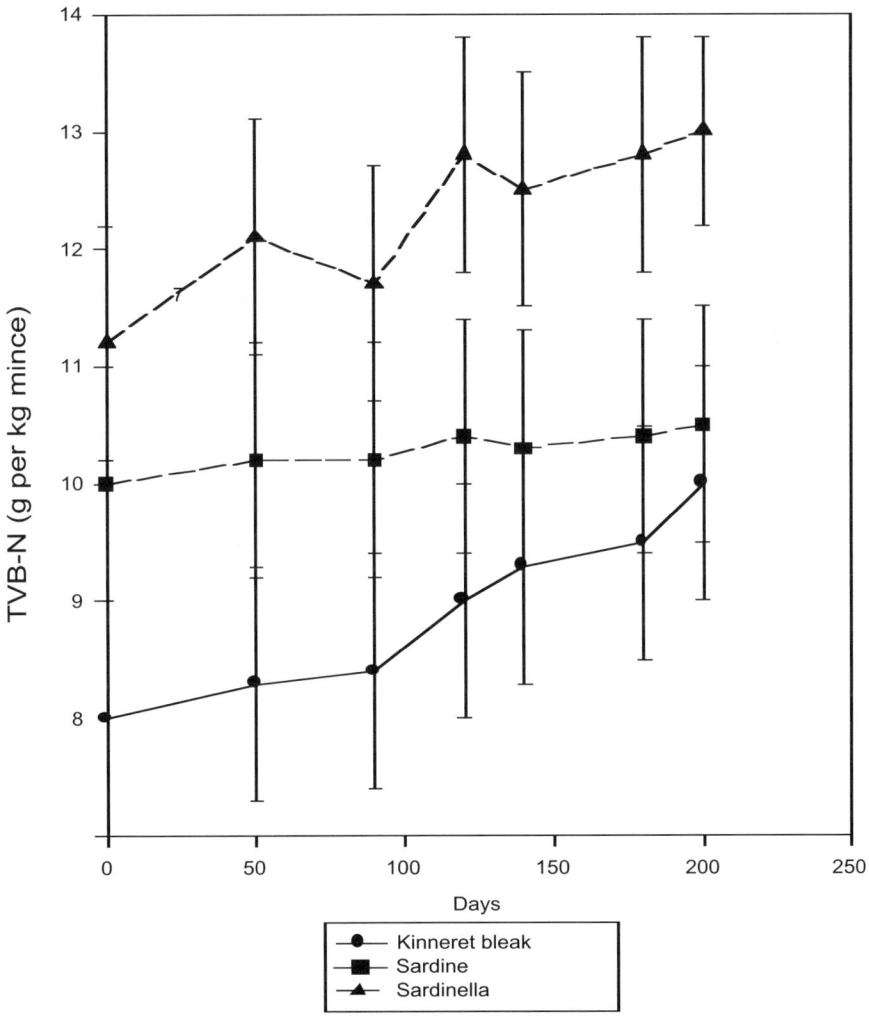

Figure 5. Changes in the TVB-N (total volatile basic nitrogen) values during storage.

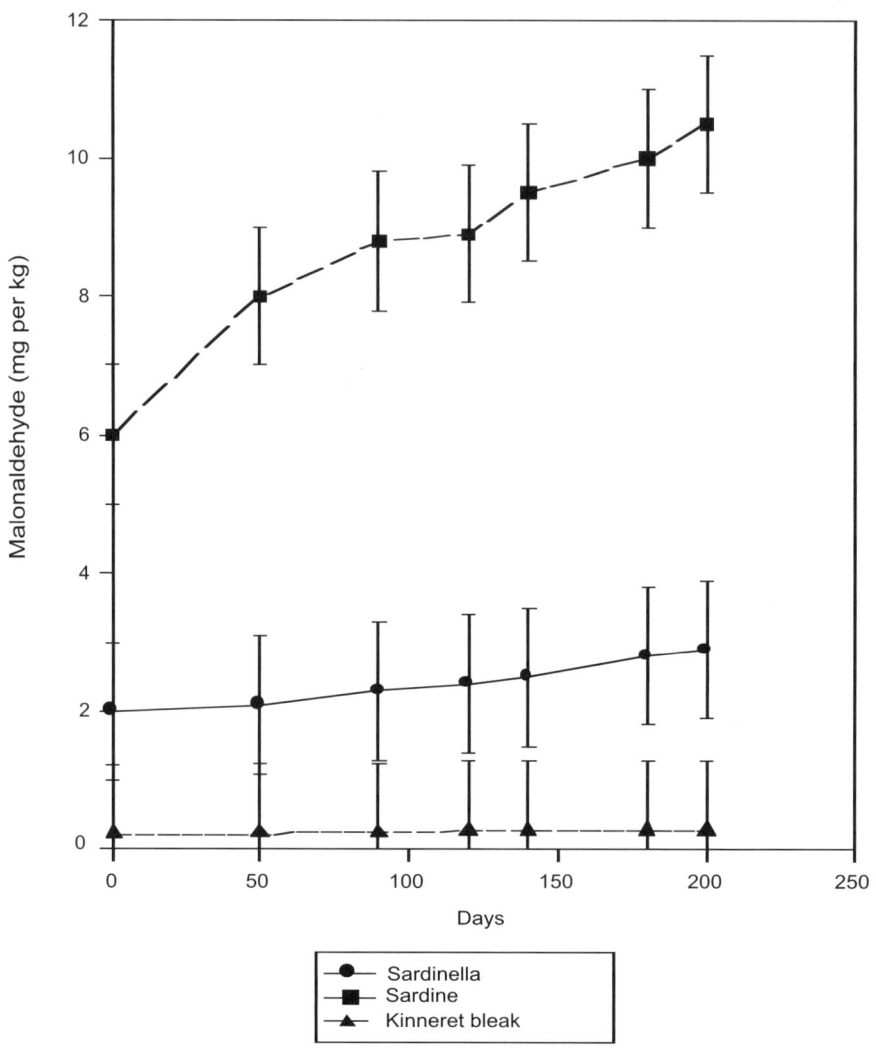

Figure 6. Changes in the TBA values during storage.

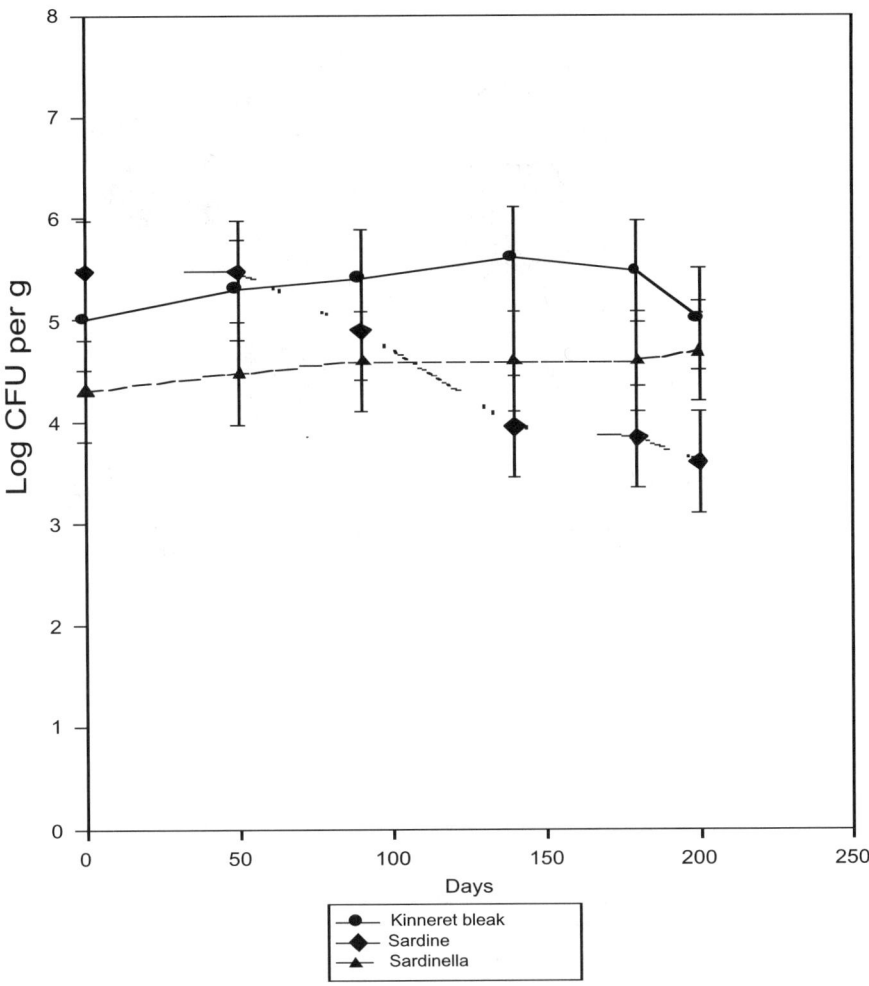

Figure 7. Changes in the microbial counts in the mince during storage.

Figure 8. Various products from fish mince. A = ravioli, B = frankfurters, C = fish balls, D = fish chips.

recover flesh from whole fish of underutilized species, which previously were considered too small or bony for consumers.

Attempts to utilize small fish without preliminary dressing were made with some fish species (Bussman 1977, Ishikawa 1978, Strom 1980), and interest in the recovery and utilization of minced fish flesh increased dramatically during the early 1970s (Ruckes 1973, Babbitt 1986). The utilization of bycatch from capture fisheries and aquaculture has progressed over recent years in the South China Sea (Yang 1999), Finnish inland fisheries (Setaelae et al. 1995), and in Venezuela (Bello 1998). Nevertheless, some problems remain unsolved: the need for more rational ways of improving the utilization of underutilized fish species; development of appropriate processing technology and equipment; rationalization of marketing processes, etc.

Our previous studies showed that production of edible material from a whole small pelagic fish (Kinneret bleak), without preliminary dressing, could be a feasible solution to the underutilization of this fish (Gelman et al. 1985). It should be noted that the marketing of the Kinneret bleak meets with many difficulties: it is not popular with the canners or the consumers because of its small size, mediocre taste, soft texture, unattractive appearance, and susceptibility to deterioration.

Table 2.   Organoleptic evaluation of the fish mince–based products.

	Noodles	Ravioli	Fish balls	Fish balls with sesame	Frank-furters	Fish-chips
Taste						
Scores	9.1	8.6	7.4	6.3	7.2	7.6
Dispersion	6-10	6-10	4-10	3-8	5-10	4-10
Standard deviation	2.18	2.28	1.7	1.91	1.45	3.04
Smell						
Scores	9.6	8.2	8.3	7.2	7.0	8.5
Dispersion	6-10	4-10	7-10	4-10	4-10	4-10
Standard deviation	2.3	2.4	0.81	1.93	1.82	2.0
Appearance						
Scores	7.6	9.1	8.7	7.9	8.1	8.3
Dispersion	6-10	8-10	7-10	5-10	5-10	6-10
Standard deviation	3.46	1.46	0.99	1.39	1.42	2.06
Texture						
Scores	7.1	7.8	6.5	6.8	6.5	5.3
Dispersion	2-10	4-10	4-9	4-9	4-8	3-10
Standard deviation	2.58	2.58	1.73	1.66	2.3	2.70

The main goals of the present study were to develop, on the pilot scale, the technology of an effective and not prohibitively expensive process for production of fish mince, and to follow this by preparing fish-mince-based food products for human consumption. Three fish species of different origins, characterized by relatively hard and thick bones, were used.

The results obtained showed that the successful scaling up of the process was rather simple, and no special problems were encountered. The yield of fish mince reached 60-65%, i.e., it was higher than those reported previously for both traditional and nontraditional methods. Typical yields of mince obtained from small fish by traditional methods average 22% for capelin, 30% for silver hake, and 40-45% for herring and flounder (CIFT 2002). A yield of 35-40% was obtained from whole capelin or sardine by nontraditional methods, compared with a yield of 25% obtained by conventional evisceration and mechanical separation (Strom 1980).

Digestion of fish tissues by endogenous enzymes at rather high temperatures and low pH led to a slight deterioration of gel strength properties in our experiments, but this shortcoming is compensated by the high yield.

Cold storage stability of the fish mince was also studied. The results obtained showed that the soluble protein content dropped gradually, and after 50 days (approximately the average storage time of raw materials before production) of storage at –18°C had decreased by 9% in Kinneret bleak and by 4% in sardinella. In sardine the soluble protein content of the mince decreased by 20% and there was an increase in rancidity after 50 days of storage; this could be attributed to the 2 months in preliminary storage at –18°C before the preparation of the mince. In the Kinneret bleak and sardinella minces, rancidity hardly increased during 200 days of storage, and this low rancidity and the small decreases in the soluble protein contents in these fish could also be attributed to the freshness of the raw materials (the minces were prepared from the chilled fish), as well as to the use of antioxidants and vacuum packing. Microbial counts did not change in any of the three species during storage. These data are in good agreement with the findings of other studies (Marti-de-Castro et al. 1997, Suvanich et al. 2000).

Various substances have been used for whitening fish mince (Yin et al. 1993, Meacock et al. 1997). The hydrogen peroxide concentration used by Young et al. (1979) and Himonides et al. (1999) for whitening was slightly decreased in our experiments, but the result obtained was quite satisfactory. No residual H_2O_2 was detected in the mince after treatment.

The traditional products prepared from fish mince are burgers, balls, patties, and sausages (Pruthiarenun et al. 1985, Bertullo 1989, Yin et al. 1993, Yo and Siah 1997). Fish mince or pulp of mixed species may be used as a partial substitute for meat in the preparation of hamburger, sausage, and mordatella (Miranda 1985; Fiddler et al. 1992, 1993). The mince obtained in our present experiments was used in the preparation of diverse traditional (frankfurters and fish balls) as well as nontraditional products, such as fish chips, noodles, and ravioli. The nontraditional products had a more attractive appearance and better taste than the traditional products. The mince was used as a partial substitute for carp flesh in gefilte fish (up to 50%) and for meat in sausages (up to 20%), and had no negative influence on the organoleptic properties of the products (Gelman et al. 1985).

Thus, this effective and not prohibitively expensive technological process for the production of fish mince from whole small fish, without preliminary dressing, was scaled up to the pilot-plant level. The mince was characterized by high yield, good storage stability, and suitability for production of fish-mince-based food products, both traditional and nontraditional, for human consumption.

References

Babbitt, J.K. 1986. Suitability of seafood species as raw materials. Food Technol. 40:97-100.

Bello, R.A. 1989. Use of under-utilized fish species by obtaining the flesh. Report of the technical consultation on fish utilization and marketing in Latin America. In: FAO Fish. Rep. 421, Suppl., pp.49-54. (In Spanish.)

Bello, R.A. 1998. Investigation on the possibilities of exploitation of shrimp by-catch in Venezuela. In: E. Garcia-Rodriguez and F. Teutscher (eds.), FAO Fisheries Department, Rome. Fish. Industries Div., Cuba Centro de Investigaciones Pesqueras (CIP), pp. 22-26. (In Spanish.)

Bertullo, E. 1989. Industry of minced fish and derived products in Latin America. Report of the technical consultation on fish utilization and marketing in Latin America. In: FAO Fish. Rep. 421, Suppl., pp. 55-61. (In Spanish.)

Bussman, B. 1977. Studies on the processing of blue whiting and horse mackerel. Inf. Fischwirtsch. 24:77-80.

CIFT. 2002. Canadian Institute of Fisheries Technology. http://is.dal.ca/~cift/.

Cunniff, P. (ed.). 1995. Official methods of analysis, 16th edn. Association of Official Analytical Chemists, Gaithersburg, Maryland.

FAO. 2000. FAO yearbook fishery statistics, 86/1, pp. 95-137.

Fiddler, W., J.W. Pensabene, R.A. Gates, M. Hale, and M. Jahncke. 1992. N-nitrosodimethylamine formation in cooked frankfurters containing Alaska pollock (*Theragra chalcogramma*) mince and surimi. J. Food Sci. 57:569-571.

Fiddler, W., J.W. Pensabene, R.A. Gates, M. Jahncke, and M. Hale. 1993. Atlantic menhaden (*Brevoortia tyrannus*) mince and surimi as partial meat substitutes in frankfurters: Effect on N-nitrosamine formation. J. Agric. Food Chem. 41: 2238-2241.

Gelman, A., E. Benjamin, and A. Herzberg. 1985. Production of mince from whole small pelagic fish. Trop. Sci. 25:253-264.

Gelman, A., E. Benjamin, A. Herzberg, and M. Kramer. 1989. Process for the preparation of boneless flesh from small fish. Israeli Patent 75139. (In Hebrew.)

Grantham, G.J. 1981. Minced fish technology: A review. FAO Fish. Tech. Pap. 216. pp. 1-14.

Himonides, A.T., K.D.A. Taylor, and M.J. Knowles. 1999. The improved whitening of cod and haddock flaps using hydrogen peroxide. J. Sci. Food Agric. 79: 485-850.

Ishikawa, S. 1978. Fish jelly products (kamaboko) and frozen minced meat (frozen surimi) made of sardine. 2. Influence of the temperature during manufacturing process on the jelly strength of kamaboko. Bull. Tokai Reg. Fish. Res. Lab. 94:37-44.

Lee, C.M., and R.T. Toledo. 1976. Factors affecting textural characteristics of cooked comminuted fish muscle. J. Food Sci. 41:391-397.

Marti-de-Castro, M.A., M.C. Gomez-Gillen, and P. Montero. 1997. Influence of frozen storage on textural properties of sardine (*Sardina pilchardus*) mince gels. Food Chem. 60:85-93.

Meacock, G., K.D.A. Taylor, M.J. Knowles, and A. Himonides. 1997. The improved whitening of minced cod flesh using dispersed titanium dioxide. 1997. J. Sci. Food Agric. 73:221-225.

Miranda, S.L. 1985. Elaboration of new products based on fish pulp. Rev. Latinoam. Technol. Aliment. Pesq. 2:5-10. (In Spanish.)

Prabin and Co. 1987. Kjeltec 1026 Dist. Unit, Manual Part no. 1002790, T8806, Application Note, Prabin and Co. AB, Klippan, pp.1-9.

Pruthiarenun, R., J. Yamprayoon, P. Suwansakornkul, P. Kiatkungwalkrai, and S. Suwansakornkul. 1985. Utilization of fish by-catch and fish ball manufacture. In: A. Reilly (ed.), Spoilage of tropical fish and product development. In: FAO Fish. Rep. 317, Suppl., pp. 428-431.

Rasekh, J., and A. Metz. 1973. Acid precipitated fish protein isolate exhibits good functional properties. Food Prod. Dev. 7:18-24.

Ruckes, E. 1973. Fish utilization, marketing and trade in countries bordering the South China Sea: Status and programme proposals. FAO Rep. 73/8. 33 pp.

Setaelae, J., P. Salmi, P. Muje, and A. Kaeyhkoe. 1996. Annales Zoologici Fennici 33:547-551.

Stodolnik, L. 1979. Effect of liquid nitrogen and contact freezing of fish upon changes in frozen storage. Bull. Inst. Int. Froid 59:1182-1185.

Strom, T. 1980. Biological and processing factors affecting functional properties. 4. Development of a process for minced fish production from capelin (*Mallotus villosus*). In: J.R. Brooker and R.E. Martin (eds.), Third National Technical Seminar on the Mechanical Recovery and Utilization of Fish, Raleigh, North Carolina. Abstracts, no. 22.

Suvanich, V., D.L. Marshall, and M.L. Jahncke. 2000. J. Food Sci. 65:151-154.

Vyncke, W. 1970. Direct determination of TBA value in extracts of fish as a measure of oxidative rancidity. Fette. Seifen. Anstrichm. 72:1084-1087.

Yang, L. 1999. A review of the bycatch and its utilization for capture fisheries in the South China Sea. In: I. Clucas and F. Teutscher (eds.), Report and Proceedings of FAO. UK Natural Resources Institute, Chatham, pp. 136-149.

Yin, B., J. Wang, and Y. Jiang. 1993. Preparation of fishball made from various kinds of small marine fishes. Shandong Fish. Qilu Yuye 5:38-40. (In Chinese.)

Yo, S.Y., and W.M. Siah. 1997. Development of fish burgers from *S. leptolepis, A. monoceros* and *A. nobilis*. In: FAO Fish. Rep. 563, pp. 257-264.

Young, K.W., S.L. Newmann, A.S. McGill, and R. Hardy. 1979. The use of dilute solution of hydrogen peroxide to whiten fish flesh. Papers presented at the Jubilee Conference at the Torry Research Station, Aberdeen, Scotland, pp. 242-250.

Byproduct Utilization: Available and Innovative Technologies

Lars Langhoff
ABD-Consulting, Lejre, Denmark

Abstract

One kilogram of byproducts not used commercial is one kilogram of mistakes. Many processors only focus on their main product and forget to take care of the byproducts in a correct manner. This influences the overall processing cost and operations in a negative way. It is desirable to process all byproducts into valuable products, first for human foods, second for pet food, and third as animal feed, pharmaceuticals, or fertilizer. Around the world a lot of scientist have developed new technologies and uses for fish byproducts but not too much has changed. The reason is simply that no one has successfully combined the use of innovative technology, processing method, and marketing of the product. During the last ten years a lot of new possibilities for processing fish and crustacean byproducts have been developed. These include not only mechanical tools but also new tools like enzymes. The combination of available fish byproducts, available technologies, and available markets is addressed.

Introduction

I have always looked at food industry byproducts as an opportunity. One kilogram of byproducts not used commercially is one kilogram of mistakes.

The daily business of most fish processors is 99% concentrated on the "main product" coming from that species of fish processed that day such as fillet, roe, or surimi. In fact 99% of the organization is concentrating all their efforts on processing, marketing, sales and distribution of the "main product" even though that the main product might only be 10-50% of the initial raw material. The market for fillets, roe, or surimi is known with established sales channels and customers, so why are we taking up so much of our time and money on the main product and neglecting the other 50-90% of the raw material? Why not do the opposite and focus 90% of our time and resources on the byproducts, which then could then become money-makers for the business?

Today's fish byproduct industry can be characterized as follows:

1. Much valuable material is lost in today's fish industry.

2. Almost all cut off material can be separated and used for human consumption, pet food, or at least as a feed or fertilizer.

3. The processing of fish byproducts can become a profitable activity.

4. The world's requirement for fish protein can be met from the present wild and aquaculture harvests, if the raw material is utilized in a correct manner.

It would be nice if one could reduce fish raw material to its components, and thereafter put the components together to formulate the desired end product. This cannot be done at present, but it gets closer and closer every year. Until this "utopia" happens one has to look carefully at all technologies available to our industry. It is not enough to look at what technologies are used in the industry today and fine-tune them. It is desirable to look into other industries and see what can be modified and utilized in our fish industry. This is "innovation."

One good example here is when Jerry Babbitt of the National Marine Fisheries Service utilized the Brown refiner for surimi and fish byproduct processing. The Brown refiner was developed for the avocado industry to separate the shell and stones from the meat, but it was successfully modified for use in the fish industry with very good results.

Separation of fish byproduct

Sedimentation

The oldest process uses sedimentation of material in tanks, and employs the same basic principles used in modem decanter and centrifuge equipment. The decanter is a machine which separates liquids and/or solids with different specific weights comparatively quickly and accurately. Decanters are especially well suited where much dry matter is present to be separated from a liquid. A newer invention, the so-called 3-phase decanter, is now used to separate materials at an early stage of the process into oil, water, and dry matter.

A centrifuge is a fast-rotating vertical decanter, which is best suited for the separation of liquids, for example fish oil and water. The centrifuge works with a high g-force and usually at a liquid temperature up to 100°C.

The hydrocyclone makes use of high inlet velocities and the normal g-force. In principle it can be said that a hydrocyclone is a stationary centrifuge in which the liquid rotates. Hydrocyclones are specially suited for the separation of solid particles with specific weights within a certain range.

Filtration

Another common process is the use of filters, which include coarse-meshed to fine-meshed wire filters, vibration screens with or without air lifting, etc., and reverse osmosis (RO). Reverse osmosis is based on a filter of a specific tight structure, whereby it is possible to predict which parts of a moving liquid will remain behind the filter, and which parts will pass through the filter. Reverse osmosis uses specific membranes as filters and is most suitable for separation of pure liquids without large amounts of suspended dry matter or high oil content.

Enzymes

One newer tool in the separation processes is the use of enzyme to free protein from other components and/or hydrolyze the protein into smaller fragments and amino acids. The smaller proteins and amino acids can then be separated from each other.

Mechanical tissue separators

Meat and bone separators work on different principles depending on the manufacturer. The mechanical separators are able separate the softer parts (often called the meat) from the more solid parts (bone and skin fraction).

Raw materials

In today's fish industry most of the leftover materials (byproducts) are combined and, if not dumped, processed one more step and turned into fish meal. Fish meal is a commodity product for which a market already exists.

In most plants water is used for transporting the cut off material to the collection bin. During water transportation protein can be lost. A potential breakdown of the protein loss on a percentage of total loss basis is given below.

- 17% lost in the drain and catch-water cistern.

- 42% loss can take place in the pump/transfer after the catch-water cistern.

- 1% loss occurs per 25-meter pipeline after the pump.

- 33% loss takes place in a rotary strainer used for separation of solid material from the water phase.

An alternative to using water as a transport medium is to install filter conveyors in connection with outlets from processing units. This will immediately separate free water from the transported product and thereby avoid transferring the protein to the water phase. Filter conveyors are often more hygienic, because less material and water will be on the floor and the material will quickly be transported out of the processing room. Using filter conveyors it is relatively easy to collect different byproducts and to process these materials separately. The separated water can under some circumstances be concentrated by means of reverse osmosis or in a compact unit evaporator.

In the process of roe stripping only the roe that is in sacs is used, and all the loose roe is mixed with other byproducts and dumped or used for fish meal. There is equipment available for separation of loose roe from other belly material, whereby it is possible to make a "single roe" product for which a market already exists.

Fish industry byproduct from source to market

Figure 1 shows the sources of fish processing byproducts, raw materials made from the byproducts, and ingredients made for the raw byproducts and markets.

Head, tail, skin, and frames are byproducts that can be used as raw material for soup stock and flavor products. Also they can be used for fish meals, hydrolysates, and silages that can be used as aquaculture and weaned pig feeds. And like other byproducts they can be use for organic fertilizers.

Head

Depending on the fish type there could be meat to recover, or the head could be processed as a product itself. As an example, heads from salmon are sold in Russia for human consumption.

Tail

The tail can be used as raw material for fish meal.

Skin

Skin can be used as a raw material for the leather industry, gelatin manufacture, and as a pet food ingredient.

Frames

The *meat* can be separated from the bone structure and used as raw material for mincemeat, surimi, pizza topping, and in pet foods. The separated *bone* material can be used as raw material in the pharmaceutical industry or as organic fertilizer byproduct from source to market

Viscera

The *roe* can be utilized in sacs or as loose roe. *Milt* can be a raw material for the pharmaceutical industry and used in pet foods. *Liver* can be a

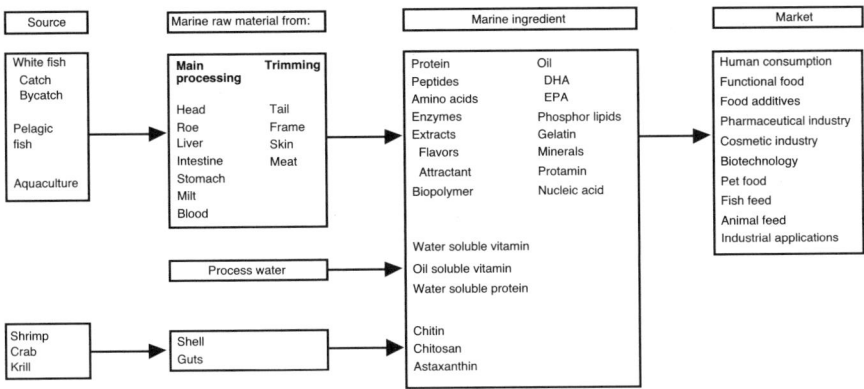

Figure 1. Fish processing byproduct from source to market.

raw material for liver oil, canned liver, the pharmaceutical industry, and used in pet foods. *Stomachs* can be raw materials for the pharmaceutical industry, and for extraction of enzymes.

Waste water

In most of today's fish processing plants the waste water from all processing areas is collected in one final pipe, and if not directly drained to the sea, processed in a DAF (dissolved air flotation) system or other waste water system. When different types of water are mixed together it is often necessary to add chemicals for the water treatment system. If a solids fraction is subsequently made the material can't be used in the food or feed chain. However, if not all waste water streams are combined it is possible to avoid use of chemicals on most of the water and thus utilize the solids in other ways.

Increasing the utilization of byproducts

The main reason for not utilizing byproducts for products other than hydrolysates, silages, or fish meal is the market. The industry does not devote enough time to marketing new products processed from byproducts. Another reason is that very few, if any, equipment manufacturers market a complete process line for byproduct upgrading. A third reason is that the amount of raw material from each processor and the capital investment do not correspond. The fourth reason is that in the industry few identify the marketing part of processing byproducts as an important core component of their business.

To increase byproduct utilization it would be desirable to establish a network of processors working together or establish companies specializing in byproduct processing. Needs would include marketing people, consultants, scientists and research and development (R&D) institutions, equipment manufactures, and turnkey equipment supplier and new venture capital. If these needs are met, awareness of byproduct utilization will be increased and more organizations will identify the processing and marketing of byproducts as their main business.

Success in Byproduct Utilization and Marketing

Bruce Buckmaster
Bio-Oregon, Inc., Warrenton, Oregon

Abstract

The classic adage states, "The devil is in the details." There are few, if any, business challenges where this is truer than in making a profitable venture out of processing fish byproducts. In a simple survey of worldwide fish meal and/or fish oil producers we find that most of the long-term players obtain their raw material from directed and dedicated fisheries. Those operations, which rely on a byproduct stream, tend to fit one of two categories: either they serve a fishery that operates at least 180 days per year or they are subsidized by the primary fish processing facility.

Traditional fish meal technology produces traditional fish meal. Advances in process have provided improved products that warrant prices at the high end of commodity protein ranges, i.e., low-temperature meal and low-ash meal. However, the capital resources required to manufacture these products are generally not supported by an adequate "byproduct time/volume ratio." Attempts to produce higher value products from processing waste streams have met with limited success due to three main issues. They are: (1) insufficient finished product volume to invite commercial interest, (2) lack of data that clearly removes the new products from the lower-priced commodity arena, and (3) an entrenched attitude among commercial interests that fish-based products are inherently variable in quality and supply. At Bio-Oregon we have begun to address each of these issues and have met with initial success.

Success at Bio-Oregon

Success is a relative term. For purposes of this presentation, success shall be deemed as a venture that is profitable enough to consistently pay all operational expenses and cover depreciation, and provide enough earnings to encourage reinvestment. Bio-Oregon, Inc. has been able to achieve the first two elements of success for over eighty years. Achieving the third element, reinvestment capital, has proven more difficult in a consistent manner.

Bio-Oregon, Inc., located at the mouth of the Columbia River on the Oregon coast, has operated as a fisheries byproduct processor since 1917. The evolution of the business and its products has mirrored the local fishery and the available markets through the decades. Examples of products include salmon eggs for sport-fishing bait; salmon, tuna, and whale meals for livestock, pet, and mink feeds; whale oil for industrial uses; aquaculture feeds; vitamin A oil supplements for human nutrition; flavors for pet food; and various fish-based agricultural fertilizers. Each product was developed in support of a primary fishery and an identified customer need. Rarely did any existing fishery last long enough to fully finance technology to serve the next emerging fishery.

Consequently, the management and technical staff of Bio-Oregon have extensive experience in quickly identifying appropriate and cost-effective methods of turning new sources of byproduct into salable products. Unfortunately, the speed of transition demanded by the realities of finance usually resulted in end-products that can be classified as low-priced commodities. This commodity-based model allowed no excess profit to fund necessary marketing and technical research in order to increase long term value to the byproduct stream.

Over the last twenty years the midwater groundfish industry has afforded Bio-Oregon a relatively stable raw material stream. Internal usage of concentrated fish hydrolysate was increasingly incorporated in aquaculture feeds manufactured at the Bio-Oregon plant. The relative cost advantage over high quality fish meal provided funding for capital projects which increased capacity, improved processing, and allowed increasing percentages of hydrolysate to be included in each feed formula.

In the mid-1990s a series of developments conjoined to cause a complete re-evaluation of the core objectives at Bio-Oregon. Groundfish allowable catches off of the Oregon-Washington coast continued to be reduced with no bottom in sight. The development of new surimi technology had opened a new fishery that produced large quantities of byproduct in a 60- to 90-day window. World aquaculture growth and the consequent growth of giant multinational aquaculture integrators were changing the competitive arena in our major market. And finally, variable costs, primarily energy and health care, were squeezing already tight margins.

Key technical and management personnel at Bio-Oregon began an analysis of obstacles and opportunities for the future direction of the company. To the extent that we had any data, we looked at markets that might value the fish-based proteins and oils inherent in our raw material. We reviewed current literature for insights and ideas concerning new processes and uses of fish-derived products. Early in the analysis it became apparent that we faced three major issues that had to be resolved if we hoped to be successful. The three issues were:

1. Insufficient finished product volume to invite commercial interest.

2. Lack of data that clearly differentiated new products from the commodity arena.

3. An entrenched attitude among commercial interests that fish-based products are inherently variable in quality and supply.

A realization that there is a clear inverse relationship between the inclusion rate of ingredients and their unit value allowed us to begin the process of identifying our new direction.

The path to success is filled with false starts and pitfalls. We soon discovered that the lure of selling small packages to retail consumers ultimately did not improve the bottom line due to the manufacturing and distribution costs. It also became clear that a commitment to efficiency over effectiveness slowed the development process. Our view of equipment manufacturers also had to change. We now see them as a valuable resource but not as a supplier of important answers.

Ultimately, we have arrived at new goals:

1. Bio-Oregon's products will be at the lower end of the descending order of ingredients (small volume, high value).

2. Bio-Oregon will produce a functional ingredient that can drive the retail purchase.

And we have adopted new rules:

1. We will accept inefficiency to build markets.

2. All processes will be portable.

3. All marketing will be "pull-through," which first identifies need.

4. All unidentified factors will be identified.

In order to achieve our new goals we have had to identify and obtain new resources. Building relationships and collaborative agreements can be time consuming, but it is essential if success is to be achieved. Bio-Oregon has accessed assistance from:

1. Academic partners.

2. Technological support from suppliers.

3. Government and NGO (non-government organization) research grants.

4. Customer and potential customer "needs" evaluations.

5. "Certified Smart People" (experts).

Currently Bio-Oregon is marketing three branded products that partially meet our new goals. SeaPro 75, SeaPro 75 (hydrolyzed), and SeaPro 45 are being sold at significant premiums over the most expensive commercially available specialty fish meals. We are currently seeking additional manufacturing capacity to increase supply in an effort to meet existing demand. Initial discussions with primary and byproduct processors are encouraging.

Bio-Oregon has identified a large number of opportunities that are consistent with our new goals, both technically and financially. We have begun exploration on the use of fish-derived ingredients that will address problems and challenges in neo-natal animals. We are exploring the role of fish products in gut-health. Fish ingredients also appear to have commercial possibilities in affecting immune response, palatability, and reproductive efficiency. Research to determine synergy with other specialty proteins offers innumerable opportunities.

Life is fraught with peril and ripe with opportunity. It is no different in the life of a commercial venture. Bio-Oregon is committed to achieve its 100 year anniversary; aged but hanging out over the cutting edge.

Analysis of Groundfish Meals Made in Alaska

Scott Smiley
University of Alaska Fairbanks, Fishery Industrial Technology Center,
Kodiak, Alaska

Jerry Babbitt
National Marine Fisheries Service, Kodiak, Alaska

Subramaniam Divakaran and Ian Forster
Oceanic Institute, Waimanalo, Hawaii

Alexandra de Oliveira
University of Alaska Fairbanks, Fishery Industrial Technology Center,
Kodiak, Alaska

Abstract

As part of an ongoing project with the principal aim of increasing the value of byproducts derived from fish processing in Alaska, we have analyzed a variety of commercially available groundfish meals. Here we report on our analyses of these meals and discuss the major types of processing systems for handling this material. Most shoreside Alaskan fish meals are made with the wet reduction method and employ Atlas Stord machinery. These plants can process upwards of 1,200 metric tons per day. The products of these plants include fish meal, fish oil, stickwater, and bone. Alaskan fish meal is generally sold to Asian aquaculturists. Alaskan fish oil is often used as a boiler fuel. Stickwater is handled variously depending on its salt concentration. Bone meal has a relatively low value. Options for making higher protein–lower ash fish meals are available but require modifications to existing equipment. Such meals have greater value for freshwater aquaculturists concerned with phosphate levels.

Introduction

Commercial fish landings in Alaska have averaged roughly 2 million metric tons (t) over the past five years (NMFS 2001). All commercial landings in Alaska are for human food. There are no "industrial" fisheries in Alaska where whole fish are made into fish meal. The most important single species in Alaska is walleye pollock (*Theragra chalcogramma*) with roughly 1 million t harvested in the Bering Sea and more than 100,000 t harvested in the Gulf of Alaska annually. Given moderate recovery rates of roughly 25%, this harvest level means that some 825,000 t of processing byproducts are generated annually in the Alaskan pollock fishery.

Implementation of the 1974 Clean Water Act by the Environmental Protection Agency (EPA) and the Alaska Department of Environmental Conservation (ADEC) led to requirements for effective handling of the byproducts of seafood processing either by grinding and dumping or, for "seafood processing centers," use of methods that reduce the amount of organic debris released into local waters. The Web site home page for EPA Region 10, covering Alaska, is http://www.epa.gov/region10; the ADEC Web address is http://www.state.ak.us/local/akpages/ENV.CONSERV/home.htm.

During the Americanization of the Bering Sea pollock fishery, several large seafood processing plants were built in Dutch Harbor, Akutan, King Cove, and Sand Point, Alaska. Some of these plants, and the community of Kodiak, can process upwards of 3 million pounds of fish per day, in landed round weight. EPA and ADEC mandated that these new processors install waste plants capable of handling the byproducts from their seafood processing operations in these locations.

With more than 2 million metric tons of fish processed into human food annually, there is a considerable volume of processing waste. In this paper we report on the quality of eleven fish meals made from the byproducts of groundfish seafood processing in Alaska. We have performed proximate analyses and mineral analyses, determined the amino acid composition of the proteins, analyzed the fatty acids in the residual oil in the fish meal, and tested the meals for nutritional effectiveness with an aquaculture finfish species. Together these data constitute a baseline of detailed information on the fish meal made in Alaska from the byproducts of seafood processing.

Materials and methods

We contacted ten different fish meal producers in Alaska and procured a single sample of roughly 100 pounds of their meal for testing. The at-sea producers included vessels operated by: American Seafoods Inc., Arctic Storm Inc., Icicle Seafoods Inc., and Premier Pacific Seafoods Inc. The shoreside processors were Alyeska Seafoods Inc., Kodiak Fishmeal Co., Peter Pan Seafoods Inc., Trident Seafood Corporation, Unisea, and Westward

Seafoods. All meals were made with the byproducts of groundfish processing; no salmon meals have been included in this study.

Proximate analyses, and analyses of amino acid and fatty acid composition, were all performed at the Fishery Industrial Technology Center in Kodiak, Alaska. Small aliquots from each sample were sent to National Marine Fisheries Service, Northwest Fisheries Science Center, in Seattle to determine mineral composition of each meal. Other samples were sent to Evergreen Analytical Services Inc., Redmond, Washington, for the ethoxyquin analysis. Then roughly 50 pounds of each meal were sent to the Oceanic Institute in Hawaii for feeding the aquacultured marine finfish Pacific threadfin (*Polydactylus sexfilis*), also called moi. The Oceanic Institute also forwarded samples from each of the meals for commercial contaminants analysis to the New Ulm Organics Laboratory, New Ulm, Minnesota.

Proximate analysis

Protein

We used a Leco model FP 2000 protein analyzer to determine total nitrogen for each meal, and this value was converted to a measure of crude protein. In this AOAC (1996) approved method, a weighed 0.4 to 0.6 g sample of each meal was placed in a combustion crucible and set into the machine. The autosampler on the device ran each sample through the system and a computer collected the data. Samples were measured in triplicate and the data are the percentages of total weight. The number reported is the average of the three replicates.

Lipid

We used a Leco FA-100 Fats Analyzer to determine percent lipid for each sample. A 1.2 to 1.6 g weighed sample of the meal was mixed with diatomaceous earth and inserted into a Leco sample thimble. The thimble was placed into the FA-100 and the lipids extracted at 100°C and 9,000 psi using critical point CO_2. The extract was dripped into a Leco collection vial and this weighed on a standard analytical scale. This method is pending AOAC approval. Samples were measured in triplicate and the data are percentages of total weight. The number reported is the average of the three replicates.

Ash

We followed standard AOAC (1996) procedures for ashing the samples. A 1-2 g sample was placed in a preweighed combustion crucible. The crucible was then placed in a Fisher Isotemp Muffle Furnace and baked at 560°C overnight. The residue not burned off was weighed and this was defined as ash. Samples were measured in triplicate and the data are percentages of total weight. The number reported is the average of the three replicates.

Moisture

We followed standard AOAC (1996) procedures for the determination of moisture. We used a Lab-Line Imperial IV Microprocessor air drying oven set at 102°C. A weighed 4 to 10 g sample was baked in the oven until constant weight was achieved. The sample was then weighed on an analytical balance and moisture determined by subtraction. Samples were measured in triplicate and the data are percentages of total weight. The number reported is the average of the three replicates.

Ethoxyquin

Although they are not technically part of a proximate analysis, we have included measurements of ethoxyquin levels in the fish meals. We sent samples of each of the eleven meals to Evergreen Analytical Services Inc. to test for the presence of ethoxyquin, an antioxidant added to the meals at the time of their manufacture. The data are mean concentrations is parts per million.

Contaminant analysis

The eleven Alaskan groundfish meals were tested for contaminants. All were analyzed for the presence of pesticide residues such as organophosphates, chlorinated hydrocarbons, and polychlorinated biphenyls (PCB) at the New Ulm Organics Laboratory, using a combination of AOAC (1996) methods with the EPA SW-486 method 3640, a gel permeation technique that removes fats and oils. The samples were extracted with a 3:1 solution of ethyl acetate:toluene. The extracts were loaded onto a gel permeation column and the eluant was flash evaporated to near dryness. The samples were then brought to volume with iso-octane and aliquots introduced into the gas chromatograph. The PCBs and pesticides were quantified using a Shimadzu 14A gas chromatograph with an electron capture detector and a Restek, RTX-35 analytical GC column.

Mineral analyses

Samples were sent to National Marine Fisheries Service, Northwest Fisheries Science Center, for mineral analysis. Samples were prepared by ashing 1 g of each meal overnight at 550°C in a muffle furnace. The ashing residue was digested overnight at room temperature in an aqueous solution containing 10% HCl and 10% H_2NO_3. Digested solutions were diluted as needed and analyzed for silver, calcium, cadmium, copper, iron, potassium, manganese, sodium, nickel, phosphorus, lead, strontium, and zinc by inductively coupled plasma optical emission spectroscopy on a Perkin Elmer Optima 3000 Radial ICP-OES. The numerical values represent the averages of three replicates. The data are percentages for calcium, potassium, magnesium, sodium, and phosphorus. For silver, cadmium, copper, iron, manganese, lead, strontium, and zinc the data are reported as parts per million (ppm).

Amino acid analyses

We followed published methods provided by the HPLC manufacturer (Waters 1993). Samples containing about 40 mg of protein from each of the eleven meals were weighed and hydrolyzed in 6 N HCl for 24 hours under nitrogen at 110°C. The hydrolysate was brought to volume with distilled water, and 1 ml was filtered through a filter with a 0.45 μm pore diameter. Samples were derivatized using the Waters AccQ Tag method. Approximately 4-6 μl of the filtered sample was introduced into an HPLC vial, dried down, and rehydrated with addition of the alpha-aminobutyric acid (AABA) internal standard, buffer, and the AccFluor reagent. The vial was heated for 10 minutes at 55°C before introduction into a Waters 2690 HPLC equipped with a Waters AccQ Tag column and a Waters 474 Scanning Fluorescence Detector. For calibration, Waters Amino Acid Hydrolysate Standard Mixture with the AABA internal standard was derivatized and run with the samples. After passing through a Waters A to D board, the data were collected by computer. Analyses were run in triplicate and the data are the means of the three replicates.

Fatty acid analyses

Lipid extraction

Extraction of lipids was carried out as described by Radin (1981). Eighteen milliliters of a 2:3 solution of isopropanol:hexane were added for each gram of sample. The mixture was homogenized to a slurry for one minute under a nitrogen gas stream. The contents were then vacuum filtered through a 150 ml sintered funnel lined with Whatman #2 filter paper. The filtrate was collected. To extract any remaining lipid, the retained solids were washed with the isopropyl:hexane solvent using the first 3 ml per gram of sample, and subsequently an additional 2 ml per sample gram, and washes added to the filtrate. The final filtrate was decanted into a flask, and the solvent removed at 49°C on a Büchi rotary evaporator to yield approximately 5 ml of lipids in solvent. This mixture was decanted into a 50 ml screw top centrifuge tube, and the flask rinsed with two 10 ml volumes of hexane and the hexane retained. Ten ml of distilled water was added to the centrifuge tube and contents mixed for 1-2 minutes. The tube was centrifuged for 20 minutes at 3,400 rpm. The top layer of supernatant was transferred into a small flask containing 2.5-3g of anhydrous sodium sulfate. An additional 10 ml of hexane was added to the centrifuge tube, and the centrifugation repeated and the top layer again added to the flask. The final lipid extract in hexane was dried over anhydrous sodium sulfate for 30 minutes, and then filtered through Whatman #1 filter paper. The solvent was evaporated at 49°C on the rotary evaporator and the remaining extracted lipids were transferred into a pre-weighed 10 ml amber screw top vial. Any retained solvent was volatilized under a nitrogen gas stream.

Fatty acid methyl esters (FAME) analysis

Fatty acid methyl esters were prepared according to the method of Maxwell and Marmer (1983). Twenty mg of extracted lipid was placed into a 10 ml centrifuge tube and the exact weight recorded. The lipid was dissolved in 1.9 ml iso-octane. Added to the tube were 100 µl of a solution of 10 mg per ml carpic acid methyl ester and 200 µl of 2N KOH (dissolved in anhydrous methanol). This solution was homogenized for 60 seconds, then centrifuged for 3 minutes at 3,000 rpm, and the lower layer discarded. This procedure was repeated twice using 500 µl of a saturated solution of ammonium acetate in water, followed by 500 µl deionized water. Methyl esters in iso-octane were then dried at room temperature with approximately 200-300 mg of anhydrous sodium sulfate for 20 minutes and then centrifuged at 3,000 rpm for 20 minutes. Fatty acid methyl esters were transferred to a 1.5 ml snap-cap amber GC vial (Agilant Technologies, Wilmington, Delaware) for chromatographic analysis.

Gas chromatography

We used an Agilant Gas Chromatograph model 6850 equipped with an autosampler, flame ionization detector, and a DB-225 capillary column to quantify the fatty acid methyl esters. Our data were collected and analyzed using the Agilant GC ChemStation program. Helium, at constant flow of 1.0 ml per min, was the carrier gas. We held the injector and detector temperatures at 250°C. The split ratio was set to 25:1, and the oven programming allowed us to raise the temperature from 140°C to 220°C at a rate of 3°C per minute. Total run time was 47 minutes. We used ChemStation to integrate the chromatogram peaks. We built a calibration curve for every compound in the Supelco 37 standard (Supelco, Bellefonte, Pennsylvania) using their calibration table and Supelco 37 standard at five different concentrations. Supelco Bacterial Acid Methyl Esters Mix, PUFA-1, and PUFA-3 at concentrations of 25% and 50% were run as additional standards in order to identify fatty acid methyl esters not found in the Supelco 37. All fatty acid methyl ester samples were run in duplicate and the data reported as the average of the two runs.

Nutritional analyses

Eleven diets were formulated and balanced to contain 44% crude protein and 11.5% crude lipid on an as fed basis. The control diet contained a high quality Norwegian fish meal (LT-94, SSF, Norway). The other diets contained an Alaskan groundfish meal in complete replacement of the control fish meal. The experimental diets were manufactured at the facilities of the Oceanic Institute. For the growth trials, Pacific threadfin *Polydactylus sexfilis* (4.3 g each) were stocked into 42 aquaria and fed to satiation twice daily for nine weeks. Individual fish were weighed at the start and the end of the trial, and at three-week intervals in between (Forster et al. 2003).

Results

Harvest levels

The most plentiful finfish harvested and processed in Alaska includes the species in Table 1. Groundfish, as used here, refers to Alaskan pollock, Pacific cod, and the flatfish in the shallow water and deep water assemblages, but excludes salmon, halibut, and blackcod.

Proximate analyses

In the tabular data presented below, the sample codes stay constant across all the tables. Therefore, one can identify the proximate composition, mineral composition, amino acid composition, fatty acid composition, and nutritional efficiency for each groundfish meal. However, our collaborators that produced the meals have requested that our analyses not identify the source of each meal, and we honor that request.

Results from the proximate analyses of the eleven different Alaskan groundfish meals are reported in Table 2. The source code refers to the company that supplied the fish meal. We have reported the data as averages of triplicate measurements for each sample meal. The means for these analyses showed that Alaskan groundfish meal contains about 6.1% moisture, 69.3% protein, 17.0% ash, and 7.6% oil. Ethoxyquin ranged in concentration from 46 to 150 parts per million (ppm) with the average being 87.5 ppm. The means and standard deviations listed at the bottom of the table refer to the average of the reported concentration of each proximate factor for the eleven meals sampled. Customer requirements govern the protein concentrations of these meals. The meals must have more than 65% protein by weight, and all meals tested are higher than this value. To those involved in the marine aquaculture of eels, ash values are not critical; however, ash values can be significant to freshwater salmonid aquaculturists as a water quality issue.

Contaminants

All eleven samples of groundfish meal made in Alaska had contaminant levels at or below the detection limit for common organophosphates, chlorinated hydrocarbons, and polychlorinated biphenyls or PCBs. Since all meals showed identical data, these values for each contaminant are reported only once in Table 3.

Mineral composition

We report on the results of our analysis of the mineral composition of the eleven Alaskan fish meal samples in Table 4. Measurements of calcium, potassium, magnesium, sodium, and phosphorus are reported as percent dry weight. Measurements of silver, cadmium, copper, iron, manganese, nickel, lead, strontium, and zinc are reported as parts per million. The levels of cadmium and lead were below detection limits and represented

Table 1.   Annual commercial landings of the most important fish species in Alaska.

Common name	(Scientific name)	Harvest (metric tons)
Alaskan pollock[a]	(*Theragra chalcogramma*)	1,165,007
Pacific cod	(*Gadus macrocephalus*)	255,155
Pink salmon[b]	(*Oncorhynchus gorbuscha*)	120,668
Yellowfin sole[a]	(*Limanda aspera*)	101,201
Dog salmon[b]	(*Oncorhynchus keta*)	64,309
Atka mackerel[a]	(*Pleurogrammus monopterygius*)	56,277
Red salmon[b]	(*Oncorhynchus nerka*)	53,890
Pacific herring[b]	(*Clupea pallasii*)	39,608
Rock sole[a]	(*Lepidopsetta bilineata*)	33,645
Arrowtooth flounder[a]	(*Atheresthes stomias*)	28,216
Flathead sole[a]	(*Hippoglossoides elassodon*)	26,139
Pacific halibut[c]	(*Hippoglossus stenolepis*)	20,068
Pacific ocean perch[a]	(*Sebastes alutus*)	19,029
Silver salmon[b]	(*Oncorhynchus kisutch*)	10,704
Greenland turbot[a]	(*Reinhardtius hippoglossoides*)	8,875
Sablefish (blackcod)[a]	(*Anoplopoma fimbria*)	8,542
Northern rockfish[a]	(*Sebastes polyspinis*)	3,639
Rex sole[a]	(*Glyptocephalus zachirus*)	2,669
Dover sole[a]	(*Microstomus pacificus*)	2,286
Total		2,024,331

[a]National Marine Fisheries Service 1999.
[b]1998 harvest statistics, Herman Savikko, Alaska Department of Fish and Game, pers. comm.
[c]International Halibut Commission 1998.

only a trace in the meals. The composition data are the averages of triplicate samples of each fish meal. The means and standard deviations listed at the bottom of the table refer to the average of the reported concentration of each mineral for the eleven meals sampled.

Amino acid composition

We report on our measurement of the amino acid composition in ten of the eleven Alaskan fish meal samples in Table 5. The data reported are percentages of total amino acids measured on a weight basis, and are averages of triplicate measurements for each meal sample. The methodology we employed would not easily allow us to measure methionine or cysteine levels. The means and standard deviations listed at the bottom of the table refer to the average of the reported concentration of each amino acid for the ten meals sampled.

Table 2. Alaskan fish meal proximate composition.

Meal code	% Moisture	% Protein	% Ash	% Oil	Ethox. ppm
1	7.6	72.3	15.4	8.2	79.0
2	2.8	66.8	23.5	5.0	150.0
3	8.0	69.7	13.5	6.0	130.0
4	7.8	69.0	15.5	7.4	100.0
5	4.7	68.8	17.8	7.0	91.0
6	8.3	71.8	14.4	7.4	75.0
7	5.8	65.2	21.8	6.3	56.0
8	5.5	69.8	15.8	9.0	57.0
9	7.9	67.2	20.2	6.3	46.0
10	3.4	67.1	19.1	9.5	94.0
11	5.7	74.3	10.1	11.3	84.0
Mean %	6.1	69.3	17.0	7.6	87.5
SD	1.9	2.7	3.9	1.8	31.3

Summary of results from proximate analyses of eleven Alaskan fish meals, plus results from ethoxyquin analyses. The meal code refers to the company that supplied the fish meal. The average percentages of dry weight are derived from triplicate measurements for each sample meal. The means and standard deviations refer to the average of the reported concentration of each proximate factor for the eleven meals.

Table 3. Contaminant detection limits in Alaskan fish meals.

Compound	Detection limit	Compound	Detection limit
Aldrin	<0.05 ppm	Heptachlor epoxide	<0.05 ppm
Alpha-BHC	<0.05 ppm	Hexachlorobenzene	<0.05 ppm
Chlordane	<0.05 ppm	Malathion	<0.05 ppm
Diazinon	<0.05 ppm	Methyl parathion	<0.05 ppm
Dieldrin	<0.05 ppm	Mirex	<0.05 ppm
2,4'-DDD	<0.05 ppm	Methoxychlor	<0.2 ppm
4,4'-DDD	<0.05 ppm	Parathion	<0.05 ppm
2,4'-DDE	<0.05 ppm	PCB-1016	<1 ppm
4,4'-DDE	<0.05 ppm	PCE-1221	<1 ppm
2,4'-DDT	<0.05 ppm	PCB-1232	<1 ppm
4,4'-DDT	<0.05 ppm	PCB-1242	<1 ppm
Endrin	<0.05 ppm	PCE-1248	<1 ppm
Ethion	<0.05 ppm	PCB-1254	<1 ppm
Gamma-BHC (Lindane)	<0.05 ppm	PCB-1260	<1 ppm
Heptachlor	<0.05 ppm	Ronnell	<0.05 ppm

Eleven samples of groundfish meal made in Alaska showed identical levels of common contaminants, including organophosphates, chlorinated hydrocarbons, and polychlorinated biphenyls (PCBs), which were below the detection limits for the methodology used by New Ulm Organics Laboratory.

Table 4.   Alaskan fish meal mineral composition.

	Percent by weight					ppm								
Code	Ca	K	Mg	Na	P	Ag	Cd	Cu	Fe	Mn	Ni	Pb	Sr	Zn
1	4.63	0.57	0.20	1.01	2.85	0.5	a	4.2	48.8	5.0	0.7	a	240	79
2	7.50	0.30	0.33	1.21	4.05	0.9	a	2.1	39.4	8.4	1.4	a	390	90
3	4.58	0.35	0.26	0.65	2.63	0.4	a	3.1	98.4	4.3	0.7	a	236	101
4	5.51	0.24	0.30	0.88	3.00	0.8	a	2.9	64.0	3.9	0.8	a	285	108
5	5.88	0.18	0.26	0.78	3.10	0.8	a	4.4	94.5	4.2	0.8	a	363	111
6	3.64	0.87	0.18	1.05	2.35	0.5	a	4.0	72.5	10.3	0.8	a	209	82
7	8.51	0.18	0.40	0.98	4.39	0.9	a	2.5	74.9	6.3	0.9	a	380	96
8	5.21	0.21	0.30	0.86	2.96	0.7	a	4.2	65.2	3.8	0.7	a	288	119
9	7.50	0.33	0.31	1.06	3.84	0.8	a	1.9	37.1	3.6	0.8	a	389	89
10	6.70	0.42	0.24	0.74	3.79	0.8	a	4.7	49.9	4.0	0.9	a	327	107
11	2.67	0.31	0.23	0.92	1.70	0.5	a	3.5	66.9	2.4	0.5	a	155	95
Aver-age	5.67	0.36	0.27	0.92	3.15	0.69	a	3.41	64.69	5.11	0.8	a	297	97.9
SD	1.8	0.2	0.1	0.2	0.8	0.2		1.0	20.2	2.3	0.2		80	12.5

[a]Trace amount.

The levels of cadmium and lead were below reporting levels and represented only a trace in the meals. The means and standard deviations refer to the average of the reported concentration of each mineral for the eleven meals sampled.

Fatty acid composition

The fatty acid composition of fish oil, extracted from the eleven different Alaskan fish meals, is listed in Table 6. We have not listed those fatty acids that showed no presence even though these were included in our standards. Table 7 lists our summary analyses based on the fatty acid composition of extracted oils from the meals. Several important results show up, the most important is the ratio of polyunsaturated fatty acids to saturated fatty acids, which is relatively high for the extracted oil in all meal samples. Another interesting value is the ratio of omega-3 fatty acids to omega-6 fatty acids in these fish meals, averaging 16.6 for the eleven meals.

Nutritional efficiency

The Pacific threadfin (moi) were fed diets containing various Alaskan fish meals for nine weeks and averaged an increase of more than eleven times their original 4.3 g weight during the trial (Table 8). Our control was Norsk LT-94. The fish meal and lipid levels were adjusted to yield a crude protein concentration of 44% and a crude lipid composition of 11.5%. Table 8 lists the final weight and feed efficiency (wet weight gain for each diet fed) of these Pacific threadfin. The weight gains for the treatments were not significantly different across all the meals tested ($P > 0.5$). However, when feed efficiency was calculated and analyzed, there was a

Table 5. Alaskan fish meal protein amino acid composition as percent.

Meal Code	ALA	ARG	ASP	CYS	GLU	GLY	HIS	ILEU	LEU	LYS	MET	PHE	PRO	SER	THR	TYR	VAL
1	8.48	5.39	14.10	a	17.80	13.28	1.63	3.56	6.40	6.09	a	2.66	5.03	4.76	4.27	2.01	4.56
2	b																
3	7.97	5.37	14.24	a	17.09	10.22	1.90	4.13	7.35	6.11	a	3.27	4.63	4.90	4.98	2.48	5.35
4	8.13	5.29	15.41	a	18.42	10.31	1.86	4.02	7.17	6.72	a	3.03	4.57	4.94	4.80	ND	5.22
5	8.02	5.00	15.10	a	18.02	10.30	1.82	3.88	6.99	6.45	a	2.96	4.44	4.78	4.68	2.39	5.06
6	8.54	5.72	13.48	a	17.02	14.05	1.67	3.52	6.30	5.95	a	2.71	5.22	4.89	4.33	2.04	4.56
7	7.96	5.14	14.44	a	17.61	11.30	1.80	3.93	7.05	5.86	a	3.00	4.84	4.86	4.78	2.40	5.03
8	7.95	5.25	14.33	a	17.36	10.76	1.90	3.92	7.09	6.36	a	3.07	4.57	4.94	4.77	2.54	5.15
9	8.28	5.07	14.85	a	17.66	11.29	1.71	3.87	6.99	6.18	a	2.94	4.86	4.82	4.53	2.25	4.98
10	8.24	5.25	14.39	a	17.36	11.66	1.75	3.90	6.97	5.94	a	2.95	4.98	4.73	4.54	2.29	5.05
11	7.88	5.12	15.66	a	18.91	8.58	1.78	4.04	7.15	6.44	a	3.04	4.14	4.65	4.88	2.53	5.20
Mean	8.15	5.26	14.60	a	17.73	11.18	1.78	3.87	6.05	6.21	a	2.96	4.73	4.83	4.66	2.10	5.02
SD	0.23	0.21	0.66	a	0.59	1.57	0.09	0.19	0.32	0.26	a	0.17	0.32	0.10	0.23	0.73	0.26

[a]Not measured.
[b]Not tested.

ND = not determined.

Meal 2 was not received in time for this analysis. Numbers are averages of triplicate measurements for each meal sample. The sum of the amounts of all the amino acids measured was adjusted to 100 to yield the percent of each measured amino acid in the meal. The means and standard deviations refer to the average of the reported concentration of each amino acid, as a percent, for the meals.

Table 6. Alaska fish meal fatty acid composition.

Fatty acid	1	2	3	4	5	6	7	8	9	10	11	Name	Mean	SD
C14:0	32.2	22.7	23.1	19.8	22.5	27.7	29.4	24.0	24.5	27.5	30.7	Myristic	25.8	3.92
C14:1ω5	1.1	1.4	1.0	0.8	0.9	1.1	1.0	0.9	0.5	1.0	1.0			
Iso-C15:0	0.7	0.0	0.6	0.7	0.8	1.0	0.8	0.8	0.8	0.5	1.0			
C15:0	2.0	1.4	1.7	1.8	1.8	2.2	2.2	1.8	2.0	2.0	2.0			
C16:0	181.6	64.3	146.0	111.0	126.7	123.3	130.8	125.4	130.5	123.1	100.1	Palmitic	123.9	28.66
C16:1ω7	47.2	25.4	35.6	31.5	37.7	50.3	44.8	40.7	41.8	44.8	38.7	Palmitoleic	39.9	7.20
C16:2ω4	1.2	1.6	1.3	2.3	2.0	3.9	2.0	1.8	1.6	1.7	2.3			
C16:3ω4	5.0	5.3	3.5	4.1	4.6	7.3	4.0	4.6	5.2	5.6	10.6			
Iso-C17:0	0.9	1.2	0.7	0.9	0.5	1.5	1.1	1.0	1.0	1.0	0.9			
C17:1ω9	19.5	6.8	14.9	11.2	13.0	11.6	15.0	13.8	16.0	12.8	5.8			
C18:0	27.2	10.2	24.0	21.9	22.4	25.2	23.6	22.7	22.1	10.8	16.0	Stearic	20.5	5.67
C18:1ω9t	0.0	0.0	0.0	0.0	0.0	0.0	0.0	0.0	0.0	0.0	11.5	Elaidic		
C18:1ω9c	134.1	18.0	115.0	95.3	107.6	116.6	110.6	110.3	115.8	108.0	55.6	cis-Oleic	98.8	33.11
C18:1ω7	77.0	5.3	60.2	51.0	59.3	63.4	65.8	61.5	61.9	59.2	36.7	Vaccenic	54.7	19.10
C18:1ω5	1.3	0.0	1.1	2.0	1.9	2.5	1.7	1.8	1.7	1.9	3.8			
C18:2ω6t	0.5	0.0	0.8	0.7	0.7	0.8	0.8	0.4	0.0	0.8	0.3	t-Linoleic		
C18:2ω6c	4.7	3.0	5.1	6.1	6.8	5.6	8.3	6.8	6.6	7.0	5.2	cis-Linoleic		
C18:3ω4	2.6	1.9	2.1	2.1	2.5	3.1	2.4	2.5	2.6	2.6	2.9			
C18:3ω3	2.3	2.1	2.5	3.2	3.1	2.9	3.8	3.1	3.2	3.5	2.4	Linolenic		
C18:4ω3	7.6	7.0	9.5	13.1	13.9	9.9	13.3	13.9	14.5	8.2	11.3			
C20:0	1.2	0.0	1.0	1.2	0.6	1.0	0.9	0.5	1.7	1.0	1.5	Arachidic		
C20:1ω11	10.5	61.1	17.6	29.5	18.3	18.0	14.2	29.4	25.5	34.6	79.8	Gadoleic	30.8	21.32
C20:1ω9	9.1	19.8	9.5	14.0	3.2	14.3	3.6	16.6	15.5	16.3	29.1			
C20:1ω7	4.6	1.4	3.0	2.7	0.0	5.3	0.6	3.1	3.8	3.0	0.0			
C20:2ω6	1.4	0.0	1.3	1.5	1.7	1.5	1.8	1.7	1.7	1.6	1.3			
C20:3ω6	0.5	0.0	0.4	1.0	3.3	1.2	5.7	0.5	0.0	1.0	1.2			
C20:4ω6	5.8	2.5	5.9	7.2	3.2	0.0	0.0	6.4	5.3	4.5	5.7	Arachidonic		
C20:4ω3	3.1	1.8	2.5	3.1	3.8	3.5	3.5	3.8	3.8	3.4	4.1			

Table 6. **(Continued). Alaska fish meal fatty acid composition.**

Fatty acid	Meal code											Name	Mean	SD
	1	2	3	4	5	6	7	8	9	10	11			
C20:5ω3	**142.5**	**70.2**	**114.7**	**112.5**	**129.7**	**136.9**	**111.3**	**99.1**	**128.4**	**116.2**	**111.7**	**EPA**	**115.8**	**19.79**
C21:0	1.3	0.0	1.3	1.4	0.0	1.3	0.0	0.6	1.6	1.3	1.6			
C22:1ω11	**11.2**	**69.8**	**15.2**	**30.8**	**33.1**	**17.3**	**19.8**	**34.7**	**31.4**	**43.5**	**94.6**	**Cetoleic**	**36.5**	**25.19**
C22:1ω9	2.8	5.3	2.6	4.4	4.9	3.2	4.2	4.7	4.4	5.8	6.7			
C22:1ω7	1.7	0.0	1.6	1.7	1.8	1.6	1.6	1.7	1.7	1.7	2.2			
C21:5ω3	5.5	3.1	4.4	4.1	4.9	5.6	4.2	4.8	5.0	4.8	4.9			
C22:5ω3	7.9	6.9	5.9	7.4	7.8	13.8	7.1	7.3	6.8	6.3	10.7	DPA		
C22:6ω3	**55.4**	**54.3**	**83.0**	**91.2**	**87.7**	**78.1**	**80.9**	**80.7**	**71.8**	**63.2**	**92.5**	**DHA**	**76.3**	**13.47**
C24:1ω9	2.5	3.3	2.5	3.9	4.0	1.9	3.4	3.5	3.4	1.5	4.1			

Data expressed as mg per g of oil extracted from meals. Each record is the average of two gas chromatography runs for the fatty acids present in the oil extracted from the samples. For the most abundant fatty acids (in bold), means and standard deviations are also listed.

Table 7. Gas chromatographic analysis of fatty acid composition in Alaskan fish meals.

	Meal code											Mean	SD
	1	2	3	4	5	6	7	8	9	10	11		
Σ saturated FA	247	100	198	159	175	183	189	177	184	167	154	176	26
Σ mono-unsat'd FA	323	223	280	279	286	307	286	323	323	334	369	310	29
Σ polyunsat'd FA	246	160	243	260	276	274	249	237	257	230	267	255	15
Polyunsat'd/sat'd	1.0	1.6	1.2	1.6	1.6	1.5	1.3	1.3	1.4	1.4	1.7	1.5	0.2
ω3	224	145	222	235	251	251	224	213	234	206	238	230	15
ω6	12.8	5.5	13.5	16.5	15.8	9.1	16.6	15.8	13.6	14.9	13.6	14.4	2.3
ω3/ω6	17.6	26.5	16.5	14.2	16.0	27.5	13.5	13.6	17.2	13.8	17.5	16.6	4.1
Σ FA identified	816	479	721	697	737	764	724	737	764	732	790	741	36
Σ FA saponifiable	840	534	744	717	783	811	759	790	785	767	853	779	42
Σ Non-id FA	24	55	23	20	46	47	35	54	21	36	63	38	15

Units are mg per g of oil extracted from meal. FA = fatty acid.

Numbers are sums in mg per g of extracted oil for each category of fatty acids. The sum of FA identified indicates the amount of fatty acid we were able to identify in mg per g of extracted oil. The sum of FA saponifiable indicates the amount of lipid, in mg per g extracted, that was subject to saponification, i.e., as a mono-, di-, or triglyceride. The sum of non-identifiable FA is equal to the total FA identified and subtracted from the total saponifiable.

statistical difference between the efficiency of one Alaskan fish meal (no. 3) when compared to a second meal (no. 7). Except for meal no. 7, none of the Alaskan fish meals were statistically different in effectiveness from the Norsk LT-94 control.

Discussion
Quality of Alaskan groundfish meal
We have documented the high quality of Alaskan groundfish meals. This report substantiates the conclusion that these fish meals are high quality products useful in formulating aquaculture feeds. Alaskan meals have proximate compositions that are equal to or exceed those reported for other available meals (Babbitt 1990, Hardy and Matsumoto 1990, Tacon 1994). The contaminants tests showed nothing above detectable limits for organophosphates, chlorinated hydrocarbons, or polychlorinated biphenyls. We have included detailed analyses of the mineral content and amino acids in Alaskan fish meal. Only a trace of either lead or cadmium were found, well below reporting levels. This is a reasonable conclusion given that Alaska is not heavily industrialized and commercial fish harvests occur in some of the most pristine waters available.

Alaskan fish meals contain a balance of amino acids, including lysine and threonine, two essential amino acids for many carnivorous species (Tacon 1994). The most abundant fatty acids are consistent with previous reports on fish and fish meal from the North Pacific (Gruger at al. 1964, Hardy and Matsumoto 1990). The relatively high ratio of omega-3 fatty acids to omega-6 is probably of significant value, as the relatively high ratio of polyunsaturated fatty acids to saturated. Alaska fish meals have been reported to suffer little from problems in rancidity (Hardy and Matsumoto 1990). Alaskan fish meal producers use the antioxidant ethoxyquin, which was detected at an average level of 87 ppm in the meals. However, the low rancidity is also probably related to the fact that the processing byproducts are derived from human food grade fish and that they are handled quickly. Finally, the nutritional test of the eleven Alaska fish meals showed clearly that when compared to an industry standard meal, fish growth of Pacific threadfin is as good or better in terms of weight gain and feed efficiency.

In part we have analyzed these Alaskan groundfish meals rigorously in order to document a baseline for future work. In the dozen years that have elapsed since the first International Conference on Fish By-Products, the average protein levels in Alaskan groundfish meal have increased from roughly 63% (Babbitt 1990) to the current average of 69.3% as documented in our proximate analyses. We anticipate that in the future, the byproducts of seafood processing in Alaska will provide fertile ground for new technologies and added values. Predicting these changes is challenging. They are especially challenging given the fluctuating nature of Alaskan

Table 8. Alaskan fish meal nutritional effectiveness.

Meal code	Final weight[1] Mean	SD	Feed efficiency[2] Mean	SD
Norsk LT-94	49.7[a]	4.9	0.89[ab]	0.07
1	50.0[a]	5.2	0.86[ab]	0.04
2	47.6[a]	0.8	0.82[ab]	0.02
3	51.2[a]	0.7	0.93[a]	0.01
4	50.8[a]	3.9	0.91[ab]	0.02
5	49.5[a]	0.8	0.90[ab]	0.02
6	52.5[a]	2.6	0.92[ab]	0.03
7	44.6[a]	5.6	0.79[b]	0.1
8	47.8[a]	2.1	0.88[ab]	0.02
9	46.9[a]	5.3	0.84[ab]	0.07
10	49.4[a]	1.5	0.86[ab]	0.02
11	45.0[a]	2.7	0.82[ab]	0.02

[1]Starting weight = 4.3 g.

[2]Feed efficiency = wet weight gain/diet, fed to Pacific threadfin.

[ab]Means within a column sharing a common letter are not significantly different ($P \geq 0.05$) using Tukey's multiple comparison test.

Values are means and standard deviations for three groups.

wild fish harvests as well as the fluctuating nature of the processing waste stream. We discuss some of the reasons for these fluctuations below.

Harvest volume fluctuations

Harvest volumes can change over time, affecting the amount of processing byproduct available for Alaskan fish meal. One reason for harvest fluctuations is catch quotas. Alaskan fish landings are well regulated and most species have population levels that are relatively stable. Should population levels decrease, the management agencies responsible lower the harvest quotas. This can have an effect on the volume of processing byproducts available for making Alaskan fish meal, but to date any changes in the allowable catch have been relatively small. The North Pacific Fishery Management Council (http://www.fakr.noaa.gov/npfmc), on the basis of recommendations by their Scientific and Statistical Committee, sets harvest quotas for groundfish species. The Alaska Department of Fish and Game sets harvest levels for Pacific salmon species and several others. The International Pacific Halibut Commission sets harvest quotas for halibut. Because of the active management of these organizations, Alaska's fisheries have been declared sustainable.

Harvest levels for some species have changed over a decadal time scale. Kodiak annually landed roughly 100,000 t of crab and shrimp in the

1970s. By 1983, king crab (*Paralithodes camtschaticus*) and pink shrimp (*Pandalus borealis*) were no longer commercially harvested in Kodiak, but roughly 100,000 t of pollock was annually harvested through the 1980s. By the 1990s, Kodiak had expanded the number of species of groundfish harvested to include Pacific cod, but total groundfish harvest levels still hovered at near 100,000 t per year (NMFS harvest data available on the Web). The reasons for these faunal shifts probably include the Pacific multi-decadal oscillation (Anderson and Piatt 1999). The North Pacific appears to be the northern-most limit to the range of some species and the southern limit of others. As water temperatures fluctuate, fish move according to their physiological preferences.

EPA regulations

Alaskan fish processors make fish meal, bone meal, fish oil, and stick-water from the byproducts of seafood processing because of state and federal requirements. In implementing the Clean Water Act of 1974, the Environmental Protection Agency (EPA) and the Alaska Department of Environmental Conservation (ADEC) determined that Kodiak, Alaska, was a "seafood processing center" and, as such, processors in Kodiak needed to employ effective methods, other than dumping into local waters, for handling the byproducts of their seafood processing. Other operations in Alaska at the time were much smaller in scale and they were allowed to handle their seafood water by "grinding and dumping" the byproducts of processing into local adjacent waters, with some controls under a General Discharge Permit. The current General Discharge Permit, released in 2002, while still allowing the grind and dump operations, contains new rules that include reduced discharge volumes, reduced debris field size, and more restrictive water quality standards.

Prior to the Clean Water Act of 1974, seafood processors in Alaska generally dumped unprocessed the byproducts of seafood processing into local waters. Anecdotal information suggests that this dumping increased the populations of local megafauna scavengers such as crab, yet solid scientific documentation of this is a challenge to discover. EPA and ADEC determined that the degradation of fish processing waste would fall primarily to microbial communities, not to megafauna, such as crab. This led to the adoption of the "grind and dump" requirements to maximize the surface area for the most effective microbial action that applies to the smaller and seasonal processors in Alaska. In general, most of these smaller and seasonal operations focus on salmon processing, and more specifically salmon canning, where the volumes of waste ground and dumped are relatively small.

All the newer higher volume seafood processing operations, as well as the community of Kodiak as a Seafood Processing Center, have had to meet the requirements of individual Special Discharge Permits, regulating their implemented waste handling systems. This is to avoid being in violation of EPA and ADEC rules. If these operations did not have Special Discharge Permits and fish meal plants, they would not be allowed to process fish.

Waste stream composition

The nature of the byproducts of seafood processing often change; this can be a conundrum for the manufacture of fish oil, fish meal, and bone meal. The composition of the waste stream in Alaska fish processing plants depends heavily, not only on the species being processed, but also on the product form the processor is making (Crapo et al. 1993). When pollock is made into surimi at shoreside plants, roughly 20% of the round weight of the fish is converted to food, leaving 80% as processing byproduct. When pollock is made into skinless fillets, roughly 34% of the round weight becomes food. Skinless fillets from Pacific cod account for about 40% of the fish round weight, whereas skinless boneless fillets (J-cut) average only 26%. When pink salmon are canned, about 65% of the round weight is made into food, leaving only 35% in the waste stream. Yet, when pink salmon are made into skinless fillets, these account for 33% of the round weight, and the waste stream accounts for two-thirds of landed fish weight. Therefore, the volume of fish processing byproducts is dependent on the customer requirements at the specific processing facility and is challenging to predict accurately.

Seasonal fluctuations

Lipid levels can show surprising fluctuations depending on the time of year the fish is caught. Fish species regulate the amount of lipid in their bodies over their annual cycle (Shirai et al. 2002). Generally among more northern species, somatic lipid levels are lowest at spawning and highest about 6 months later. This is probably due, at least in part, to sequestration of the lipoprotein vitellogenin in oocytes during the reproductive buildup and storing energy as lipid for nutrition during the poorly productive winter season. The differences can be spectacular. Atlantic herring (*Clupea harengus*) can have somatic lipid levels as low as less than 2% at spawning to as high as more than 20% six months later (McGurk et al. 1980). There are also marked changes in the fatty acid composition of fish oils on a seasonal basis (Aidos et al. 2002). Significant seasonal change in somatic lipid levels also occurs in Alaskan pollock and Pacific cod. The data indicate that the season of harvest can have a measurable effect on the types of consumer products made and the composition of the waste stream produced in Alaskan seafood processing plants.

Unit operations in handling fish waste

The unit operations involved in handling seafood processing wastes are somewhat dependent on the kind of equipment used and the types of secondary products made. In general, the fish processing waste is ground up, cooked at close to the boiling point of water, dewatered either with a screw press or a decanter centrifuge, and the resultant solid phase dried. Bone can be separated from the protein by screening after drying, or it can be separated after mincing and prior to cooking. Deboned fish meal is principally protein. The liquid phase can be centrifuged to remove suspended solids. To separate oil from the aqueous phase, the liquid is sent through separators, sometimes referred to as refiners. The deoiled aqueous phase is stickwater, which contains substantial amounts of water, and this is usually concentrated by dewatering in some manner. There are numerous different approaches to completing these unit operations.

Secondary products made from fish waste

The efficient utilization of the waste stream from fish processing plants has two major quality considerations. First, the source materials are from high quality human food processing facilities and are handled expeditiously allowing for little microbial degradation. Second, there is a much higher ratio of skin and bone to flesh protein in processing byproducts compared with whole fish reduction operations. This last issue has led to a flawed perception that fish meal made from the groundfish processing byproducts in Alaska is of lower quality than fish meals from other sources—specifically that ash levels are higher. In reality, ash levels in Alaskan fish meals are closely regulated by the fish meal producers to meet customer demands and world fish meal standards.

There are three or four secondary product types made from the byproducts of seafood processing. These are fish protein meal, fish bone meal, fish oil, and stickwater. The stickwater contains a high concentration of the water soluble components and can have a biological oxygen demand (BOD)—a measure of the concentration of organics in solution—averaging from 30,000 to 60,000, and running as high as 150,000 (G. Peters, Alyeska Seafoods, Unalaska, Alaska, pers. comm.). Stickwater from those waste processing operations that employ seawater as the fluming agent have relatively high levels of sodium chloride. Customer requirements mean that stickwater from these operations cannot easily be added back to the dried meal. In those operations where fresh water is used for fluming, the organic content of the stickwater can be used to augment the fish meal.

High volume wet reduction systems

No single technology for making profitable products from the waste stream will address the needs of all seafood processing operations in Alaska cost effectively. The major distinguishing criterion is the volume of waste to be handled. Some of the major shoreside seafood processing operations on the Bering Sea can process as much as 1,500 t of fish per day. If the plant were making surimi with a 20% conversion efficiency (round weight into human food weight), this would equate with roughly 1,200 t of processing byproduct per day. The wet reduction method used in equipment made by Atlas Stord can effectively handle these volumes. Unless there are volume projections that range near this 1,200 t per day limit, there may be more cost effective alternatives available. The capital expense of an Atlas Stord meal plant can be considerable. The fish meal made from these systems generally is sold at roughly 65% protein, 17% ash, 8% oil, and 10% moisture. Virtually all the operating fish meal plants in Alaska are Atlas Stord systems. This includes the seven shoreside operations and the several on-board operations in the at-sea factory trawl fleet.

Moderate volume wet reduction systems

Alfa Laval pioneered a smaller capacity plant in Kodiak a few years ago called the ConKix system (Babbitt at al. 1994). A wet reduction system, the unit operations are similar to that of the Atlas Stord systems. Designed to produce a high quality protein powder from fish processing waste, the system employed a Brown finisher early in the process to remove bone prior to cooking. When installed in the processing plant in Alaska, the devices in the ConKix system were linked together with stainless steel pipe assuring the highest quality possible. The system had a capacity of roughly 100 t per day, and the protein meal made with it was used to feed children in South America. This technology, with modifications, can probably handle up to 250 t per day cost effectively. However, it would probably not be cost effective to operate this system with much less than 25 t per day. The proximate composition of the fish protein meal made by the Alfa-Laval ConKix system was about 78% protein, 6% ash, 10% oil, and less than 10% moisture. This specific plant is no longer in operation in Alaska, but some of the same equipment is now installed in a fish meal operation outside Astoria, Oregon. None of our test fish meals were made with the Alfa-Laval ConKix system.

Low volume enzyme hydrolysis

There have been several attempts at operating an enzymatic hydrolysis–based system to handle seafood processing wastes in Alaska. Secondary products produced in a hydrolysis system can include the liquid hydrolysate, protein meal, fish oil, and bone meal. Available information suggests these will be most cost effective with smaller volumes of processing waste, in the range of 5-25 t per day. One such system, employing equipment made by Alfa-Laval, is installed in Alaska now. Unit operations in a hydrolysis system are somewhat different from those found in the wet reduction method. Fish waste is ground up, the hydrolytic enzyme metered in at a specific concentration, and hydrolysis accomplished with gentle heating and mixing. The tissues become liquid after hydrolysis and are subsequently heated to denature the enzyme and then sieved to remove bone. This liquid can then be sent to a refiner to separate the oil, while the aqueous phase must be dewatered to produce the protein meal.

The advantage of hydrolysis systems is their lower capital cost, but this is balanced by their lower capacity. A wide variety of different commercial enzymes is available to accomplish the liquification of the fish processing waste (Hall and Ahmed 1992, Flick and Martin 2000), and in addition local fish species contain heat-activated proteases of considerable aggressiveness which could be used to hydrolyze fish wastes (Wasson et al. 1992, Tschersich and Choudhury 1998).

Other handling methods

There are other methods to effectively handle the byproducts of seafood processing, but they generally involve making fertilizers rather than fish meal. One used in Alaska involves the use of mineral acids to make acid silage—a fish fertilizer (Hall et al. 1985, Arason et al. 1990). Similar to enzymatic hydrolysis in chemical mechanism, acid hydrolysis differs in generally being allowed to proceed to completion assuring that the proteins are fully broken down to their constituent amino acids. In the United States where phosphoric acid is often used, the sum of bone calcium phosphate plus the phosphate in the acid yields the phosphate concentration of the final silage. Acidified hydrolysate is shelf stable against microbial degradation if the pH is lower than 4.5. Another option involves the EPA's mandate which extends to the 12 mile limit. Should processors choose to they may legally barge the byproducts of seafood processing past the 12 mile limit and dump them into the offshore waters without penalty.

Conclusions

Our data show that groundfish meal, made from the byproducts of seafood processing in Alaska, is a high quality commodity easily comparable to other fish meals available in world markets (Babbitt 1990, Hardy and Matsumoto 1990, Tacon 1994). The fluctuations in the volume of processing byproducts caused by changing product form and small changes in harvest quota levels both are generally overwhelmed by the volume of fish harvested in Alaska. Seasonal fluctuations in the proximate composition of the landed species will be most apparent in the volume of fish oil recovered, but this has little impact on the composition of the fish meals. Until seasonal analyses of the fatty acid composition of fish lipids are completed we will not know whether there are significant changes in the nature of the fish oil recovered. The data presented show that ash levels are managed effectively even though the greater part of the fish flesh has been removed for the human food market. The real test of a fish meal is how it performs nutritionally; the data from the moi feeding trials show that Alaskan fish meal was comparable to the Norsk LT-94 control meal. Alaskan fish meal contains relatively high levels of the amino acids, such as lysine and threonine, critical for aquaculture of carnivorous fish. The ratio of polyunsaturated to saturated fatty acids averages greater than 1.0 for the meals tested. In addition, the ratio of omega-3 to omega-6 fatty acids is relatively high at between 12 and 15 to 1. Since these criteria are of significance in human diets, they are valuable in feeds formulated for other fish species that will become human food.

If all the byproducts of seafood processing in Alaska were available to be made into fish meal, bone meal, and fish oil, we could generate a sizable volume of each, but the fluctuations discussed above make greater precision difficult. If one accepts an average conversion efficiency of 25% (round weight into human food), then processing annual harvests of 2,000,000 t would result in 1,500,000 t of wet byproduct. Assuming that the solids comprise 25% of the wet byproduct, this yields 375,000 t of dried protein, oil, stickwater solubles, and bone. Our last survey showed that only about 65,000 t of 65-70% protein fish meal are being currently made in shoreside and at-sea reduction plants in Alaska. Figures for stickwater, bone meal, and oil are more difficult to discover. There is significant opportunity for the conversion of these unutilized processing byproducts in making high quality fish meal.

Acknowledgments

We wish to acknowledge the assistance of the seafood processors of Alaska in completing this work. Without their help we would not have been able to compile as much significant information as we did. We also acknowledge the help of the U.S. Department of Agriculture, and specifically Peter Bechtel of the Agricultural Research Service. ARS provided financial support for this work.

References

Aidos, I., A. van der Padt, J.B. Luten, and R.M. Boom. 2002. Seasonal changes in crude and lipid composition of herring fillets, byproducts and respective produced oils. J. Agric. Food Chem. 50(16):4589-4599.

Alaska Department of Fish and Game. 2002. Alaska's commerical salmon fishery. http://www.state.ak.us/adfg/geninfo/special/sustain/fishery.pdf.

Anderson, P.J., and J.F. Piatt. 1999. Community reorganization in the Gulf of Alaska following ocean climate regime shift. Mar. Ecol. Prog. Ser 189:117-123.

Arason, S., G. Thoroddsson, and G. Valdimarsson. 1990. The production of silage from waste and industrial fish: The Icelandic experience. In: Making profits out of seafood wastes: Proceedings of the International Conference on Fish By-Products. Alaska Sea Grant, University of Alaska Fairbanks, pp. 79-85.

AOAC. 1996. Official methods of analysis of AOAC International, vol. 1-2. AOAC (Association of Official Agricultural Chemists) International, Gaithersburg, Maryland.

Babbitt, J.K. 1990. Intrinsic quality and species of North Pacific fish. In: Making profits out of seafood wastes: Proceedings of the International Conference on Fish By-Products. Alaska Sea Grant, University of Alaska Fairbanks, pp. 39-43.

Babbitt, J.K., R.W. Hardy, K.D. Reppond, and T.M. Scott. 1994. Processes for improving the quality of whitefish meal. J. Aquatic Food Product Technol. 3(3): 59-68.

Crapo, C., B. Paust, and J.K. Babbitt. 1993. Recoveries and yields from Pacific fish and shellfish. Alaska Sea Grant, University of Alaska Fairbanks. 32 pp.

EPA (U.S. Environmental Protection Agency) Region 10. http://www.epa.gov/region10.

Flick Jr., G.J., and R.E. Martin. 2000. Industrial products: Hydrolysates. In: R.E. Martin, E.P. Carter, G.J. Flick Jr., and L.M. Davis (eds.), Marine and freshwater products handbook. Technomic Publications Co., Inc., Lancaster, Pennsylvania, pp. 619-625.

Gruger, E.H., R.W. Nelson, and M.E. Stansby. 1964. Fatty acid composition of oils from 21 species of marine fish, freshwater fish and shellfish. J. Am. Oil Chemists Soc. 41(10):662-667.

Hall, G.M., and N.H. Ahmad. 1992. Functional properties of fish-protein hydrolysates. In: G.M. Hall (ed.), Fish processing technology. Blackie Academic and Professional, Glasgow, pp 249-274.

Hall, G.M., D. Keeble, D.A. Ledward, and R.A. Lawrie. 1985. Silage from tropical fish 1. Proteolysis. J. Food Technol. 20:561-572.

Hardy, R.W., and T. Masumoto. 1990. Specifications for marine by-products for aquaculture. In: Making profits out of seafood wastes: Proceedings of the International Conference on Fish By-Products. Alaska Sea Grant, University of Alaska Fairbanks, pp. 109-120.

International Pacific Halibut Commission. 1998. 1998 harvest statistics. http://www.iphc.washington.edu.

Maxwell, R.J., and W.N. Marmer. 1983. Fatty acid analysis: Phospholipid rich samples. Lipids 18:453-459.

McGurk, M.D., J.M. Green, W.D. McKone, and K. Spencer. 1980. Condition indices, energy density and water and lipid content of Atlantic herring (*Clupea harengus*) of southern Newfoundland. Can. Tech. Rep. Fish. Aquat. Sci. 958. 41 pp.

National Marine Fisheries Service. 1999. Fisheries of the United States, 1998. Current fishery statistics no. 1998. NOAA, NMFS, Bethesda, Maryland.

National Marine Fisheries Service. 2001. Fisheries of the United States, 2000. Current fishery statistics no. 2000. NOAA, NMFS, Bethesda, Maryland.

North Pacific Fishery Management Council. http://www.fakr.noaa.gov/npfmc.

Radin, N.S. 1981. Extraction of tissue lipids with a solvent of low toxicity. In: J. Lowenstein (ed.), Methods of enzymology, vol. 72. Academic Press, New York, pp. 5-7.

Shirai, N., M. Terayama, and H. Takeda. 2002. Effect of season on the fatty acid composition and free amino acid content of the sardine *Sardinops melano-stictus*. Comp. Biochem. Physiol. B. Biochem. 131(3):387-393.

Tacon, A.G.J. 1994. Feed ingredients for carnivorous fish species: Alternatives to fishmeal and other fishery resources. FAO Fish. Circ. No. 881. 35 pp.

Tschersich, P., and G.S. Choudhury. 1998. Arrowtooth flounder (*Atheresthes stomias*) protease as a processing aid. J. Aquat. Food Product Technol. 7(1):77-89.

Wasson, D.H., J.K. Babbitt, and J.S. French. 1992. Characterization of a heat stable protease from arrowtooth flounder (*Atheresthes stomias*). J. Aquat. Food Product Technol. 1(3/4):167-182.

Waters Inc. 1993. Waters AccQ Tag chemistry package instruction manual: WAT052874. Millipore Corporation, Milford, Massachusetts. 88 pp.

Ketchikan Seafood Byproduct Studies: Partnerships in Problem Solving

Stephanie Madsen
Pacific Seafood Processors Association, Juneau, Alaska

Pacific Seafood Processors Association (PSPA)

I have been a resident of Alaska for 30 years, having lived in Cordova, Kodiak, Dutch Harbor, and Juneau. I have always been involved in seafood in some way.

PSPA is a trade association of seafood processors and has been in existence since 1914. Representing most of the major seafood producers from salmon-only processors to white fish processors, the PSPA also has three "mothership" operations: in the Bering Sea, off the coast of Oregon, and in the Washington fisheries. PSPA is located onshore in coastal Alaska. Because the seafood industry is the number-one private employer and the second largest taxpayer in Alaska, we are, in our opinion, an important part of the economy.

Processors' perspectives on seafood byproducts

Turning to the processors' perspective of seafood byproducts, we first must admit that if there is money in it, we do it. Second, Alaska's geography and terrain limits building construction. Most communities are remote, and transportation is limited between processors and coastal communities. Populations are centralized, which creates logistical and geographical limitations. Flat land is generally unavailable except near the water. Thus, obtainable space to build facilities is at a high premium.

Third, seafood processors are very diverse. Large operations in Kodiak, for example, handle multiple species, primarily pollock. These fisheries management systems can fish for four to six months, making their profit margins greater than those for the salmon-only processors.

Salmon-only processors have a real challenge. Depending on the area, a season is only four to six weeks. Furthermore, these processors cannot control the quality, timing, and volume of their raw material. If they had control, they could take time to handle product differently and experiment with it. The food management structure has changed the variables. Most fisheries make money by volume. Though the salmon-only fisheries want well-valued fish and high volume, they must operate as quickly as possible because their time is so limited. Those fisheries that work with white fish and halibut are fortunate to have their fishing season from March to December.

Finally, regulatory compliance problems mostly impact salmon-only fisheries. Many times, if a solution is economically feasible, a company cannot implement it to meet environmental or other regulations. PSPA thinks that it is fine to put fish back into water as parts, depending on the volume, location, and timing. We don't believe grinding is practical; this is addressed under the "Ketchikan seafood waste study" section of this paper.

I find it difficult to talk to you about all of your studies, because I understand that there is some reluctance for processors to get involved. Even though I represent 14 processors, they compete with each other. They will not discuss things with their neighbors, including their research and development with byproducts. During different times of profitability, they have looked at diverse things like value-added processes, but this often produces more waste—that is, more byproducts from their primary product.

Ketchikan seafood waste study

The Ketchikan Narrows is a problem area for salmon processing. There are five processors in a small area, and discharging fish processing waste is an issue. They used to grind and discharge their byproducts, and the animals would help by eating it. They put more money into grinders and other machinery. But the discharge soon compacts, crusts over, puffs and smells, and then burps. We struggle with how to repair this and have considered alternatives, such as producing fish meal. In this scenario, all processors would take their byproducts to the fish meal plant, giving the Ketchikan residents new odors, discharges, and different consequences to deal with. Furthermore, the fish meal plant would need to burn more diesel fuel to operate. Let us now move on to problem solving.

Solving problems

Two people, present at the 2nd International Seafood Byproducts Conference, know more about our problem solving strategy than I do: Kevin George with the Alaska Department of Environmental Conservation and

Lance Miller of the Juneau Economic Development Council. Some of the processors that are actually involved in the studies attended the conference.

Specifically, we must consider the following items:

1. Test the regulation. Is discharging ground heads better than discharging whole heads?

2. The environment around discharge piles has changed. Prior to the current grinding regulation, halibut and blackfish roamed the area, but now sharks do. In a public hearing, residents and fishermen stated that they did not remember the burping, crusting, and sliding problem, etc., before the current grind and discharge regulation.

3. What would be the impact on the zone of deposit? Can we use mechanical mitigation/aeration of the piles, so that we can move them around? If we are allowed to do that, then piles would shrink and we could comply with regulations. This would incur short-term water quality violations, because it can't move a pile without resuspending some of the fine solids.

4. What issues affect the size of the mixing zone? Not everything will stay within that zone, due to currents and the like. What impact would there be to the environment if regulators gave us a larger mixing zone? What are the trade-offs?

We need to consider the zone of deposit, grind-size studies, and pile remediation project. Furthermore, the PSPA hopes to implement alternative technologies in Ketchikan to discover what the trade-offs may be and whether such would be economically feasible.

A true partnership composed of the Environmental Protection Agency, the Alaska Department of Environmental Conservation, seafood processors, and consultants now exists. The group is looking at study design. With the current fisheries management system, there is less time for the processors to find markets for byproducts. PSPA encourages you to keep researching, but be sensitive to bottom line and some of our difficulties in Alaska, including those related to the remote locations.

Increasing the Value of Alaska Pollock Byproducts

Joe Regenstein
Cornell University, Ithaca, New York

Susan Goldhor
Center for Applied Regional Studies, Cambridge, Massachusetts

Don Graves
Unisea Inc., Dutch Harbor, Alaska

Introduction

The initial work was done in Dutch Harbor at the UniSea plant during the pollock B fishing season (i.e., when roe is not ripe and not taken as a major product). The researchers had complete access to every aspect of waste handling and byproduct utilization. A lab was made available inside the fish meal plant and the researchers were briefed extensively on all aspects of byproducts, including economics. This was important, and let us know what consequences would ensue from the extraction of particular organs from the fish meal line. UniSea also provided us with an experienced assistant, who was an invaluable partner (Ray Harris). By working with him, we were also confident that requests for materials from the plant would be obtained accurately and promptly.

After our initial orientation to UniSea's operations, and with the agreement of Don Graves, UniSea's head of research and development, and our primary contact, we decided to explore the following research avenues:

- Pollock liver oil, and the possibility of low temperature extraction.

- Pollock visceral oil as a potential astaxanthin source.

- Pollock scales and their potential use to flocculate solids out of stickwater, either alone or with the help of an edible gum as a coagulant/flocculant.

- Pollock skin and bones as gelatin sources.

- Pollock viscera prior to and subsequent to stomach removal as feedstocks for hydrolysis.

- Pollock frame mince as a food ingredient.

Each of these goals had a rationale either because of potential value of the product in the marketplace, or because the raw material in question had a negative value to UniSea in its current operation and the new use would increase its value. Pollock liver oil and visceral oils are currently burned as fuel on site. This gives them a value of approximately $1.20 per gallon, the price of diesel delivered into Dutch Harbor/Unalaska. This is competitive with what the company can get, selling them as undifferentiated fish oil, given the high cost of transport to and from Unalaska.

We were curious as to whether they would command higher prices as separated liver and visceral oils. We isolated each oil at the lowest possible temperature, and both oils were sent out for analysis; liver oil for fatty acid composition and vitamin content, and visceral oil, which is a deep red color, for astaxanthin content. A high astaxanthin oil would be very valuable for some aquaculture operations, combining the benefits of fish oil with pigment.

The use of scales as a flocculant was pioneered by Robert R. Zall and his group at Cornell University about 20 years ago. Because stickwater in Dutch Harbor is pumped out to sea, and because BOD (biological oxygen demand) is always an issue, UniSea would like to reduce their BOD discharge. Flocculating some of the suspended solids in the stickwater seems to be a rational way of initially accomplishing this goal. The advantages are obvious: using an essentially free material (scales) as the flocculating agent and using a material that is organic and could be put back into the meal. Basically we'd be intercepting it, increasing its protein content by flocculating some of the stickwater solids, and, thus, yielding a net increase in fish meal protein content.

The markets for fish gelatin are increasing. Not only is fish gelatin especially useful for products that melt at relatively low temperatures (frozen desserts, yogurts, etc.), but fish gelatin is the only gelatin that can easily meet the most stringent kosher and halal requirements with sufficient quantities to meet commercial needs. In addition, the Japanese and other countries are moving to fish gelatin as a way to avoid consumer concerns about BSE (mad-cow disease). Fish bones can be screened out of the meal, and fish skin actually has negative value in the meal since it (counter-intuitively) lowers the protein level and raises the ash level. Using skin and bones for gelatin was one of the driving forces behind this project. Because of the low extraction temperatures for fish gelatins, this can be a relatively economical process even in remote locations. However, fish bones need to be obtained at temperatures below those used for fish meal, so hydrolysis of adhering flesh will be necessary to obtain clean bones (see below).

Hydrolysates (or digests) can be engineered to enter a variety of markets, and can have values ranging from low to extremely high. During this first round of work, our concern was simply to make autolysates, and to see if the material left after the removal of the pollock stomachs (which are sold as a specialty item) would make an autolysate comparable to that made of whole viscera.

Frame mince is meat taken off the bones and, in particular, the backbone. Because this meat is dark in color, containing blood and other pigments, and because these components also add strong flavors (including some distinct "meaty" notes), frame mince has been undervalued and usually discarded along with the bones, while the white trim mince has been used. However, it is our contention that frame mince has greater nutritional value than trim mince, precisely because it offers bio-available heme iron, and it has greater culinary value in dishes where strong, meaty flavors are desirable. We were able to use equipment already in place in UniSea to collect frame mince, to taste it, and to successfully present it to UniSea's executives as a finished dish, which, much to their surprise, they enjoyed.

Our work is described below in the form of four reports on specific phases of the work with the researchers identified.

Project 1: Extraction of gelatin from pollock skin

Peng Zhou[1] and Joe M. Regenstein

Introduction

The waste from fish processing after filleting can account for as much as 75% of the total catch weight (Shahidi 1994). About 30% of such waste consists of skin and bone with high collagen content that can be used to produce fish gelatin (Gómez-Guillén et al. 2002). The quality of the prepared gelatin depends on its physicochemical properties, which are greatly influenced not only by the species or tissue from which it is extracted, but also by the severity of the extraction process, which mainly depends on pH, temperature, and extraction time (Montero and Gómez-Guillén 2000). Thus, an optimization of the extraction procedure may be helpful in rationalizing the use of fish residues. The aim of the present study is to determine the appropriate process for the manufacture of gelatin from pollock skin. Specifically, we want to determine such processing conditions as pretreatment temperature and chemistry, and extraction temperature, time and chemistry, which may affect the total yield and the functional properties of gelatin.

[1] Department of Food Science, Cornell University, Ithaca, New York.

A. Extraction of gelatin with distilled water

Materials and methods

Gelatin extraction by distilled water. For extraction temperature studies, 25 g of fish skin was mixed with 100 ml distilled water (1:4 w/v), and incubated at different temperatures for 60 min. For extraction time studies, 37.5 g of fish skin was mixed with 150 ml distilled water (1:4, w/v), and incubated at 40 or 45°C. Solutions (15 ml) were collected after 15, 30, 45, 60, 75, and 90 min.

Gelatin fractionation. The gelatin extract was centrifuged for 10 min at 5,000 rpm, and the supernatant was collected and is referred to as the total gelatin faction. Saturated ammonium sulfate solution was added to the extract to obtain 22% saturation. After slowly stirring at room temperature for 1 hour, the solution was centrifuged for 10 min at 5,000 rpm. The precipitate was collected as fraction 1 (F1), and the supernatant was then increased to 30% saturation by adding more saturated ammonium sulfate solution. The solution was stirred at room temperature for 1 hr and centrifuged for 10 min at 5,000 rpm. The precipitate was collected as fraction 2 (F2). The fractions were dissolved and dialyzed against distilled water, and then the total volume, protein concentration, and gel strength of each fraction were determined.

Protein concentration determination. The protein concentration of the solutions was determined by the Biuret method with bovine serum albumin (BSA) as the standard (Gornall et al. 1949). The absorption of a 1% BSA solution at 280 mm was taken as 6.6 (Sober 1970).

Hydroxyproline determination. The hydroxyproline content was determined by the method of Woessner (1961) with L-hydroxyproline as a standard. The hydroxyproline content of each sample was calculated using the following equation:

$$\text{Hydroxyproline content (\%)} = \text{Concentration of hydroxyproline} \\ \text{(w/v)} / \text{Concentration of protein (w/v)} \times 100$$

Gel strength determination. Gelatin extracts were diluted to 3.0% with distilled water. Then 10 ml of diluted protein solutions were placed in special bottles. The bottles were moved from room temperature to 10°C and incubated at 10°C for 17 hours. The gel strength was determined by the Bloom method using the Stevens-L.F.R.A. Texture Analyzer (Wainewright 1977). Calibration curves with commercial gelatins permitted extrapolation to standard Bloom conditions (6.67% protein in standard Bloom jars, 10°C).

Results and discussion

The influence of extraction temperature. The result showed that with increasing extraction temperature, more protein was extracted. Below 35°C, hardly any protein can be extracted (Fig. 1). However, with

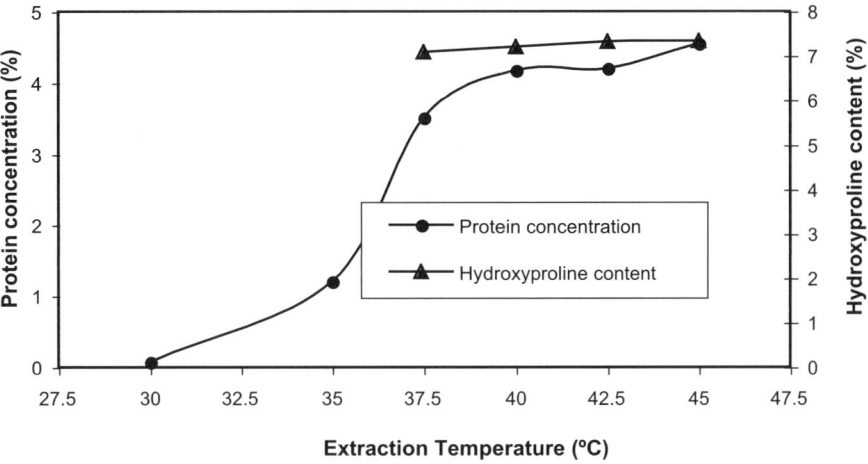

Figure 1. Extraction of gelatin at different temperatures.

temperatures between 37.5 and 45°C, the changes in hydroxyproline content of the extracts were not significant (Fig. 1). The gel strength of gelatin extracts at 45°C was lower than others extracted at lower temperature, which may be due to the degradation of gelatin at higher temperature during the extraction process (Fig. 2).

The influence of extraction time. With increasing extraction time more protein was extracted, and the most protein was extracted after 60 min incubation at 40 or 45°C (Fig. 3). However, neither temperature (40 or 45°C) nor time affected the hydroxyproline contents of the extracts. Although within 90 min, the extraction time has no effect on the gel strength of the extracts, the gel strength of the extracts at 45°C were lower than those with the same extraction time at 40°C (Fig. 4).

Gelatin fractionation by ammonium sulfate precipitation. Nearly half of the gelatin extract (F1) protein was precipitated by the 22% saturated ammonium sulfate (Fig. 5), while the fraction that precipitated between 22% and 30% saturated ammonium (F2) contributed 31% of the total extract protein. F1 had a darker color than the total gelatin while F2 had a lighter color compared with the total gelatin extract. Although there is no significant difference in the hydroxyproline content between the fractions (Fig. 6), the gel strength of F1 was higher than F2 (Fig. 7).

B. Extraction of gelatin with alternative extraction methods
Materials and methods
Gelatin extraction with pretreatment at 20°C. Fish skin (25 g) was washed with tap water (1:6 w/v) at 20°C for 10 min and then was rinsed

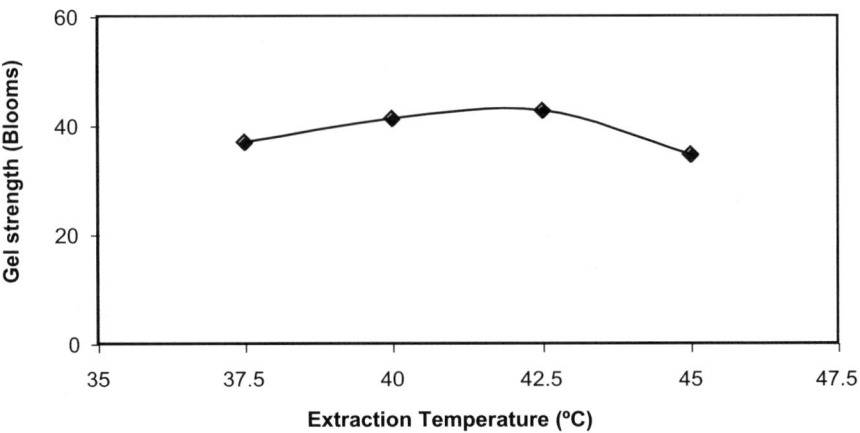

Figure 2. Gel strength of gelatin extracted at different temperatures.

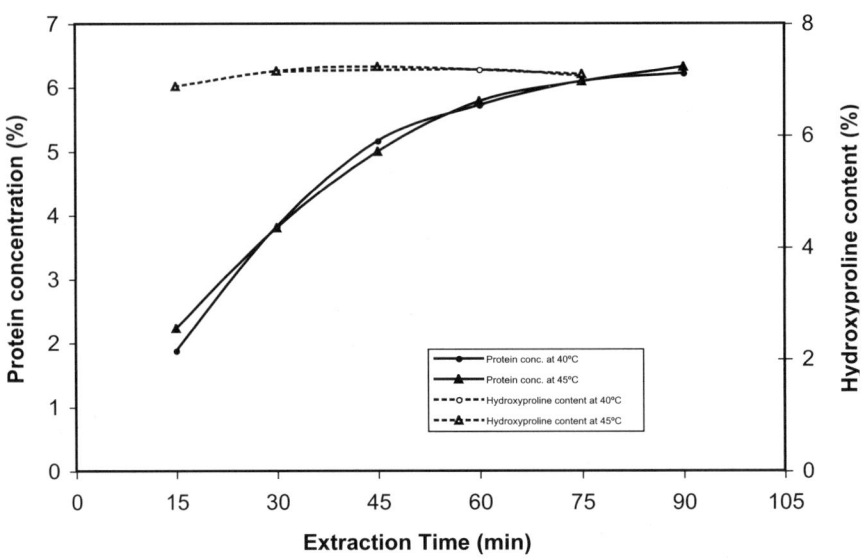

Figure 3. Extraction of gelatin for different times at 40°C and 45°C.

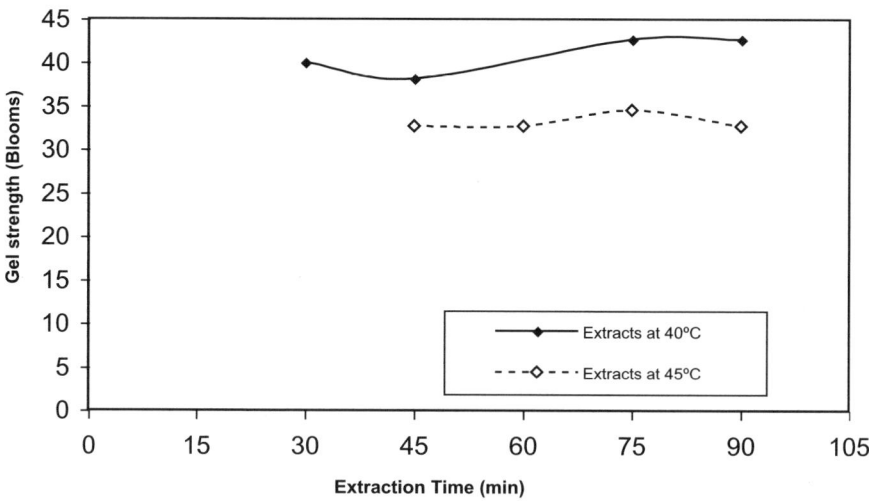

Figure 4. Gel strength of gelatin extracted for different times at 40°C and 45°C.

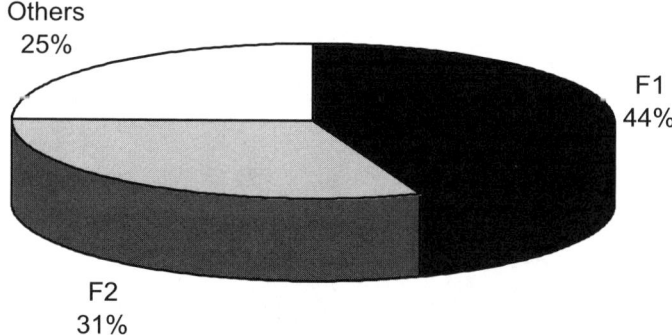

Figure 5. Pollock skin gelatin fractions by ammonium sulfate precipitation. F1 is 0-22% ammonium sulfate; F2 is 22-30% ammonium sulfate.

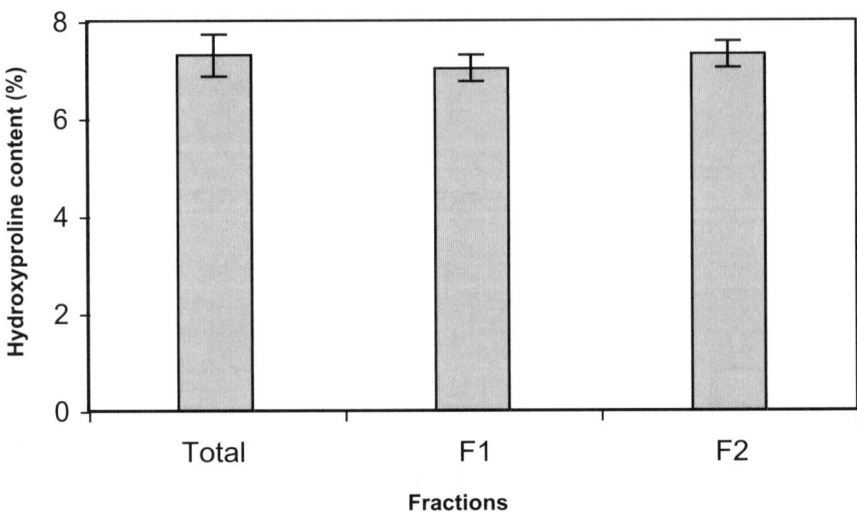

Figure 6. Hydroxyproline content of pollock skin gelatin fractions.

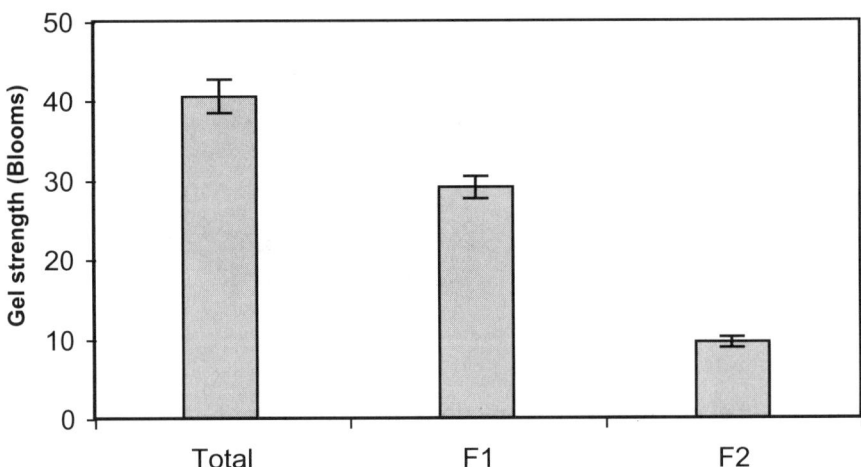

Figure 7. Gel strength of pollock skin gelatin fractions.

with abundant running tap water. This step was repeated three times. Excess water was removed by draining the cleaned skins and manual squeezing.

Cleaned skin was treated with 0.05 M sodium hydroxide (1:6 w/v) for 45 min, and then the sample was drained and rinsed with tap water. This step was repeated two times.

Samples were then treated with 0.025 M sulfuric acid (1:6 w/v) for 45 min and then 0.35 M citric acid (1:6 w/v) for 45 min (NaOH-H_2SO_4-Citric acid method, Na-H-C); 0.05 M acetic acid (1:6 w/v) for 75 min (NaOH-acetic acid method, Na-A), or 0.35 M citric acid (1:6 w/v) for 75 min (NaOH-citric acid method, Na-C). The samples were then drained and rinsed with tap water.

All the above were done at 20°C. After that, samples were mixed with distilled water (total volume 100 ml), and first extracted at 40°C for 180 min and then at 60°C for 180 min. Solutions (1 ml) were collected during extraction for protein concentration determination.

Gelatin extraction with pretreatment at 2°C. Fish skin (40 g) was washed with tap water (1:6 w/v) at 2°C for 10 min and then was rinsed with abundant running tap water. This step was repeated three times. Excess water was removed by draining the cleaned skins and manual squeezing.

Cleaned skin was treated with 0.2 M sodium hydroxide or calcium hydroxide (1:6 w/v) for 45 min, and then the sample was drained and rinsed with tap water (1:6 w/v). This step was repeated two times.

Samples were then treated with 0.1 M sulfuric acid (1:6 w/v) for 45 min and then 0.035 M citric acid (1:6 w/v) for 45 min (Na-H-C), 0.05 M acetic acid (1:6 w/v) for 75 min [Na-A; or Ca(OH)$_2$-acetic acid method, Ca-A], or 0.35 M citric acid (1:6 w/v) for 75 min (Na-C). Finally, the sample was washed with distilled water (1:6 w/v) for 10 min, and then the sample was drained and rinsed with tap water.

All the above were done at 2°C. After that, samples were mixed with distilled water (total volume 160 ml), and extracted at 40°C for 360 min. Solutions (10 ml) were collected during extraction for protein concentration, hydroxyproline content, and gel strength determination.

Yield of gelatin extraction. The yield of extraction was calculated using the following equation:

Yield (%) = Protein concentration (w/v) × Volume of extract / 40 g × 100

Results and discussion
Results showed that pretreatment at room temperature (20°C), which was recommended for warm water fish skin gelatin extraction (Grossman and Bergman 1992), is not suitable for the gelatin extraction from the skin of pollock, which is a cold water fish. Because most of the proteins were lost during the pretreatment at 20°C, the protein concentration of

extracts with pretreatment at 20°C (Fig. 8) was much lower than those with pretreatment at 2°C (Fig. 9).

With pretreatment at 2°C, protein can be extracted after incubation at 40°C for 180 min (Fig. 9), and the total yields after incubation for 360 min were more than 12% (Fig. 10). The hydroxyproline determination (Fig. 11) showed that there were no significant differences between different methods and extraction times, which agreed with our previous results. However, the result of gel strength depended on the methods (Fig. 12), among which the Ca-A method could get the extract with the highest gel strength while the extract of Na-H-C did not gel. The difference between the extraction methods may be due to the final pH following pretreatment, and further research will be done to get a better understanding of these observations.

Conclusions

Our study showed that the processing conditions, such as pretreatment temperature and method, and extraction temperature and time influenced the yield and rheological properties of pollock skin gelatin. The extraction process of gelatin from pollock skin, which is a cold-water fish, was less harsh than that for other fish species (Grossman and Bergman 1992, Montero and Gómez-Guillén 2000). However, the pollock skins prepared still have some fish odor, which would limit the application of this gelatin product in food industry. These issues will be the focus of future research.

Project 2: Extraction and characterization of gelatin from pollock bone
Ali Motamedzadegan[2] and Joe M. Regenstein

Introduction

Collagen is the precursor of gelatin, the principal protein found in the skin and bone of animals and fish. The largest use of gelatin is in gel desserts. Estimated world usage is 200,000 metric tons per year with U.S. usage being 30,000 metric tons per year for food and about 10,000 metric tons per year for pharmaceutical applications (Choi and Regenstein 2000). Due to its lower melting temperature, fish gelatin can be used in some cold-served food products. It also is used by some chemical and electronic industries. Fish gelatin, despite a higher cost, is able to serve as a halal (Muslim) or kosher (Jewish) food ingredient (Choi and Regenstein 2000), and some feel it is safer because it cannot carry bovine spongiform encephalopathy (mad-cow disease) (Fernandez-Diaz et al. 2001).

[2] Department of Food Science, Cornell University, Ithaca, New York.

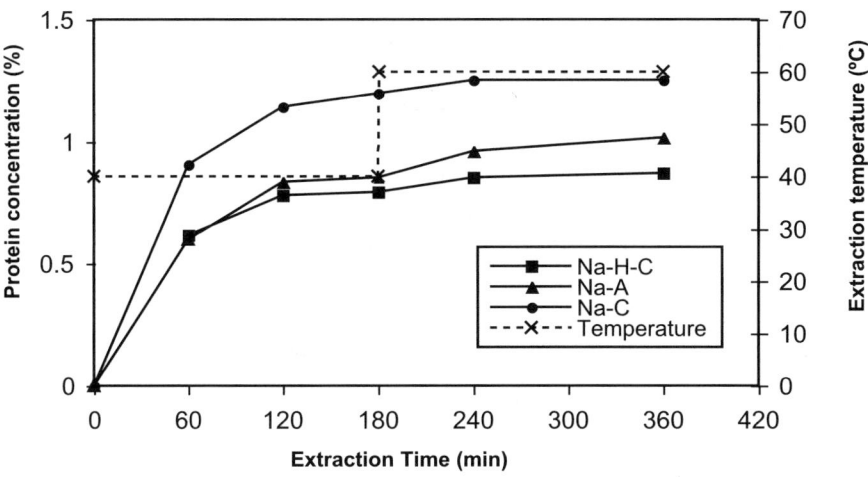

Figure 8. Extraction of gelatin by different methods with pretreatment at 20°C.

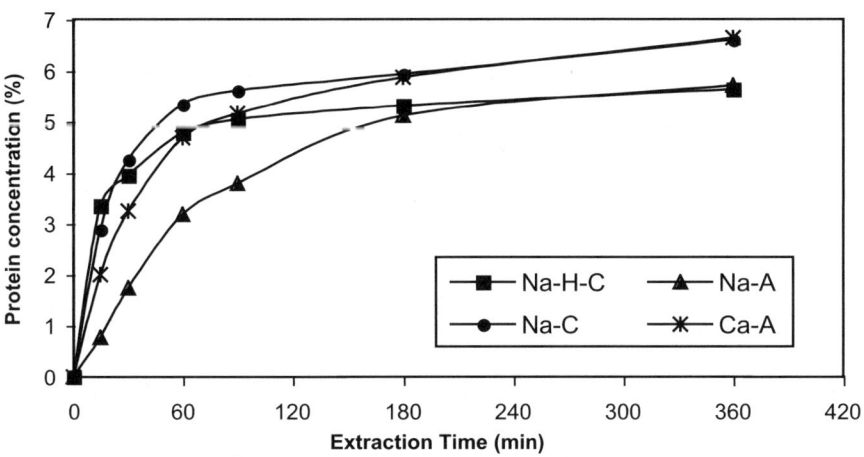

Figure 9. Extraction of gelatin by different methods with pretreatment at 2°C.

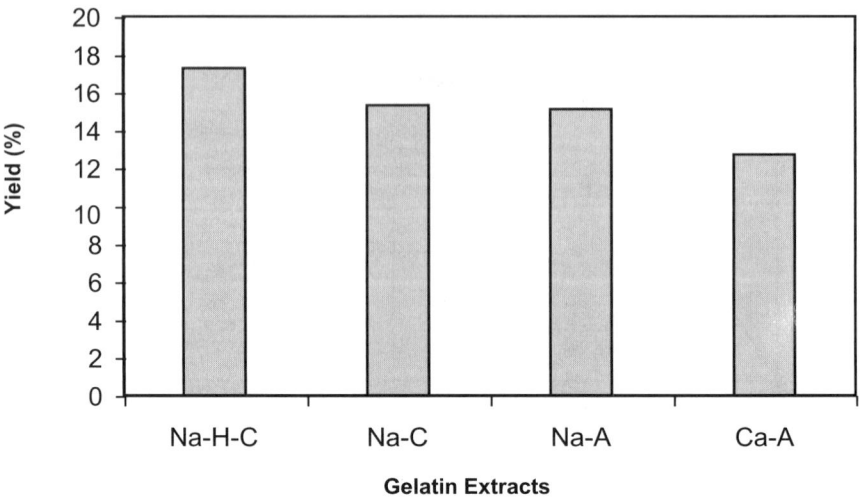

Figure 10. Yields of gelatin extracts after extraction for 360 min.

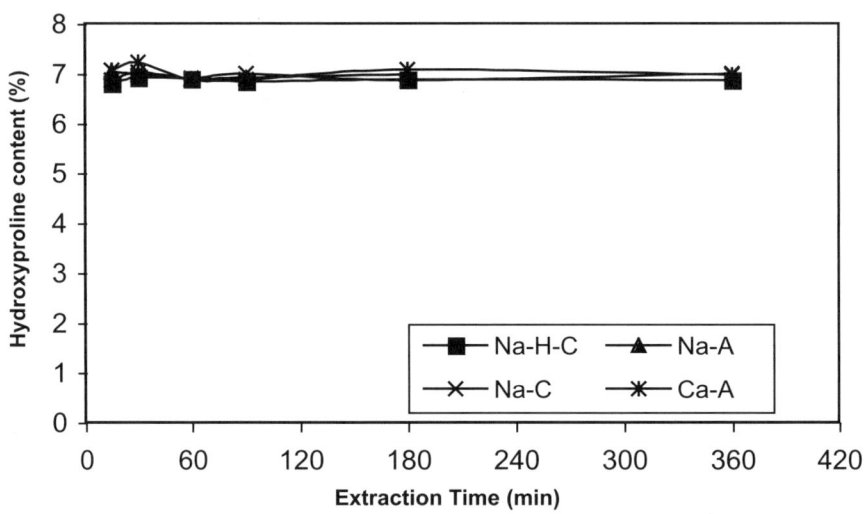

Figure 11. Hydroxyproline content of gelatin extracts.

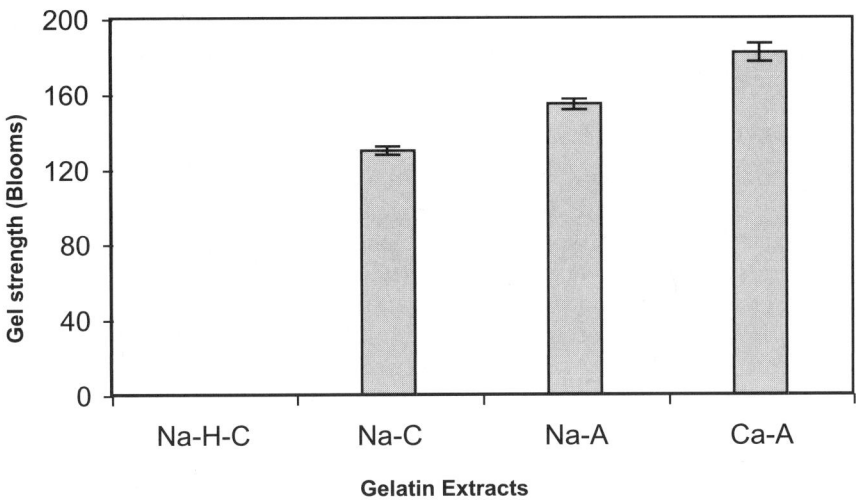

Figure 12. Gel strength of gelatin extracts after extraction for 180 min.

During the last decade the annual world fish catch has stabilized at about 90 million metric tons, and limited growth is expected in the future (Gildberg et al. 2002). Thus, the use of fish waste to generate additional value-added products has become more important. In the production of fish fillets more than 50% of the fish weight is waste; bone is one of the major byproducts, making up about 15% of the fish weight. On a dry matter basis, the pure bone contains about 60-70% minerals, mainly calcium, phosphate, and hydroxyapatite, and about 30% protein, mainly collagen (Nagai and Nobutaka 1999, Gildberg et al. 2002, Morimora et al. 2002). However, a minor amount of skin collagen from the tail and fins is also a part of the commercial fish backbone fraction.

Effective degradation or hydrolysis of materials composed of collagen is difficult due to its unique protein structure (Morimora et al. 2002). Most investigations to date have been done on fish skin gelatin; fish bone gelatin has not been as well characterized. Several authors have studied protein recovery from fish bone. The process of protein recovery depends on several parameters, of which temperature, time, and pH are the most important. Using an appropriate enzyme for bone cleaning (to remove adhering meat) is important, both to have a clean material for gelatin extraction and to produce a usable fish protein hydrolysate for either food or some other industrial use. The present study is related to the optimization of gelatin recovery from pollock bone, using Corolase L10 (AB Enzymes GmbH, Darmstadt, Germany) as the meat digesting enzyme.

Corolase, with an optimum temperature at 40°C, is a commercial alkaline protease, which has the potential of digesting and separating the meat from bone

Materials and methods

Frozen pollock bone was obtained from a fish processing plant in Dutch Harbor, Alaska, and kept at –30°C before use. It was defrosted overnight at 2.5°C. Most further work was done at this refrigerated temperature. The bones were crushed to 3-5 cm particles and mixed in a 1% Corolase solution for 20 min. The solution temperature had been raised to 60°C, so that submerging the refrigerated bones in the solution gave an ultimate temperature of 35-37°C with 2-3 min come up time.

The enzyme treated bones were washed several times with distilled water and crushed in a blender to a 5-7 mm particle size. Twice the amount of distilled water was added to the bones (v/w) and extraction was done in a water bath at elevated temperatures (± 1°C). For the extraction at different pH, sodium citrate buffer, sodium phosphate buffer, and tris buffer were used for the range of pH 2.6-5.8, 5.4-7.6, and 7.6-10.1, respectively, with 0.5 ± 0.2 unit increments. The extract was removed and centrifuged at 3,500 rpm for 10 min to separate insoluble particles. Total protein was measured as a simple indicator of gelatin extraction yield with the method of Lowry et al. (1951) using BSA (Sigma) for the standard curve. Total protein of bone and bone residue was measured using the Lowry method after solubilizing all the proteins of the dried sample with 1N NaOH solution at 100°C for 10 min. The bone residue was extracted two more times using a water to bone ratio of 2:1 and 1:1 in the second and third steps, respectively. The same procedure was followed for the second and third extracts. Hydroxyproline content of the samples was measured with the Woessner method (1961), using L-hydroxyproline as the standard.

$$\text{Hydroxyproline content (\%)} = \text{Concentration of hydroxyproline} \\ \text{(w/v) / Concentration of protein (w/v)} \times 100$$

Due to the small quality and low gelatin concentration in the extract, skin gelatin was used as a control for measuring the gel strength and 5% of the total skin protein was replaced with bone protein. Gel strength was determined using 10 ml of 3% (at 10°C) or 2.3% (at 2.5°C) protein solution in a glass container after 17-18 hours using the Stevens-L.F.R.A. Texture Analyzer. The results are shown as Bloom units (Wainewright 1977).

Statistical method

MSTATC software was used for comparing the means with Multiple Duncan's Range Test ($\alpha = 0.05$). Each treatment had at least two replications.

Results

Cleaning of the bone with Corolase at 40°C probably solubilized a part of the collagen, thus reducing the yield of gelatin during extraction. However, the gel strength of both enzyme and hand cleaned bone was examined for gel strength with 3% total protein at 10°C with substitution of pollock skin gelatin with pollock bone gelatin (Fig. 13). The results shows that even though cleaning of the bone with Corolase gives a better gel strength because of a higher purity gelatin, the bone gelatin still decreases the gel strength significantly when added to fish skin gelatin.

The loss of bone weight after enzyme treatment and the loss of total solids content of the cleaned bone were 62.0% and 40.2%, respectively. On a dry matter basis, enzyme cleaned bone and bone residues (after three extractions) have 12.9% and 10.0% protein, respectively, which contain 10.4% and 13-14.2% hydroxyproline, respectively. The solubilized gelatin hydroxyproline percentage varies between 5 and 8%, mostly 6-7.5%. Results for the bone residues are shown in Fig. 14 as a function of extraction time.

Extraction of gelatin in three steps increases the yield of protein recovery up to 45%. However, the second and third solutions are more dilute than the first one, and are of lower quality (Fig. 15).

The effect of temperature, time, and pH

Raising the extraction temperature up to 70°C will increase the protein recovery significantly. Raising the temperature more than 70°C has no significant effect on the extraction process. However, elevated temperatures may have an adverse effect on gelling properties of the gelatin. Neither the hydroxyproline content, nor the gel strength at 2.5°C varied much with the different temperatures. The hydroxyproline content varies between 5 and 6% (Fig. 16).

By increasing the extraction time up to 90 min, the yield of gelatin extracted increases at 70°C. The hydroxyproline content varies between 7 and 8% and the gel strength both at 2.5°C and 10°C does not vary significantly (Fig. 17).

Using a sodium citrate buffer, the protein extraction shows a slightly higher yield at pH around 3-4 (Fig. 18). However, except for pH around 3.3 and 4.7, all the samples using citrate buffer with the pH values less than 5.8 had significantly lower protein recovery than at higher pH values (Fig. 19). However, by increasing the pH to the alkaline range using tris buffer, the yield significantly increases. By elevating the pH to 10.2, the yield again decreases. The gel strength (in Bloom units) of all the samples was less than the control (83 Blooms) (Fig.20). At pH 4.7-5.8, it was significantly lower than control and the other samples at 2.5°C. The gel strength at these pHs were more than 10 Bloom units less than the control. Although the other samples had no significant difference with

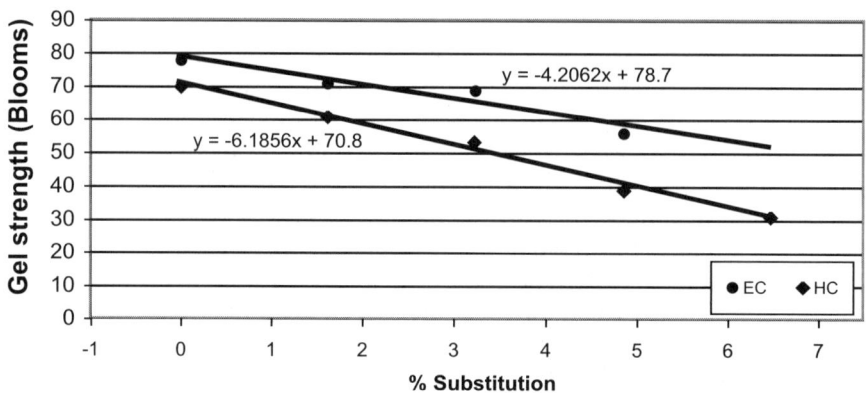

Figure 13. Gel strength affected by substitution of pollock skin protein with bone protein at 10°C, hand cleaned (HC) and enzyme cleaned (EC) bones.

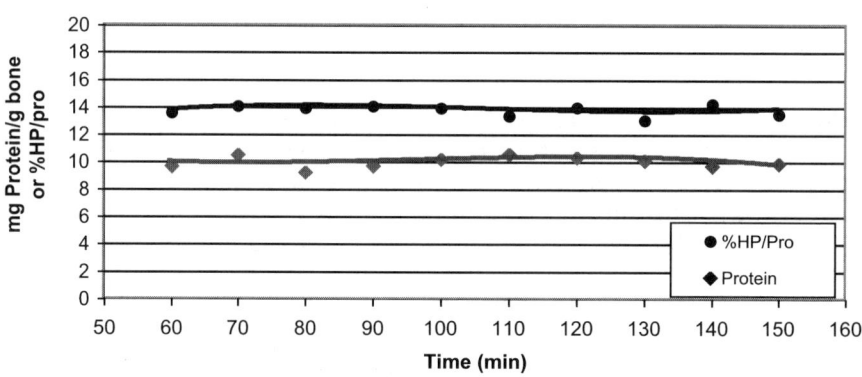

Figure 14. Protein and hydroxyproline content of bone residues after three extractions at 70°C.

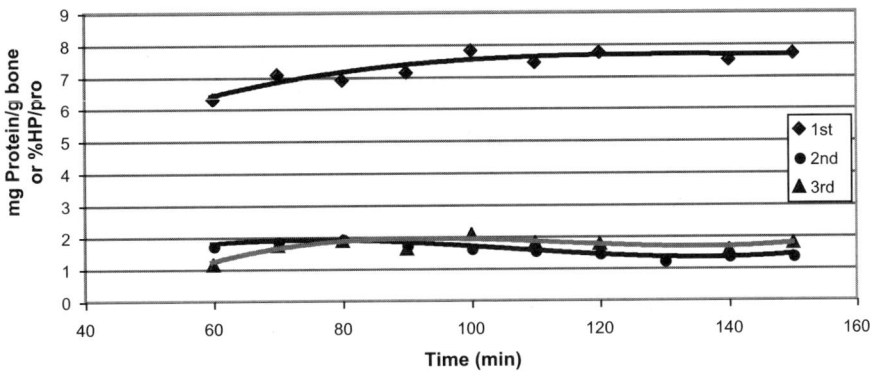

Figure 15. Three successive protein extractions at 70°C.

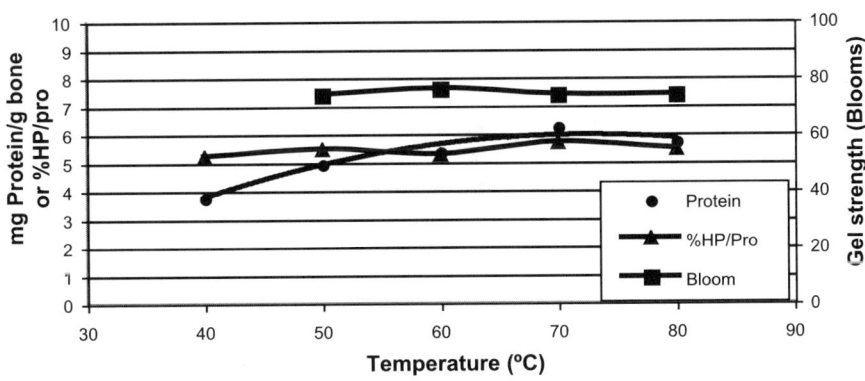

Figure 16. Protein concentration, hydroxyproline content, and gel strength at different extraction temperatures.

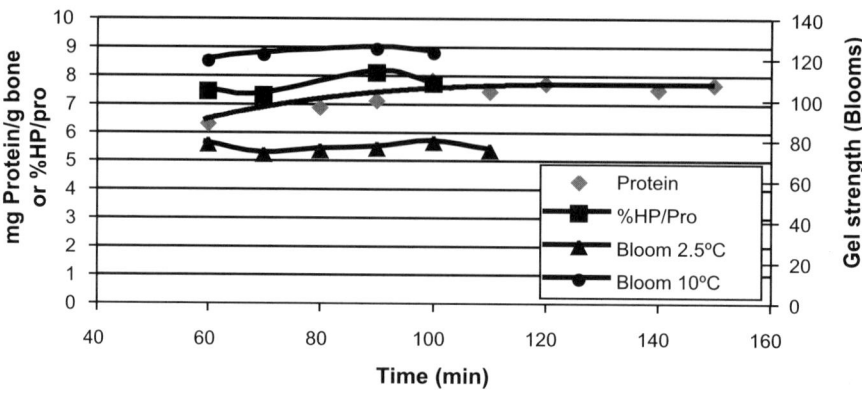

Figure 17. Protein extraction, hydroxyproline content, and gel strength at 70°C in different times.

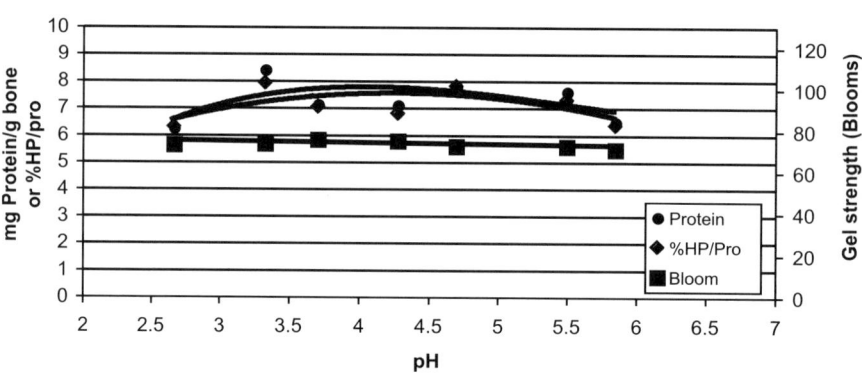

Figure 18. Protein extraction, hydroxyproline content, and gel strength at different pH.

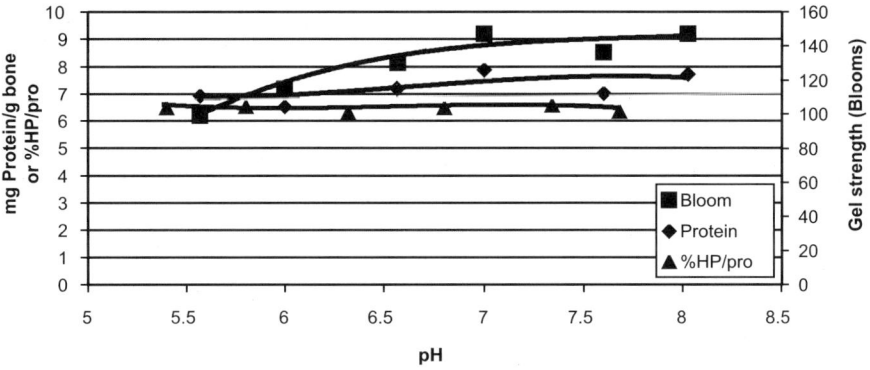

Figure 19. Protein extraction, hydroxyproline content, and gel strength at different pH

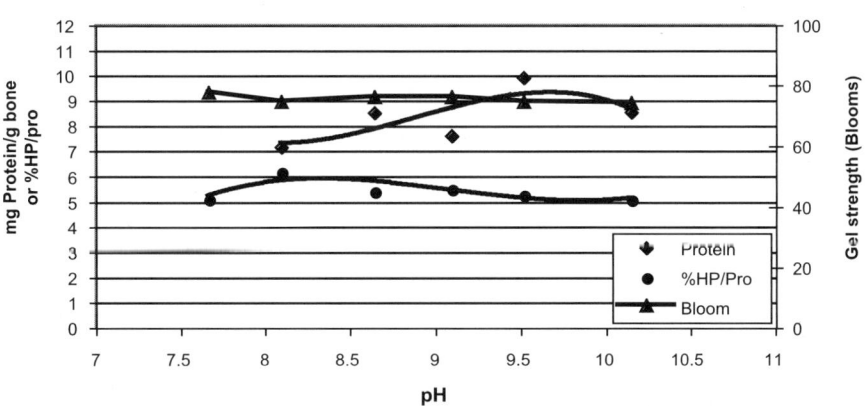

Figure 20. Protein extraction, hydroxyproline content, and gel strength at different pH. See Materials and methods for composition of buffers at different pH values.

the control, the closest sample to the control at pH 7.7 (tris buffer) was 5.5 Bloom units lower. All the gels in the acidic range were more fragile than the gels in the alkaline range when tested for gel strength. The fragility decreased by increasing the pH. However, this might be due to different pH of gels. The pH of gels varied from 5.6 to 7.6 while the pH of the control was 6.5. The control gel (2.3% pollock skin gelatin) was fairly fragile. When interpreting these Bloom values it should be noted that gels consisted of 95% skin gelatin and 5% bone gelatin.

Conclusion

The Corolase will solubilize a part of the gelatin and decrease the protein content of the cleaned bone. Using an enzyme with lower optimum temperature, we hope, leaves a higher protein concentrate in the extract. The hydroxyproline content of the extracted protein is the same as for cod fish, but the hydroxyproline content of the bone protein residue (13-14%) is about double that of cod (Gildberg et al. 2002). The 10.3% hydroxyproline content of un-extracted pollock bone suggests that a fraction of collagen with high hydroxyproline has not been extracted from the bone. However, the hydroxyproline content of the extract seems to be almost constant during the extraction process. The extract solution with around 1% protein is not as viscous as the equivalent fish skin extract, but is more viscous at refrigeration temperature (2.5°C). The quality of the fish bone gelatin is not as good as the fish skin gelatin. Although the fish bone gelatin may not be suitable for some application, it may be used in some situations where a viscous solution is needed at refrigerated temperatures. Precipitating the gelatin from the bone extract and characterizing it, along with separating calcium from the bone residues, are the goals for our future research.

Project 3: Pollock liver oil
Susan H. Goldhor

Pollock, like cod and haddock, use their livers as oil storage organs. The livers are large (8-11% of total body weight in the B fishing season, which were used in this work), and contain a high percent of oil, generally over 50%. Current practice is to include the livers in the fish meal process. The oil that is derived from the livers gets combined with oil from cuttings and viscera, and is decanted from the press water. This combined oil is a deep red color, due to the presence of pigment in the visceral (non-liver) oil. Most of the oil produced is used as fuel by the plant's burners and replaces diesel, which currently costs about $1.20 per gallon. The remainder is sold into the feed market. The price in this market (currently $0.17 per lb) is comparable, but when packaging and shipping are factored in, burning the oil as boiler fuel is more profitable. Our interest was in seeing if either pollock liver oil or visceral oil could enter more profitable markets. One limitation on the value of oil produced by a meal plant comes from almost invariably having been overcooked and oxidized. AFDF's 1988 study on pollock liver oil, carried out with Eagle Fisheries, pinpointed oil oxidation as one major impediment to market acceptance.

Ideally, oils would be produced with as little exposure to heat and oxygen as possible. To this end, we carried out a series of small-scale experiments, using different levels of heat to render oil from livers. All livers were pureed in a Waring Blender prior to heating, and then 50 ml of liver was spun in a Beckman GS-15R centrifuge for 10 minutes at 5,500 rpm at room temperature.

Pureed livers were allowed to sit at room temperature (21°C) or were quickly heated to 50°C or 80°C prior to centrifugation. At the highest temperature, oil color altered and the oil had a definite cooked odor. At the intermediate temperature (50°C), oil separated well and appeared to be in good shape. However, the surprise was that at room temperature, oil separation in most of the livers tested was as good as it was in those livers that were heated, with up to 65% oil separating. It would appear that rendering is certainly unnecessary and any heating may be unnecessary for liver oil production in B season pollock. However, one caveat is necessary. When we tested individual livers, certain livers emulsified and would not separate at room temperature. We do not know why those livers behaved differently. Also, because we did not do large scale testing, we cannot say if those livers would influence batch behavior. The emulsion that was formed by those livers could be broken only by heat, although the relatively low temperature of 50°C for less than a minute was sufficient.

Liver oil was sent out for fatty acid analysis. The red visceral oil, which requires more heat and a longer cook time to separate out, was sent for astaxanthin analysis. Astaxanthin was present at 19 ppm. This seems low; however, there are other peaks in the viscera oil chromatogram that could be esterified forms of astaxanthin and need to be saponified to show that they are in fact astaxanthin. During the next year, we hope to quantitate the amount of astaxanthin present in the viscera oil, check for potential variability during the season, and—if technical problems can be overcome—get an understanding of its bioavailability, which will affect its value in the feed market. Fatty acid analyses show that B season oil has about 15% EPA and 5% DHA. While not enough to make this oil a desirable human food supplement, pollock liver oil should have value in the feed market. These PUFA levels are comparable to those found by AFDF in their single analysis of B season oil: most of their work was done on A season oil, which has about 6% DHA and 9% EPA. We wish to thank Dr. Mahmoud Rowshandelim for performing the astaxanthin and fatty acid analysis.

Project 4: Fish protein hydrolysates/digests
Susan H. Goldhor

Fish protein hydrolysis refers to any process in which the fish flesh is broken down by protein-digesting enzymes. These enzymes may be present naturally in the fish viscera, or they may be purchased commercially and added during the process. Digestion may go sufficiently far to turn the flesh into liquid, or it may be partial and highly controlled, such as that used to cure herring, softening the flesh and developing flavors.

Liquid seafood digests have been used as fertilizers, aquaculture feed ingredients and attractants, animal feed ingredients, seafood flavorants, and peptones for microbial growth media (Goldhor 1992). Asian fish sauce is a well-known product manufactured by slow endogenous hydrolysis occurring over a six- to twelve-month period (Raksakulthai 1986).

The choice of enzymes and conditions for hydrolysis dictates the final characteristics and, thus, markets for the product. The enzymes used will affect not only the fineness of the product (measured by its ability to pass through a screen), but also the flavors generated. For example, if one wanted to produce a liquid fertilizer product, the cheapest enzyme capable of rapid digestion to a fine particle size would be indicated. On the other hand, if one wanted to produce seafood flavors for human consumption, the enzymes used would be chosen because they did not produce bitter peptides, and because of subtle flavor enhancements. Since flavors are expensive products entering demanding markets, the cost of the enzymes would be essentially irrelevant.

In our initial work at UniSea on pollock wastes during the B fishing season, our interest was in crude hydrolysates of viscera and viscera and heads, using only endogenous enzymes. In these first experiments, whole viscera were collected from a processing line and ground in a silent cutter, or collected as a macerated mass after stomach removal. Although this work was done in August, when roes were still immature, the larger roes had been removed from the viscera slated for stomach removal. The ground viscera were heated gradually, with constant stirring, in a double boiler. Temperature was monitored. Significant liquefaction did not occur until the temperature reached 60°C. All digestion was due to endogenous enzymes; no commercial enzymes were added.

Although significant quantities of oil were present, no oil separated out from the hydrolysate until the temperature reached about 63-65°C. At that point, oil separated out and stayed separate. Large quantities of small, soft, whitish particles remained undigested. We do not know if these were collagen or parasites or some other entity. These were resistant to digestion, and raising the temperature to 70°C did not affect them.

Heads could be digested alone, under the same conditions, or they could be added to the visceral mass and digested along with it. The digestion of heads added large quantities of bone, which were screened out at the end of the process. The crude autolysis that we performed has limited value in the marketplace. Further hydrolyses with selected commercial enzymes will be performed during the 2003 B season to produce samples of higher value hydrolysates.

Endogenous enzymes were effective at cleaning bones; however, they were also effective at digesting gelatin. Therefore, while crude hydrolysis with endogenous enzymes may be useful in itself, in the manufacture of a product that can use essentially all viscera, heads, etc., and provides an alternative to fish meal, it does not seem to be useful in cleaning bones for gelatin production. Cleaning bones for gelatin production may require digestion with an enzyme capable of working at lower temperatures. Proteolytic enzymes isolated from cold water fish have been shown to have low temperature optima, e.g., around 40°C (Haard 1992). It is interesting that the crude mix of visceral enzymes has such a high temperature optimum. Further work is needed to see if proteolytic enzymes with lower temperature optima can be isolated from this mix.

References

AFDF. 1988. U.S. market prospects for Alaskan pollock liver oil. Alaska Fisheries Development Foundation (AFDF), Anchorage, Alaska.

Choi, S.-S., and J.M. Regenstein. 2000. Physicochemical and sensory characteristics of fish gelatin. J. Food Sci. 65:194-199.

Férnández-Díaz, M.D., P. Montero, and M.C. Goméz-Guillén. 2001. Gel properties of collagens from skins of cod and hake and their modification by the coenhancers magnesium sulphate, glycerol and transglutaminase. Food Chem. 74:161-167.

Gildberg, A., J.A. Arnesen, and M. Carlehog. 2002. Utilization of cod backbone by biochemical fractionation. Process Biochem. 38:475-480.

Goldhor, S.H. 1992. Fish protein hydrolysates and their uses. In: The 1991 Seafood Environmental Summit Proceedings. UNC-SG-92-06, University of North Carolina Sea Grant, Raleigh, p. 73.

Gómez-Guillén, M.C., J. Turnay, M.D. Fernández-Díaz, N. Ulmo, M.A. Lizarbe, and P. Montero. 2002. Structural and physical properties of gelatin extracted from different marine species: A comparative study. Food Hydrocolloids 16:25-34.

Gornall, A.G., C.J. Bardawill, and M.M. David. 1949. Determination of serum proteins by means of the biuret reaction. J. Biol. Chem. 177(2):751-766.

Grossman, S., and M. Bergman. 1992. Process for the production of gelatin from fish skins. U.S. Patent 5,093,474.

Haard, N. 1992. A review of proteolytic enzymes from marine organisms and their application in the food industry. J. Aquatic Food Product Tech. 1:17.

Lowry, O.H., N.J. Rosebrough, A.L. Farr, and R.J. Randall. 1951. Protein measurement with the folin phenol reagent. J. Biol. Chem. 193:265-275.

Montero, P., and M.C. Gómez-Guillén. 2000. Extraction conditions for megrim (*Lepidorhombus boscii*) skin collagen affect functional properties of the resulting gelatin. J. Food Sci. 65:434-438.

Morimora, S., H. Nagata, Y. Uemura, A. Fahmi, T. Shigematsu, and K. Kida. 2002. Development of an effective process for utilization of collagen from livestock and fish waste. Process Biochem. 37:1403-1412.

Nagai, T., and S. Nobutaka 1999. Isolation of collagen from fish waste material: Skin, bone and fins. Food Chem. 68:277-281.

Raksakulthai, N. 1986. Role of protein degradation in fermentation of fish sauce. Ph.D. thesis, Memorial University of Newfoundland, St. John's, Canada.

Shahidi, F. 1994. Seafood processing by-product. In: F. Shahidi and J.R. Botta (eds.), Seafood chemistry, processing, technology and quality. Blackie Academic & Professional, Glasgow, pp. 320-334.

Sober, H.A. 1970. Handbook of biochemistry. Selected data for molecular biology. The Chemical Rubber Co., Cleveland, Ohio.

Wainewright, F.W. 1977. Physical tests for gelatin and gelatin products, In: A.G. Ward and A. Courts (eds.), The science and technology of gelatin. Academic Press, New York, pp. 507-534.

Woessner Jr., J.F. 1961. The determination of hydroxyproline in tissue and protein samples containing small proportions of this imino acid. Arch. Biochem. Biophys. 93:440-447.

Cleaner Production Practices in the Seafood Industry Can Add Profits to Your Operation

Anthony P. Bimbo
Technical Consultant, Kilmarnock, Virginia

Abstract

Over the 1950-2000 period the United States landed an average 3.6 million metric tons of fish annually. Alaska landed an average of 1.3 million tons of fish over this same period. Alaskan landings began to accelerate in 1976 when the pollock fishery was first entered and during the period 1976 to 2000 Alaska landings have averaged about 38% of the U.S. landings, a large jump from 9% during the 1950-1975 period. Statistical figures for the disposition of the U.S. fish catch indicate that about 74% of the catch is utilized for food and 26% for reduction to fish meal and oil. Since 100% of the edible fish are not consumed as canned, cured, or frozen, a relatively large portion of the fish (heads, tails, and viscera) will be generated as waste. The most obvious outlet for this waste is the production of fish meal and oil, and this option is being exercised where the waste is concentrated. But what about the small- or medium-sized fish processors who do not have a friendly fish meal neighbor to take their waste? What are they to do? The options might only be to landfill or dispose of the waste at sea. This paper discusses options that the small to medium-sized fish processor can consider for the utilization of waste. These options cover a wide range of possibilities with the much more profitable ones requiring more initial capital and greater effort.

Introduction

In order to put us in the right perspective let's look at some statistics. First of all, the world landings of fish appear to be increasing and reached 130 million metric tons in 2000. However, a closer look at the data shows that the marine capture segment has been relatively static for a long period of time and all of the growth is coming from aquaculture development (Fig. 1) (O'Bannon 2002, FAO 2002b).

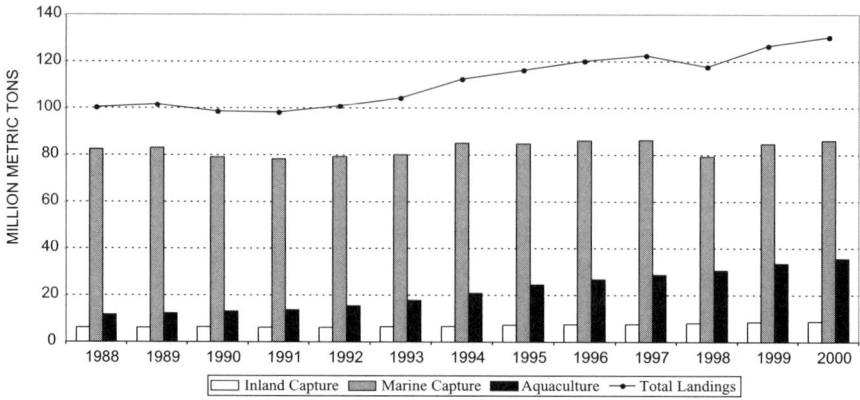

Figure 1. Composition of world landings, excluding mammals and aquatic plants. Sources: O'Bannon 2002 and FAO 2002b.

The distribution of the world catch indicates that about one-fourth is specifically for fish meal and oil production while the remaining 76% is distributed as fresh (32%), canned (10%), frozen (22%), cured (9%), and other (3%). The uses, other than fish meal and oil, offer opportunities for additional products that can be produced from the waste which conservatively accounts for about half of the edible landings.

The United States is responsible for 4% of the world's fish landings and Alaska represents almost 50% of the U.S. landings (O'Bannon 2002). The rise in Alaska landings coincides with the Alaska pollock (*Theragra chalcogramma*) fishery which accelerated in the mid-1980s. Pollock now accounts for about 27% of the U.S. landings of fish (O'Bannon 2002). A breakdown of the composition of the U.S. landings is shown in Fig. 2.

Many fish species are caught specifically for fish meal and oil production, including anchovy (*Engraulis ringens*), jack mackerel (*Trachurus murphyi*), capelin (*Mallotus villosus*), menhaden (*Brevoortia* sp.), and sand eel (*Ammodytes personatus*). Edible fish trimmings are also used for fish meal and oil production in a number of countries. These include skipjack tuna (*Katsuwonus pelamis*), salmon (*Salmo salar, Oncorhynchus* sp.), catfish (*Ictalurus punctatus*), sardine/pilchards (*Sardinops sagax*), various white fish, pollock (*Theragra chalcogramma*), mackerel (*Scomber japonicus, Scomber scombrus*), and herring (*Clupea harengus*). Unfortunately not all of the trimmings are utilized and the disposition of these materials has been a constant source of friction and problems over the years. These materials certainly offer unique opportunities for future products including fish meal and oil production.

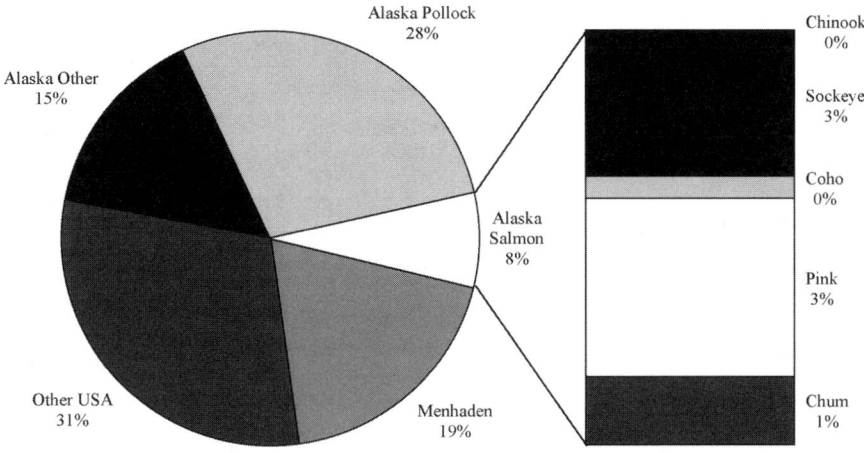

Figure 2. Composition of U.S. catch, 1993-2001 averages. Source: O'Bannon 2002.

Pollution prevention (cleaner production)

We have all seen the term pollution prevention (P2) but we probably all have different interpretations of what that means. First of all to me, pollution prevention does not mean an end of pipe treatment. In other words, it is not a step at the end of the process that attempts to treat the entire residual product that was not recovered during the primary processing of the fish or cuttings. End of pipe treatment characterizes the product in terms of BOD_5 (biological oxygen demand), COD (chemical oxygen demand), oil and grease, suspended solids, etc., and attempts to remove or at least reduce them. These terms, BOD_5, COD, etc., do not adequately describe the composition of the waste streams that come from typical fish processing plants. A better interpretation is that pollution prevention is a concept designed to maximize product recovery and reduce the conditions that cause product losses, which increases revenues; by maximizing product recovery you reduce or eliminate the source of the pollution. In order to visualize this we must think of the material being discharged or discarded as protein and oil (fish meal and oil) rather than BOD_5 or oil and grease (effluent contaminants). Corporate management tends to identify the term BOD_5 with expense that has no possible generation of revenue, a psychological turnoff. But when you put this in terms of protein and oil (especially for the fish meal industry), management can easily identify with the terms since it translates into marketable products with a defined value. Waste streams, both liquid and solid, should then always be characterized in terms of protein and oil. Once the effluents are characterized in these terms then we will be less likely to entertain the "end of pipe" concept. Table 1 illustrates this concept using the Peruvian fish meal industry as an example. If we characterize the components of the various liquid streams in terms of protein and oil then we can calculate how much product is being thrown away if these liquid streams are not processed. On average, almost half of the dry solids from the fish are contained in the three liquid streams generated in the plants in Peru. Failure to recover the solids results in loss of yield and loss of revenue.

The "end of pipe" concept does not address the cause of the problem. It normally only deals with today's conditions and therefore requires continuous add-on steps as the effluent regulations evolve. It does not offer the option of product recovery or improving process efficiency. End of pipe treatment does have its place, however, but only after all of the potential product that can be recovered has been recovered. Investments made within the process to improve efficiency will always pay off with increased yields and a higher rate of return. This is easily demonstrated when we look at Fig. 3 which demonstrates how the addition of technology within the process not only reduces the environmental load in the discharge but actually improves yields and revenue. This data is for the pumpwater only from the Peruvian fish meal industry.

Cleaner production practices are important in your daily operations. Bad publicity about pollution or waste leads to consumer pressures on your customers. Your customers are reluctant to purchase products that are associated with damage to the environment because this is perceived as supporting environmental mismanagement. Lending institutions are reluctant to loan money to industries and/or companies with poor environmental records. And bad publicity stays on the Internet forever. The information is easily retrieved by inserting key words into a search engine and the information is always there for the public to access.

Finally, it does not matter whether we like it or whether we agree with it. The current effluent regulations are the best we are going to see and these will change and will get stricter in the future—just ask the Peruvian and Chilean fish meal industry. Eventually at some point in the future we will not be allowed to discharge into the sea. So it makes good sense to plan for the time when this will come about and to get ahead of the curve.

Raw material

The first key to pollution prevention or increased revenues is the raw material quality. It doesn't matter whether your raw material is whole fish or cuttings. The condition of the raw material determines yields and your bottom line. As the raw material ages a series of biochemical reactions take place which, depending on the length of time and temperature, can result in low quality products and odors in the plant or drier stacks

Table 1.   Summary of solids (dry matter basis) in the various liquid and solid streams from a typical fish meal operation in Peru.

	Kg solids, dry matter	Value for 1,000 kg raw material
Fresh fish, everything recovered	250 kg meal 40 kg oil	$114.75
Pumpwater discarded	–51	–$23.41(20.4%)
Bloodwater discarded	–20	–$9.18(8%)
Stickwater discarded	–48	–$22.03(19%)
Presscake	121	$55.54(48.4%)
Other	10	$4.59(4%)

Raw material contains 200 kg of dry solids plus 90 kg fat. Oil yield 40 kg per ton of fish. Based on $459 per ton fish meal. Source: Bimbo 1996.

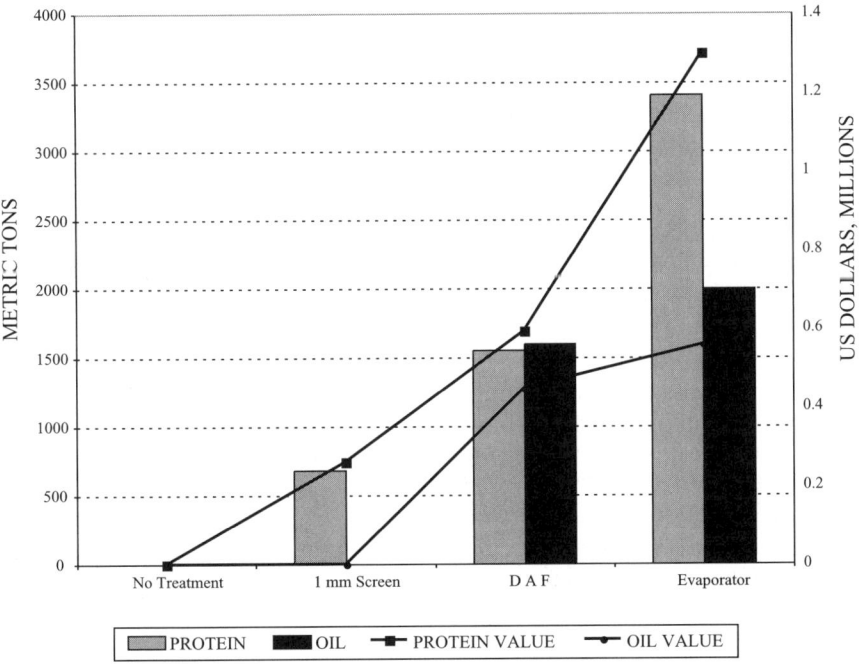

Figure 3. Cumulative recovery of protein and oil (dry matter basis) in pump-water from the Peruvian fish meal industry. Source: Bimbo 1996, Mueller-Vollmer et al. 1998. Fish meal calculated at US$381 per t, fish oil calculated at US$279 per t. DAF = dissolved air flotation.

Table 2.   Change in the TVN mg per 100 g of herring and small cod over a 13-day period.

	Cod			Herring		
Days	0°C	6°C	12°C	0°C	6°C	12°C
1	23.5	20.8	21.5	18	26	26
3	30.9	37	73.9		94	140
5	32		141.8	66		216
6	41.7				199	
7		88				
9				100	212	370
10		119.6				
11	61.8					
12					278	420
13				100		

TVN = total volatile nitrogen. Source: Petersen 1971.

if you are making fish meal. More of your raw material ends up in the liquid streams instead of the solid stream which then requires more processing for recovery. If the liquid streams are not recovered then a yield loss results. When yields are reduced, production costs go up and you generate less revenue. From a technical perspective, volatile amines are formed resulting in high total volatile nitrogen (TVN). This can be seen in Table 2 for capelin and herring from the North Sea area, and in Table 3 for anchovy from South America stored with and without ice.

As the fish age, the amino acids break down into biogenic amines such as histamine, and the fat is hydrolyzed and forms high free fatty acids (FFA) in the oil. The change in the FFA in the fat in capelin and the TVN in the flesh over time is shown in Table 4. All of these measurements (TVN, FFA, and histamine) are indicators of both the quality of the fish meal and the quality of the raw material.

Reduced yields during seasons when the catch is low mean that less product is recovered and therefore less revenue is generated than during seasons when the catch is good. For example in Fig. 4 we see that over the period 1995-2002 when the catch in Peru and Chile was high, world fish meal and oil prices were low. But when these countries experienced catch problems as in the 1998-1999 El Niño, prices soared and the value of the fish meal was almost two times higher than in the high catch period. Efficient recovery of product pays off during these low catch periods and it maximizes use of the resource, which is really what sustainable or responsible fishing is all about. It makes no sense to attempt to have a fishery defined as sustainable when you are discarding a major portion of the catch as waste.

Table 3. Comparison of the TVN mg per 100 g of anchovy stored with and without ice.

Hours	Anchovy	Anchovy on ice[a]
5	4.7	4.7
12		5.11
18	33.4	32
29	85.9	41.3
36	112	59.07

[a]Ice added at 160 kg per 1,000 kg of fish.
TVN = total volatile nitrogen. Source: Mueller-Vollmer et al. 1998.

So the first key to reducing waste and enhancing yields (pollution prevention) is the raw material quality. It is like money in the bank. Any investment in maintaining or improving the quality of the raw material pays dividends at each stage of the process. Reducing the storage temperature of the fish by 5-6°C extends the storage time by 100%. And when the storage time is extended there is less loss of raw material in the blood water. Figure 5 shows the cumulative increase in the amount of product lost as blood water for fish offal (U.K.) stored at ambient temperature over 26 hours.

Opportunities for profit

The fish meal and oil industry and the edible seafood industry have many things in common. In the fish meal process you concentrate the components you want: protein, oil, and minerals; and discharge what you don't want: water. There really should not be any waste except for the water. In edible seafood processing you recover the components you want: fillets, dressed fish, surimi, etc.; and discard the components you don't want: blood, viscera, heads, backbones, tails, shells, etc. In performing the processing steps necessary to produce these edible products water to fish ratios as high as 30:1 have been used. If the solid waste is mixed with the water you have a major recovery problem on your hands. A solid waste problem can be converted into a major water effluent problem. A comparison of raw material, finished product, and waste streams for each type of industry is shown in Table 5. The data is based on a plant handling 100,000 tons of fish per year. Water management therefore becomes a major issue when talking about the processing of fish.

Table 4.   Change in capelin TVN and capelin oil FFA over time.

Days	Temp. °C	Fish TVN mg/100g	Oil FFA %
0		12	0.4
1	4.0	11	0.7
2	4.3	16	0.9
3	6.1	58	1.4
4	7.3	86	2.1
5	7.9	103	2.4
6	8.4	118	3.4
7	8.8	134	5.4
8	9.0	151	5.8
9	9.1	157	7.5
11	9.1	202	9.2
12	9.2	202	9.2

TVA = total volatile nitrogen; FFA = free fatty acids. Source: Strøm 1971.

Challenges can lead to opportunities

A report issued in 1988 (University of Alaska 1988) characterized the Alaska seafood wastes at that time. Some of their data is reproduced in Table 6. Landings for 1987 and 2001 were taken from U.S. government statistics (O'Bannon 2002) and the waste calculations have been added. The species of fish listed represent about 85-90% of the total Alaskan landings.

Assuming that the waste to fish meal conversion is 20%, in 1987 Alaska could have produced 61,259 metric tons of fish meal. In 2001 the fish meal production could have been 266,182 tons. For information purposes, according to Mielke 2002, if we use the average price for fish meal for the period 1986-1987: $350, and for the period 1992-2001: $486 delivered price to Hamburg, and subtract $60 for freight, we find that the potential fish meal value in 1987 was $17.8 million while in 2001 it would have been $113.4 million. The "Hamburg" price is based on 66% protein fair average quality (FAQ) fish meal so it would be necessary to adjust these figures depending on the protein content of the Alaska fish meal.

Based on U.S government statistics for Alaska (NOAA 2002), cod and pollock fish meal and oil production during the period 1994-2002 has been in the 50-70,000 metric ton range for fish meal and in the 8,000-19,000 ton range for fish oil with major increases taking place in 2001-2002 (Fig. 6). There are no figures for the volume of salmon fish meal produced in Alaska but with an average annual landing in the 275,000-425,000 ton range, the potential is quite substantial.

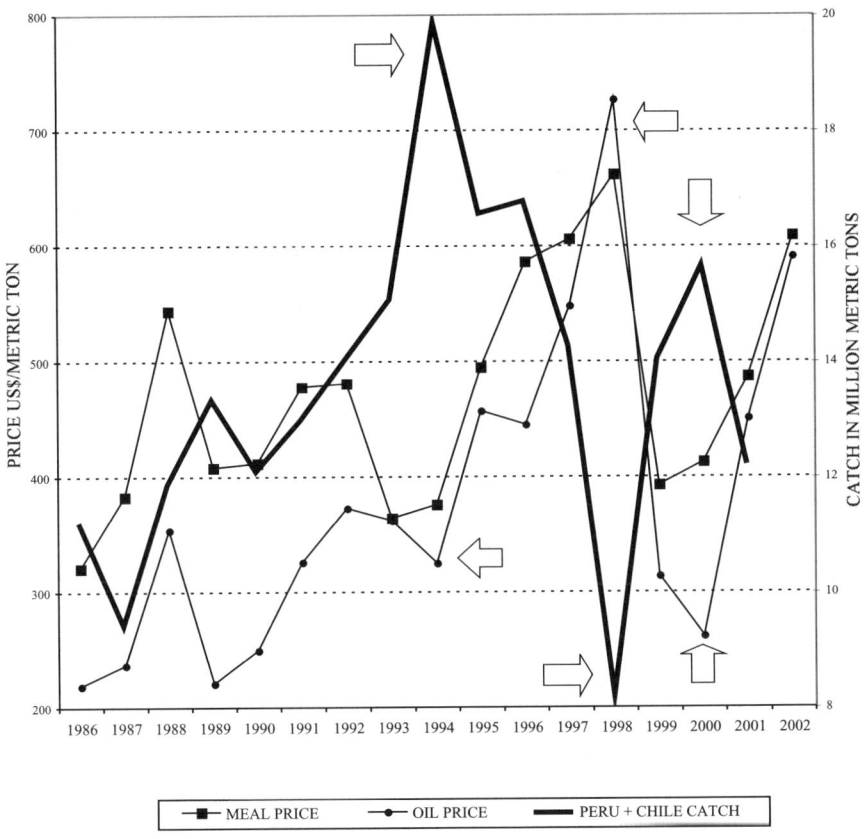

Figure 4. Relationship of fish meal and fish oil prices to the catch in Peru and Chile. Source: Mielke 2002, O'Bannon 2002.

Future considerations

Fish protein

According to the Food and Agriculture Organization (FAO 2002a) the average per capita consumption of fish could reach as high as 22.5 kg per person by the year 2030. This suggests a total annual demand for fish of 186 million tons by 2030—almost double the present level. However, since supply will probably be limited by environmental factors, FAO suggests a more likely range of 150 to 160 million tons, or between 19 and 20 kg per person.

The FAO report goes on to say that there is a growing trend to market fresh fish for human consumption. This is because the costs of delivering

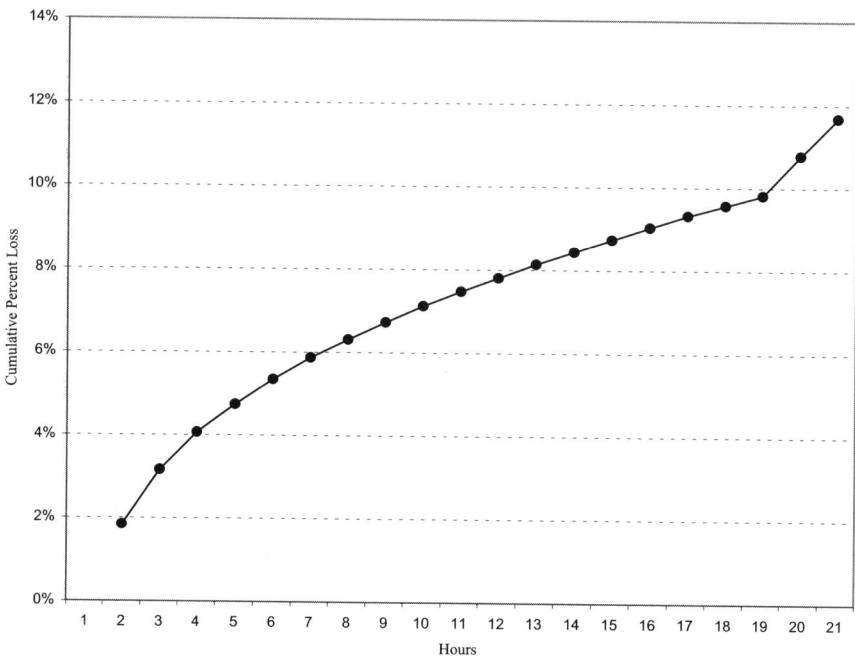

Figure 5. Cumulative weight loss percent of U.K. fish offal stored at ambient temperature. Source: Carpenter 1961.

Table 5.   **Comparison of raw material, various waste streams, and finished product for fish meal and oil and edible seafood processing.**

	Fish meal and oil process	Edible seafood process
Raw material	100,000	100,000
Pumpwater range	250,000-500,000	a
Blood water	10,000	30,000[b]
Process water range	c	400,000-1,000,000
Stickwater	60,000	NA
Salmon blood	NA	5,000-10,000
Solid waste	NA	70,000
Finished product(s)	26,000	30,000

[a]The edible seafood processing sector may or may not have a pumpwater stream.
[b]Bloodwater in the edible seafood process may include melting ice.
[c]Normally process water in the fish meal plant is wash-up water and condensate water.
Numbers are in metric tons. Source: Bimbo 2002.

Table 6. **Estimation of the amount of solid fish waste generated in the Alaska seafood industry, 1987 and 2001 compared.**

	Cod	Pollock	Halibut	Salmon[a]	Salmon[a]	Crab	Flatfish	Total
Landings, tons 1987	74,191	255,344	34,040	221,404	221,404	69,242	7,631	780,740
Landings, tons 2001	213,539	1,441,904	33,739	311,351	311,351	21,406	4,501	2,284,468
Edible product	Fillets	Surimi and fillets	Headless	Canned	Frozen	Meat	Fillets	
% Edible	25	15-25	76-82	62-69	75-90	17-25	18-25	15-90%
% Waste	75	75-85	18-24	31-38	10-25	75-83	75-82	10-85%
1987 waste, tons	5,618	191,508	6,127	34,318	11,070	51,932	5,723	306,296
2001 waste, tons	160,154	1,081,428	6,073	48,259	15,568	16,055	3,376	1,330,912

[a]Assume 50% of salmon used for canning and 50% as frozen.
Source: University of Alaska 1988, NMFS 2002b.

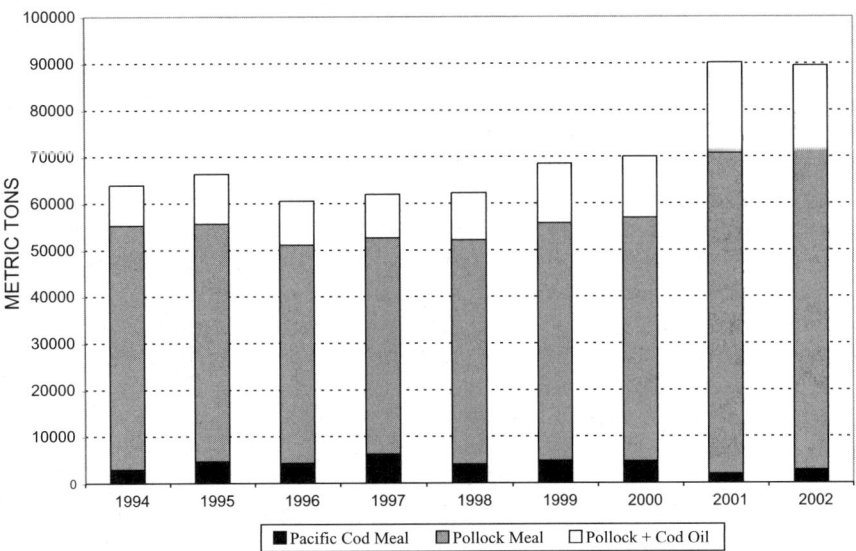

Figure 6. Fish meal and oil production in Alaska, 1994-2002 (pollock and Pacific cod only). Source: NMFS 2002a.

fresh fish to markets are falling and consumers are willing to pay a premium for this product. Demand for fish meal and fish oil will continue to grow rapidly. So far the raw material for fish meal and oil has been supplied by capture fisheries, and in all likelihood this will continue. However, the competition for small surface fish will become more intense, and the fish meal and oil industry will need to exploit other raw materials, such as mesopelagic fish (fish living in the middle depths) and krill. Rising prices will also drive a switch to substitute aquaculture feeds. However, a satisfactory replacement for fish oil has not yet been found according to FAO.

Over the next three decades, the world's fisheries will meet demand by continuing the same shift from fish capture to fish cultivation that gained momentum in the 1990s. The share of capture fisheries in world production will continue to decline while aquaculture will continue to expand. The maximum sustainable marine production has been estimated at around 100 million tons a year, which is higher than the annual catches of 80 to 85 million tons achieved during the 1990s. The report assumes that large quantities of previously underexploited aquatic resources will be used, including krill, mesopelagic fish, and oceanic squids. As in the 1990s, most of the shortfall will be made up by aquaculture, which will probably continue to grow at rates of 5 to 7% per year, at least until 2015.

If we project the world demand (per capita consumption x population) for fish proteins out to the year 2030 we get a figure of about 165 million tons of demand. If we project the current world landings of ocean fish out to the same time there is a very slight increase when we force the ocean fish in at 100 million tons. When we project aquaculture out to the same period we find aquaculture overtaking the worldwide edible fish availability around the year 2015, which would be sufficient to meet the world's demand for fish protein. However, that quantity of cultured fish will require feeds, and feeds will require fish meal and oil. Not all of the aquaculture reared fish are carnivorous so the demands for fish meal and oil are not as great, but the aquaculture-raised fish could also supply a level of fish meal and oil from the cuttings and byproducts. Figure 7 shows the projected demand for fish protein against the supply from aquaculture and marine fish.

Fish oil

There has been a great deal of interest in fish oils for their omega-3 fatty acids. The need for fish oils to supply these fatty acids has been well discussed at this conference and at many other meetings. But not all fish oils meet the requirements for the market. In the first place, all fish oils do not supply the same levels of these highly regarded omega-3 fatty acids. Table 7 gives a comparison of various commercial fish oils graded according to the level of the omega-3 fatty acids. These are only average figures as there are variations within species over a season.

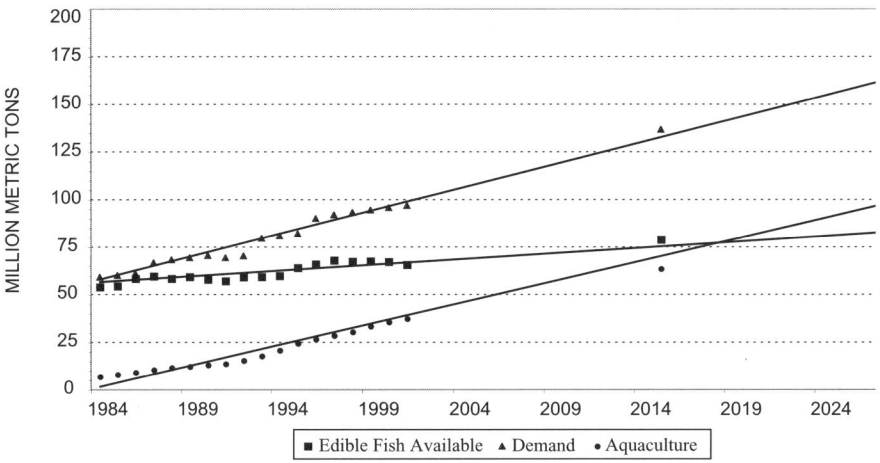

Figure 7. Projected world demand for fish protein. Sources: O'Bannon 2002 and FAO 2002a.

Fish oils are valued in the health market for their omega-3 fatty acid content and oils with higher levels of these fatty acids will command a premium over oils with lower levels. If a variety of fish species are being processed at a facility it would be of value to know what the omega-3 content of the individual oils really is. Fish oil volume is not an issue for this market so even small facilities can compete if they can segregate and protect oils with the desired properties.

The market for fish oil in the omega-3 arena actually has four levels leading to the consumer:

1. The crude fish oil producer. This level controls the normal quality of the fish oil such as free fatty acids, moisture and impurities, and oxidation products. They can also screen for omega-3 fatty acids and if there is storage diversification they can segregate oils based on the omega-3 fatty acids. This will command a premium price. Also, wild salmon oil is a desired oil over aquaculture salmon oil for this market.

2. The semi-refined or refined fish oil producer. This level purchases desired fish oils on the world market from producers above or from brokers. They then process the oil to remove the "gross" impurities, free fatty acids, moisture and impurities, color, stearines, and heavy metals. They stabilize the oil and package it in drums or containers.

3. The pharmaceutical fish oil producer. At this level the refined or semi-refined fish oil is further processed to remove oxidation products,

Table 7.   Comparative levels of EPA and DHA in commercial fish oils.

EPA>DHA	Anchovy (31), sardine/pilchard (25), Atlantic menhaden (22), gulf menhaden (19)
EPA≈DHA	Jack mackerel (28), sand eel (22), Norway pout (20), salmon (19), pollock liver (16), capelin (14), mackerel (12), herring (11), red fish (8)
DHA>EPA	Tuna (26), blue whiting (23), white fish (23), sprat (15), dogfish liver (12)
Low EPA and DHA	Catfish (1)

DHA = docosahexaenoic acid; EPA = eicosapentaenoic acid.
The values in ( ) = EPA C20:5 n-3 + DHA C22:6 n-3. Source: Bimbo 1998.

cholesterol, chemical contaminants such as pesticides, PCBs (polychlorinated biphenyls) and dioxins, and heavy metals. The oils can then be further processed to split the triglycerides into fatty acids, convert the acids to either esters or glycerides, and concentrate the omega-3 fatty acids to 50%, 75%, 90%, or possibly 95% + EPA (eicosapentaenoic acid) and DHA (docosahexaenoic acid). The oil is then stabilized and packaged for further distribution.

4.   Health food supplement producer. This can be part of the third level. At this stage the oil is packaged in capsules or tablets or emulsions, or bottled. It can be micro-encapsulated to a powder. It is then ready for the last step.

5.   Consumer.

As you move through these levels, the volume of the final product is greatly reduced and the value is increased. The higher value producers (level 3 and 4) do not want crude fish oils because their refining processes are not geared up for removal of large quantities of impurities. They therefore rely heavily on the group at level 2 to be their supplier.

To illustrate this market, Fig. 8 shows the imports of fish oil and fish oil fractions into the United States over the period 1989-2002 (through August). Because of nomenclature issues I have found it useful to categorize the products by value per kilogram.

The small seafood processor

The small seafood processor who is not fortunate enough to have a local friendly fish meal plant to take his cuttings is faced with a number of options. At first glance it might appear that he is at a disadvantage to the

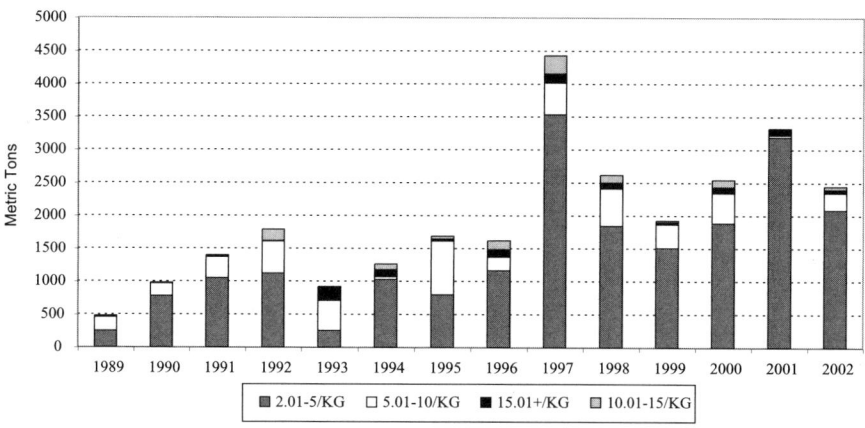

Figure 8. U.S. imports of fish oil and fish oil fractions (2002 data through August only). Source: NMFS 2002a.

larger processors but that is not necessarily the case. The small processor can make quick decisions and change directions, options normally not available to the larger processors. His options range from those that will cost something in order to dispose of the material, through intermediate options that are very market-dependent, and finally to options where the value of the finished products may exceed the value of the edible portions of the fish (Fig. 9). The very high-end options require large capital investments to recover small volumes of high value products but the intermediate options only require some capital and lots of ingenuity to achieve success. In the middle of these options is the production of fish meal and oil. It should be emphasized that these products are market driven; in other words, find the market before producing the product. One final thought, as you move up the scale toward the more highly valued products you will produce more waste products and generate more water which will then have to be processed or handled in some way. That brings us back around to the production of fish meal and oil, where there should be no waste except the water in the fish.

Conclusions

There are numerous opportunities for Alaska to utilize its large fish catch. In 1987 it was estimated that there were opportunities to produce fish meal and oil at volumes equal to the present U.S. production. Since 1987 the landings in Alaska have increased almost threefold, yet according to the available statistics the fish meal plus oil production is only now reach-

Immunoglobulins, enzymes, and other biochemicals
Pigment products (astaxanthin) from crustacean shells
Chitin and chitosan from crab shell waste
Specialty sauces for Asian markets
Mollusk shells in calcium health food supplements
Leather and gelatin from fish (kosher) and shark skins (non-kosher)
Shark, halibut, and cod liver oils
Hydrolysate production
Fish meal and oil production
Use solid waste in advanced silage production
Use solid waste in cold or crude silage production
Use solid waste in composting—Need vegetable waste source
Use liquid and solid wastes directly as fertilizer
Use solid waste for bait and chum fishing
Use crushed mollusk shells for road bed construction
Dispose of liquid and solid wastes in the sea
Landfill solid wastes   Liquid waste to sewer system

Figure 9. Possible uses for fish waste from most valuable requiring the highest capital investment (top) to least valuable requiring lowest investment (bottom). Source: Bimbo 2002.

ing the 90,000 ton level from 60,000 tons in 1994 (1994 is the earliest year statistics are available). It is estimated that about 1.33 million tons of waste was generated in 2001 which would equate (conservatively) to about 266,000 tons of fish meal. At 10-year average fish meal prices ($426 per ton) that is about $113 million dollars. The demand for fish meal will continue to increase as the growth in aquaculture continues.

Over the last seven years the U.S. government (U.S. Agency for International Development, and Environmental Protection Agency) has found the funds to support waste audits for the seafood industry in Peru. Private European funding agencies have supported waste audits in the Chilean seafood industry. The Food and Agricultural Organization (FAO) has supported waste audits in the various developing countries of Africa, Asia, and South America. The audits provide a one-on-one evaluation of the operation with recommendations for improvement and increased efficiency. It would therefore seem that such audits in the Alaskan seafood industry would be most beneficial and would pay dividends for many years to come. All of the involved agencies in Alaska should come together and work to fully utilize the resource in this state.

References

Bimbo, A.P. 1996. Pollution prevention and control in the seafood industry and particularly for small and medium sized fishmeal plants. http://www.cepis.ops-oms.org/muwww/fulltext/epa/pcsi/pcsi.html. U.S. EPA/CEPIS Seminar: Prevención de la contaminación en la pequeña y mediana industria, Lima, Perú. Pan American Health Organization, Lima, Peru.

Bimbo, A.P. 1998. Paper presented at the Lysi HF International Seminar on Fish Oil & Human Health. Beijing and Shanghai, China, November 1998. Unpublished.

Bimbo, A.P. 2002. Cleaner production practices in the Chilean seafood industry. INTEC/GTZ Report, Valparaiso, Chile, February 2002. Unpublished.

Carpenter, G.A. 1961. Drainage of offal. In: International Association of Fish Meal Manufacturers News Summary 4, pp. 15-20.

FAO. 2002a. Towards 2015/2030 world agriculture: Summary report. Food and Agriculture Organization of the United Nations, Rome, Italy.

FAO. 2002b. Fisheries Global Information Service. http://www.fao.org/fi/figis/index.jsp.

Mielke, T. (ed.). 2002. Oil World Annual 2002. ISTA Mielke GmbH, Hamburg, Germany.

Mueller-Vollmer, J., A.P. Bimbo, F. Basurco, and L. Egocheaga. 1998. Prácticas recomendadas para mejorar la eficiencia de los procesos en la industria de harina de pescado. Guía Técnica. Consejo Nacional del Ambiente (CONAM), Lima, Perú.

NMFS. 2002a. Fisheries Statistics and Economics. NOAA, NMFS Office of Science and Technology. http://www.st.nmfs.gov/st1/commercial/index.html.

NMFS. 2002b. National Marine Fisheries Service. Alaska Regional Office. Catch Statistics. http://www.fakr.noaa.gov/2002/2002.htm.

O'Bannon, B. (ed.). 2002. Fisheries of the United States 2001. NMFS, Silver Spring, Maryland.

Petersen, T.E. 1971. The degradation and loss of protein and oil from herring stored at temperatures between 0° and 12° C. In: International Association of Fish Meal Manufacturers News Summary 31, p. 12. St. Albans, Hertfordshire, U.K.

Strøm, T. 1971. Post mortem changes during the storage of capelin. In: International Association of Fish Meal Manufacturers News Summary 31, p. 78. St. Albans, Hertfordshire, U.K.

University of Alaska. 1988. Final report on the characterization of Alaska seafood wastes. Report submitted to the Alaska Fisheries Development Foundation by the University of Alaska Fishery Industrial Technology Center, Kodiak, Alaska, October 12, 1988.

Bycatch Utilization:
The Asian Experience

S. Subasinghe
INFOFISH, Kuala Lumpur, Malaysia

Abstract

Asia exhibits a relatively high level of bycatch utilization. Much of the bycatch is generated from multispecies trawl fisheries. Depending on the utilization pattern of various species, bycatch may include all of "trash fish" and varying amounts of low-valued species, mostly small pelagics. In four Southeast Asian countries examined, bycatch landings mainly consisted of twelve species of fish, which constituted nearly 30-85% of the total bycatch. The total trash fish or bycatch landed in the four countries amounted to over 3.4 million metric tons or over 35% of total fish landed. The main product forms using bycatch include fish meal, dry/salted/fermented products, jelly products, fish crackers, fish extractives, nutraceuticals, and feed additives.

In the last few years there has been an increase in the demand for bycatch, especially for meal and direct human consumption, resulting in almost a doubling of raw material prices in some instances. Profit margin on processing, which has become very competitive, varies significantly depending on the end product, market, quality, and end user. Much effort is now being focused to maintain bycatch quality. Present day processors have to address a host of special criteria including quality and safety of the product, packaging, labeling criteria, resource sustainability, and environmental friendly production techniques. Bycatch is fast losing its status as "a source of animal protein for the masses." Globalization of world trade also has implications in product marketing.

Introduction

Southeast Asia exhibits a relatively high level of fisheries bycatch utilization. Much of the bycatch in the region is generated from multispecies trawl fisheries. These fisheries target a mixture of fish species and often very little is discarded making it difficult to distinguish between target and non-target species. On the other hand, due to low catches per unit effort and not so attractive returns from shrimping, even dedicated shrimp trawlers now often carry out non-shrimp fishing operations, depending on season and location. Despite this, there has been a reduction in fish supply to local markets due to rising populations, reduction in traditional catches, and an increase in the export trade. All these factors have reduced the amount of fish available from traditional sources. This has resulted in a wide range of bycatch species being marketed in fresh form to bridge the shortfall. In some countries domestic marketing of fresh bycatch has increased to such an extent that often it has become difficult to distinguish between bycatch and target species in a fishery.

Thus, it is difficult to give a very clear-cut definition to the term "bycatch" in the Southeast Asian scenario. Depending on the utilization pattern of various species in different countries, it may include all of "trash fish" and varying amounts of low-valued species, mostly small pelagic. The term trash fish is generally used to denote that portion of the catch that is composed of young or juvenile fishes, including damaged fish. Thus exact quantification of the bycatch component of a particular fishery becomes difficult. This paper is largely based on a study carried out by INFOFISH on behalf of the Food and Agriculture Organization (FAO) on bycatch utilization pattern in four countries in the region, namely, Indonesia, Malaysia, Philippines, and Thailand.

Total landing of bycatch in the four countries is around 2.2 million metric tons (t) (Table 1). Bycatch landings mainly consist of twelve species of fish, which constitute 30-55% of the total bycatch. In addition, varying amounts of the landings are categorized as trash fish. Trash fish landings total around 1.2 million t. Interestingly, Indonesia does not allow landings of trash-fish or rather the use of the terminology for any of the catch, and all are categorized as bycatch. Categorization of fish as trash fish or bycatch fish is a fairly loose one and hence when discussing utilization of low-value species, it is more prudent to group together both trash fish and bycatch. The percentage of the total trash fish and bycatch component of the total fish landed in the four countries varies from nearly 20% in Indonesia to a high of 49% in the Philippines.

Processing and marketing of bycatch-based products

Bycatch marketing channels in the region show significant similarities (Fig. 1). Even though the range of products using bycatch/trash fish is very large, the main product forms using bycatch include dry/salted fish, fish jelly products, fermented products, fish crackers, and fish meal. However, over the last decade there has been an increase in the usage of bycatch fish and fish waste utilization in the production of specialty products, feed additives, and a wide range of nutraceuticals (Table 2). Table 3 shows the number of plants producing traditional bycatch fish products in Malaysia and Thailand.

Table 1.   Bycatch composition (1997).

Bycatch species	Indonesia Quantity (t)	% of total bycatch	Malaysia Quantity (t)	% of total bycatch	Philippines Quantity (t)	% of total bycatch	Thailand Quantity (t)	% of total bycatch
1. Slipmouth (Leiognathidae)	89,403	12.00	2,362	1.1	61,254	6.9	–	–
2. Lizard fish (Synodontidae)	15,158	2.00	14,490	6.6	6,671	0.8	71,315	3.9
3. Flounders (Bothidae)	15,075	2.00	2,193	1.0	627	Negligible	7,423	0.4
4. Red bulls eye (Priacanthidae)	4,448	0.7	13,434	6.2	18,378	2.1	77,825	4.3
5. Threadfin (Nemipteridae)	29,340	3.9	31,052	14.2	29,839	3.4	87,717	4.8
6. Goat fish (Mullidae)	24,203	3.3	10,302	4.7	15,884	1.8	420	Negligible
7. Trevalle (Carangidae)	32,097	4.3	–	–	32,175	3.6	49,747	2.7
8. Conger eel (Muraenesocidae)	–	–	6,844	3.1	2,053	0.2	1,600	0.1
9. Catfish (Siluriformes)	78,578	10.6	13,599	6.2	6,324	0.7	763	0.00
10. Ribbon fish (Lampridiformes)	–	–	–	–	9,412	1.1	17,586	1
11. Sardine (Clupeidae)	295,550	39.8	–	–	302,341	34.2	202,792	11.1
12. Croakers (Theraponidae)	44,837	6.0	20,368	9.3	3,992	0.5	29,961	1.6
1. Total (1-12)	628,689	85	114,644	53	488,950	55	547,149	30
2. Total bycatch	742,997		218,200		884,950		399,890	
3. Trash fish	–	–	333,668		7,869		822,110	
4. Trash fish + bycatch	742,997		551,868		892,520		1,222,000	
5. Total landing	3,791,025		1,172,922		1,805,806		2,877,622	
6. Bycatch + trash fish % of total landing	20		47		49		42	

Fish meal

Fish meal production consumes a very significant quantity of landings, varying from around 25% in the Philippines to nearly 40% in Thailand. Almost all of trash fish and varying amounts of bycatch are used in fish meal. Fish meal prices in the global market have increased due to increased aquaculture effort, coupled with the low production of major global producers Peru and Chile and the poor demand for bovine meals due to mad cow disease. Over the last few years demand for fish meal, and hence the raw material fish, has risen locally.

The processing technology of meal varies according to the sophistication of the industry and the type of raw material used. Protein content of crude meals is relatively low and may vary from 45 to 55%, as compared to over 70% for good quality meal. Crude production operations may

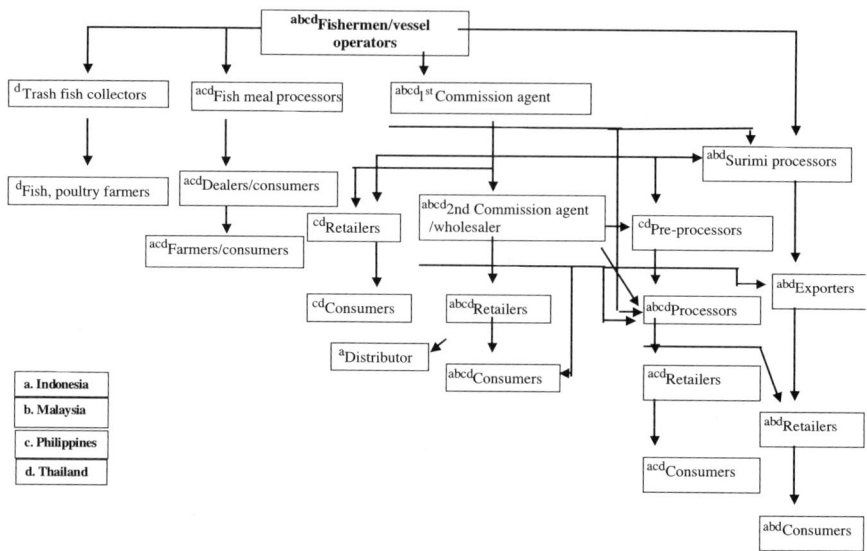

Figure 1. Bycatch marketing channels.

even incorporate or use waste from fish processing factories. Such meals have a very high fat content and low (40-45%) protein content. Fish meal is marketed in 25 kg or 50 kg polyethylene lined poly-sacks. In many countries, heavy competition for raw material and diminishing supplies have resulted in many operators operating below capacity or even winding up business. For example, in Thailand the number of fish meal plants decreased over the three-year period 1995-1998 from 123 to 97. Thai production of 342,438 t of fish meal in 1998 used nearly 758,465 t of trash fish, 53,841 t of other bycatch fish, and nearly 511,581 t of food factory waste, mostly from surimi factories and tuna canning waste.

Dried Fish

Production of dried fish is one of the most widespread traditional industries in the region. The amount of bycatch used by the industry shows wide variation depending on the seasonal availability. Among the bycatch species, those most commonly used for processing into dry fish are sardine (Clupeidae), croaker (Theraponidae), ribbonfish (Lampridiformes), threadfin (Nemipteridae), flounder (Bothidae) and lizardfish (Synodontidae). The process involves gutting, brining, and either sun or kiln drying of the fish depending on the facilities available. The dried product has a moisture content of 30-40% and is thus drier than dry fish made of high value species such as Spanish mackerel (*Scomberomorus* sp.), threadfin

Table 2.   Bycatch and fish waste–based products in Asia.

Product	Species used[a]	Part of the body used
Main products		
Fish meal	Nonspecific	Whole
Dry/salted fish	Nonspecific	Whole
Fish sauce	Sardine/anchovy	Whole
Fish cracker	Sardine/croaker/shrimp	Meat
Fish floss	Rays	Meat
Jerky	Tuna	Meat
Surimi	Croaker/threadfin/lizard fish	Meat
Extractives	Tuna/shrimp/oyster	Cook water
Sausage/burgers	Tuna	Frame meat/off-cuts
Others		
Chitin/chitosan	Shrimp	Shell
Astaxanthin	Shrimp	Shell
Squid oil	Squid	Offal
Squid meal	Squid/cuttlefish	Offal
Squid ink	Squid/cuttlefish	Offal/viscera
Tuna oil (DHA/DPA)[b]	Yellowfin/bigeye	Head/eye muscle
Fish gelatin	Tuna	Skin
Leather	Eel/shark/barramundi	Skin
Squalene	Shark	Liver
Cartilage	Shark	Skeleton

[a]Scientific names given in text.
[b]Docosahexaenoic acid/docosapentaenoic acid.

Table 3.   Number of plants producing traditional fishery products (1998-1999) in Malaysia and Thailand.

	Malaysia	Thailand	
Product	No. plants	No. plants	Raw material used
1. Fish ball/fish cake/surimi	28	95	6,410 t
2. Fish meal	22	97	1,323,887 t
3. Fish/shrimp cracker	24	122	2,417 t
4. Dried salted fish	38	630	40,376 t
5. Fish sauce/paste	24	158	100,193 t
6. Smoked fish	–	20	1,405 t
7. Steamed fish	–	93	7,271 t
Total	136	1,215	1,481,959 t

bream (Nemipteridae), or snapper (Lutjanidae). Dried fish is bulk packed in polyethylene lined carton boxes. The product is sold in supermarkets packed in polyethylene bags or is retailed at open markets or night markets in urban and suburban areas.

Fish cracker

Fish cracker is a snack food widely consumed in the region. Depending on the country, crackers may contain varying amounts of fish flesh, ranging from 5 to 15% or more. In Malaysia, fish cracker should contain more than 15% fish flesh by law. The fish cracker dough is made by mixing fish mince with tapioca flour, salt, warm water (and various other ingredients such as garlic and pepper, depending on the recipe), and kneading well. The dough is made into a cylindrical shape and steamed for 1-2 hours. The cooked dough is cooled for 6-12 hours or more, cut into thin slices manually or mechanically using a slicer, and kiln dried for 4-6 hours or sun dried for 1-2 days. The dried product is deep fried until it becomes light and crisp. Fried, ready to eat crackers are packed in 50-200 gm consumer packs for the retail market. Some cottage scale operators in Malaysia and Indonesia sell the dried product, which may take various shapes, direct to the consumers.

In Thailand the main species used for fish cracker is mince of bigeye (Priacanthidae), a quality mince which gives good color and texture to the final product. Though shrimp cracker is more popular in the region, the high price of the minced shrimp raw material (US $1.5-3.0 per kg) is a hindrance to its growth. The industry is mostly based on bigeye (*Priacanthus* spp.) the mince of which costs around US $1.0 per kg. There are nearly 122 shrimp cracker producers in Thailand using nearly 2,500 t of raw material annually. As the production process is labor intensive, cost of labor accounts for over 25% of the production cost. Freshness of fish used and the quality of the other major ingredient, tapioca flour, determines the quality of the cracker. For example in Thailand, poor quality cracker which has very little fish meat sells at around Baht 75 per kg at retail level, while good quality product is sold at around Baht 100-110 per kg.

Fish ball

Fish ball is a favorite snack food in the region. It may be served boiled or fried by street vendors or may be incorporated into soups and even cooked dishes. The main ingredients in fish ball are fish mince, tapioca or corn starch, salt, baking powder, and various seasonings. Manufacturers may make their own fish mince from fresh fish or buy filleted fish or ready-made fish mince blocks from suppliers. The ingredients are made into balls manually or mechanically. The fish balls are allowed to settle in chilled water (10-20°C) for 20-30 minutes and are steamed or cooked. Cooked product is cooled under a fan to room temperature and packed in bulk packs (2 kg) or smaller consumer packs (250 g, 500 g, 1 kg). The product is marketed chilled and is sold in both supermarkets and wet markets. In Thailand nearly 100 licensed fish ball plants are in operation using around 6,500 t of fish annually.

Fish sauce

Traditionally fish sauce production has been a cottage scale operation. However, the modern processing operations are heavily automated. The basic process involves fermentation of fish species such as anchovies, sardines, or mackerel in about 25% salt by weight of fish used. The fish is mixed well with salt and left covered in earthen pots or jars for 3-12 months until a clear yellow-colored liquid sauce is obtained. Sauce made of anchovy generally demands a better market price. After the first extraction, leftover fish is further fermented up to two or three times with the addition of salt to give poorer quality sauces. In Malaysia and Thailand there are 24 and 158 licensed fish sauce/paste manufacturing facilities respectively. Nearly 100,193 t of fish are used by the plants in Thailand.

Surimi

White fleshed fish species with good gel-forming abilities are used in the production of surimi. Species often used in the region are bigeye, croaker, and threadfin bream. Minced fish is leached 2-4 times with chilled water to remove soluble components, mainly proteins and blood, and then kneaded with sugar and polyphosphates and frozen into blocks. Surimi with a lighter, more whitish color and good gel strength commands a better price. For example in Thailand the higher grades with a gel strength of over 1,000 g per cm^2 fetches around Baht 85-90 per kg (US $2.0 per kg) while the lower grades fetch only Baht 65-80 per kg (US $1.45-1.75 per kg). Japan Imported 18,578 t of surimi from Thailand during 2000.

The processors produce various grades of surimi. Even species such as sardines, tilapia, and mackerel may be used to produce type C surimi which is also used for the manufacture of products such as fish ball and fish cake for the local market. In fact, the local population prefers products made of type B and type C surimi as the products retain some fish flavor and juiciness as compared to the bland type A surimi.

Raw material cost varied from US $355 per ton in Thailand to US $576 in the Philippines, while the adjusted cost stood at US $1,422 and US $1,153 respectively, indicating a higher extraction rate in the Philippines industry. This is expected as the figures quoted are for two different grades of surimi. Thai product is meant for high end analog products such as crab sticks for export market while the Philippines surimi is mostly used by the local industry in the production of fish ball or fish cake for the domestic market.

Processing cost of surimi shows wide variation depending on the grade of surimi produced. High grade surimi prepared from white fleshed fish with good gel strength is processed into specialty analogue products or is exported. When supplying quality raw material for the production of high grade surimi for export, commission agents keep a markup of about 5%, selling the product to processors. Processors sell the product at around US $1,800 per t keeping a margin of about 20%. Retail/export price of type A surimi is around US $2,000 per t with a retail profit margin of around 11%.

Production and profitability

Bycatch based industries show significant profitability. In spite of the gradual increase in the price of fish over the years, and relatively tight competition for raw material as a result, in many countries in the region processors of traditional fish products figure prominently in the fishery sector. Though the numbers of processors may have dropped significantly due to stiff competition, economic downturn experienced by many countries in the region, and stricter quality requirements demanded by the industry, those who survived have consolidated themselves into economically viable entities (Table 4). In fact, though the number of processors has come down in some product categories, the total production by weight has shown a significant increase (Table 5). Application of novel methods of quality control under the guidance of government regulatory bodies has helped these processors enter foreign ethnic markets.

Table 6 shows the profitability of some of the bycatch based products discussed here. In Thailand, for example, the price of trash fish has risen from Baht 2.8-3.6 (US $0.06-0.08) per kg during 1994-1998 to Baht 7 (US $0.15) per kg in July 2001. Fish meal price during the same period has increased from Baht 14,700 per t to Baht 24,000 per t (US $326 to 533 per t). It is interesting to note that in 2000, while Thailand exported 25,199 t of fish meal at approximately US $480 per t, the country imported 11,821 t of high quality meal at around US $598 per t. The profit margin per metric ton of product varied from around US $8 (Thailand) to a high of US $75 (Malaysia). Ex-factory sales price of meal was lowest in Thailand (US $327) and highest in the Philippines (US $385).

In the case of dry fish, a strict comparison of profit margins is difficult as the market price and the profitability of the operation largely depends on the raw material used, the price of which may vary several fold, depending on the species, seasonality, and quality. Cost of production in the Philippines (raw material priced at US $1,175 per t) and Thailand (at US $1,368 per t) amounted to US $1,260 per t and US $1,387 per t respectively. Corresponding selling prices were US $1,603 per t (Philippines) and US $1,444 per t (Thailand) respectively, with processors maintaining a profit margin of 27% and a low 4% respectively.

Fish cracker production seems to be a very profitable bycatch-based industry. Variable costs show wide variation from US $1,701 per t in Thailand to a low of US $304 per t in Malaysia. Thailand pays nearly US $1 per

Table 4. **Number of fish processing factories by type (1980-1998) in Thailand.**

	1980	1985	1990	1995	1998	Increase (1980-1998)	Increase (1985-1998)
Freezing	90	80	108	144	131	46%	64%
Canning	9	39	42	52	42	×4	8%
Fish sauce	98	114	116	102	88	−10%	−23%
Budu sauce	14	33	29	54	70	×5	112%
Steamed fish	81	115	55	80	93	15%	−19%
Smoked fish	1	171	36	26	20	×20	−88%
Salted fish	327	978	750	727	630	92%	−36%
Dried shrimp	139	148	205	158	128	−8%	−14%
Dried squid	317	879	712	561	534	−68%	−39%
Dried shellfish	16	674	613	237	197	×12	−71%
Fish ball	11	54	94	98	95	×9	76%
Fish cracker	4	76	90	125	122	×42	61%
Fish meal	95	92	104	122	97	2%	5%

Source: Compiled using data from Department of Fisheries, Thailand.

Table 5. **Use of marine landings including bycatch by the traditional fish processing sector (1994-1998) in Thailand.**

Type of plant	1994	1995	1996	1997	1998	% change 1994-1998
Fish sauce	32,855	50,745	65,190	68,409	99,546	+200%
Budu sauce	325	318	364	410	647	+100%
Steamed fish	4,282	4,554	7,272	7,125	7,271	+70%
Smoking	1,499	1,374	1,652	4,726	1,405	−6%
Salted fish	48,581	55,000	66,354	51,811	40,376	−17%
Dried shrimp	26,637	23,318	27,231	20,922	18,329	−31%
Dried squid	31,690	29,767	30,006	21,717	16,286	−49%
Dried shellfish	2,522	2,684	2,525	1,715	2,350	−7%
Fish ball	6,005	5,664	7,453	7,764	6,410	+7%
Fish/shrimp cracker	1,118	2,086	2,280	2,266	2,417	+116%
Total	155,514	175,510	210,327	186,865	195,037	+25%

Source: Compiled using data from Department of Fisheries, Thailand.

Table 6.   Raw materials and processing costs per metric ton (in US$).

Products	Indonesia	Malaysia	Philippines	Thailand
Dried Fish				
1. Variable cost	NA	NA	1,175	1,368
2. Fixed expenses	NA	NA	36	19
3. Selling expenses	NA	NA	50	0
4. Total 1-3	NA	NA	1,260	1,387
5. Ex-factory price	NA	NA	1,604	1,444
6. Profit margin (%)	NA	NA	344 (27)	58 (4)
Fermented product				
1. Variable cost	401	363	411	NA
2. Fixed expenses	42	74	39	NA
3. Selling expenses	80	16	23	NA
4. Total 1-3	523	453	473	NA
5. Ex-factory price	675	524	529	NA
6. Profit margin (%)	152 (23)	71 (14)	56 (12)	NA
Fish ball				
1. Variable cost	399	720	1,679	732
2. Fixed expenses	64	117	90	12
3. Selling expenses	87	32	156	7
4. Total 1-3	550	868	1,925	752
5. Ex-factory price	575	1,053	2,885	778
6. Profit margin (%)	25 (5)	184 (21)	960 (50)	26 (3)
Fish meal				
1. Variable cost	257	211	306	292
2. Fixed expenses	30	47	23	18
3. Selling expenses	50	9	9	9
4. Total 1-3	337	268	338	319
5. Ex-factory price	375	342	385	327
6. Profit margin (%)	38 (11)	74 (23)	47 (14)	8 (2)
Fish cracker				
1. Variable cost	497	304	NA	1,702
2. Fixed expenses	37	131	NA	86
3. Selling expenses	108	11	NA	0
4. Total 1-3	641	446	NA	1,788
5. Ex-factory price	950	1,053	NA	2,000
6. Profit margin (%)	309 (48)	607 (136)	NA	213 (12)

Table 6.   (Continued.) Raw materials and processing costs per metric ton (in US$).

Products	Indonesia	Malaysia	Philippines	Thailand
Fish floss				
1. Variable cost	NA	NA	NA	4,943
2. Fixed expenses	NA	NA	NA	96
3. Selling expenses	NA	NA	NA	0
4. Total 1-3	NA	NA	NA	5,040
5. Ex-factory price	NA	NA	NA	5,333
6. Profit margin (%)	NA	NA	NA	294 (6)
Surimi				
1. Variable cost	NA	NA	NA	1,622
2. Fixed expenses	NA	NA	NA	93
3. Selling expenses	NA	NA	NA	87
4. Total 1-3	NA	NA	NA	1,801
5. Ex-factory price	NA	NA	NA	2,000
6. Profit margin (%)	NA	NA	NA	199 (11)

kg for raw material (or nearly 45% of the production cost) while the cost of raw material used in Malaysia is around US $0.20 per kg. The marked difference in raw material cost is due to the exclusive use of mince of big-eye by the Thai industry to get a product with good color and texture. In Malaysia the manufacturing process is mostly a cottage scale operation, which uses a wide range of bycatch species available at very competitive prices. Furthermore, cracker sold in the domestic market should have a minimum of 15% fish which leaves very limited room and incentive for the producers to improve color and texture of the product.

The fish cracker product in Thailand, which is manufactured in large commercial scale operations, has a lower content of fish and is lighter. Pack size and method of packaging and presentation contributes significantly at the retail level. The processor's price showed wide variation ranging from US $446 per t in Malaysia to US $1,788 per t in Thailand, with a markup of 136% and 12% respectively.

The fish ball industry has grown over the last decade. The cost of mince raw material shows wide variation. In the Philippines and Thailand the producers buy the ready mince at prices ranging from US $1.00 in Thailand to around US $1.60 in the Philippines. In Malaysia and Indonesia the raw material cost of US $0.59 per kg and US $0.36 per kg respectively applies to industries directly using fish as raw material, and not ready mince as in the Philippines and Thailand. In Thailand, fish ball made of good quality sea fish such as threadfin bream and croaker may fetch a retail price of Baht 60-65 per kg (around US $1.35 per kg) while those from demersal species may fetch around Baht 45-55 per kg (US $1.00-1.20 per kg). Products with tapioca and less fish flesh fetch a lower price and are mostly bought by noodle sellers. In the Philippines the product is retailed at a price range of US $2.90-3.10 per kg whereas in Malaysia the product is retailed around US $1.30-2.00 per kg. Raw material used for fish ball production shows very wide price differences ranging from around US $338 per t in Indonesia to a high of around US $1,446 per t in the Philippines.

In the fish sauce industry raw material cost is US $217 per t in Malaysia, US $317 per t in the Philippines, and US $340 per t in Indonesia, amounting to nearly 60-85% of the variable cost and around 49-59% of the cost of production of fish sauce. The processor's selling price of the product ranges from US $524 per t (Philippines) to US $675 per t (Indonesia). The profit margin for processors varies from US $56 per t in the Philippines to US $152 per t in Indonesia, a markup of 12% and 23% respectively. The cost of production of lower grades of fish sauce is lower due to the fact that after the first extraction, the leftover fish is further fermented up to two or three times with the addition of salt to give poorer quality sauces.

Prospects for improved utilization and marketing

The introduction of large scale commercial fisheries resulted not only in the rapid commercialization of the traditional processing sector but also the introduction of new products based on bycatch species, such as surimi, a wide range of fish jelly products, fish cake, and fish floss. On the other hand, the advent of industrial fisheries increased the amount of fish extracted from the waters, often to the extent that in some areas the resources dwindled to such a degree that conflicts arose among stakeholders. Increased demand for fish has led some countries to import fish and fishery products to cater to the domestic market and the processing sector, which re-exports some after value addition. At present very little bycatch fish is discarded by the fisheries in the region. Even boats operating in distant waters try to save the bycatch or sell it to collector vessels. Sometimes the bycatch is shared by the crew. Considering the present utilization pattern, in order to promote sustainable use of resources, it is important to preserve the quality of trash fish and bycatch onboard to facilitate improved utilization for human food purposes. In this respect the following approaches could be recommended:

1. *Improved onboard facilities and infrastructure*: Improved onboard facilities for sorting, washing, storing, and icing of bycatch would help to improve the quality of bycatch fish. It may not be always

possible to achieve this due to various limitations such as lack of storage space for ice, bycatch fish, or labor. It is encouraging to note that the incentive of better price has led some trawl operators to preserve bycatch quality. Infrastructure at landing sites too plays an important role in improved bycatch utilization. Availability of clean water, ice for re-icing the catch if necessary, storage space, mechanical facilities for unloading bycatch without damaging, transportation, wholesaling, and retailing are important in this respect.

2. *Promote waste utilization*: It is heartening to note that much progress has been made in the utilization and marketing of fish waste, especially shrimp waste, in the region. Production of chitin and chitosan and extraction of astaxanthin and other pigments and feed additives has become an integral part of the shrimp industries of most countries of the region. Government-industry cooperation in promoting such activities by way of technology transfer, market promotion, as well as streamlining collection of waste from processing centers, would be helpful in further promoting waste utilization.

3. *Product development*: Product and market diversification is another important aspect to be addressed in improving utilization of bycatch. The products presently being marketed could be further improved to extend their shelf life, safety, quality and marketability, which effectively amounts to improved utilization. On the other hand, modern methods of packaging, transportation, and storage could also be used to market some species of bycatch dressed, in consumer packs, or in a ready-to-prepare form.

4. *Transfer of technology and improving efficiency of processing*: Over the last decade the number of small-to-medium-scale bycatch processing operations have declined due to competition from large-scale operators with capital and advanced technological know-how. Broad basing the industry with transfer of such technology and training and improving entrepreneurship could be considered an important step in improving bycatch utilization and marketability.

5. *Improved quality and safety of products*: Improving quality and assurance of product quality and safety is a vital step in present-day product marketing. Considering the diffused nature of raw material supply in the bycatch processing sector and traditional nature of the production technologies involved it is important to have HACCP-based production operations in place if the products are to win the confidence of the present-day discerning consumers. In this respect much support and assistance could be provided by regulatory authorities.

6. *Financial assistance*: Governments in the region have a general policy to promote small-to-medium-scale entrepreneurship by loan and/or

funding agencies, which are widely applied to the agriculture and fisheries sector through various agricultural banks. They often offer concessionary terms and conditions through government assisted schemes. However, in spite of such governmental intervention many opt to go to sources outside the banking scheme at much higher interest rates, since they find the government and institutional funding too complicated due to much paperwork involved and the need for collateral. Thus it would be prudent to simplify such assistance packages and where possible to channel them through specialized fisheries or cooperative banks with user friendly schemes.

Even though the demand for bycatch is expected to continue in the years to come, many worry about the long-term sustainability of the resource. Mesh size regulation, tow time limitation, and implementing closed seasons are considered useful approaches for minimizing bycatch in fisheries. However, considering conflicting interests among stakeholders, successful sustained implementation of the measures becomes difficult. In this respect, community-based management of trawling, including zoning, is promoted by some as a cost effective way of controlling and conserving the resource.

On the other hand, traditional approaches to improved bycatch utilization need to be reassessed in the light of new developments in the global seafood market. The concept of utilizing bycatch to bridge the protein-deficit in populations and to provide additional income and employment opportunities to artisanal sectors in production and processing may not be applicable in the present day aggressive market oriented scenario. In the past most traditional products, which are the key bycatch based products of today, were processed under cottage scale operations with very little mechanization. The present day product processing and marketing has to address certain special criteria, not only with respect to quality and safety of products, but also on aspects such as packaging, including eco-labeling, presentation, and resource sustainability. In this respect, it is also important for the marketers to have a good knowledge of market requirements and consumer expectations, including ready access to market and price information.

In such a scenario, we have to be mindful of the high human and material cost of product processing and marketing, by way of trained manpower, market/product information, upgraded processing facilities, quality control, and monitoring. Globalization of world trade and development of regional trading agreements such as the ASEAN (Association of Southeast Asian Nations) Free Trade Agreement (AFTA) also have implications in product marketing. Even though there is provision for providing a certain degree of protection for indigenous industries in these agreements, they will undoubtedly bring in heavy competition for a share of the market.

Microbiological Aspects of Animal Feed Manufacturing: Safety and Quality

Brian H. Himelbloom
University of Alaska Fairbanks, Fishery Industrial Technology Center, Kodiak, Alaska

Abstract

Safety and quality of animal feeds are under scrutiny due to the potential transmission of unwanted microorganisms and other agents. Feed manufactured as a seafood byproduct are reviewed with regard to bacteria, molds, and viruses. Controls used to prevent contamination and post-processing recontamination are addressed. New technologies to stabilize seafood wastes for extended shelf life, to enhance animal nutrition, and to test manufactured products are presented.

Introduction

The usefulness of post-mortem fish and fishery byproducts is dependent on the conditions used for preservation of the material. Great care is needed if the food is destined for human consumption. Animal feeds produced from fishery waste have not been held to such a high standard. However, there is concern about the potential transmission of microorganisms and end products of microbial metabolism into feeds for animals, which ultimately become consumed by humans. In this review, the microscopic world is addressed with respect to spoilage and pathogens in fish waste processing. Processing options for maintaining quality and safety are explored along with analytical testing and the application of probiotic bacteria as feed adjuncts in animal nutrition.

Discussion

Fish spoilage microbiology

Bacteria

Several types of bacteria can subsist on the nutrients found in fish. Muscle, skin, mucus, gills, guts, and blood provide components for growth and multiplication of bacteria. The microbial content of Alaska fish waste was determined and contained approximately 10^7 bacteria per gram, which included 10^4 coliforms per gram and 10^2 *E. coli* per gram (Himelbloom and Stevens 1994).

The end products of microbial metabolism in fish tissues are typically odoriferous chemicals that are easy to detect sensorially. Compounds that may be present are amine- and sulfhydryl-derivatives, such as ammonia, trimethylamine, dimethylamine, hydrogen sulfide, and dimethyl sulfide, occurring from the breakdown of amino acids and other low molecular weight chemicals. These spoilage compounds will be present in various concentrations depending on the fish species, the level of microbial growth, the microflora composition, and the environmental conditions of the stored fish or fishery waste.

Fungi

Molds and yeasts are very minor players compared to bacteria in spoiling of fish. These "higher microorganisms" do not compete well because of the absence of freely available carbohydrates that are necessary for fermentation. However, menhaden stickwater can be fermented by pure cultures of two fungi metabolizing the fish lipids even though no carbohydrates are present (Green et al. 1976). Due to their tolerance to water activities below the limit for most bacteria, the molds may spoil semimoist fish meal. Since fungi are acid-tolerant, these microorganisms can persist in fish silage tanks with molds and oxidative yeasts occupying the aerobic surfaces (Levin et al. 1989).

Fish meal quality—spoilage effect on feeding fish

Freshness of fish meals is a major criterion for quality. Advanced spoilage of raw herring prior to press cake meal manufacturing resulted in depressed digestibility in feeding trials of chinook salmon and rainbow trout (Clancy et al. 1995). Much research has indicated that spoiled fish prior to being manufactured into fish meal contains high levels of biogenic amines and other volatile amines. Histamine can be found in fish solubles, such as mackerel and sardine stickwater, and whole meal in which solubles have been added to fish press cake (Toyama et al. 1981, Okuzumi et al. 1984). Some researchers have speculated that the biogenic amines such as histamine, cadaverine, tyramine, and putrescine cause animal feeding problems. Cadaverine levels greater than 600 ppm reduced growth and reduced feed efficiencies in Atlantic halibut (Aksnes and Mundheim 1997). However, a recent report indicates that the high biogenic amine concentrations do

**Table 1. Bacterial decarboxyl-
ation of amino acids
into biogenic amines.**

Amino acid	Biogenic amine
Arginine	Putrescine
Histidine	Histamine
Lysine	Cadaverine
Tryptophan	Tryptamine
Tyrosine	Tyramine

From Pike and Hardy 1997, Silla Santos 1996.

not affect animal production or lead to pathological conditions in the gastrointestinal tract of Atlantic salmon smolts (Opstvedt et al. 2000). More research is needed to determine if the effects of biogenic amines in feeds are specific to certain fish species at a particular life stage.

Biogenic amine formation

Biogenic amines occur from the decarboxylation of certain amino acids (Table 1). Biogenic amines are nonvolatile, water-soluble, and stable to heat processing (Ricque-Marie et al. 1998). The conversion from histidine to histamine is the most studied biogenic amine and causes chemical intoxication known as scombroid poisoning. The causative bacteria that contribute the enzymes are in the family Enterobacteriaceae, and the most potent producers are *Morganella morganii* and *Klebsiella* species. Fish that contain high levels of the amino acids capable of being metabolized by bacteria to biogenic amines are tuna, mackerel, and mahi-mahi. Temperatures above refrigeration are necessary for the growth of biogenic amine–producing bacteria. Some species of lactic acid bacteria produce biogenic amines; however, these bacteria require a carbohydrate source for growth and metabolism. Fermented or ensiled fish may be a problem if biogenic amine–forming lactic acid bacteria dominate the microflora such as *Lactobacillus* species in ensiled Atlantic mackerel (Dapkevicius et al. 2000).

Instances of overcooking fish meal followed by acid hydrolysis caused the toxic compound gizzerosine, a derivative of histamine, to be formed and result in a serious disease in chicks (Okazaki et al. 1983). Morphological abnormalities were observed in rainbow trout intestines during a short-term feeding trial of gizzerosine-positive fish meal (Fairgrieve et al. 1994).

Microbial safety—pathogenic bacteria

Only a few types of bacteria are of concern to seafood processors. These are *Clostridium botulinum*, *Listeria monocytogenes*, *Staphylococcus aureus*,

Table 2.   Pathogenic bacteria potentially associated with fish meal production.

Bacterium	Characteristics and minimum growth requirements
Clostridium botulinum Type E	Gram-positive, sporeformer, strict anaerobe, a_W >0.97, pH > 5, temp > 3.3°C, causes botulism (neurotoxigenic)
Listeria monocytogenes	Gram-positive, facultative anaerobe, a_W > 0.85, pH > 4.5, temp > 2.5°C, causes listeriosis (systemic infection)
Salmonella sp.	Gram-negative, facultative anaerobe, a_W > 0.93, pH > 4.0, temp > 5°C, causes salmonellosis (enterotoxigenic)
Staphylococcus aureus	Gram-positive, facultative anaerobe, a_W > 0.85, pH > 4.5, temp > 10°C, causes staphylococcal poisoning (enterotoxigenic)

Adapted from Himelbloom 1998, Doyle et al. 1997.

and *Salmonella* species (Table 2). These bacteria can enter the processing plant via raw materials, humans, and pests. The anaerobic sporeformer, *C. botulinum* Type E, is associated with soils and marine sediments. *Listeria monocytogenes* is considered ubiquitous and resides in cold, wet environments. If cleaning and sanitation are substandard, these bacteria can create reservoirs from which they can continue to contaminate pre- and post-processed products. The mesophilic non-sporeformers, *Staphylococcus aureus* and *Salmonella* species, are found typically on humans and other warm-blooded animals. Bacterial toxins can be produced under particular conditions for these four types of pathogenic bacteria. Of these, only *S. aureus* produces a heat-stable enterotoxin.

Salmonella

Since *Salmonella* is a leading bacterial cause of foodborne illnesses, primary producers of poultry, pork, beef, and fish have been trying to restrict the bacterium from farms. One of the first steps is to ensure that the feeds are *Salmonella*-free. This would eliminate a major factor in contamination of agricultural and aquatic farms. *Salmonella* species were introduced in many countries through contaminated fish meal used for animal feeds (ICMSF 1998). Standards were applied forty years ago to improve the safety of fish meals (Dreosti et al. 1962). Since then, fish meal used as a component in animal feeds has tested positive for *Salmonella* in 7-48% of the samples and various serotypes have been isolated (Jacobs et al. 1963, Williams et al. 1969, Morris et al. 1970, Trust 1971, Fassi-Fehri and Kochanski 1971, Roumani et al. 1981, Nabbut et al. 1982, Veldman et al. 1995). A more recent investigation of poultry feed ingredients in northern Greece found no *Salmonella* in thirty samples of fish meals (Zdragas et al. 2001).

Antibiotic-resistant pathogenic bacteria

The environmental prevalence of antibiotic-resistant bacteria and the upsurge of resistant pathogenic bacteria in hospital patients is a major health issue. Some of the resistant bacteria are theorized to arise from industrial feeding of animals raised for human food and transmission of the pathogens into the food supply. These animals are being supplemented with antibiotics to ward off infections and increase body weight rapidly. Pathogens such as *Salmonella* species are prevalent on animal farms and readily pick up the genetic determinants, i.e., plasmid deoxyribonucleic acid (DNA), for resisting antibiotics. Antibiotics are also used in aquaculture; however, there has been no evidence of antibiotic-resistant bacteria surviving the pasteurization of farmed fish waste into fish meal.

Non-bacterial safety concerns

Other safety aspects of fishery waste may involve non-bacterial vectors. Marine biotoxins, such as paralytic shellfish poison, produced by dinoflagellates (microscopic marine algae) may occur in molluscan shellfish. There have been no reports of these heat-stable toxins in fish meals. Tetrodotoxin from puffer fish would be a concern if fish silage includes these species (Disney and James 1980, Raa and Gildberg 1982). A number of fish kills and fishermen ailments on the East Coast of the United States have been attributed to *Pfiesteria piscicida*. This dinoflagellate would not survive the processing of fish waste into fish meal. Although the stability of the toxin is unknown, fish killed by the dinoflagellate would not be used for fish meal. Intracellular parasites or viruses require live hosts for propagation. Although infectious viruses can occur in aquaculture and hatcheries, the viruses cannot survive the conversion of fish tissues into fish meal. Parasites such as roundworms and cod worms and their encysted states would not be expected to survive the heating and drying operations used in fish meal production. Prions, short for proteinaceous infectious agent, have emerged as disease-causing agents in humans and animals. The most well-known prion causes bovine spongiform encephalopathy (BSE) or "mad cow disease," which has raised havoc in European farms (FDA 2001). Many herds were destroyed after a link was established to a degenerative brain ailment in humans consuming beef products. The herds were infected by consuming feeds containing central nervous tissue (brain and spinal cord byproducts) from BSE-infected cattle (FDA 2001). A similar chronic wasting disease is causing problems in game animals in the United States. However, to date there is no equivalent disease found for fish. Chemical pollutants such as pesticides, polychlorinated hydrocarbons, and heavy metals in the food chain are an increasing concern, but the topic is not covered in this review.

Processing controls

Fish meal

Pasteurization via indirect steam is the simplest and most common method for ridding fish waste of viable vegetative bacterial and mold cells, yeasts, and viruses. At temperatures above their maximum growth ranges, these microscopic organisms will have ruptured cell walls and viral capsids which cause leakage of intracellular liquid and inactivation of cells and viruses. Death rates for salmonellae during heating of fish meal can vary due to the strain involved, its temperature sensitivity, water activity, storage temperature, and atmosphere (Doesburg et al. 1970a,b). *Salmonella*-free poultry feed pellets are processed to an internal temperature of 80°C for one minute, although thermoresistant strains of *S. typhimurium* may exist (Zdragas et al. 2001). Spores from bacteria and molds are not affected by sub-sterilization temperatures. Several samples of fish meal, used in animal feeds, were found to contain spores from enterotoxigenic *Clostridium perfringens* Type A, which was implicated in the deaths of livestock and birds (Chakrabarty and Boro 1981).

After dehydration of the pasteurized fish waste, the material is ground and bagged for storage and shipment to end users. Bacterial contamination and post-processing contamination can occur at various manufacturing steps, depending on the operations and quality control used in the facility (Fig. 1). Most fish meal has a low enough water activity that the bacterial contaminants will not pose a quality or safety problem unless the material becomes rehydrated before its intended use.

Fish feeds can contain numerous viable microorganisms, including *Salmonella* species, which may persist during extended storage (Trust 1971). Two strains of *Salmonella enteriditis*, inoculated in fish meal, survived storage at room temperature for 60-90 days (Pelagić et al. 1998). Bacteria and fungi were viable in fish feeds stored at ambient, refrigerated, and freezing temperatures (Zmysłowska and Lewandowska 1999, 2000). Experimental fish feeds made into semi-moist pellets and stored at 25°C and 2°C resulted in bacterial and yeast growth, attributed to high water activity and neutral pH (B.H. Himelbloom, unpubl. data, 1993).

Silage from acidified fish waste

Acidified silage, developed in Scandinavia, is a non-thermal alternative to thermal processing of fishery waste (Fig. 2). Ground material is mixed with adequate portions of an inorganic acid, i.e., sulfuric acid and hydrochloric acid, to about pH 2 (Raa and Gildberg 1982, Arason 1994). The highly acidic environment is sufficient to prevent bacterial growth during storage but allows autolysis to occur if digestive enzymes are present. However, more recently formic acid has been used, which lowers the silage to pH 4 and does not require neutralization before use (Raa et al. 1983, Arason 1994). Silage made from fish offal with the addition of formic acid to an initial pH 3.5 causes a 4-log reduction in bacterial counts

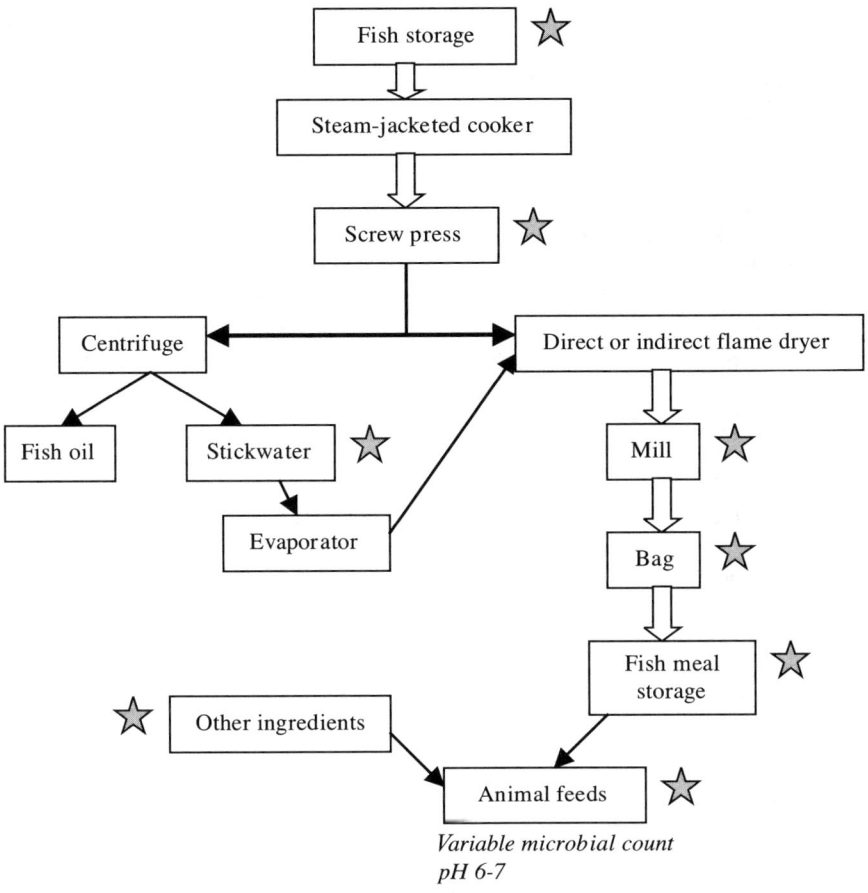

Figure 1. Fish feed processing (adapted from Bimbo 2000, Windsor and Barlow 1981) and potential microbiological contamination sites (denoted by stars).

within two days, followed by negligible counts by two weeks (Alwan et al. 1993). Coliform counts are reduced from 10^4 per gram to negligible levels in fish silage after acidification to pH 3.5 (Mahendrakar et al. 1991). However, potentially beneficial lactic acid bacteria are inhibited at pH 3.5 (Alwan et al. 1993). The bacterial fish pathogens, *Aeromonas salmonicida*, *Yersinia ruckeri*, and *Renibacterium salmoninarum*, are destroyed in fish silage made with formic and propionic acids; however, infectious pancreatic necrosis virus can survive unless the silage is treated using heat or a virucide (Smail et al. 1993). Aflatoxin-producing fungi caused by *Aspergillus flavus* can grow in moist mixtures of fish silage combined with carbohydrate meals unless propionic acid is included as a mold inhibitor (Disney and James 1980, Raa and Gildberg 1982).

Silage from fermented fish waste

An economic alternative to acidified fish silage is fermented fish silage, which is produced in various countries. Fermentation of fish waste requires the addition of a readily usable and cheap carbohydrate such as molasses along with a source of starter culture, usually composed of lactic acid bacteria alone or with yeast (Fig. 3, Table 3). High culture concentrations of 10^8 cells per gram of fish waste are used to start the process and assure a successful fermentation. The starter culture metabolizes the carbohydrate, i.e., glucose into organic acids such as lactic acid, which acidifies the material to about pH 4. The quality and stability of naturally fermented fish silage depends on the species of *Lactobacillus* used, the quality of the fish material, the type of carbohydrate, and the quantity added (van Wyk and Heydenrych 1985). Spoilage bacteria present at high levels, 10^5 colony-forming units (CFU) per gram, in raw fish waste can be reduced to negligible levels after fish fermentation is complete (Faid et al. 1997).

Rapid fermentation is essential to eliminate harmful bacteria and improve the storage stability of fish waste (Lindgren 1999). Hygiene indicator bacteria (coliforms) are reduced from an initial load of 10^6 CFU per gram to less than one CFU per gram during fish fermentation (Samuels et al. 1992, Ahmed et al. 1996, Ahmed and Mahendrakar 1997, Faid et al. 1997). As with inorganic acid-stabilized silage, fermented silage will prevent any contaminating pathogens from growing in this acidic environment. *Clostridium botulinum* Type E will not grow below pH 5 at 29°C (Owens and Mendoza 1985) and fish pathogens, *Vibrio anguillarum* and *Aeromonas salmonicida*, will be eliminated (Lindgren and Pleje 1983). However, fermented or ensiled fish waste in which a carbohydrate source has been added can be spoiled by competing yeasts and molds. Preservatives such as sorbic, propionic, and benzoic acids are used to inhibit the fungi in fermented wastes (Ahmed et al. 1996, Lindgren 1999). As a feed ingredient, fermented or acid preserved fish silage is as good as or better than fish meal or soy meal in aquaculture feeding trials (Raghunath and Gopakumar 2002).

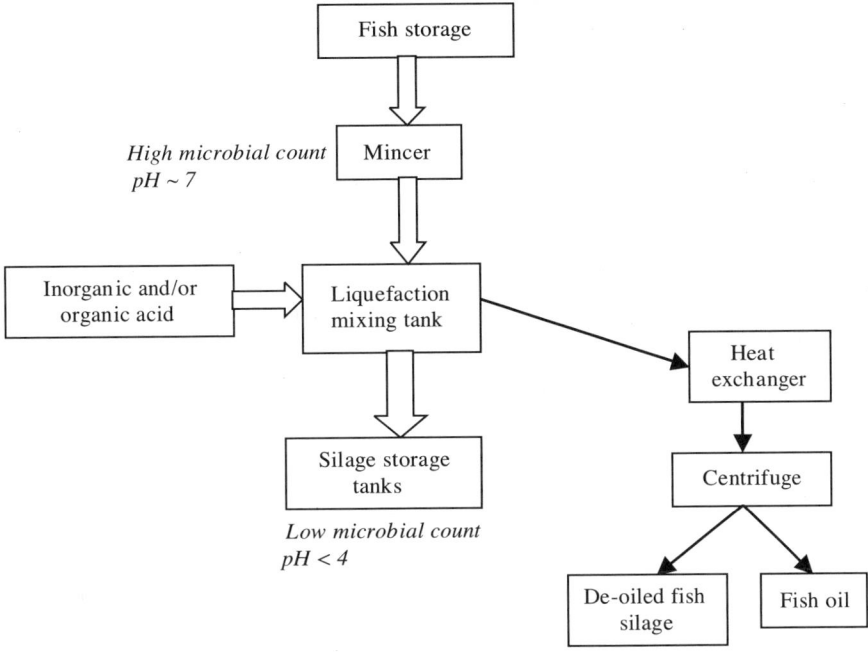

Figure 2. Acidified fish silage process (adapted from Bimbo 2000, Windsor and Barlow 1981) with respect to microbial content.

Other processing controls

Processes for inactivating *Salmonella,* that have been tested but are not necessarily favored, include ethylene oxide (Trust and Wood 1973, Saheki et al. 1988a), gamma-irradiation (Mossel et al. 1967, Reusse et al. 1976, Saheki et al. 1988b) and antibiotics. Newer technologies such as non-isotopic irradiation through the use of electron beams (SureBeam Corporation, San Diego, Calif.) for sterilizing fish meals may be a future possibility.

Analytical testing
Bacteriology

Unlike human foods, microbial standards for animal feeds have not been as demanding and can vary depending on the buyer's specifications. Microbiological analysis requires access to a laboratory or at least a separate room for diluting samples and incubating petri plates or Petrifilm (3M Microbiology Products, St. Paul, Minn.). A trained technician is essential for conducting the analysis. Alternatively, samples can be tested by contracting

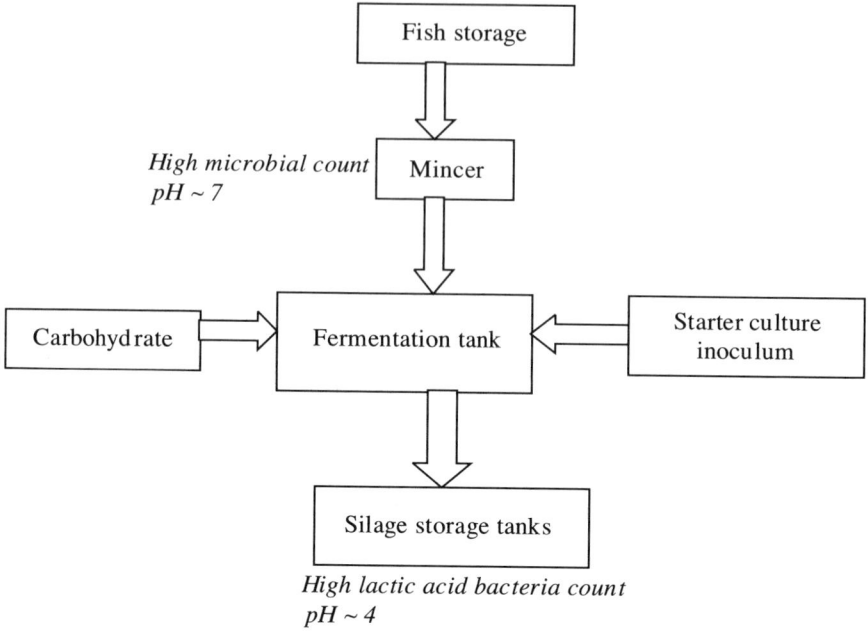

Figure 3. Fish silage produced using bacterial fermentation.

with an outside laboratory. The initial analysis for fish meal or silage may cover total bacterial counts, coliforms, and fungi (Table 4). Pathogen testing for *Salmonella* species is more intricate and most likely would be performed by a contract laboratory. New methods of testing for *Salmonella* include rapid immunological test kits (*Salmonella* 1-2 Test, BioControl, Bellevue, Wash.; REVEAL, Neogen Corporation, Lansing, Mich.) and molecular methods such as the polymerase chain reaction (PCR, PROBELIA, Bio-Rad Laboratories, Hercules, Calif.). Pulse-field gel electophoresis (PFGE) was recently used to show persistence of *Salmonella enterica* serovars in four Norwegian fish meal plants (Nesse et al. 2003).

    Salmonella testing of fish meals is mandatory in the European Union (Pike and Hardy 1997). The standard would be zero *Salmonella* cells per sample, typically 25 grams. A statistical sampling plan would be required for analyzing large production runs of fish meal for *Salmonella* contamination. *Salmonella* species belong to the family Enterobacteriaceae, which is used as an indicator of the pathogen's possible presence in fish meal plants (Quevedo et al. 1965). Enterobacteriaceae counts correlate with *Salmonella* presence in poultry feeds (Veldman et al. 1995). As an index of sanitary quality, a limit of 300 Enterobacteriaceae per gram of animal feed is enforced by the European Union (Zdragas et al. 2001).

Table 3. Fish and shellfish silages produced using bacterial fermentation.

Bacteria inoculant	Fermentable substrate	Fish or shellfish	Time, temp	Other factors	Final pH	Reference
Lactobacillus plantarum	15% molasses	Shrimp	1 wk, 30°C	Anaerobic	4.3	Fagbenro and Bello-Olusoji 1997
Lactobacillus plantarum	15% molasses	Tuna viscera	5 days, 35°C		4.3-4.4	Yoon et al. 1997
Lactobacillus plantarum	15% molasses	Mixed species	6 days, 30°C		4.2	Ottati et al. 1990
Lactobacillus plantarum	7.5% molasses	Cod gurry	3 days, 37°C	Hydrolyzed	4.0	Levin 1994, Giurca and Levin 1992
Lactobacillus plantarum	5% sucrose and 5% molasses	Sardines	1 wk, 28°C		4.5	Zuberi et al. 1993
Lactobacillus plantarum	5% dextrose	Herring offal	1 wk, 25°C		4.1-4.3	Lassén 1995a,b
Lactobacillus plantarum	5% lactose	Perch	2 days, 35°C		4.2-4.4	Hassan and Heath 1986
Lactobacillus plantarum	4% corn syrup	Cod gurry	3 days, 37°C	Hydrolyzed	4.0	Giurca and Levin 1993
Pediococcus acidilactici and *Lactobacillus plantarum*	10% barley and oats and 10% molasses	Herring	2 days, 24°C		4.4	Lindgren and Pleje 1983
Streptococcus faecium and *Lactobacillus plantarum*	4% molasses or 3% dextrose	Salmon viscera	1 wk, 25°C	Pre-acidified	4.5	Dong et al. 1993
Lactobacillus plantarum and *Saccharomyces cereviseae*	30% molasses	Sardines	10 days, 26°C	Pre-acidified	4.0-4.4	Faid et al. 1994
Lactobacillus plantarum and *Saccharomyces* species	25% molasses	Pilchard	1 wk, 22°C		4.2-4.5	Faid et al. 1997
Wild	40% molasses	Sardines	1 wk, 25°C	Anaerobic, stirred	4.4	Zahar et al. 2002
Wild	10% molasses	Freshwater species	2 days, 37°C		4.0	Ahmed and Mahendrakar 1997
Wild	10% molasses	Freshwater species	4 days, 26°C	Micro-aerophilic	4.2-4.3	Ahmed and Mahendrakar 1995
Wild	4% malt meal and 20% cereal meal	Cod	1 wk, 28°C		4.5	Nilsson and Rydin 1965

Biogenic amines

Biogenic amine analysis would require access to a high performance liquid chromatograph (HPLC) for operation by an advanced laboratory technician. As an alternative to HPLC, test kits for histamine only are available commercially (Alert, Neogen Corp.). Histamine, cadaverine, putrescine, and tyramine concentrations are indicators of raw fish freshness and serve as quality indicators for fish meal (Opstvedt et al. 2000). A suggested standard for Atlantic salmon and shrimp feeds made from fish meal is a biogenic amine content of less than 3,000-3,400 ppm (Anderson et al. 1997, Pike and Hardy 1997).

A feed quality index can be expressed by dividing the combined concentrations of the amino acids lysine, arginine, and tyrosine, by the combined respective biogenic amines (Lassén 1995c). In addition, the total volatile nitrogen content, determined using distillation and titration, for raw fish should not exceed 50 mg N per 100 g when used for fish meal (Arason 1994, Pike and Hardy 1997). Chick bioassays are used to determine if gizzerosine levels are toxic (Fairgrieve et al. 1994, Romero et al. 1994).

Other tests

Additionally, pH and water activity (a_w) measurements would be necessary to determine if the fish waste byproducts will be shelf-stable. Fish meal is microbiologically stable due to an a_w of 0.33-0.65, unless it is rehydrated (ICMSF 1998). Newer instrumentation has allowed faster results for determining a_w (AquaLab, Decagon Devices, Inc., Pullman, Wash.).

New developments in animal nutrition— probiotics

Over the last decade, there have been efforts to understand the role of protective bacteria (probiotics) in animal nutrition. These are primarily lactic acid bacteria which are considered prime candidates as supplements in feeding regimes. Rather than rid the animals of disease-causing bacteria through therapeutics, i.e., antibiotic-containing feeds, the probiotic bacteria may provide the same effect by restricting pathogen access to sensitive areas of the animals' gastrointestinal systems. Probiotic bacteria positively affect the composition of the intestinal microflora (Strøm and Raa 1992, Tannock 1999).

Some probiotics may directly inhibit the pathogens through the production of antimicrobial compounds. These may include the production of lactic acid, hydrogen peroxide, and antimicrobial peptides (bacteriocins). Thus, if the level of potential pathogens could be reduced in fish then it is expected there will be less risk of transmitting the pathogens through fishery waste byproducts into animal feeds.

Much of the primary work has been done using terrestrial animals while limited efforts have been made in aquaculture (Gatesoupe 1999,

Table 4.   Microbial content of fish meal and fish silages.

Product	Aerobic plate count per gram	*Salmonella*	Other	Reference
Fish meal	200	negative		Furuta et al. 1980
Fish meal	10^4	nd		Okuzumi et al. 1984
Fish meal (low temperature dried)	10^8	negative		Wood et al. 1985
Acidified fish silage	<500	nd	pH 3.7-3.9	Backhoff 1976
Acidified fish silage	200	negative	pH 3.6	Wood et al. 1985
Fermented fish silage	10^6-10^8	nd	Starter culture used	van Wyk and Heydenrych 1985

nd = not determined.

Lyndon 1999, Skjermo and Vadstein 1999, Hansen 2000, Vijayakumaran 2001, Irianto and Austin 2002). Larvae of finfish, mollusks, and crustaceans benefit from disease prevention and growth promotion through consuming probiotics added to live foods such as rotifers (Douillet and Langdon 1994, Skjermo and Vadstein 1999, Hansen 2000). Flounder fed a probiotic *Lactobacillus* species included in commercial feed had higher weight gains than the controls (Byun et al. 1997). More research is needed in utilizing probiotic bacteria for aquatic applications before they are accepted for use by the commercial industry.

Conclusions

Quality and safety are important issues for utilizing fish processing waste. Increased quality occurs when fish waste is protected from overt spoilage prior to fish byproduct processing and from post-processing contamination. Demands from buyers are expected to require processors to ensure the byproducts are safe to use in various animal feeds. Probiotic-enhanced fish silage may accomplish the safety requirement while providing additional health benefits to the farmed animals and, ultimately, humans.

Acknowledgments

The author thanks the University of Alaska Fairbanks (UAF) Biosciences Library for assistance in procuring many of the articles cited in this review. Travel funds were provided from the U.S. Department of Agriculture, Agricultural Research Service, through a special projects grant to UAF.

References

Ahmed, J., and N.S. Mahendrakar. 1995. Effect of different levels of molasses and salt on acid production and volume of fermenting mass during ensiling of tropical freshwater fish viscera. J. Food Sci. Technol. 32:115-118.

Ahmed, J., and N.S. Mahendrakar. 1997. Chemical and microbial changes in fish viscera during fermentation ensiling at different temperatures. Bioresour. Technol. 59:45-46.

Ahmed, J., B.S. Ramesh, and N.S. Mahendrakar. 1996. Changes in microbial population during fermentation of tropical freshwater fish viscera. J. Appl. Bacteriol. 80:153-156.

Aksnes, A., and H. Mundheim. 1997. The impact of raw material freshness and processing temperature for fish meal on growth, feed efficiency and chemical composition of Atlantic halibut (*Hippoglossus hippoglossus*). Aquaculture 149:87-106.

Alwan, S.R., D.J. Buckley, and T.P. O'Connor. 1993. Silage from fish waste: Chemical and microbiological aspects. Ir. J. Agric. Food Res. 32:75-81.

Anderson, J.S., D.A. Higgs, R.M. Beames, and M. Rowshandeli. 1997. Fish meal quality assessment for Atlantic salmon (*Salmo salar* L.) reared in sea water. Aquac. Nutr. 3:25-38.

Arason, S. 1994. Production of fish silage. In: A.M. Martin (ed.), Fisheries processing: Biotechnological applications. Chapman and Hall, New York, pp. 244-272.

Backhoff, H.P. 1976. Some chemical changes in fish silage. J. Food Technol. 11: 353-363.

Bimbo, A.P. 2000. Fish meal and oil. In: R.E. Martin, E.P. Carter, G.J. Flick Jr., and L.M. Davis (eds.), Marine and freshwater products handbook. Technomic Publishing Co., Inc., Lancaster, Pennsylvania, pp. 541-581.

Byun, J-W., S-C. Park, Y. Benno, and T-K. Oh. 1997. Probiotic effect of *Lactobacillus* sp. DS-12 in flounder (*Paralichthys olivaceus*). J. Gen. Appl. Microbiol. 43: 305-308.

Chakrabarty, A.K., and B.R. Boro. 1981. Prevalence of food-poisoning (enterotoxigenic) *Clostridium perfringens* type A in blood and fish meal. Zentbl. Bakteriol. Mikrobiol. Hyg. 172:427-433.

Clancy, S., R. Beames, D. Higgs, B. Dosanjh, N. Haard, and B. Toy. 1995. Influence of spoilage and processing temperature on the quality of marine fish protein sources for salmonids. Aquac. Nutr. 1:169-177.

Dapkevicius, M.L.N.E., M.J.R. Nout, F.M. Rombouts, J.H. Houben, and W. Wymenga. 2000. Biogenic amine formation and degradation by potential fish silage starter microorganisms. Int. J. Food Microbiol. 57:107-114.

Disney, J.G., and D. James (eds.). 1980. Fish silage production and its use. FAO Fish. Rep. 230.

Doesburg, J.J., E.C. Lamprecht, and M. Elliott. 1970a. Death rates of salmonellae in fishmeals with different water activities. I. During storage. J. Sci. Food Agric. 21:632-635.

Doesburg, J.J., E.C. Lamprecht, and M. Elliott. 1970b. Death rates of salmonellae in fishmeals with different water activities. II. During heat treatment. J. Sci. Food Agric. 21:636-640.

Dong, F.M., W.T. Fairgrieve, D.I. Skonberg, and B.A. Rasco. 1993. Preparation and nutrient analysis of lactic acid bacterial ensiled salmon viscera. Aquaculture 109:351-366.

Douillet, P.A., and C.J. Langdon. 1994. Use of a probiotic for the culture of larvae of the Pacific oyster (*Crassostrea gigas* Thunberg). Aquaculture 119:25-40.

Doyle, M.P., L.R. Beuchat, and T.J. Montville (eds.). 1997. Food microbiology: Fundamentals and frontiers. American Society for Microbiology Press, Washington, D.C.

Dreosti, G.M., S.G. Wiechers, and R.J. Nachenius. 1962. Quality standards for fish meal and related problems. Fishing Industry Research Institute Memorandum No. 127, University of Cape Town, Rondebosch, South Africa. 8 pp.

Fagbenro, O.A., and O.A. Bello-Olusoji. 1997. Preparation, nutrient composition and digestibility of fermented shrimp head silage. Food Chem. 60:489-493.

Faid, M., H. Karani, A. Elmarrakchi, and A. Achkari-Begdouri. 1994. A biotechnological process for the valorization of fish waste. Bioresour. Technol. 49: 237-241.

Faid, M., A. Zouiten, A. Elmarrakchi, and A. Achkari-Begdouri. 1997. Biotransformation of fish waste into a stable feed ingredient. Food Chem. 60:13-18.

Fairgrieve, W.T., M.S. Myers, R.W. Hardy, and F.M. Dong. 1994. Gastric abnormalities in rainbow trout (*Oncorhynchus mykiss*) fed amine-supplemented diets or chicken gizzard-erosion-positive fish meal. Aquaculture 127:219-232.

Fassi-Fehri, M., and J. Kochanski. 1971. Results of research on *Salmonella* in fish meal. Maroc Med. 548:460-464. (In French.)

FDA. 2001. Prions and transmissible spongiform encephalopathies. In: Foodborne pathogenic microorganisms and natural toxins handbook. <http://www.cfsan.fda.gov/~mow/prion.html>. Center for Food Safety and Applied Nutrition, U.S. Food and Drug Administration, Rockville, Maryland.

Furuta, K., S. Morimoto, and S. Sato. 1980. Bacterial contamination in feed ingredients, formulated chicken feed and reduction of viable bacteria by pelleting. Lab. Anim. 14:221-224.

Gatesoupe, F.J. 1999. The use of probiotics in aquaculture. Aquaculture 180:147-165.

Giurca, R., and R.E. Levin. 1992. Optimization of the lactic acid fermentation of hydrolyzed cod gurry with molasses. J. Food Biochem. 16:83-97.

Giurca, R., and R.E. Levin. 1993. Optimization of the lactic acid fermentation of hydrolyzed cod (*Gadus morhua*) gurry with corn syrup as carbohydrate source. J. Food Biochem. 16:277-289.

Green, J.H., S.L. Paskell, and D. Goldmintz. 1976. Lipolytic fermentations of stickwater by *Geotrichum candidum* and *Candida lipolytica*. Appl. Environ. Microbiol. 31:569-575.

Hansen, G.H. 2000. Use of probiotics in marine aquaculture. Feed Mix 8:32-34.

Hassan, T.E., and J.L. Heath. 1986. Biological fermentation of fish waste for potential use in animal and poultry feeds. Agric. Wastes 15:1-15.

Himelbloom, B.H. 1998. Primer on food-borne pathogens for subsistence food handlers. In: R. Fortuine, G.A. Conway, C.D. Schraer, M.J. Dimino, C.M. Hild, and J. Braund-Allen (eds.), Circumpolar health 96, Proceedings of the Tenth International Congress on Circumpolar Health. American Society for Circumpolar Health, Anchorage, Alaska, pp. 228-234.

Himelbloom, B.H., and B.G. Stevens. 1994. Microbial analysis of a fish waste dump site in Alaska. Bioresour. Technol. 47:229-233.

ICMSF. 1998. Feeds and pet foods. In: International Commission on Microbiological Specifications for Foods (eds.), Microorganisms in foods 6: Microbial ecology of food commodities. Aspen Publishers, Inc., Gaithersburg, Maryland, pp. 190-214.

Irianto, A., and B. Austin. 2002. Probiotics in aquaculture. J. Fish Dis. 25:633-642.

Jacobs, J., P.A.M. Guinée, E.H. Kampelmacher, and A. van Keulen. 1963. Studies on the incidence of *Salmonella* in imported fish meal. Zentbl. Vetmed. Series A 10:542-550.

Lassén, T.M. 1995a. Evaluation of conditions for fermented of fish offal. Agric. Sci. Finl. 4:11-17.

Lassén, T.M. 1995b. Lactic acid fermentation of fish offal and chicken by-product with different starter cultures. Agric. Sci. Finl. 4:19-26.

Lassén, T.M. 1995c. Biological quality of fermented fish offal and chicken by-products. Agric. Sci. Finl. 4:27-33.

Levin, R.E. 1994. Lactic acid and propionic acid fermentations of fish hydrolysates. In: A.M. Martin (ed.), Fisheries processing: Biotechnological applications. Chapman and Hall, New York, pp. 273-310.

Levin, R.E., R. Witkowski, Y. Meirong, and S. Goldhor. 1989. Preparation of fish silage with phosphoric acid and potassium sorbate. J. Food Biochem. 12:253-259.

Lindgren, S. 1999. Storage of waste products for animal feed. In: B.J.B. Wood (ed.), The lactic acid bacteria, Vol. 1, The lactic acid bacteria in health and disease. Aspen Publishers, Gaithersburg, Maryland, pp. 387-407.

Lindgren, S., and M. Pleje. 1983. Silage fermentation of fish or fish waste products with lactic acid bacteria. J. Sci. Food Agric. 34:1057-1067.

Lyndon, A.R. 1999. Fish growth in marine culture systems: A challenge for biotechnology. Mar. Biotechnol. 1:376-379.

Mahendrakar, N.S., V.S. Khabade, K.P. Yashoda, and N.P. Dani. 1991. Chemical and microbiological changes during autolysis of fish and poultry viscera. Trop. Sci. 31:45-54.

Morris, G.K., W.T. Martin, W.H. Shelton, J.G. Wells, and P.S. Brachman. 1970. Salmonellae in fish meal plants: Relative amounts of contamination at various stages of processing and a method of control. Appl. Microbiol. 19:401-408.

Mossel, D.A.A., M. van Schothorst, and E.H. Kampelmacher. 1967. Comparative study on decontamination of mixed feeds by radicidation and by pelletisation. J. Sci. Food Agric. 18:362-367.

Nabbut, N.H., E.K. Barbour, and H.M. Al-Nakhli. 1982. Occurrence of salmonellae in animal feed ingredients in Saudi Arabia. Am. J. Vet. Res. 43:1703-1705.

Nesse, L.L., K. Nordby, E. Heir, B. Bergsjoe, T. Vardund, H. Nygaard, and G. Holstad. 2003. Molecular analyses of *Salmonella enterica* isolates from fish feed factories and fish feed ingredients. Appl. Environ. Microbiol. 69:1075-1081.

Nilsson, R., and C. Rydin. 1965. A new method of ensiling foodstuffs and feedstuffs of vegetable and animal origin. Enzymologia 11:126-142.

Okazaki, T., T Noguchi, K. Igarashi, Y. Sakagami, H. Seto, K. Mori, H. Naito, T. Masumara, and M. Sugahara. 1983. Gizzerosine, a new toxic substance in fish meal, causes severe gizzard erosion in chicks. Agric. Biol. Chem. 47:2949-2952.

Okuzumi, M., K. Toyama, H. Yamanaka, and M. Nagano. 1984. Histamine-forming bacteria in raw materials of fish meals. Bull. Jpn. Soc. Sci. Fish. 50:883-888.

Opstvedt, J., H. Mundheim, E. Nygård, H. Aase, and I.H. Pike. 2000. Reduced growth and feed consumption of Atlantic salmon (*Salmo salar* l.) fed fish meal made from stale fish is not due to increased content of biogenic amines. Aquaculture 188:323-337.

Ottati, M., M. Gutiérrez, and R. Bello. 1990. Study of microbial fish silage from underutilized fish species. Arch. Latinoam. Nutr. 40:408-425. (In Spanish.)

Owens, J.D., and L.S. Mendoza. 1985. Enzymatically hydrolyzed and bacterially fermented fishery products. J. Food Technol. 20:273-293.

Pelagić, V.R., V. Jurić, B. Radenković, and M. Ristić. 1998. The survival of *Salmonella enteriditis* in animal feed and deep litter. Acta Vet. (Belgr.) 48:317-322.

Pike, I.H., and R.W. Hardy. 1997. Standards for assaying quality of feed ingredients. In: L.R. D'Abramo, D.E. Conklin, and D.M. Akiyama (eds.), Advances in world aquaculture, Vol. 6, Crustacean nutrition. The World Aquaculture Society, Louisiana State University, Baton Rouge, pp. 473-492.

Quevedo, F., J. Nakasone, J. Calderon, J. Portocarrero, E. Alcala, G. Aguena, P. Salgado, and J. Castro. 1965. Enterobacteriaceae in fish meal. Annales de l'Institut Pasteur de Lille 16:157-162. (In French.)

Raa, J., and A. Gildberg. 1982. Fish silage: A review. CRC Crit. Rev. Food Sci. Nutr. 16:383-419.

Raa, J., A. Gildberg, and T. Strøm. 1983. Silage production: Theory and practice. In: D.A. Ledward, A.J. Taylor, and R.A. Lawrie (eds.), Upgrading waste for feeds and food. Butterworths, Boston, pp. 117-132.

Raghunath, M.R., and K. Gopakumar. 2002. Trends in production and utilization of fish silage. J. Food Sci. Technol. 39:103-110.

Reusse, U., J. Bischoff, G. Fleischhauer, and R. Geister. 1976. Pasteurization of fish meal by irradiation. I. Determination of the minimal radiation dose necessary to destroy salmonellae in contaminated fish meal. Zentbl. Vetmed. Series B 23:158-170. (In German.)

Ricque-Marie, D., M.I. Abdo-de La Parra, L.E. Cruz-Suarez, G. Cuzon, M. Cousin, Aquacop, and I.H. Pike. 1998. Raw material freshness, a quality criterion for fish meal fed to shrimp. Aquaculture 165:95-109.

Romero, J.J., E. Castro, A.M. Díaz, M. Reveco, and J. Zaldívar. 1994. Evaluation of methods to certify the "premium" quality of Chilean fish meals. Aquaculture 124:351-358.

Roumani, B.M., A.M. Abdelnoor, and C. Hilan. 1981. *Salmonella agona* isolated from fish meal and a *Salmonella* strain isolated from shrimps in Lebanon. Zentbl. Bakteriol. Mikrobiol. Hyg. 172:411-414.

Saheki, K., T. Hashimoto, and S. Shinamura. 1988a. Destruction of salmonellae in the coastal fish meals by dry heating and ethylene oxide gas fumigation. Nippon Suisan Gakkaishi 54:2099-2105. (In Japanese.)

Saheki, K., K. Konno, T. Sato, and T. Kawabata. 1988b. Destruction of *Salmonella* from the coastal fish meals by ^{60}Co γ ray irradiation. Nippon Suisan Gakkaishi 54:2107-2112. (In Japanese.)

Samuels, W.A., J.P. Fontenot, V.G. Allen, and G.J. Flick. 1992. Fermentation characteristics of ensiled seafood wastes and low-quality roughages. Anim. Feed Sci. Technol. 38:305-317.

Silla Santos, M.H. 1996. Biogenic amines: Their importance in foods. Int. J. Food Microbiol. 29:213-231.

Skjermo, J., and O. Vadstein. 1999. Techniques for microbial control in the intensive rearing of marine larvae. Aquaculture 177:333-343.

Smail, D.A., P.J. Huntly, and A.L.S. Munro. 1993. Fate of four fish pathogens after exposure to fish silage containing fish farm mortalities and conditions for the inactivation of infectious pancreatic necrosis virus. Aquaculture 113:173-181.

Strøm, T., and J. Raa. 1992. Ensiling of fish by lactic acid bacterial fermentation for feed production. FAO Fish. Rep. 470:142-147.

Tannock, G.W. 1999. Introduction. In: G.W. Tannock (ed.), Probiotics: A critical review. Horizon Scientific Press, Portland, Oregon, pp. 1-4.

Toyama, K., M. Okuzumi, T. Yokoi, and H. Aoe. 1981. Histamine content of fish meal. Bull. Jpn. Soc. Sci. Fish. 47:415-419. (In Japanese.)

Trust, T.J. 1971. Bacterial counts of commercial fish diets. J. Fish. Res. Board Can. 28:1185-1189.

Trust, T.J., and A.J. Wood. 1973. An initial evaluation of ethylene oxide for the sterilization of formulated and pelleted fish feeds. J. Fish. Res. Board Can. 30:269-274.

van Wyk, H.J., and C.M.S. Heydenrych. 1985. The production of naturally fermented fish silage using various lactobacilli and different carbohydrate sources. J. Sci. Food Agric. 36:1093-1103.

Veldman, A., H.A. Vahl, G.J. Borggreve, and D.C. Fuller. 1995. A survey of the incidence of *Salmonella* species and Enterobacteriaceae in poultry feeds and feed components. Vet. Rec. 136:169-172.

Vijayakumaran, M. 2001. Probiotics in aquaculture. In: N.G. Menon and P.P. Pillai (eds.), Perspectives in mariculture. Marine Biological Association of India, Cochin, pp. 369-380.

Williams, L.P., J.B. Vaughn, A. Scott, and V. Blanton. 1969. A ten-month study of *Salmonella* contamination in animal protein meals. J. Am. Vet. Med. Assoc. 155:167-174.

Windsor, M.L., and S.M. Barlow. 1981. Introduction to fishery by-products. Fishing News Books Ltd, Farnham, Surrey, England.

Wood, J.F., B.S. Capper, and L. Nicolaides. 1985. Preparation and evaluation of diets containing fish silage, cooked fish preserved with formic acid and low-temperature-dried fish meal as protein sources for mirror carp (*Cyprinus carpio*). Aquaculture 44:27-40.

Yoon, H-D., D-S. Lee, C-I. Ji, and S-B. Suh. 1997. Studies on the utilization of wastes from fish processing. I. Characteristics of lactic acid bacteria for preparing skipjack tuna viscera silage. J. Korean Fish. Soc. 30:1-7. (In Korean.)

Zahar, M., N. Benkerroum, A. Guerouali, Y. Laraki, and K. El Yakoubi. 2002. Effect of temperature, anaerobiosis, stirring and salt addition on natural fermentation silage of sardine and sardine wastes in sugarcane molasses. Bioresour. Technol. 82:171-176.

Zdragas, A., N. Iliadis, and K. Sarris. 2001. Isolation of Salmonellas and Enterobacteriaceae from poultry feed in northern Greece. Wien. Tieraerztl. Monschr. 88:54-58.

Zmysłowska, I., and D. Lewandowska. 1999. Survival of bacterial strains in fish feeds stored at different temperatures. Pol. J. Environ. Studies 8:447-451.

Zmysłowska, I., and D. Lewandowska. 2000. The effect of storage temperatures on the microbiological quality of fish feeds. Pol. J. Environ. Studies 9:223-226.

Zuberi, R., R. Fatima, S.I. Shamshad, and R.B. Qadri. 1993. Preparation of fish silage by microbial fermentation. Trop. Sci. 33:171-182.

Advances in Seafood Byproducts
Alaska Sea Grant College Program • AK-SG-03-01, 2003

The Impact of Food Safety and Competitive Markets on Byproduct Recovery Strategies

Alan Ismond
Aqua-Terra Consultants, Bellevue, Washington

The connection between seafood byproducts and consumers is not always apparent, and is dependent on the type of byproduct that is recovered. The most direct connection is in the production of neutraceuticals and ingredients for nutrient enriched foods. Byproducts recovered for use as fertilizers indirectly impact consumer fruits and vegetables. In the case of byproducts used in fish and animal feeds, consumer meats and some cultured fish are impacted. Given the connection between seafood byproducts and consumer food choices, the seafood industry would benefit from tracking consumer preferences.

Issues relating to the positive and negative aspects of food safety and quality have attracted abundant media attention in the last 10 years. Consumers are aware of livestock health issues, and they have reacted negatively to them. Bovine spongiform encephalopathy (BSE) in cattle has had a devastating impact on producers and meat sales. Consumers are concerned with the potential link between BSE and CJD, a similar human ailment. Antibiotic use and residues in livestock have raised fears of human pathogen resistance. On a regular basis, the media reports on food contaminants such as pesticides, mercury, and dioxins. Pathogens in food are a recurring problem resulting in approximately 5,000 to 9,000 deaths in the United States annually. Some of the organisms of concern are *Salmonella, Listeria*, and *E. coli* O157:H7. On the other end of the spectrum, consumers are increasingly willing to pay for foods with positive attributes. Organic foods grown under strict guidelines for farming and husbandry methods are one example. Neutraceuticals and functional foods are two more examples.

Understanding the link between seafood byproducts and consumer preferences requires a more detailed look at the food chain. Figure 1 illustrates the pathways by which seafood byproducts can impact consumers. In the case of seafood-byproduct-derived fertilizers, the quality and type of seafood waste affects the fertilizer quality and attributes. This, in turn, affects the soil quality, which affects crop quality and safety. The quality

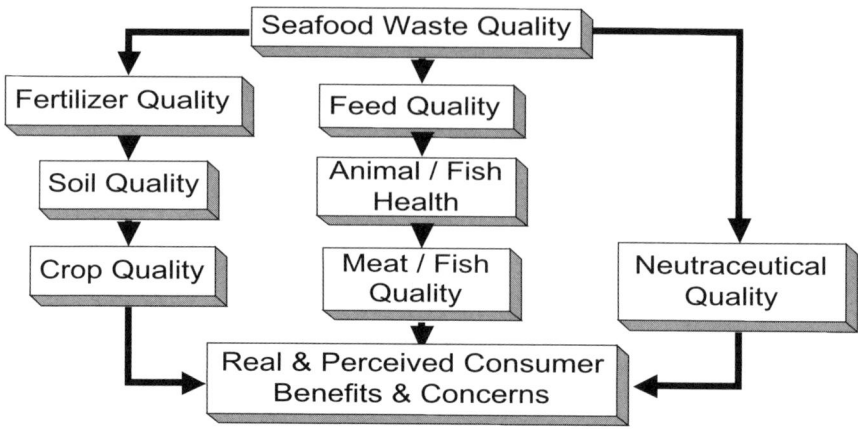

Figure 1. Seafood waste and the food chain.

of seafood waste used for producing animal and fish feeds certainly impacts animal and fish health. This, in turn, affects the quality and safety of the cultured fish and meat products. Last, the quality of seafood waste directly affects the quality and safety of neutraceuticals.

The process by which seafood waste is converted to byproducts directly affects the quality of the byproducts. The method for storing and preserving the waste prior to processing into byproducts is also critical. Improper storage and preservation can lead to the formation of undesirable chemical and microbial decomposition. The processing method must preserve the desirable qualities and attributes while separating or destroying unwanted attributes. Finally, the storage and preservation of the byproducts must protect the product quality.

Historically, the byproduct recovery strategy in Alaska has fallen short of addressing all of the issues outlined. Seafood waste quality standards and testing have been limited or nonexistent. In many cases, materials of varying freshness and quality have been commingled. Storage conditions and handling methods have resulted in mixed waste quality. While byproduct quality has been a consideration, the dominant concerns have been costs and margins. Processing methods have been predominately focused on capital and operating costs with product quality as an outcome. Many byproduct production facilities have been considered as waste disposal facilities and are production driven. The goal has been to find markets for the products produced. In many cases, variable quality has necessitated selling to commodity markets that buy on price and are dominated by supply and demand. This has resulted in variable margins and profitability.

Table 1.   Seafood byproduct business strategies.

Production driven	Market driven
One product	Several products
Variable product quality	Consistent product quality
Limited product testing	More extensive raw material and product testing
Sell to commodity markets	Sell to commodity and niche markets
Variable margins	Variable and high margins

The future profitability of the seafood industry in Alaska will require a shift to a more market driven strategy. Food safety issues at the consumer level will increasingly result in liabilities throughout the food production chain. Product claims and lawsuits will likely extend to the producers of food production inputs. This will place greater demands on the quality of seafood byproducts. On the positive side, consumer spending for beneficial foods will potentially increase margins throughout the food production chain. However, this will also place greater demands on seafood byproduct quality and specifications. Two production related considerations that impact the seafood industry currently will become increasingly important in the near future. Seafood processing wastewater discharge limits will likely be more stringent over time. This will necessitate recovering additional byproducts and rationalizing existing facilities. Another production-related issue is the impact of air quality permits. These permits limit the amount of power and BTUs available for recovering byproducts.

Given all of the market and production limitations, the challenge is to determine the best byproduct production strategy. Table I summarizes two potential modes of operation. The financial survival of the seafood processing industry will necessitate a greater shift to market driven models. Historically, seafood waste in Alaska has not been sorted by quality and type prior to processing. In order to better control the quality and consistency of seafood byproducts, there will be a need to presort seafood waste by quality and attributes. While this could add to the cost of producing byproducts, this will also permit the marketing of products to more profitable market segments.

Seafood byproducts that are inputs for other food production systems will come under greater scrutiny for their effect on the food product. Presently, the focus has been on the physical and chemical attributes of the seafood byproduct and assurances of what the byproduct does and does not contain. The future will require a move toward more functional testing and the impact of the seafood byproduct on the food production system. For example, in the case of fertilizers, it is the vigor of the plants that is ultimately of concern. Phytotoxicty testing is now being used to assess the impact of a fertilizer on plants. This involves germination tests utilizing the fertilizer to determine if it has an inhibitory effect on the test plants. While this kind of testing can take days rather than minutes, it provides a more complete picture of the potential benefits or risks of using a given fertilizer.

Focusing on niche markets is a potentially profitable strategy; however, it is not a static one. The education and the sophistication of consumers will continue to evolve. Today's niche markets will become tomorrow's commodity markets. Today's commodity markets may no longer exist tomorrow. The feeding of mammalian protein to mammals is a good example in the meat industry. Concerns over animal health have dramatically curtailed this market.

The traditional seafood byproduct production and marketing model in Alaska will need to heed consumer and environmental concerns in order to survive financially in the future. Increasing the recovery of byproducts may entail greater costs and potentially greater profits. Facilities that currently store all of the seafood waste in one containment will need to consider sorting the waste by quality. The higher quality materials can be converted to higher margin products for niche markets, and the lower quality materials will be destined for commodity markets. There is no doubt that increasing consumer demands for safer and more wholesome foods will affect all aspects of the food industry. The key will be to keep one step ahead of the quality continuum.

Seafood Byproduct Research Priorities and Opportunities

The 2nd International Seafood Byproduct Conference was held in Anchorage, Alaska, on November 10-13, 2002. More than 125 attended, from 15 states and 18 countries. The final session of the meeting was a 1.5 hour panel discussion moderated by Chris Mitchell titled "Panel Discussion: Future Research and Product Needs."

Panel discussion
P.J. Bechtel,[1] Chris Mitchell,[2] and Ian Forster[3]

Report from the panel discussion
Panel members included Ronald Hardy, University of Idaho Hagerman Fish Culture Experiment Station, Hagerman, Idaho; Albert Tacon, Aquatic Farms, Kaneohe, Hawaii; Lars Langhoff, ABD Consulting, Lejre, Denmark; Ron Anderson, Bio-Oregon Inc., Warren, Oregon; John Kilpatrick, marine protein and oil consultant, West Vancouver, British Columbia; Tony Bimbo, technical consultant, Kilmarnock, Virginia; and Fereidoon Shahidi, Memorial University of Newfoundland, St. John's, Newfoundland. Each panel member briefly discussed the research priorities and opportunities they thought important. Following the panel presentations Chris Mitchell moderated the discussion, open to all conference participants.

A very brief summary of ideas expressed during the discussion follows. Not all ideas were expressed in the context of research priorities and opportunities. While the ideas are not those of the authors, the authors have interpreted comments and have combined some ideas. Comments are listed in the order discussed.

1. One reason for waste material not being used in commerce today is due to logistics. It is often difficult to collect small amounts of byproduct from remote locations, and often byproducts are available

[1]University of Alaska Fairbanks, USDA Agricultural Research Service Laboratory, Fairbanks, Alaska.
[2]Seafood Market Developers, Bellevue, Washington.
[3]The Oceanic Institute, Waimanalo, Hawaii.

for only a short period of time making it difficult to process them in a timely manner. There is a need to find methods to preserve and stabilize byproducts, so they can be stored and collected at a later time and then further processed at a central location. Preservation methods available today include silage and hydrolysis of the product, but it would be desirable to develop a more "natural" way of preserving this material for long-term storage. A processor could then collect the material and further process it into marketable products.

2. Transportation in and out of Alaska is expensive, placing a burden on development of industries that rely on exports. A new logistic system needs to be developed to avoid these high transport costs.

3. Fish viscera contain a number of valuable materials, which are usually not separated during fish processing and utilized to the full extent possible. One example is liver, which is possibly one of the most valuable products from fish processing due to its oil and protein content.

4. In processing plants a tremendous amount of water is used to transport material that is collected for fish meal. Better methods of collecting the material, which uses less water, should be devised.

5. More of the materials from food grade byproducts can be incorporated into fish sausage products. These products can be valuable and markets already exist.

6. There is a need to continue to look for ways to increase the value of bone and connective tissue components, which are becoming an increasing portion of byproducts. For example, the bones from food grade byproducts can be used to make human grade products and as a mineral premix. Fish bone contains calcium, phosphorus, selenium, molybdenum, chromium, iodine, and fluorine, and is well suited as a mineral premix for human consumption. Processes such as pressure-cooking and grinding can be utilized.

7. In addition to general inexpensive commodities (fish meal), feed grade hydrolysates or feeding attractant, etc., could be made. One use of these products would be to enhance the palatability of vegetable proteins in feed formulations. Making standardized attractants or palatability enhancers from fish byproducts can be complex depending on the properties desired. In addition to hydrolysates and silage technologies, fermentation of byproducts is another possibility. *Lactobacillus* fermentation of byproducts could provide a good source of protein, attractants, and probiotics.

8. There is a need for standardization of the definition and criteria for byproduct freshness that can be clearly communicated, understood,

and applied to current processing systems. Better maintenance of freshness of the raw material can increase the value of the final product.

9. Industry should be involved when priorities are developed and projects are being initiated. Research project leaders should try to bring in industry partners as advisors, funders, and markets for the products. Industry knows the markets and will capitalize on perceived opportunities.

10. Often industry is reluctant to invest in projects but has an interest in obtaining government support for the efforts. Should government support be withdrawn, there is concern about having an industry that is self-sustaining.

11. There is a need to have sound scientific basis for nutraceuticals and functional food health claims.

12. There is a need to have a better understanding of the amounts, location, and types of fish processing wastes, both liquid and solid. An audit of a representative number of processing plants in various segments of the seafood industry could identify where the most immediate improvements in byproduct utilization can be made. Different processing plants have different problems. An environmental audit with representation of the entire seafood industry would be useful.

13. Devise methods of reducing salt content in stickwater where there is a problem.

14. There is an immediate need in the aquaculture industry for low-ash fish meals that can be used in the manufacture of low-pollution diets.

15. All elements of the waste stream should be captured for specific uses. If one part is taken out for high valued products there is still a need to utilize the other components.

16. The use of plant and other protein sources are projected to increase as replacements for fish meal, and the same is true for fish oil. The supply of fish oil used in aquaculture feeds needs to be extended in a way that maintains product quality at the end of the food chain. There is a need to know more about the dynamics of uptake and retention of omega-3 fatty acids in fish flesh to ensure final product quality.

17. Awareness of the dioxin problem is high in the European Union. This problem has been solved for fish oil, but there is concern in the industry about fish meal for which tighter regulations are possible.

18. It is possible to make a fish equivalent of a meat extract that can be concentrated to become a high-value, nutritional food ingredient.

19. It is important to project a better image of the seafood byproduct processing industry as quality seafood innovators. The term "co-products" instead of "waste" or "byproducts" could be part of creating a better image.

20. A more in depth evaluation of the economics of byproduct utilization is needed, including the economics of shipment.

21. Processors should have a person directly responsible for byproduct quality, processing, marketing, etc.

22. Changes in fish harvesting to longer seasons results in a more uniform fish-processing tempo over a longer period of time. This results in increased efficiencies, and fish processors having time to extract more value from all parts of the fish.

23. Older processing technologies can have application today if one pays attention to changes in the markets.

24. There is a need to evaluate the potential for extraction of extremely high-value products from primary and secondary fish byproducts in light of current market conditions and demands.

25. If wild fish can meet "organic" labeling requirements there will be additional marketing opportunities for the byproducts.

26. There is a need to have a future conference that will focus on the market side of byproducts rather than the technology aspects.

27. The most obvious human food use is making fish minces from byproducts.

28. There is a potential conflict in utilization of fish byproducts for human food and utilization by the aquaculture industry. Many would like to see all byproducts being used directly by humans, which could have a negative impact on aquaculture.

29. When making products such as hydrolysates and meals there is a need to look closely at what raw materials are going through the plant at different seasons, changes in the composition of the raw materials, and how these changes alter the final products.

30. Nutraceuticals and biomedical products can be made from fish processing byproducts; however, pharmaceutical industries are less interested in products that cannot be patented or protected in some manner. Research in these areas will have to be initiated and supported by other means.

31. Issues of importance to the entire industry could be worked on together such as analytical methods for heavy metals and other compounds. Current methods are lengthy and expensive.

32. There is an opportunity for a global byproduct Web site.

33. Processors need to have long-term commitment to purchase byproducts at a given price before making major investments in byproduct processing operations.

34. Differences in cultural and traditional customs need to be understood in efforts to successfully utilize and market byproducts.

35. Market research is needed to successfully produce products that buyers want and are willing to pay for. There is the need to be practical, and some of these ideas are excellent, but are they economical and will they result in product being sold and profits generated? There is a need for pilot operations that are supported by strong science, and that result in physical items to market.

36. Making a new product is more than having plant space and the process. New products often require more attention than the company is willing to provide. When making products, don't fall into these traps:

    A. "If we make it, they will come and buy it." In the end, getting a product to market is only 10% technology.

    B. Trying to make "final" consumer products. There is opportunity in producing high-value ingredients that are one step away from final products. Making final consumer products for the shelf is a lot to take on and it is often prudent to make high-value ingredients with special functional values or composition. These ingredients need to be highly characterized and, where possible, have demonstrated where the opportunities are for the human or animal markets.

37. Until the fish processing industry begins to realize and think of byproducts as valuable raw materials, they won't become such.

38. Aquaculture commodity ingredients such as meal and oil are on the low end of the value scale, whereas ingredients for the human side are on the high end. More progress in developing materials for specific markets that represent reasonable value is needed (much higher than the average aquaculture value, but not up to the human value) such as horticulture fertilizers. The horticulture fertilizer market is more than a commodity market. There is a need to have an understanding of these products and their application in organic agriculture, as well as some aspects of horticulture such as the production of flowers and ornamentals for household use.

39. Utilization of fish processing byproducts to make high value ingredients for very young animals such as pigs is an area that warrants further investigation. There are also opportunities in developing pet food ingredients.

Research priorities and opportunities from ranking by conference participants
P.J. Bechtel, Chris Mitchell, and Ian Forster

Introduction

Panel members and all conference participants were asked to identify three research needs, priorities, and opportunities for the future and leave them in a box before the closing of the sessions. These were then collated into a list of 29 research priorities (data not shown) and the list was distributed to all participants immediately prior to the panel discussion. After the panel discussion each conference participant was asked to rank their top five priorities and leave these with the conference organizers. The following six groups were then synthesized and are presented in ranked order, with number one being ranked high most often by the conference participants in this exercise.

1. Research priorities and needs should be identified in consultation with industry and other stakeholders and should include a holistic approach to problem solving, including scientific, legal, and economic considerations. Marketing and identifying the market potentials of fish processing byproducts should be the topic of a future conference.

2. Economic analysis of the fish processing byproduct areas is needed.

   A. Develop standard economic modeling methods for the analysis of all aspects of fish processing byproduct operations and products.

   B. Develop economic analysis of the options for handling small amounts of fish processing byproducts, i.e., fish processing byproducts produced by small isolated processors.

   C. Develop models to aid in forecasting the value of all byproducts.

3. Develop new or improved products with increased value from fish processing byproducts.

   A. Develop procedures for making hydrolysates from different fish processing byproducts as unique feed and food ingredients, attractants, and products with unique functional properties, bioactive peptides, and other uses.

B. Develop new and improved uses for products from fish byprod-ucts and separated components of the byproduct stream, such as separation of high valued viscera components for human use, higher valued protein and oil aquaculture feed components, unique nutritional ingredients, ingredients for different seg-ments of the life cycle for aquaculture, farm animals and pets, and supplements from minerals, protein, and oil, etc.

C. Extract interesting biomolecules and fractions from fish process-ing byproducts.

4. Develop processes and methods for enhancing the value of fish byproducts and increasing efficiencies.

A. Develop storage and preservation procedures, including natural preservatives, for fish processing byproducts, which increase dur-ing storage of fish processing primary and secondary byproducts.

B. Develop processes for reducing water consumption and process-ing of wastewater and its components.

C. Develop and standardize methods for accessing the quality and freshness of fish processing byproducts.

D. Develop systems for "full utilization" of fish processing byproducts.

E. Develop logistical solutions for small volume or seasonal proces-sors that cannot support traditional byproduct operations.

5. Provide science-based research for health claims and for communi-cating issues of importance to the public concerning the environ-ment and sustainability of the fishing, fish byproduct utilization, and aquaculture industries.

6. Develop HACCP programs to ensure the safety and quality of fish processing byproducts for use as animal feed ingredients and as components of the human diet.

Seafood byproducts: A call to action
Chris Mitchell

Introduction
The final session of the 2nd International Seafood Byproduct Conference was a participant panel entitled "Panel Discussion: Future Research and Product Needs." The 125 conference attendees were asked to help deter-mine future industry research needs and to prioritize such work.

Panel priority

Overwhelmingly, the panel and participating attendees said that though much good work was being done, most of the research is being funded and undertaken with little participation and/or interaction by the industry the research is supposed to assist. Most attendees noted that few breakthroughs had occurred in the previous decade, likely because many research projects operated in an industry vacuum. Therefore, the simple technology of meal manufacture remained the mainstay because the technology was already established and markets simply awaited production. Many of the other simple technologies, such as commercializing hydrolysate and silage production, had rarely progressed beyond pilot projects undertaken by university scientists and research institutes. These technologies were the main themes 10 years ago at the first International Conference on Seafood By-Products and have yet to advance much beyond the conceptual.

The Alaska seafood industry throws away (or minimally processes into fish meal, where economically viable to do so) most of what it catches as unrecovered byproducts. This discard is unlike almost every other protein category. By throwing away so much we drive all production and marketing costs against that small percentage retained for human consumption. No wonder seafood is thought to be expensive and many harvesters and processors barely survive. This does not have to be the case. Solutions exist. We must only overcome the inertia of thinking, treating, and acting as though these recoverable byproducts are waste, a universal term for materials deemed not worth the effort of recovery.

Rather than wait another ten years and rehash the same things over once again at the next International Seafood Byproduct Conference without much progress during the interim, this most recent conference should be a catalyst for ACTION. An action plan can bring researchers, industry, government, markets, and funding together to focus on demonstrating (in a commercial way) some of the most promising technologies—and connecting these technologies to real markets.

Three-phase action plan

Phase 1. The panel discussion listed as its highest priority that researchers and industry should work together to bring the best ideas to realization. It went on to suggest that a byproduct conference is needed to focus solely on markets—and we can't afford to wait another ten years before this conference occurs. The idea behind this urgent call was that markets should drive the research effort—not vice versa as is most often the case.

Phase 2. The conference should be organized (and presenters selected) with the understanding that during the conference, a select few would be tasked with the responsibility of selecting two or three of the most promising technologies to be trialed thereafter in small (but commercially viable) demonstration projects. Each project would be designed to have its own "development/implementation team," including a cooperating community, processing plant, researchers, and marketers. This is necessary to ground-truth the technology, strengthen vested interests, and tie the project to real production and markets.

Phase 3. Demonstration projects would include a 3-6 month preparatory and planning phase and a 12-18 month operational and marketing phase. The objective would be to set up trials/demonstration projects that, if successful, would evolve into actual commercial operations—albeit initially on a small scale. The design team would remain involved throughout the project as an advisory group to support the operational component and to provide guidance should unexpected problems arise.

Setting

It is suggested that Alaska be the site of these demonstration projects, because more than 60% of America's seafood harvests take place in Alaska waters far from existing markets, because much of that production takes place in remote communities dependent upon fishery production for employment and community sustenance, because the fisheries in many Alaska communities are high volume but very seasonal, and because Alaska's remoteness from markets dictates that traditional solutions may not work.

Funding

Annually tens of millions of dollars are found for funding disaster relief to disaffected fishing communities, fishery management needs, habitat understanding, fisheries/mammal interactions, seafood marketing, etc., while next to nothing is directed toward unlocking the solutions to the recovery of the discarded 60% of America's fishery harvests. Various federal and state institutions provide millions of dollars annually for research on enhancing utilization of byproducts and to find uses for secondary raw materials—and as such are logical sources for this call to ACTION.

The direct costs of the effort as outlined would be less than $2 million earmarked as follows: Phase 1 $250,000; Phase 2 $100,000; Phase 3 $1,500,000 (an average of $500,000 for each of three demonstration projects). If the action plan is designed by the industry, in collaboration with researchers, equipment, and technology suppliers and markets, one could expect significant industry financial support (likely exceeding the federal funds) via equipment, technology transfer, and market support.

Index

*(Page numbers in italic typface
denote illustrations.)*